Human Evolution

M000118391

Human Evolution
Trails from the Past

Camilo J. Cela-Conde and Francisco J. Ayala

OXFORD
UNIVERSITY PRESS

OXFORD

UNIVERSITY PRESS

Great Clarendon Street, Oxford OX2 6DP

Oxford University Press is a department of the University of Oxford.
It furthers the University's objective of excellence in research, scholarship,
and education by publishing worldwide in

Oxford New York

Auckland Cape Town Dar es Salaam Hong Kong Karachi
Kuala Lumpur Madrid Melbourne Mexico City Nairobi
New Delhi Shanghai Taipei Toronto

With offices in

Argentina Austria Brazil Chile Czech Republic France Greece
Guatemala Hungary Italy Japan Poland Portugal Singapore
South Korea Switzerland Thailand Turkey Ukraine Vietnam

Oxford is a registered trade mark of Oxford University Press
in the UK and in certain other countries

Published in the United States
by Oxford University Press Inc., New York

© Oxford University Press 2007

The moral rights of the authors have been asserted
Database right Oxford University Press (maker)

First published 2007
Reprinted 2008

All rights reserved. No part of this publication may be reproduced,
stored in a retrieval system, or transmitted, in any form or by any means,
without the prior permission in writing of Oxford University Press,
or as expressly permitted by law, or under terms agreed with the appropriate
reprographics rights organization. Enquiries concerning reproduction
outside the scope of the above should be sent to the Rights Department,
Oxford University Press, at the address above

You must not circulate this book in any other binding or cover
and you must impose the same condition on any acquirer

British Library Cataloging in Publication Data
Data available

Library of Congress Cataloging in Publication Data
Data available

Typeset by Newgen Imaging Systems (P) Ltd., Chennai, India
Printed in Great Britain
on acid-free paper by
Antony Rowe, Chippenham, Wiltshire

ISBN 978–0–19–856779–0 978–0–19–856780–6 (Pbk.)

10 9 8 7 6 5 4 3 2

Contents

Preface: *Homo sapiens*, the rational animal

Questions about human nature, about what it is that makes us distinctively human, have been raised since antiquity, at least since the dawn of the philosophical way of thinking in Classical Greece, four centuries before the Christian Era. Aristotle, the greatest biologist of antiquity, identified humans first and foremost as "animals," but he called them "rational animals": the power of reason as the distinctive feature that separates us from the rest of the living world. It would be only many centuries later, with the publication of Darwin's theory of evolution by natural selection, that an understanding could emerge of why we share so much of what we are with other animals, particularly with those closest to us, the primates.

One and a half centuries after the publication of Darwin's *The Origin of Species* we know much about the evolutionary history that brought about the distinctive features that make us human. Surely, Darwin would have been pleased by the numerous discoveries of hominid ancestors and by the progress of knowledge concerning human origins. We now know that Darwin's conjectures were correct about the fundamental anatomical event in human origins, namely the evolution of bipedalism, and about the place, Africa, where humanity came about. We also know the time when our lineage separated from the chimpanzee lineage, some 7 million years ago. Scientists are currently hard and fast at work, chromosome by chromosome, and nucleotide by nucleotide, seeking to ascertain the features of our genetic code that make us different from, as well as very similar to, our simian relatives. Further, work is underway to decipher the DNA of our closest fossil relatives, the Neanderthals.

In the midst of the dramatic advances of current paleoanthropology, we are aware that many questions remain unanswered. How many genera, and which ones, belong to the human lineage? And which among the hominid species are our direct ancestors? How did they come about? What are the critical adaptive events in humankind's evolutionary sojourn? Who were the very first ancestors of the human lineage and what were their most prominent features? And what about the first migrants out of Africa? What lithic tools had they and how were they modified as they wandered through other continents? What about the Neanderthals? Were they *Homo sapiens*, members of our species? Did they speak? Did they appreciate art?

Step by step these questions will be answered or, at least, formulated more precisely. These questions and the partial answers available are part of this book, which tells the story of human evolution as it is known today, with as many questions, or more, as there are answers. This is how we must approach our evolutionary lineage, by exploring with rigor the medley of fossil and archeological records, the complexity of the genetic data, and taking advantage of all sorts of available tools, conceptual and technical, which may cast some light on the understanding of our origins. We must formulate testable theories and explore the empirical observations that corroborate or falsify them.

The resulting panorama will not be simple, but need not be confusing. If we are rational beings, we may as well use our powers of rationality to the maximum extent possible as we seek to discover the history of our own evolution.

Camilo J. Cela-Conde
Francisco J. Ayala

CHAPTER 1

Evolution, genetics, and systematics

1.1 The theory of evolution

All organisms are related by descent from common ancestors. Humans and other mammals descend from shrewlike creatures that lived more than 150 million years ago (Ma); mammals, birds, reptiles, amphibians, and fishes share as ancestors aquatic worms that lived 600 Ma; and all plants and animals derive from bacteria-like microorganisms that originated more than 3 billion years ago. Biological evolution is a process of descent with modification. The process consists of two components. Lineages of organisms change through the generations (*anagenesis* or phyletic evolution); diversity arises because the lineages that descend from common ancestors diverge through time (*cladogenesis* or lineage splitting, the process by which new species arise).

Human cultures have advanced explanations for the origin of the world and of humans and other creatures. Traditional Judaism and Christianity explain the origin of living beings and their adaptations to life in their environments—legs and wings, gills and lungs, leaves and flowers—as the handiwork of the Creator. Myths proposing that different kinds of organisms can be transformed one into another are found in diverse cultures since antiquity. Among the philosophers of ancient Greece, Anaximander proposed that animals could metamorphose from one kind into another, and Empedocles speculated that organisms were made up of various combinations of preexisting parts.

The notion that organisms may change by natural processes was considered, usually incidentally, as a possibility by Christian scholars of the Middle Ages, such as Albertus Magnus (1200–1280) and his student Thomas Aquinas (1224–1274). Aquinas concluded, after detailed discussion, that the development of living creatures, such as maggots and flies, from nonliving matter, such as decaying meat, was not incompatible with Christian faith or philosophy. But he left it to others (to scientists, in current parlance) to determine whether this actually happened.

The first broad theory of evolution was proposed by the French naturalist Jean-Baptist de Monet, chevalier de Lamarck (1744–1829). In his *Philosophie zoologique* (1809; translated as *Zoological Philosophy*), Lamarck held the enlightened view, shared by the intellectuals of his age, that living organisms represent a progression, with humans as the highest form. Lamarck's theory of evolution asserts that organisms evolve through eons of time from lower to higher forms, a process still going on, always culminating in human beings. The remote ancestors of humans were worms and other inferior creatures, which gradually evolved into more and more advanced organisms, ultimately humans.

The inheritance of acquired characters is the theory most often associated with Lamarck's name. Yet this theory was actually a subsidiary construct of his theory of evolution: that evolution is a continuous process so that today's worms will yield humans as their remote descendants. As animals become adapted to their environments through their habits, modifications occur by "use and disuse." Use of an organ or structure reinforces it; disuse leads to obliteration. The characteristics acquired by use and disuse, according to Lamarck, would be inherited. This assumption that would later be called the inheritance of acquired characteristics (or Lamarckism) was disproved in the twentieth century.

Lamarck's evolution theory was metaphysical rather than scientific. Lamarck postulated that life possesses an innate tendency to improve over time, so that progression from lower to higher organisms would continually occur, and always following the same path of transformation from lower organisms to increasingly higher and more complex organisms. A somewhat similar evolutionary theory was formulated one century later by another Frenchman, the philosopher Henri Bergson (1859–1940) in his *L'Evolution créatrice* (1907; *Creative Evolution*).

Erasmus Darwin (1731–1802), a physician and poet, and the grandfather of Charles Darwin, proposed, in poetic rather than scientific language, a theory of the transmutation of life forms through eons of time (*Zoonomia, or the Laws of Organic Life*; 1794–1796). More significant for Charles Darwin was the influence of his older contemporary and friend, the eminent geologist Sir Charles Lyell (1797–1875). In his *Principles of Geology* (1830–1833), Lyell proposed that the Earth's physical features were the outcome of major geological processes acting over immense periods of time, incomparably greater than the few thousand years since Creation generally assumed at the time.

1.1.1 Charles Darwin

The founder of the modern theory of evolution was Charles Darwin (1809–1882; Figure 1.1), the son and grandson of physicians. He enrolled as a medical student at the University of Edinburgh. After 2 years, however, he left Edinburgh and moved to the University of Cambridge to pursue his studies and prepare to become a clergyman. Darwin was not an exceptional student, but he was deeply interested in natural history. On December 27, 1831, a few months after his graduation from Cambridge, he sailed as a naturalist aboard the *HMS Beagle* on a round-the-world trip that lasted until October 1836. Darwin was often able to disembark for extended trips ashore to collect natural specimens. The discovery of fossil bones from large extinct mammals in Argentina and the

Figure 1.1 Charles Darwin (1809–1882), *c.*1854, discoverer of natural selection.

observation of numerous species of finches in the Galápagos Islands were among the events credited with stimulating Darwin's interest in how species originate.

The observations he made in the Galápagos Islands may have been the most influential on Darwin's thinking. The islands, on the equator 600 miles off the west coast of South America, had been named Galápagos (the Spanish word for tortoises) by the Spanish discoverers because of the abundance of giant tortoises, different on different islands and different from those known anywhere else in the world (Figure 1.2). The tortoises clanked their way around sluggishly, feeding on the vegetation and seeking the few pools of fresh water. They would have been vulnerable to predators, but these were conspicuously absent on the islands. In the Galápagos, Darwin found large lizards, feeding unlike any others of their kind on seaweed, and mockingbirds, quite different from

Testudo microphyes, Isabela I.

Testudo abingdonii, Pinta I.

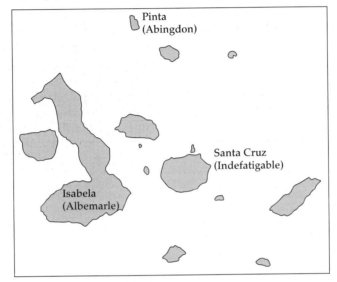

Testudo ephippium, Santa Cruz I.

Figure 1.2 The Galápagos Islands, with drawings of three tortoises found in different islands.

those found on the South American mainland. Well known is that he found several kinds of finches, varying from island to island in various features, notably their distinctive beaks, adapted to disparate feeding habits: crushing nuts, probing for insects, and grasping worms.

In addition to *The Origin of Species* (1959), Darwin published many other books, notably *The Descent of Man and Selection in Relation to Sex* (1871), which extends the theory of natural selection to human evolution. Darwin's theory of natural selection is summarized in *The Origin of Species* as follows.

Can it, then, be thought improbable, seeing that variations useful to man have undoubtedly occurred, that other variations useful in some way to each being in the great and complex battle of life, should sometimes occur in the course of thousands of generations? If such do occur, can we doubt (remembering that more individuals are born than can possibly survive) that individuals having any advantage, however slight, over others, would have the best chance of surviving and of procreating their kind? On the other hand, we may feel sure that any variation in the least degree injurious would be rigidly destroyed. This preservation of favourable variations and the rejection of injurious variations, I call Natural Selection.

The argument consists of three parts: (1) hereditary variations occur, some more favorable than others to the organisms; (2) more organisms are produced than can possibly survive and reproduce; and (3) organisms with more favorable variations will survive and reproduce more successfully. Two consequences follow: (1) organisms are adapted to the environments where they live because of the successful reproduction of favorable variations and (2) evolutionary change will occur over time.

Natural selection was proposed by Darwin primarily to account for the adaptive organization of living beings; it is a process that promotes and maintains adaptation. Evolutionary change through time and evolutionary diversification (multiplication of species) are not directly promoted by natural selection, but they often ensue as by-products of natural selection as it fosters adaptation to different environments.

1.1.2 Alfred Russel Wallace

Alfred Russel Wallace (1823–1913) is famously given credit for discovering, independently from Darwin, natural selection as the process accounting for the evolution of species. On June 18, 1858 Darwin wrote to Charles Lyell that he had received by mail a short essay from Wallace such that "if Wallace had my [manuscript] sketch written in [1844] he could not have made a better abstract." Darwin was thunderstruck.

Darwin and Wallace, who was at the time in the Malay archipelago collecting biological specimens, had started occasional correspondence in late 1855,

with Darwin at times offering sympathy and encouragement to the occasionally dispirited Wallace for his "laborious undertaking." In 1858, Wallace had come upon the idea of natural selection as the explanation for evolutionary change and he wanted to know Darwin's opinion about this hypothesis, since Wallace, as well as many others, knew that Darwin had been working on the subject for years, had shared his ideas with other scientists, and was considered by them as the eminent expert on issues concerning biological evolution.

Darwin hesitated as to how to proceed about Wallace's letter. He wanted to credit Wallace's discovery of natural selection, but he did not want altogether to give up his own earlier independent discovery. Eventually two of Darwin's friends, the geologist Sir Charles Lyell and the botanist Joseph Hooker, proposed, with Darwin's consent, that Wallace's letter and two earlier writings of Darwin would be presented at a meeting of the Linnaean Society of London. On July 1, 1858, three papers were read by the society's undersecretary, George Busk, in the order of their date of composition: two short essays that Darwin had written in 1844 and 1857 and Wallace's essay, "On the Tendency of Varieties to Depart Indefinitely from Original Type." The meeting was attended by some 30 people, who did not include Darwin or Wallace. The papers generated little response and virtually no discussion; their significance apparently was not discerned by those in attendance.

Wallace's independent discovery of natural selection is remarkable. But Wallace's interest and motivation was not the explanation of the adaptation of organisms to their environments and the adaptive design of their organs and other features, but rather how to account for the evolution of species, as indicated in his paper's title: "On the Tendency of Varieties to Depart Indefinitely from Original Type." Wallace thought that evolution proceeds indefinitely and is progressive. Darwin, on the contrary, did not accept that evolution would necessarily represent progress or advancement. Nor did he believe that evolution would always result in morphological change over time; rather, he knew of the existence of "living fossils," organisms that had remained unchanged for millions of years. For example, "some of the most ancient Silurian animals, as the Nautilus, Lingula, etc., do not differ much from living species."

1.1.3 *The Origin of Species*

The publication of *The Origin of Species* took the British scientific community of the mid-nineteenth century by storm. It also caused considerable public excitement. Scientists, politicians, clergymen, and notables of all kinds read and discussed the book, defending or deriding Darwin's ideas. The most visible actor in the controversies immediately following publication was the English biologist T.H. Huxley, known as Darwin's bulldog, who defended the theory of evolution with articulate and sometimes mordant words, on public occasions as well as in numerous writings.

A younger English contemporary of Darwin, with considerable influence during the latter part of the nineteenth and in the early twentieth century, was Herbert Spencer. A philosopher rather than a biologist, he became an energetic proponent of evolutionary ideas, popularized a number of slogans, such as "survival of the fittest" (which was taken up by Darwin in later editions of *Origin*), and engaged in social and metaphysical speculations (which Darwin thoroughly disliked). His ideas considerably damaged proper understanding and acceptance of the theory of evolution by natural selection. Most pernicious was the crude extension by Spencer and others of the notion of the "struggle for existence" to human economic and social life that became known as Social Darwinism.

Darwinism in the latter part of the nineteenth century faced an alternative evolutionary theory known as neo-Lamarckism. This hypothesis shared with Lamarck's the importance of use and disuse in the development and obliteration of organs, and it added the notion that the environment acts directly on organic structures, which explained their adaptation to the way of life and environment of the organism. Adherents of this theory discarded natural selection as an explanation for adaptation to the environment.

Prominent among the defenders of natural selection was the German biologist August Weismann, who in the 1880s published his germ-plasm theory. He distinguished two substances that make up an organism: the soma, which comprises most body parts and organs, and the germ plasm, which contains the cells that give rise to the gametes and hence to progeny. Early in the development of an embryo, the germ plasm becomes segregated from the somatic cells that give rise to the rest of the body. This notion of a radical separation between germ plasm and soma—that is, between the reproductive tissues and all other body tissues—prompted Weismann to assert that inheritance of acquired characteristics was impossible, and it opened the way for his championship of natural selection as the only major process that would account for biological evolution. Weismann's ideas became known after 1896 as neo-Darwinism.

One important reason why Darwin's theory of evolution by natural selection encountered resistance among Darwin's contemporaries and beyond was the lack of an adequate theory of inheritance that would account for the preservation through the generations of the variations on which natural selection was supposed to act. Contemporary theories of "blending inheritance" proposed that offspring merely struck an average between the characteristics of their parents. Darwin's own theory of "pangenesis" proposed that each organ and tissue of an organism throws off tiny contributions of itself that are collected in the sex organs and determine the configuration of the offspring. These theories of blending inheritance could not account for the conservation of variations, because differences between variant offspring would be halved each generation, rapidly reducing the original variation to the average of the preexisting characteristics.

1.1.4 Gregor Mendel

The missing link in Darwin's argument was provided by Mendelian genetics. About the time *The Origin of Species* was published, the Augustinian monk Gregor Mendel was starting a long series of experiments with peas in the garden of his monastery in Brünn, Austria-Hungary (now Brno, Czech Republic). Mendel's paper, published in

1866 in the *Proceedings* of the Natural Science Society of Brünn, formulated the fundamental principles of the theory of heredity that is still current. His theory accounts for biological inheritance through particulate factors (now known as genes) inherited one from each parent, which do not mix or blend but segregate in the formation of the sex cells, or gametes.

Mendel's discoveries remained unknown to Darwin, however, and indeed they did not become generally known until 1900, when they were simultaneously rediscovered by a number of scientists in Europe.

Mendel's theory of heredity was rediscovered in 1900 by the Dutch botanist Hugo de Vries, the German Carl Correns, and others. Mendel's theory provided a suitable mechanism for the natural selection of hereditary traits. But a controversy arose between those who thought that the kind of characters transmitted by Mendelian heredity were not significant for natural selection (because this concerned very small, "continuous," variations among individuals) and those who thought Mendelian heredity was all, or most, that there was in evolution, with natural selection relegated to a minor role, or no role at all.

de Vries himself proposed a new theory of evolution known as mutationism, which essentially did away with natural selection as a major evolutionary process. According to de Vries (and other geneticists such as William Bateson in England), two kinds of variation take place in organisms. One is the "ordinary" variability observed among individuals of a species, such as small differences in color, shape, and size. This variability would have no lasting consequence in evolution because, according to de Vries, it could not "lead to a transgression of the species border even under conditions of the most stringent and continued selection." The variation that is significant for evolution is the changes brought about by mutations, spontaneous alterations of genes that result in large modifications of the organism, which may give rise to new species: "The new species thus originates suddenly, it is produced by the existing one without any visible preparation and without transition."

Many naturalists and some mathematicians, particularly in Britain but also on the European continent, rejected mutationism, and even Mendelian heredity, as irrelevant to natural selection, because mutations produced only large, even monstrous, morphological variations, whereas natural selection depends on minor variations impacting, most of all, life span and fertility. These scientists, among them the English statistician Karl Pearson, defended Darwinian natural selection as the major cause of evolution through the cumulative effects of small, continuous, individual variations (which they assumed passed from one generation to the next without being limited by Mendel's laws of inheritance).

1.1.5 The synthetic theory of evolution

The controversy between the two groups approached a resolution in the 1920s and 1930s through the theoretical work of geneticists, such as R.A. Fisher and J.B.S. Haldane in Britain and Sewall Wright in the USA. These scientists used mathematical arguments to show that (1) continuous variation (in such characteristics as body size, number of progeny, and the like) could be explained by Mendel's laws and (2) natural selection acting cumulatively on small variations could yield major evolutionary changes in form and function. Their work provided a theoretical framework for the integration of genetics into Darwin's theory of natural selection, but it had a limited impact on contemporary biologists because it was formulated in a mathematical language that most biologists could not understand and was presented with little empirical corroboration.

The synthesis of Darwin's theory of natural selection and Mendelian genetics became generally accepted by biologists only in the mid-twentieth century, after the publication of several important books by biologists who provided observations and experimental results that supported the formulations of the mathematical theorists. One important publication, in 1937, was *Genetics and the Origin of Species* by Theodosius Dobzhansky, a Russian-born American naturalist and experimental geneticist. Dobzhansky's book advanced a

Figure 1.3 G. Ledyard Stebbins, George Gaylord Simpson, and Theodosius Dobzhansky (left to right), three main authors of the modern theory of evolution, in 1970, at a conference at the University of California, Davis, organized by Francisco J. Ayala.

reasonably comprehensive account of the evolutionary process in genetic terms, laced with experimental evidence supporting the theoretical argument. *Genetics and the Origin of Species* had an enormous impact on naturalists and experimental biologists, who rapidly embraced the new understanding of the evolutionary process as one of genetic change in populations. Other significant contributions were *Systematics and the Origin of Species* (1942) by the German-born American zoologist Ernst Mayr, *Evolution: the Modern Synthesis* (1942) by the English zoologist Julian Huxley, *Tempo and Mode in Evolution* (1944) by the American paleontologist George Gaylord Simpson, and *Variation and Evolution in Plants* (1950) by the American botanist George Ledyard Stebbins (Figure 1.3). The synthetic theory of evolution, as it became known, elaborated by these scientists contributed to a burst of evolutionary studies in the traditional biological and paleontological disciplines and stimulated the development of new disciplines, such as population and evolutionary genetics, evolutionary ecology, and paleobiology.

1.1.6 The second half of the twentieth century

In the second half of the twentieth century, population genetics and evolutionary genetics became very active disciplines that eventually incorporated

molecular biology, a new discipline which emerged from the 1953 discovery by James Watson and Francis Crick of the molecular structure of DNA, the hereditary chemical contained in the chromosomes of every cell nucleus. The genetic information is encoded within the sequence of nucleotides that make up the chainlike DNA molecules. This information determines the sequence of amino acid building blocks of protein molecules, which include structural proteins as well as the numerous enzymes that carry out the organism's fundamental life processes. Genetic information could now be investigated by examining the sequences of amino acids in the proteins, and eventually the sequences of the nucleotides that make up the DNA.

In the mid-1960s laboratory techniques such as electrophoresis and selective assay of enzymes became available for the rapid and inexpensive study of differences among enzymes and other proteins. These techniques made possible the pursuit of evolutionary issues, such as quantifying genetic variation in natural populations (such variation sets bounds on the evolutionary potential of a population) and determining the amount of genetic change that occurs during the formation of new species. Comparisons of the amino acid sequences of corresponding proteins in different species provided precise measures of the divergence among species evolved from common ancestors, a considerable improvement over

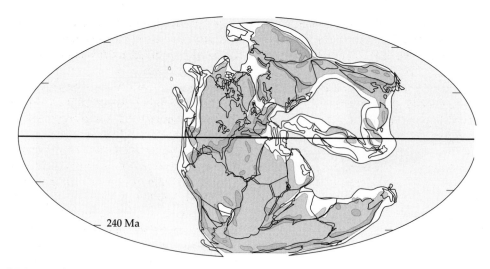

240 Ma

Figure 1.4 Pangea. About 240 Ma, in the early Triassic, most of the continents land was aggregated into a single mass.

the typically qualitative evaluations obtained by comparative anatomy and other evolutionary subdisciplines.

The laboratory techniques of DNA cloning and sequencing have provided a new and powerful means of investigating evolution at the molecular level. The fruits of this technology began to accumulate during the 1980s following the development of automated DNA-sequencing machines and the invention of the polymerase chain reaction (PCR), a simple and inexpensive technique that obtains, in a few hours, billions or trillions of copies of a specific DNA sequence or gene. Major research efforts such as the Human Genome Project further improved the technology for obtaining long DNA sequences rapidly and inexpensively. By the first few years of the twenty-first century, the full DNA sequence—that is, the full genetic complement, or genome—had been obtained for more than 20 higher organisms, including humans, the house mouse (*Mus musculus*), the rat *Rattus norvegicus*, the vinegar fly (also known as the fruit fly) *Drosophila melanogaster*, the mosquito *Anopheles gambiae*, the nematode worm *Caenorhabditis elegans*, the malaria parasite *Plasmodium falciparum*, and the mustard weed *Arabidopsis thaliana*, as well as for the yeast *Saccharomyces cerevisiae* and numerous

microorganisms. A draft of the chimpanzee genome was published in 2005. Rapid advances have also occurred in the study of evolutionary developmental biology, which has become known as evo-devo.

In the second half of the twentieth century, the earth sciences also experienced a conceptual revolution of great consequence for the study of evolution. The theory of plate tectonics, which was formulated in the late 1960s, revealed that the configuration and position of the continents and oceans are dynamic features of Earth. Oceans grow and shrink, while continents break into fragments or coalesce into larger masses, altering the face of the planet and causing major climatic changes along the way. The consequences for the evolutionary history of life are enormous. Thus, biogeography, the evolutionary study of plant and animal distribution, has been revolutionized by the knowledge, for example, that Africa and South America were part of a single landmass some 200 Ma and that the Indian subcontinent was not connected with Asia until geologically recent times (Figure 1.4).

New methods for dating fossils, rocks, and other materials have made it possible to determine, with much greater precision than ever before, the age of the geological periods and of the fossils themselves. This has greatly contributed to advances in paleontology, and to emergence of the new field of paleobiology. Increased interest and investment have favored in particular paleoanthropology, which has experienced a notable acceleration in the rate of discovery and investigation of hominid remains and their associated faunas and habitats. We will review these discoveries throughout this book.

Finally, ecology, the study of the interactions of organisms with their environments, has evolved from descriptive studies—natural history—into a vigorous biological discipline with a strong mathematical component, both in the development of theoretical models and in the collection and analysis of quantitative data. Evolutionary ecology has become a very active field of evolutionary studies. Major advances have also occurred in evolutionary ethology, the study of the evolution of animal behavior. Sociobiology, the evolutionary study of social behavior, is perhaps the most active subfield of ethology and surely the most controversial, because it seeks to explain human behavior and human societies similarly as animal social behavior, as largely determined by their genetic make-up.

1.2 Population and evolutionary genetics

1.2.1 Evolution by natural selection

We introduce in this section some fundamental concepts of genetics, particularly those that are relevant for understanding evolution. Students in general, but particularly those who have had a college-level course in genetics, may want to skip the section, or refer to it only when seeking understanding of some particular issues. There are numerous genetics textbooks, as well as texts focused on population and evolutionary genetics, where the concepts introduced here are developed in greater detail.

Biological evolution is the process of change and diversification of living things over time, and it affects all aspects of their lives: morphology (form and structure), physiology, behavior, and ecology. Underlying these changes are genetic changes.

In genetic terms, the process of evolution consists of changes through time in the genetic make-up of populations. Evolution can be seen as a two-step process. First, hereditary variation arises; second, selection is made of those genetic variants that will be passed on most effectively to the following generations. The origin of hereditary variation also entails two mechanisms: the spontaneous mutation of one variant into another and the sexual process that recombines those variants to form a multitude of new arrangements of the variations. Selection, the second step of the evolution process, occurs because the variants that arise by mutation and recombination are not transmitted equally from one generation to another. Some may appear more frequently in the progeny because they are favorable to the organisms carrying them, which thereby leave more progeny. Other factors affect the transmission frequency of hereditary variations, particularly chance, a process called genetic drift.

Darwin's argument of evolution by natural selection starts with the existence of hereditary variation. Experience with animal and plant breeding had demonstrated to Darwin that variations that are "useful to man" can be found in organisms. So, he reasoned, variations must occur in nature that are favorable or useful in some way to the organism itself in the struggle for existence. Favorable variations are ones that increase chances for survival and procreation. Those advantageous variations are preserved and multiplied from generation to generation at the expense of less-advantageous ones. This is the process known as natural selection. The outcome of the process is an organism that is well adapted to its environment, and evolution often occurs as a consequence.

Natural selection, then, can be defined as the differential reproduction of alternative hereditary variants, determined by the fact that some variants increase the likelihood that the organisms having them will survive and reproduce more successfully than will organisms carrying alternative variants. Selection may occur as a result of differences in survival, in fertility, in rate of development, in mating success, or in any other aspect of the life cycle. All of these differences can be

incorporated under the phrase differential reproduction because all result in natural selection to
the extent that they affect the number of progeny
an organism leaves.

Darwin maintained that competition for limited
resources results in the survival of the most-
effective competitors. Nevertheless, natural selection may occur not only as a result of competition
but also as a result of some aspect of the physical
environment, such as inclement weather. Moreover, natural selection would occur even if all the
members of a population died at the same age,
simply because some of them would have produced more offspring than others. Natural selection is quantified by a measure called Darwinian
fitness or relative fitness. Fitness in this sense is the
relative probability that a hereditary characteristic
will be reproduced; that is, the degree of fitness is
a measure of the reproductive efficiency of the
characteristic.

1.2.2 Deoxyribonucleic acid: DNA

Two related polynucleotides in organisms are
deoxyribonucleic acid (DNA) and ribonucleic acid
(RNA). The hereditary chemical in most organisms
is DNA; in some viruses, such as human immuno-
deficiency virus (HIV), it is RNA. But RNA
fulfills important functions in all organisms, such
as being the *messenger* (messenger RNA, or
mRNA) conveying the information encoded in
DNA from the nucleus into the body of the cell,
where it directs protein synthesis, as well as the
agent for *transfer* (transfer RNA, or tRNA), which
brings the individual amino acids that are succes-
sively added to protein (polypeptide) chains following the instructions conveyed by mRNA. There
also are other kinds of RNA molecule, such as
microRNA (miRNA), very short molecules, typically consisting of 22 nucleotides, which are directly
transcribed from the DNA, and perform important
functions in gene regulation, including early
development in mammals. Hundreds of genes
encoding miRNAs have been identified in animals
and many more are predicted (Berezikov *et al.*,
2006); in plants the number of known miRNAs is
smaller (Mallory and Vaucheret, 2006). In animals,

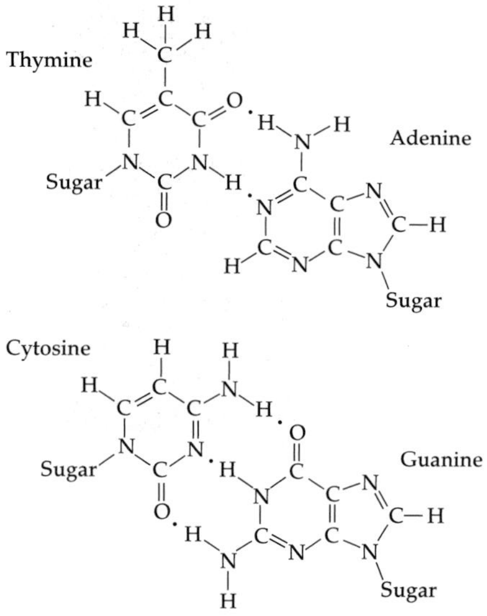

Figure 1.5 The four nitrogen bases of DNA: adenine (A), cytosine (C), guanine (G), and thymine (T). In the double helix, the bases of the two complementary strands are held together by hydrogen bonds: two between A and T, and three between C and G.

including humans, it is estimated that the expression of more than one-third of all genes is controlled by miRNAs.

Nucleic acids are long polymers of a basic unit,
the *nucleotide*. A nucleotide is composed of three
distinct chemical parts joined by covalent bonds.
One part is a pentose sugar: doxyribose in DNA
and ribose in RNA. The second part is a nitrogenous base, which in DNA can be either a
purine—adenine (A) or guanine (G)—or a
pyrimidine—cytosine (C) or thymine (T; Figure 1.5).
RNA contains the same bases as DNA, except that
it has uracil (U) rather than T. The third part of the
nucleotide is a phosphate group, which forms the
joint between successive nucleotides by phospho-
diester bridges between the 5′-carbon of one sugar
moiety and the 3′-carbon of another. The 5′–3′
links establish directionality in the nucleic acids.

DNA molecules consist of two chains of
nucleotides paired in a double helix. The pairing is
effected by hydrogen bonds between the nucleotides of the two strands, so that the pairing is
always between A and T or between G and C
(Figure 1.5). The genetic information is conveyed
by linear sequences of these letters, similarly as

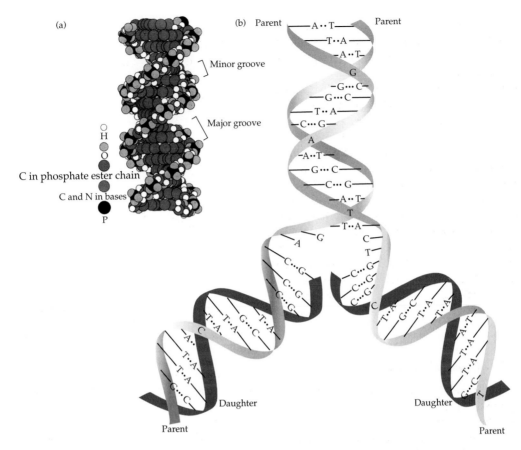

Figure 1.6 The DNA double helix. (a) A space-filling model. (b) Mode of replication: the two strands separate and each one serves as a template for the synthesis of a complementary strand, so that the two daughter double helices are identical to each other and to the original molecule.

semantic information is conveyed by sequences of the 26 letters of the English alphabet.

During replication the two strands of the DNA double helix separate and each becomes a template for a complementary strand (Figure 1.6). Because of the strict rules of pairing, the two daughter molecules are identical to the mother molecule and to each other. Hence, the fidelity of biological heredity.

The DNA of eukaryotic organisms is organized into chromosomes, which consist of several kinds of histone protein associated with the DNA. The chromosomes occur in pairs, one inherited from each parent. The number of chromosomes, characteristic of each species, varies broadly from only

one pair, as in some parasitic nematodes, to more than 100, as in some species of butterflies, and to more than 600, as in some ferns. Humans have 23 pairs of chromosomes (Figure 1.7). Other primates have 24 pairs; two of their chromosomes fused into one, chromosome 2, in our hominin ancestors. In all primates the two chromosomes of a certain pair are identical in females (XX), but not in males (XY).

A gene is a DNA segment that becomes *transcribed* into mRNA, which in turn becomes *translated* into a *polypeptide*; that is, a protein or part of a protein. (Some proteins consist of several polypeptides; e.g., hemoglobin A, the most common in adult humans, consists of four polypeptides, two

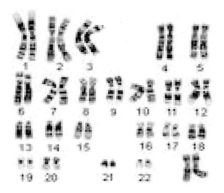

Figure 1.7 The 23 pairs of chromosomes in a human female.

	Second position				
First position	U	C	A	G	Third position
U	UUU ⎤ Phe UUC ⎦ UUA ⎤ Leu UUG ⎦	UCU ⎤ UCC UCA Ser UCG ⎦	UAU ⎤ Tyr UAC ⎦ UAA Stop UAG Stop	UGU ⎤ Cys UGC ⎦ UGA Stop UGG Trp	U C A G
C	CUU ⎤ CUC CUA Leu CUG ⎦	CCU ⎤ CCC CCA Pro CCG ⎦	CAU ⎤ His CAC ⎦ CAA ⎤ Gln CAG ⎦	CGU ⎤ CGC CGA Arg CGG ⎦	U C A G
A	AUU ⎤ AUC Ile AUA ⎦ AUG Met	ACU ⎤ ACC ACA Thr ACG ⎦	AAU ⎤ Asn AAC ⎦ AAA ⎤ Lys AAG ⎦	AGU ⎤ Ser AGC ⎦ AGA ⎤ Arg AGG ⎦	U C A G
G	GUU ⎤ GUC GUA Val GUG ⎦	GCU ⎤ GCC GCA Ala GCG ⎦	GAU ⎤ Asp GAC ⎦ GAA ⎤ Glu GAG ⎦	GGU ⎤ GGC GGA Gly GGG ⎦	U C A G

Figure 1.8 The genetic code. Each set of three consecutive letters (codon) in the DNA determines one amino acid in the encoded protein. DNA codes for RNA (transcription), which codes for amino acids (translation). RNA uses uracil (U) rather than thymine (T). The 20 amino acids making up proteins (with their three-letter and one-letter representations) are as follows: alanine (Ala, A), arginine (Arg, R), asparagine (Asn, N), aspartic acid (Asp, D), cysteine (Cys, C), glycine (Gly, G), glutamic acid (Glu, E), glutamine (Gln, Q), histidine (His, H), isoleucine (Ile, I), leucine (Leu, L), lysine (Lys, K), methionine (Met, M), phenylalanine (Phe, F), proline (Pro, P), serine (Ser, S), threonine (Thr, T), tyrosine (Tyr, Y), tryptophan (Trp, W), and valine (Val, V).

of each of two different kinds, called α and β.) The number of protein-encoding genes is about 30,000 in primates and other mammals, 13,000 in *Drosophila*, and 5,000 in yeast. Some plants, such as *Arabidopsis*, seem to have nearly as many genes as mammals. Most of the DNA of eukaryotes, which does not embrace genes, is often called *junk DNA* and a good part of it consists of sequences of various lengths, some quite small but repeated many thousands or even millions of times, such as the *Alu* sequences of the human genome. Much of the junk DNA may not be functional at all, but some sequences, such as those encoding the miRNAs mentioned above, play a role in regulating the transcription or translation of other DNA sequences.

The coding part of a gene often occurs in parts (*exons*) that are separated by segments of non-coding DNA, called *introns*. Typically a gene is preceded by untranscribed DNA sequences, usually short, that regulate its transcription. The rules that determine the translation of mRNA into proteins are known as the *genetic code* (Figure 1.8). Particular combinations of three consecutive nucleotides (*codons* or triplets) specify particular amino acids, out of the 20 that make up proteins. Tryptophan and methionine are specified each by only one codon; all others are specified by several, from two to six, which are said to be synonymous. Three codons are *stop* signals that indicate termination of the translation process.

Chemical reactions in organisms must occur in an orderly manner; organisms therefore have ways of switching genes on and off, since different sets of genes are active in different cells. Typically a gene is turned on and off by a system of several switches acting on short DNA sequences adjacent to the coding part of the gene. There are switches acting on a given gene activated or deactivated by feedback loops that involve molecules synthesized by other genes, as well as molecules present in the cell's environment. There are a variety of gene-control mechanisms, which were discovered first in bacteria and other microorganisms.

The investigation of gene-control mechanisms in mammals (and other complex organisms) became possible in the mid-1970s with the development of recombinant DNA techniques. This technology made it feasible to isolate single genes (and other DNA sequences) and to clone them, in billions of identical copies, to obtain the quantities necessary for ascertaining their nucleotide sequence. In mammals, insects, and other complex organisms, there are control circuits and master switches (such as the so-called *homeobox* genes) that operate at higher levels than the control mechanisms that activate and deactivate individual genes. These higher-level switches act on sets rather than individual genes. The details of how these sets are controlled, how many control systems there are, and how they interact remain largely to be elucidated, although great advances in evo-devo have been made in recent years.

1.2.3 Mutation

The nucleotide sequence of the DNA is, as a rule, reproduced faithfully during replication. But heredity is not a perfectly conservative process; otherwise, evolution could not have taken place. Occasionally mistakes, or mutations, occur in the DNA molecule during replication, so that daughter cells differ from the parent cells in the sequence or in the amount of DNA. A mutation first appears in a single cell of an organism, but it is passed on to all cells descended from the first. Mutations occur in all sorts of cells, but the mutations that count in evolution are those that occur in the sex cells (eggs and sperm), or in cells from which the sex cells derive, because these are the cells that produce the offspring.

Mutations can be classified into two categories: gene, or point, mutations, which affect only one or a few nucleotides within a gene, and chromosomal mutations, which either change the number of chromosomes or change the number or arrangement of genes on a chromosome.

A gene mutation may be either a substitution of one or a few nucleotides for others or an insertion or deletion of one or a few pairs of nucleotides. Substitutions in the nucleotide sequence of a structural gene may result in changes in the amino acid sequence of the protein, although this is not always the case. Consider the triplet AUA, which codes for the amino acid isoleucine. If the last A is

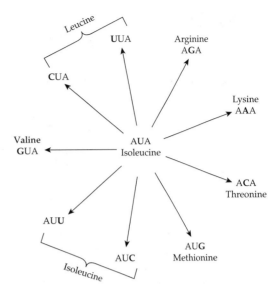

Figure 1.9 Point mutations. Substitutions at the first, second, or third position in the mRNA codon for the amino acid isoleucine can give rise to nine new codons that code for six different amino acids as well as isoleucine. The effects of a mutation depend on what change takes place: arginine, threonine, and lysine have chemical properties that differ sharply from those of isoleucine.

replaced by C, the triplet still codes for isoleucine, but if it is replaced by G, it codes for methionine instead (see Figure 1.9).

A nucleotide substitution in the DNA that results in an amino acid substitution in the corresponding protein may or may not severely affect the biological function of the protein. Some nucleotide substitutions change a codon for an amino acid into a signal that terminates translation (a stop codon). These mutations are likely to have harmful effects. If, for instance, the second U in the triplet UUA, which codes for leucine, is replaced by A, the triplet becomes UAA, a stop codon; the result is that the triplets following this codon in the DNA sequence are not translated into amino acids.

Additions or deletions of nucleotides within the DNA sequence of a structural gene often result in a greatly altered sequence of amino acids in the coded protein. The addition or deletion of one or two nucleotides shifts the reading frame of the nucleotide sequence all along the way from the point of the insertion or deletion to the end of the molecule. To illustrate, assume that the

DNA segment …CATCATCATCATCAT… is read in groups of three as …CAT-CAT-CAT-CAT-CAT…. If a nucleotide base—say, T—is inserted after the first C of the segment, the segment will then be read as …CTA-TCA-TCA-TCA-TCA…. From the point of the insertion onward, the sequence of encoded amino acids is altered. If, however, a total of three nucleotides is either added or deleted, the original reading frame will be maintained in the rest of the sequence. Additions or deletions of nucleotides in numbers other than three or multiples of three are called frameshift mutations.

1.2.4 Effects of mutation

Newly arisen mutations are more likely to be harmful than beneficial to their carriers, because mutations are random events with respect to adaptation; that is, their occurrence is independent of any possible consequences. The allelic variants present in an existing population have already been subject to natural selection. Most are present in the population because they improve the adaptation of their carriers; their alternative alleles have been eliminated or kept at low frequencies by natural selection. A newly arisen mutation is likely to have been preceded by an identical mutation in the previous history of a population. If the previous mutation no longer exists in the population, or it exists at very low frequency, it is a sign that the new mutation is not likely to be beneficial to the organism and is also likely to be eliminated.

Occasionally, however, a new mutation may increase the organism's adaptation. The probability of such an event happening is greater when organisms colonize a new territory or when environmental changes confront a population with new challenges. In these cases the established adaptation of a population is less than optimal, and there is greater opportunity for new mutations to be better adaptive. The consequences of mutations depend on the environment. Increased melanin pigmentation may be advantageous to inhabitants of tropical Africa, where dark skin protects them from the Sun's ultraviolet radiation, but it is not beneficial in Scandinavia, where the intensity of sunlight is low and light skin facilitates the synthesis of vitamin D in the deeper layers of the dermis.

Mutation rates have been measured in a great variety of organisms, mostly for mutants that exhibit conspicuous effects. Mutation rates are generally lower in bacteria and other microorganisms than in more complex species. In humans and other multicellular organisms, the rate for any given mutation typically ranges from about one per 100,000 gametes to one per 1,000,000 gametes. There is, however, considerable variation from gene to gene as well as from organism to organism. Moreover, there are different ways of measuring mutation rates; for example, rates with respect to changes in any given nucleotide of the DNA sequence of a gene, or with respect to any change in any given *gene* (which encompasses hundreds or thousands of DNA nucleotides). Also, rates are quite different for gene mutations in the strict sense and for reorganizations, duplications, and deletions of sets of genes.

Although mutation rates are low, new mutants appear continuously in nature, because there are many individuals in every species and many gene loci in every individual. The process of mutation provides the organisms of each generation with many new genetic variations. Thus, it is not surprising to see that, when new environmental challenges arise, species are able to adapt to them.

Consider the resistance of disease-causing bacteria and parasites to antibiotics and other drugs. When an individual receives an antibiotic that specifically kills the bacterium causing the disease—say, tuberculosis—the immense majority of the bacteria die, but one in a million may have a mutation that provides resistance to the antibiotic. These resistant bacteria will survive and multiply, and the antibiotic will no longer cure the disease. This is the reason why modern medicine treats bacterial diseases with cocktails of antibiotics. If the incidence of a mutation conferring resistance for a given antibiotic is one in a million, the incidence of one bacterium carrying three mutations, each conferring resistance to one of three antibiotics, is one in a million million million; such

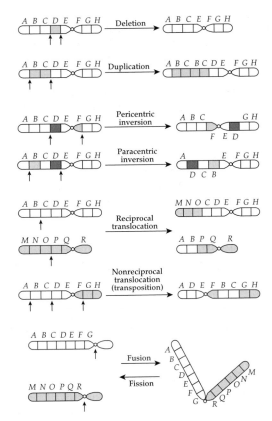

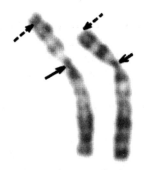

Figure 1.11 The second chromosome of a child showing a deletion. The terminal segment of the chromosome on the right is missing (broken arrows). The solid arrows indicate the position of the centromere.

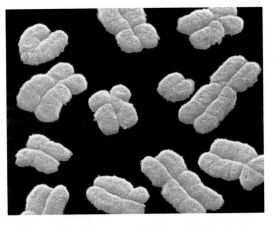

Figure 1.12 Electron photograph of human chromosomes joined at their centromere.

Figure 1.10 Chromosomal mutations. A deletion has a chromosome segment missing. A duplication has a chromosome segment represented twice. Inversions and translocations are chromosomal mutations that change the locations of genes in the chromosomes. Centric fusions are the joining of two chromosomes at the centromere to become one single chromosome. Centric fissions, or dissociations, are the reciprocal of fusions: one chromosome splits into two chromosomes.

bacteria are far less likely to exist in any infected individual.

1.2.5 Chromosomal mutations

Changes in the number, size, or organization of chromosomes within a species are termed chromosomal mutations, chromosomal abnormalities, or chromosomal aberrations (Figure 1.10). Changes in number may occur by the fusion of two chromosomes into one, by fission of one chromosome into two, or by addition or subtraction of one or more whole chromosomes or sets of chromosomes. (The condition in which an organism acquires one

or more additional sets of chromosomes is called polyploidy.) Changes in the structure of chromosomes may occur by inversion, when a chromosomal segment rotates 180° within the same location; by duplication, when a segment is added; by deletion, when a segment is lost (Figure 1.11); or by translocation, when a segment changes from one location to another in the same or a different chromosome. These are the processes by which chromosomes evolve (Figure 1.12).

Inversions, translocations, fusions, and fissions do not change the amount of DNA. The importance of these mutations in evolution is that they change the linkage relationships between genes. Genes that were closely linked to each other

Table 1.1 Genotypic and allelic frequencies for the M-N blood groups in three human populations

Population	Blood-group individuals			Total	Genotypic frequency			Allelic frequency	
	M	MN	N		$L^M L^M$	$L^M L^N$	$L^N L^N$	L^M	L^N
Australian Aborigines	22	216	492	730	0.030	0.296	0.674	0.178	0.822
Navajo Indians	305	52	4	361	0.845	0.144	0.011	0.917	0.083
White North Americans	1787	3039	1303	6129	0.292	0.496	0.213	0.539	0.461

Individuals with blood group M are homozygotes with genotype $L^M L^M$; those with blood group MN are heterozyotes, $L^M L^N$; those with blood group N are homozyotes, $L^N L^N$. The allelic frequency of L^M is the frequency of $L^M L^M$ plus half the frequency of the heterozygotes $L^M L^N$; for example, $0.030 + 0.148 = 0.178$. Similarly the frequency of L^N is the frequency of $L^N L^N$ plus half the frequency of $L^M L^N$.

become separated and vice versa; this can affect their expression because genes are often transcribed sequentially, two or more at a time. Human chromosomes differ from those of chimps and other apes in number: they have 24 whereas we have 23 pairs as a consequence of the fusion of two of their chromosomes into one, as noted above. In addition, inversions and translocations that distinguish human from ape chromosomes have been identified in several chromosomes.

1.2.6 Genetic variation in populations

The sum total of all genes and combinations of genes that occur in a population of organisms of the same species is called the *gene pool* of the population. This can be described for individual genes or sets of genes by giving the frequencies of the alternative genetic constitutions; different forms of the same gene are called alleles. Consider, for example, a particular gene, such as the one determining the M-N blood groups in humans. One allele codes for the M blood group, while the other allele codes for the N blood group. The M-N gene pool of a particular population is specified by giving the frequencies of the alleles *M* and *N*. Thus, in the USA the *M* allele occurs in people of European descent with a frequency of 0.539 and the *N* allele with a frequency of 0.461. In other populations these frequencies are different; for instance, the frequency of the *M* allele is 0.917 in Navajo Indians and 0.178 in Australian Aborigines (Table 1.1).

The genetic variation present in a population is sorted out in new ways in each generation by the process of sexual reproduction, which recombines the chromosomes inherited from the two parents during the formation of the gametes that produce the following generation. But heredity by itself does not change gene frequencies. This principle is stated by the Hardy–Weinberg law, which describes the genetic equilibrium in a population by means of an algebraic equation. It states that genotypes, the genetic constitution of individual organisms, exist in certain frequencies that are a simple function of the allelic frequencies: namely, the square expansion of the sum of the allelic frequencies (Figure 1.13).

If there are two alleles, *A* and *a*, at a gene locus, three genotypes will be possible: *AA*, *Aa*, and *aa*. If the frequencies of the alleles *A* and *a* are *p* and *q*, respectively, the equilibrium frequencies of the three genotypes will be given by $(p+q)^2 = p^2 + 2pq + q^2$ for *AA*, *Aa*, and *aa*, respectively. The genotype equilibrium frequencies for any number of alleles are derived in the same way.

The genetic equilibrium frequencies determined by the Hardy–Weinberg law assume that there is random mating; that is, the probability of a particular kind of mating is the same as the combined frequency of the genotypes of the two mating individuals. Random mating can occur with respect to most gene loci even though mates may be chosen according to particular characteristics. People, for example, choose their spouses according to all sorts of preferences concerning looks,

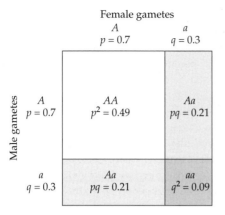

Figure 1.13 The Hardy–Weinberg law, a representation of the relationship between allele and genotype frequencies.

personality, and the like. But concerning the majority of genes, people's marriages are essentially random. People are unlikely to choose their matings according to their M-N blood-group genotypes, or according to the genotype they have with respect to a particular enzyme.

Assortative, or selective, mating takes place when the choice of mates is not random. Marriages in the USA, for example, are assortative with respect to many social factors, so that members of any one social group tend to marry members of their own group more often, and people from a different group less often, than would be expected from random mating. Consider the sensitive social issue of interracial marriage in a hypothetical community in which 80% of the population is white and 20% is black. With random mating, 32% $(2 \times 0.80 \times 0.20 = 0.32)$ of all marriages would be interracial, whereas only 4% $(0.20 \times 0.20 = 0.04)$ would be marriages between two blacks. These statistical expectations depart from typical observations even in modern society, as a result of persistent social customs.

The Hardy–Weinberg equilibrium expectations also assume that gene frequencies remain constant from generation to generation, that there is no gene mutation or natural selection and that populations are very large. But these assumptions are not correct. Organisms are subject to mutation, selection, and other processes that change gene frequencies, but the effects of these processes can be calculated

by using the Hardy–Weinberg law as the starting point.

1.2.7 Processes of genetic change

The allelic variations that make evolution possible are generated by the process of mutation, but new mutations change gene frequencies very slowly, because mutation rates are low. If mutation were the only genetic process of evolution, this would occur very slowly. Moreover, organisms would become dysfunctional over time, because most mutations are harmful rather than beneficial.

Gene flow, or gene migration, takes place when individuals migrate from one population to another and interbreed with its members. Gene frequencies are not changed for the species as a whole, but they change locally whenever different populations that have different allele frequencies exchange genes by migration or intermarriage. In general, the greater the difference in allele frequencies between the resident and the migrant individuals, and the larger the number of migrants, the greater effect the migrants have in changing the genetic constitution of the resident population.

Gene frequencies can change from one generation to another by a process of pure chance known as genetic drift. This occurs because the number of individuals in any population is finite, and thus the frequency of a gene may change in the following generation by accidents of sampling, just as it is possible to get more or fewer than 50 heads in 100 throws of a coin simply by chance.

The magnitude of the gene-frequency changes due to genetic drift is inversely related to the size of the population—the larger the number of reproducing individuals, the smaller the effects of genetic drift. The reason is similar to what happens with a coin toss. If you toss a coin 10 times, you may obtain only three heads (0.30 frequency) with a probability that is not very small. But if you toss a coin 1,000 times, it is extremely unlikely that you'll get heads with a frequency of only 0.30 (300 heads) or less. The effects of genetic drift in changing gene frequencies from one generation to the next are quite small in most natural populations, which generally consist of thousands of reproducing individuals. The effects over many generations are more important.

Genetic drift can have important evolutionary consequences when a new population becomes established by only a few individuals, a phenomenon known as the founder principle. The allelic frequencies present in these few colonizers are likely to differ at many loci from those in the population they left, and those differences have a lasting impact on the evolution of the new population. The colonization of the continents of the world, starting from Africa and between continents, by *Homo erectus* as well as by modern *Homo sapiens*, as well as the colonization of different regions of the same continent, was likely carried out at various times by relatively few individuals, which may have differed genetically by chance from the original population. For example, the absence of the B blood group among Native Americans is likely due to the chance absence of this relatively rare blood group among the original American colonizers.

1.2.8 Natural selection

Mutation, gene flow, and genetic drift change gene frequencies without regard for the consequences that such changes may have in the ability of the organisms to survive and reproduce; they are random processes with respect to adaptation. If these were the only processes of evolutionary change, the organization of living things would gradually disintegrate. The effects of such processes alone would be analogous to those of a mechanic who changed parts in an automobile engine at random, with no regard for the role of the parts in the engine.

Natural selection keeps the disorganizing effects of mutation and other processes in check because it multiplies beneficial mutations and eliminates harmful ones. Natural selection accounts not only for the preservation and improvement of the organization of living beings but also for their diversity. In different localities or in different circumstances, natural selection favors different traits, precisely those that make the organisms well adapted to their particular circumstances and ways of life.

The effects of natural selection are measured with a parameter called *fitness*. Fitness can be expressed as an absolute or as a relative value. Consider a population consisting at a certain locus of three genotypes: A_1A_1, A_1A_2, and A_2A_2. Assume that on average each A_1A_1 and each A_1A_2 individual produces one offspring but that each A_2A_2 individual produces two. One could use the average number of progeny left by each genotype as a measure of that genotype's absolute fitness over the generations. (This, of course, would require knowing how many of the progeny survive to adulthood and reproduce.) It is, however, mathematically more convenient to use relative fitness values (typically represented with the letter w). Evolutionists usually assign the value 1 to the genotype with the highest reproductive efficiency and calculate the other relative fitness values proportionally. For the example just used, the relative fitness of the A_2A_2 genotype would be $w = 1$ and that of each of the other two genotypes would be $w = 0.5$. A parameter related to fitness is the selection coefficient, often represented by the letter s, which is defined as $s = 1 - w$. The selection coefficient is a measure of the reduction in fitness of a genotype. The selection coefficients in the example are $s = 0$ for A_2A_2 and $s = 0.5$ for each A_1A_1 and A_1A_2.

Selection may favor one homozygote over the other and over the heterozygote, or may favor the heterozygote over both homozygotes. A particularly interesting example of heterozygote superiority among humans is provided by the gene responsible for sickle-cell anemia in places where malaria is rife. Human hemoglobin in adults is for the most part hemoglobin A, a four-component molecule consisting of two α and two β hemoglobin chains. The gene Hb^A codes for the normal β hemoglobin chain, which consists of 146 amino acids. A mutant allele of this gene, Hb^S, causes the β chain to have in the sixth position the amino acid valine instead of glutamic acid (Figure 1.14). This seemingly minor substitution modifies the properties of hemoglobin so that homozygotes with the mutant allele, Hb^SHb^S, suffer from a severe form of anemia that

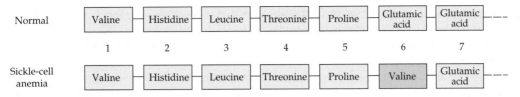

Figure 1.14 The first seven amino acids of the β chain of human hemoglobin; the β chain consists of 146 amino acids. A substitution of valine for glutamic acid at the sixth position is responsible for the severe disease known as sickle-cell anemia.

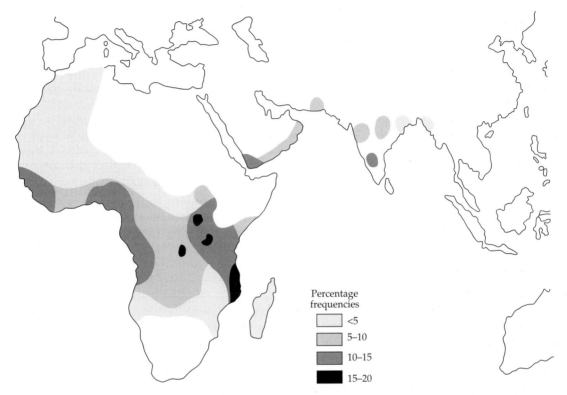

Figure 1.15 Geographic distribution of the allele Hb^S, which in the homozygous condition is responsible for sickle-cell anemia. The frequency of Hb^S is high in those regions of the world where *P. falciparum* malaria is endemic, because $Hb^A Hb^S$ individuals, heterozygous for the Hb^S and the "normal" allele, are highly resistant to malarial infection.

in most cases leads to death before the age of reproduction.

The Hb^S allele occurs in some African populations with a high frequency. This seems puzzling because of the severity of the anemia. The strong natural selection against the $Hb^S Hb^S$ homozygotes should have eliminated the defective allele. But the Hb^S allele occurs at high frequency precisely in regions of the world where a particularly severe form of malaria, caused by the parasite *Plasmodium falciparum*, is endemic (Figure 1.15). It was

hypothesized that the heterozygotes, $Hb^A Hb^S$, were resistant to malaria, whereas the homozygotes $Hb^A Hb^A$ were not. In malaria-infested regions then the heterozygotes survived better than either of the homozygotes, which were more likely to die from either malaria ($Hb^A Hb^A$ homozygotes) or anemia ($Hb^S Hb^S$ homozygotes). This hypothesis has been confirmed in various ways. Most significant is that most hospital patients suffering from severe or fatal forms of malaria are homozygotes $Hb^A Hb^A$. In a study of 100 children who died from malaria, only one was found to be a heterozygote, whereas 22 were expected to be so according to the frequency of the Hb^S allele in the population.

The malaria example illustrates the general principle that the fitness of genotypes depends on the environment. Thus, as mentioned earlier, dark skin is favored in the tropics where the incidence of ultraviolet (UV) radiation from the sun is high and may cause melanoma and other cancers. At high latitudes lighter skin may be favored because of the low-level UV radiation, which is required for synthesizing vitamin D in the lower layers of the dermis and is less likely to cause melanoma and other UV-radiation-induced diseases.

1.2.9 Modes of selection

The population density of organisms and the frequency of genotypes may impact the fitness of genotypes. Insects, for example, experience enormous yearly oscillations in density. Some genotypes may possess high fitness in the spring, when the population is rapidly expanding, because such genotypes yield more prolific individuals. Other genotypes may be favored during the summer, when populations are dense, because these genotypes make for better competitors, ones more successful at securing limited food resources.

The fitness of genotypes can also vary according to their relative numbers. Particularly interesting is the situation in which genotypic fitnesses are inversely related to their frequencies, a common situation that preserves genetic polymorphism in populations. Assume that two genotypes, A and B, have fitnesses related to their frequencies in such a way that the fitness of either genotype increases when its frequency decreases and vice versa. When A is rare, its fitness is high, and therefore A increases in frequency. As it becomes more and more common, however, the fitness of A gradually decreases, so that its increase in frequency eventually comes to a halt. A stable polymorphism occurs at the frequency where the two genotypes, A and B, have identical fitnesses.

Frequency-dependent selection may arise because the environment is heterogeneous in such

a way that different genotypes better exploit different subenvironments. When a genotype is rare, the subenvironments that it exploits better will be relatively abundant. But as the genotype becomes common, its favored subenvironment becomes saturated. Sexual preferences also may lead to frequency-dependent selection. It has been demonstrated in some insects, birds, mammals, and other organisms that the mates preferred often are those that are rare. People also seem to experience this rare-mate advantage: blonds may seem attractively exotic to brunettes, or brunettes to blonds.

Natural selection can be explored by examining its effects on the phenotypes of individuals in a population. Distribution scales of phenotypic traits such as height, weight, number of progeny, or longevity typically show greater numbers of individuals with intermediate values and fewer and fewer toward the extremes: this is the so-called normal distribution. By reference to this distribution, we may distinguish three modes of natural selection: stabilizing, directional, and diversifying (Figure 1.16).

When individuals with intermediate phenotypes are favored and extreme phenotypes are selected against, the selection is said to be stabilizing. The range and distribution of phenotypes then remains approximately the same from one generation to another. An example of selection favoring intermediate phenotypes is mortality among newborn infants, which is highest when they are either very small or very large.

Directional selection occurs when the distribution of phenotypes in a population changes systematically in a particular direction. The physical and biological aspects of the environment are changing continuously, and over long periods of time the changes may be substantial. The climate and even the configuration of the land or waters vary incessantly. Changes also take place in the biotic conditions; that is, in the other organisms present, whether predators, prey, parasites, or competitors. Genetic changes occur as a consequence, because the genotypic fitnesses may shift so that different sets of alleles are favored.

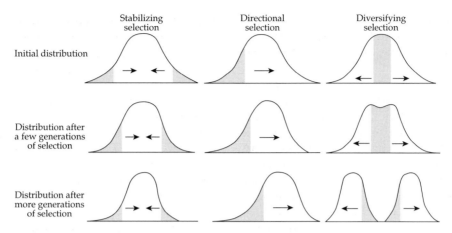

Figure 1.16 Three types of natural selection showing the effects of each on the distribution of phenotypes within a population. The shaded areas represent the phenotypes against which selection acts. Stabilizing selection acts against phenotypes at both extremes of the distribution, favoring the multiplication of intermediate phenotypes. Directional selection acts against only one extreme of phenotypes, causing a shift in distribution toward the other extreme. Diversifying selection acts against intermediate phenotypes, creating a split in distribution toward each extreme.

Over geologic time, directional selection leads to major changes in morphology and ways of life. Evolutionary changes that persist in a more or less continuous fashion over long periods of time are known as evolutionary trends. Directional evolutionary changes increased the cranial capacity of the human lineage from the small brain of *Australopithecus*—human ancestors of several million years ago—which was about $400\,cm^3$ in volume, to a brain more than three times as large in modern humans. Directional selection— particularly, long-term evolutionary trends—often does not occur in a continuous or sustained manner, but rather in spurts. Surely the increase in brain size from *Australopithecus* to *H. sapiens* did not occur at a constant rate of, say, so many cubic centimeters per thousand years.

Two or more divergent phenotypes in an environment may be favored simultaneously by diversifying selection. No natural environment is homogeneous; rather, the environment of any plant or animal population is a mosaic consisting of more or less dissimilar subenvironments. There is heterogeneity with respect to climate, food resources, and living space. Also, the heterogeneity may be temporal, with change occurring over time, as well as spatial. Species cope with environmental heterogeneity in diverse ways. One strategy is

genetic monomorphism, the selection of a generalist genotype that is well adapted to all the subenvironments encountered by the species. Another strategy is genetic polymorphism, the selection of a diversified gene pool that yields different genotypes, each adapted to a specific subenvironment.

1.2.10 Sexual selection

Sexual selection is a special form of natural selection. Other things being equal, organisms more proficient in securing mates have higher fitness. There are two general circumstances leading to sexual selection. One is the preference shown by one sex (often the females) for individuals of the other sex that exhibit certain traits. The other is increased strength (usually among the males) that yields greater success in securing mates. Sexual selection explains, for example, the presence of exorbitant antlers in male deer and the spectacular plumage of male peacocks. These traits would seem disadvantageous because of increased energy costs or exposure to predators, but have evolved by natural selection because they help to secure mates.

The presence of a particular trait among the members of one sex can make them somehow

Figure 1.17 Sexual dimorphism in the extinct Irish Elk, *Megaloceros* (left), and in a South American hummingbird, *Spathura underwoodi* (right, from Darwin). Males are on the right.

more attractive to the opposite sex. This type of sex appeal has been demonstrated experimentally in all sorts of animals, from vinegar flies to pigeons, mice, dogs, and rhesus monkeys. Sexual selection can also come about because a trait—the size of the antlers of a stag, for example—increases prowess in competition with members of the same sex. Stags, rams, and bulls use antlers or horns in contests of strength; a winning male usually secures more female mates. Therefore, sexual selection may lead to increased size and aggressiveness in males. Male baboons are more than twice as large as females, and the behavior of the docile females contrasts with that of the aggressive males. A similar dimorphism occurs in the northern sea lion, *Eumetopias jubata*, where males weigh about 1,000 kg (2,200 lb), about three times as much as females. The males fight fiercely in their competition for females; large, battle-scarred males occupy their own rocky islets, each holding a harem of as many as 20 females (Figure 1.17 shows other examples).

1.2.11 Kin selection

The apparent altruistic behavior of many animals is, like some manifestations of sexual selection, a trait that at first seems incompatible with the theory of natural selection. Altruism is a form of behavior that benefits other individuals at the expense of the one that performs the action; the fitness of the altruist is diminished by its behavior, whereas individuals that act selfishly benefit from it at no cost to themselves. Accordingly, it might be

expected that natural selection would foster the development of selfish behavior and eliminate altruism. This conclusion is not so compelling when it is noticed that the beneficiaries of altruistic behavior are usually relatives. They share part of their genes, including genes that promote altruistic behavior. Altruism may evolve by kin selection, which is simply a type of natural selection in which relatives (and therefore genes in common) are taken into consideration when evaluating an individual's fitness. The fitness of a gene or genotype that takes into account its presence in relatives is known as *inclusive fitness*.

Natural selection favors genes that increase the reproductive success of their carriers, but it is not necessary that all individuals that share a given genotype have higher reproductive success. It suffices that carriers of the genotype reproduce more successfully on average than those possessing alternative genotypes. Parental care is, therefore, a form of altruism readily explained by kin selection. The parent spends energy caring for the progeny because it increases the reproductive success of the parent's genes.

Kin selection extends beyond the relationship between parents and their offspring. It facilitates the development of altruistic behavior when the energy invested, or the risk incurred, by an individual is compensated in excess by the benefits ensuing to relatives. The closer the relationship between the beneficiaries and the altruist and the greater the number of beneficiaries, the higher the risks and efforts warranted in the altruist. Individuals that live together in a herd or troop usually are related and often behave toward each other in protective or helping ways. Adult zebras, for instance, will turn toward an attacking predator to protect the young in the herd rather than fleeing to protect themselves.

An extreme form of kin selection occurs in some species of bees, wasps, ants, and other social insects. We may use as an example the stingless bees, with hundreds of species in the tropics. These bees live in colonies, typically with a single queen, and hundreds or thousands of workers, which are morphologically different from the queen. The female workers build the hive, care for the young, and gather food, but they are sterile; the queen alone produces progeny. It would seem that the workers' behavior would in no way be promoted or maintained by natural selection. Any genes causing such behavior would seem likely to be eliminated from the population, because individuals exhibiting the behavior favor not their own reproductive success but that of the queen.

The expectations change, however, when we take into account the genetic make-up of these social insects, in which the females are diploid (have two sets of chromosomes), but the males are haploid (have only one set of chromosomes). This genetic structure is called *haplodiploidy* (Figure 1.18). Queens produce some eggs that remain unfertilized and develop into males, or drones, and are haploid, having a mother but no father. Their main role is to engage in the nuptial flight during which one of them fertilizes a new queen. Other eggs laid by queen bees are fertilized and develop into diploid females, the large majority of which are workers. In many species of social insects, the queen typically mates with a single male once during her lifetime; the male's sperm is stored in the queen's spermatheca, from which it is gradually released as she lays fertilized eggs. All the queen's female progeny therefore have the same father, so that workers are more closely related to one another and to any new sister queen than they are to the mother queen. The female workers receive one-half of their genes from the mother and one-half from the father, but they share among themselves three-quarters of their genes. This is because the half of the set from the father is the same in every worker, given that the father had only one set of genes rather than two (the male developed from an unfertilized egg, so all his sperm carry the same set of genes). The other half of the workers' genes come from the mother, and on the average half of them are identical in any two sisters. Consequently, with three-quarters of her genes present in her sisters (while only half of her genes would be passed on to a daughter), a worker's genes are transmitted one and a half times more effectively when she raises a sister (whether another worker or a new queen) than if she were to produce a daughter of

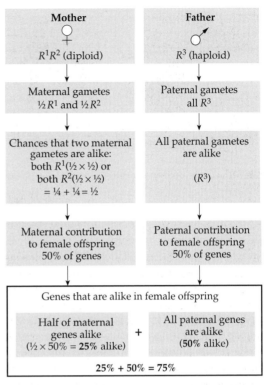

Figure 1.18 Haplodiploid reproduction of social hymenopterans (ants, bees, and wasps) with respect to any given gene, R. The gene has three allele forms, R^1 and R^2 in the diploid mother and R^3 in the haploid father. The probability that two daughters will both inherit R^1 is 1/4 and that both will inherit R^2 is also 1/4. Therefore, the probability that both daughters will inherit the same allele from the queen is 1/2. They will both inherit R^3 from the father. Each daughter inherits half of her genes from the mother and half from the father. Thus, the probabilities that the two sisters will have inherited the same genes from both parents are 1/4 for the mother genes (half of the genes with a 1/2 probability of being identical) and 1/2 for the father genes (half of the genes with probability 1) or 3/4 for both genes. Mother and daughters share only half of their genes.

her own. With such genetic population structure, natural selection will maximize the number of sterile female workers and minimize the number of reproductive females, which is accomplished by having only one queen.

1.2.12 Reciprocal altruism and group selection

Altruism also occurs among unrelated individuals when the behavior is reciprocal and the altruist's costs are smaller than the benefits to the recipient. This reciprocal altruism is found in the mutual grooming of chimpanzees and other primates as they clean each other of lice and other pests.

Altruistic behavior may also evolve by the so-called *group selection*, when populations (groups) with certain attributes (such as altruistic behaviors) will persist and multiply better than populations lacking such attributes. But group selection can occur only under very restrictive conditions. Within a population or group, an altruistic genotype will have lower fitness than a selfish genotype, because altruistic individuals incur a cost, from which selfish individuals benefit. Therefore, altruistic genotypes will tend to be eliminated from the population. But populations made up of selfish genotypes may become extinct more readily (for example, by over-exploiting food resources) than populations with altruistic genotypes. Altruism may evolve in a species if the rate of extinction of selfish populations is large compared to the rate at which selfish genotypes increase in frequency within populations. Evolutionists have shown that these restrictive conditions rarely occur in nature. We will return to the issues of altruistic behavior and group selection in section 10.4.

1.2.13 Species and speciation

Species come about as the result of gradual change prompted by natural selection. Environments differ from place to place and change in time. Natural selection favors different characteristics in different situations. The accumulation of differences between populations exposed to different environments may eventually yield different species.

External similarity is the common basis for identifying individuals as being members of the same species. Nevertheless, there is more to a species than outward appearance. A bulldog, a terrier, and a golden retriever are very different in appearance, but they are all dogs because they can interbreed. People can also interbreed with one another, and so can cats with other cats, but people cannot interbreed with dogs or cats, nor can these with each other. Although species are usually identified by appearance, there is something basic,

of great biological significance, behind similarity of appearance: individuals of a species are able to interbreed with one another but not with members of other species. Among sexual organisms, species are groups of interbreeding natural populations that are reproductively isolated from other such groups.

The ability to interbreed is of great evolutionary importance, because it determines that species are independent evolutionary units. Genetic changes originate in single individuals; they can spread by natural selection to all members of the species but not to individuals of other species. Individuals of a species share a common gene pool that is not shared by individuals of other species. Different species have independently evolving gene pools because they are reproductively isolated.

Although the criterion for deciding whether individuals belong to the same species (i.e. reproduction isolation) is clear, there may be ambiguity in practice for two reasons. One is lack of knowledge: it may not be known for certain whether individuals living in different sites belong to the same species, because it is not known whether they can interbreed naturally. The other reason for ambiguity is rooted in the nature of evolution as a gradual process. Two geographically separate populations that at one time were members of the same species may gradually diverge into two different species. Since the process is gradual, there is no particular point at which it is possible to say that the two populations have become two different species; that is, that there is one particular generation in which reproductive isolation is present, but it was not present in the previous generation.

A similar kind of ambiguity obtains when we compare ancestral and descendant populations living at different times. There is no way to test whether today's humans could interbreed with those who lived thousands of years ago. It seems reasonable that living people would be able to interbreed with people who lived a few generations earlier and look more or less like other people now living. But what about ancestors who lived thousands of generations earlier? There is no precise time at which *H. erectus* became *H. sapiens*, but it would not be appropriate to classify remote human ancestors and modern humans in the same species just because the changes from one generation to the next surely were small. It is useful to distinguish between two groups that look different and lived at different times by means of different species names, just as it is useful to give different names to childhood and adulthood even though no single moment can separate one from the other. Biologists distinguish species in organisms that lived at different times by means of a common-sense morphological criterion: if two organisms differ from each other in form and structure about as much as do two living individuals belonging to two different species, they are classified in separate species and given different names. Species that may be related as ancestral and descendant are called *chronospecies*. This is a matter to which we'll return later in this book because it is quite relevant in the study of fossils.

Given that species are groups of populations reproductively isolated from one another, asking about the origin of species is equivalent to asking how reproductive isolation arises between populations. This may occur as an incidental consequence of genetic divergence between populations that are geographically separated from one another. But reproductive isolation may be directly promoted by natural selection when populations are somewhat diverged, or adapted to different features of the environment, so that hybrids have low fitness. In the extreme, this occurs when hybrids are inviable or sterile. When hybrids have lower fitness than nonhybrids, genes will be favored by natural selection that reduce the probability of hybridization, and eventually complete reproductive isolation may ensue.

Geographic separation may result in complete reproductive isolation if it persists long enough. Consider, for example, the evolution of many endemic species of plants and animals in the Hawaiian archipelago (Table 1.2; Figure 1.19). The ancestors of these species arrived on these islands several million years ago. There they evolved as they became adapted to the environmental conditions and colonizing opportunities

Table 1.2 Species endemism in Hawaii

	Number of species	Percent endemic
Ferns	168	65
Flowering plants	1729	94
Land mollusks	1064	99+
Insects	3750	99+
Birds	71	99

Species are endemic in a place when they have evolved in that place and are not found naturally in any other locality.

present. Reproductive isolation between the populations evolving in Hawaii and the populations on continents was not as such promoted directly by natural selection; their geographic remoteness forestalled any opportunities for hybridizing. Nevertheless, reproductive isolation became complete in many cases as a result of gradual genetic divergence over thousands of generations.

1.2.14 Homology, analogy, and convergent evolution

Different species may exhibit features that are similar in appearance, structure, or function. The legs of dogs resemble the legs of leopards; bats and birds use wings for flying. Resemblances may be due to inheritance from a common ancestor or may have evolved independently as adaptations to similar functions. Correspondence of features in different organisms that is due to inheritance from a common ancestor is called *homology*. The forelimbs of humans, whales, dogs, and bats are homologous. The skeletons of these limbs are all constructed of bones arranged according to the same pattern because they derive from a common reptilian ancestor with similarly arranged forelimbs.

Correspondence of features due to similarity of function but not related to common descent is termed *analogy*. The wings of birds and of flies are analogous. These wings are not modified versions of a wing present in a common ancestor; rather they have evolved independently as adaptations to a common function, flying. Some features may be

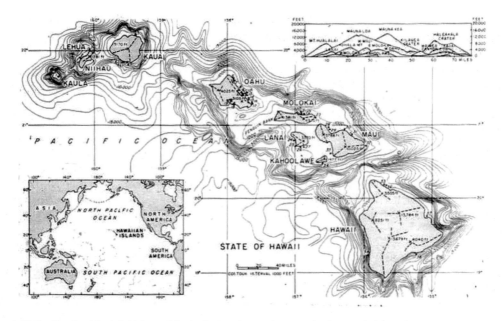

Figure 1.19 The Hawaiian Islands (with inset of the Pacific Ocean) more than 3,300 km (2,000 miles) away from the nearest continent. These volcanic islands formed between 5 Ma (Kauai) and more than half a million (Hawaii) years ago. (Some small islands or atolls northwest of Kauai are even older.)

partially homologous and partially analogous; for example, the wings of bats and birds. Their skeletal structure is homologous, due to common descent from the forelimb of a reptilian ancestor; but the modifications for flying are different and independently evolved, and in this respect they are analogous.

Features that become more rather than less similar through independent evolution are said to be convergent. *Convergence* is often associated with similarity of function, as in the evolution of wings in birds, bats, and flies. The shark (a fish) and the dolphin (a mammal) are much alike in external morphology; their similarities are due to convergence, since they have evolved independently as adaptations to aquatic life.

In section 1.3 we'll return to the distinctions between homologous, analogous, and convergent features, because they play a critical role in a prevailing theory of systematics known as cladistics. We will now turn, however, to the methods for reconstructing and representing evolutionary history, both anagenesis and cladogenesis, which are often represented as evolutionary trees.

1.2.15 Evolutionary trees

The evolution of all living organisms, or of a subset of them, can be represented as a tree, with branches that divide into two or more as time progresses, which represent the splitting of species (or higher taxonomic groups). Such trees are called phylogenies. Their branches represent evolving lineages, some of which eventually die out, while others persist in themselves or in their derived lineages down to the present time. Evolutionists are interested in the history of life and hence in the topology, or configuration, of phylogenies, which represent the splitting of taxa through time. They are concerned as well with the nature of the anagenetic changes within lineages (in morphology, function, behavior, genetic make-up, etc.) and with the timing of both anagenetic and cladogenetic events.

Evolutionary trees are hypotheses or models that seek to reconstruct the evolutionary history of taxa; that is, species or other groups of organisms, such as genera, families, or orders. The branching relationships of the trees reflect the relative relationships of ancestry, or cladogenesis.

Box 1.1 Species endemism in remote archipelagos

Species are called endemic when they have evolved in the place where they live and are not found in any other locality naturally; that is, unless they have been introduced by humans. Endemism is particularly apparent when colonizers reach geographically remote areas, such as islands, where they find few or no competitors and have an opportunity to diverge as they become adapted to the new environment.

Many examples of endemism are found in archipelagoes removed from the mainland. The Galápagos Islands are about 1,000 km (600 miles) off the west coast of South America. When Charles Darwin arrived there in 1835 during his voyage on the *HMS Beagle*, he discovered many species not found anywhere else in the world: for example, several species of finches, of which 14 are now known to exist (called Darwin's finches). These passerine birds have adapted to a diversity of habitats and diets, some feeding mostly on plants, others exclusively on insects. The various shapes of their bills are clearly adapted to probing, grasping, biting, or crushing: the diverse ways in which the different Galápagos species obtain their food.

The explanation for such diversity is that the ancestor of Galápagos finches arrived in the islands before other kinds of birds and encountered an abundance of unoccupied ecological niches. Its descendants underwent adaptive radiation, evolving a variety of finch species with ways of life capable of exploiting opportunities that on various continents are already exploited by other species.

The Hawaiian archipelago also provides striking examples of endemism. Its several volcanic islands, ranging from nearly 1 million to more than 5 million years in age, are far from any continent or even other large islands. In their relatively small total land area, an astounding number of plant and animal species exist. Most of the species have evolved on the islands, among them about two dozen species (about one-third of them now extinct) of honeycreepers, birds of the family Drepanididae, all derived from a single immigrant form. In fact, all but one of Hawaii's 71 native bird species are endemic. More than 90% of the native species of flowering plants, land mollusks, and insects are also endemic, as are two-thirds of the 168 species of ferns (see Table 1.2).

Thus, in Figure 1.20, humans and rhesus monkeys are seen to be more closely related to each other than either is to the horse. Stated another way, this tree shows that the most recent common ancestor to all three species lived in a more remote past than the most recent common ancestor to humans and monkeys.

Evolutionary trees may also indicate the changes that have occurred along each lineage, or anagenesis. Thus, in the evolution of cytochrome *c* since the last common ancestor of humans and rhesus monkeys, one amino acid changed in the lineage going to humans but none in the lineage going to rhesus monkeys (Figure 1.20). In cladistic representations, decisive anagenetic changes that account for the configuration of the tree are marked by notches or otherwise along the branch leading to a particular taxon.

There exist several methods for constructing evolutionary trees. Some were developed for interpreting morphological data, others for interpreting molecular data; some can be used with either kind of data. The main methods currently in use are called distance, maximum parsimony, and maximum likelihood.

Distance methods are used primarily with molecular data, but also with morphological information. A *distance* is the number of differences between two taxa. The differences are measured with respect to certain traits (such as morphological features) or to certain macromolecules (the sequence of amino acids in proteins or the sequence of nucleotides in DNA or RNA). The tree illustrated in Figure 1.20 was obtained by taking into account the distance, or number of amino acid differences, between three organisms with respect to a particular protein (cytochrome *c*). Table 1.3 shows the (minimum) number of nucleotide differences in the genes of 20 species that account for the amino acid differences in their cytochrome *c*. An evolutionary tree based on the data in that table, showing the numbers of nucleotide changes in each branch, is illustrated in Figure 1.21.

	1–8	9	10									20
Human	——	Gly–Asp–Val–Glu–Lys–Gly–Lys–Lys–Ile –Phe– Ile –Met–
Rhesus monkey	——	Gly–Asp–Val–Glu–Lys–Gly–Lys–Lys–Ile –Phe– Ile –Met–
Horse	——	Gly–Asp–Val–Glu–Lys–Gly–Lys–Lys–Ile –Phe– Val –Gln –

21   30   40
Lys–Cys–Ser–Gln–Cys–His–Thr–Val–Glu–Lys–Gly–Gly–Lys–His–Lys–Thr–Gly–Pro–Asn–Leu–
Lys–Cys–Ser–Gln–Cys–His–Thr–Val–Glu–Lys–Gly–Gly–Lys–His–Lys–Thr–Gly–Pro–Asn–Leu–
Lys–Cys– Ala –Gln–Cys–His–Thr–Val–Glu–Lys–Gly–Gly–Lys–His–Lys–Thr–Gly–Pro–Asn–Leu–

41   50   60
His–Gly–Leu–Phe–Gly–Arg–Lys–Thr–Gly–Gln–Ala–Pro–Gly–Tyr–Se r–Tyr–Thr–Ala–Ala–Asn–
His–Gly–Leu–Phe–Gly–Arg–Lys–Thr–Gly–Gln–Ala–Pro–Gly–Tyr–Se r–Tyr–Thr–Ala–Ala–Asn–
His–Gly–Leu–Phe–Gly–Arg–Lys–Thr–Gly–Gln–Ala–Pro–Gly– Phe– Thr –Tyr–Thr– Asp –Ala–Asn–

61   70   80
Lys–Asn–Lys–Gly–Ile– Ile –Trp–Gly–Glu–Asp–Thr–Leu–Met–Glu–Tyr–Leu–Glu–Asn–Pro–Lys–
Lys–Asn–Lys–Gly–Ile–Thr–Trp–Gly–Glu– Asp –Thr–Leu–Met–Glu–Tyr–Leu–Glu–Asn–Pro–Lys–
Lys–Asn–Lys–Gly–Ile–Thr–Trp– Lys – Glu– Glu –Thr–Leu–Met–Glu–Tyr–Leu–Glu–Asn–Pro–Lys–

81   90   100
Lys–Tyr– Ile –Pro–Gly–Thr–Lys–Met– Ile –Phe–Val–Gly – Ile –Lys–Lys–Lys–Glu–Glu–Arg–Ala–
Lys–Tyr– Ile –Pro–Gly–Thr–Lys–Met– Ile –Phe–Val–Gly – Ile –Lys–Lys–Lys–Glu–Glu–Arg–Ala–
Lys–Tyr– Ile –Pro–Gly–Thr–Lys–Met– Ile –Phe– Ala –Gly – Ile –Lys–Lys–Lys– Thr –Glu–Arg– Glu –

101   110   112
Asp–Leu–Ile–Ala–Tyr–Leu–Lys–Lys–Ala–Thr–Asn–Glu
Asp–Leu–Ile–Ala–Tyr–Leu–Lys–Lys–Ala–Thr–Asn–Glu
Asp–Leu–Ile–Ala–Tyr–Leu–Lys–Lys–Ala–Thr–Asn–Glu

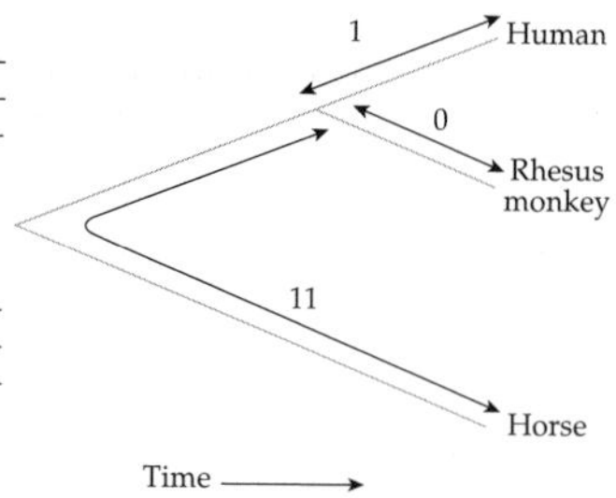

Figure 1.20 Left: the 104 amino acids in cytochrome *c* of human, rhesus monkey, and horse. The human sequence is shown at the top. Humans differ from monkeys by one amino acid and from horse by 12 amino acids; monkey and horse differ by 11 amino acids. Right: the phylogeny of human, rhesus monkey, and horse, based on their cytochrome *c* sequences. The one difference between human and monkey, at site 66, is due to a change in the human lineage, since monkey and horse are identical at this site. The numbers represent the number of amino acid changes in each lineage.

Morphological data also can be used for constructing distance trees. The first step is to obtain a distance matrix, such as that making up Table 1.3, but one based on a set of morphological comparisons between species or other taxa. For example, in some insects one can measure body length, wing length, wing width, number and length of wing veins, or another trait.

A most common procedure to transform a distance matrix into a phylogeny is called cluster analysis. The distance matrix is scanned for the smallest distance element, and the two taxa involved (say, A and B) are joined at an internal node, or branching point. The matrix is scanned again for the next smallest distance, and the two new taxa (say, C and D) are clustered.

The procedure is continued until all taxa have been joined. When a distance involves a taxon that is already part of a previous cluster (say, E and A), the average distance is obtained between the new taxon and the preexisting cluster (say, the average distance from E to A and from E to B). This simple procedure, which can be used with morphological as well as molecular data, assumes that the rate of evolution is uniform along all branches.

Some distance methods relax the condition of uniform rate and allow for unequal rates of evolution along the branches. One of the most extensively used methods of this kind is called *neighbor joining*. The method starts, as before, by identifying the smallest distance in the matrix and linking the

Table 1.3 Minimum number of nucleotide differences in the genes coding for cytochrome *c* in 20 species

Species	1	2	3	4	5	6	7	8	9	10	11	12	13	14	15	16	17	18	19	20
1. Human	–	1	13	17*	16	13	12	12	17	16	18	18	19	20	31	33	36	63	56	66
2. Monkey		–	12	16*	15	12	11	13	16	15	17	17	18	21	32	32	35	62	57	65
3. Dog			–	10	8	4	6	7	12	12	14	14	13	30	29	24	28	64	61	66
4. Horse				–	1	5	11	11	16	16	16	17	16	32	27	24	33	64	60	68
5. Donkey					–	4	10	12	15	15	15	16	15	31	26	25	32	64	59	67
6. Pig						–	6	7	13	13	13	14	13	30	25	26	31	64	59	67
7. Rabbit							–	7	10	8	11	11	11	25	26	23	29	62	59	67
8. Kangaroo								–	14	14	15	13	14	30	27	26	31	66	58	68
9. Duck									–	3	3	3	7	24	26	25	29	61	62	66
10. Pigeon										–	4	4	8	24	27	26	30	59	62	66
11. Chicken											–	2	8	28	26	26	31	61	62	66
12. Penguin												–	8	28	27	28	30	62	61	65
13. Turtle													–	30	27	30	33	65	64	67
14. Rattlesnake														–	38	40	41	61	61	69
15. Tuna															–	34	41	72	66	69
16. Screwworm fly																–	16	58	63	65
17. Moth																	–	59	60	61
18. Neurospora																		–	57	61
19. Saccharomyces																			–	41
20. Candida																				–

Source: from Fitch and Margoliash (1967).

* These numbers are for nucleotide differences, which are greater than the amino acid differences shown in Figure 1.21.

two taxa involved. The next step is to remove these two taxa and calculate a new matrix in which their distances to other taxa are replaced by the distance between the node linking the two taxa and all other taxa. The smallest distance in this new matrix is used for making the next connection, which will be between two other taxa or between the previous node and another taxon. The procedure is repeated until all taxa have been connected with one another by intervening nodes.

Maximum-parsimony methods seek to reconstruct the tree that requires the fewest number of changes (i.e. it is the most parsimonious) summed along all branches. This is a reasonable assumption, because it usually will be the most likely. But evolution may not necessarily have occurred following a minimum path, because the same change instead may have occurred independently along different branches, and some differences may have involved intermediate steps that are not apparent in the organisms now living.

Not all evolutionary changes, even those that involve a single step, may be equally probable. For example, among the four nucleotide bases in DNA, cytosine (C) and thymine (T) are members of a family of related molecules called pyrimidines; likewise, adenine (A) and guanine (G) belong to a family of molecules called purines. A change within a DNA sequence from one pyrimidine to another (C $\leftrightarrow$ T) or from one purine to another (A $\leftrightarrow$ G), called a transition, is more likely to occur than a change from a purine to a pyrimidine or the converse (G or A $\leftrightarrow$ C or T), called a transversion. Parsimony methods take into account different probabilities of occurrence if they are known.

Maximum-parsimony methods are related to cladistics (see section 1.3), a very formalistic theory of taxonomic classification, used extensively with morphological and paleontological data. The critical feature in cladistics is the identification of derived shared traits, called synapomorphic traits.

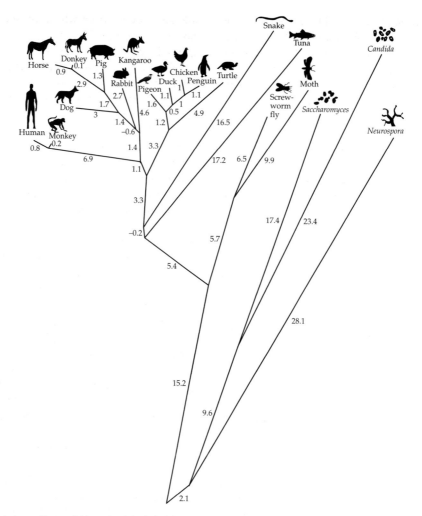

Figure 1.21 Evolutionary history of 20 species, based on the cytochrome *c* amino acid sequence. The common ancestor (at the bottom) of yeast and humans lived more than 1 billion years ago. The numbers on the branches are the estimated minimum number of nucleotide substitutions along the branch. Although fractional (or negative) numbers of nucleotide substitutions cannot occur, the numbers given are those that best fit the data.

A synapomorphic trait is shared by some taxa but not others because the former inherited it from a common ancestor that acquired the trait after its lineage separated from the lineages going to other taxa. In the evolution of carnivores, for example, domestic cats, tigers, and leopards are clustered together because of their possessing retractable claws, a trait acquired after their common ancestor branched off from the lineage leading to dogs, wolves, and coyotes. It is important to ascertain that the shared traits are homologous rather than analogous. For example, mammals and birds, but not lizards, have a four-chambered heart. Yet birds are more closely related to lizards than to mammals; the four-chambered heart evolved independently in the bird and mammal lineages, by parallel (or convergent) evolution.

Maximum-likelihood methods seek to identify the most likely tree, given the available data. They require that an evolutionary model be identified, which would make it possible to estimate the probability of each possible individual change. For example, as is mentioned above, transitions are more likely than transversions among DNA nucleotides, but a particular probability must be assigned to each. All possible trees are considered. The probabilities for each individual change are multiplied for each tree. The best tree is the one with the highest probability (or maximum likelihood) among all possible trees.

Maximum-likelihood methods are computationally expensive when the number of taxa is large, because the number of possible trees (for each of which the probability must be calculated) grows factorially with the number of taxa. With 10 taxa, there are about 3.6 million possible trees; with 20 taxa, the number of possible trees is about 2 followed by 18 zeros (2×10^{18}). Even with powerful computers, maximum-likelihood methods can be prohibitive if the number of taxa is large. Heuristic methods exist in which only a subsample of all possible trees is examined and thus an exhaustive search is avoided.

The statistical degree of confidence of a tree can be estimated for distance and maximum-likelihood trees. The most common method is called bootstrapping. It consists of taking samples of the data by removing at least one data point at random and then constructing a tree for the new data-set. This random sampling process is repeated hundreds or thousands of times. The bootstrap value for each node is defined by the percentage of cases in which all species derived from that node appear together in the trees. Bootstrap values above 90% are regarded as statistically strongly reliable; those below 70% are considered unreliable.

1.2.16 Gene duplication

Similarity between features due to common descent is called homology and the traits are called homologous, as mentioned above. Two kinds of homologous trait can be distinguished, orthologous and paralogous, a distinction that is particularly helpful with respect to genes and other genetic features. Orthologous genes are descendants of an ancestral gene that was present in the ancestral species from which the species in question have evolved. The evolution of orthologous genes therefore reflects the evolution of the species in which they are found. The cytochrome *c* molecules of the 20 organisms shown in Figure 1.21 are orthologous, because they derive from a single ancestral gene present in a species ancestral to all 20 organisms.

Paralogous genes are descendants of a duplicated ancestral gene. Paralogous genes, therefore, evolve within the same species (as well as in different species). The genes coding for the α, β, γ, and δ hemoglobin chains in humans are paralogous. The evolution of paralogous genes reflects differences that have accumulated since the genes duplicated. Homologies between paralogous genes serve to establish gene phylogenies; that is, the evolutionary history of duplicated genes within a given lineage.

Figure 1.22 is a phylogeny of the gene duplications giving rise to the myoglobin and hemoglobin genes found in modern humans. Hemoglobin molecules are tetramers, consisting of two polypeptides of one kind and two of another kind. In embryonic hemoglobin E, one of the two kinds of polypeptide is designated ε; in fetal hemoglobin F, it is γ; in adult hemoglobin A, it is β; and in adult hemoglobin A_2, it is δ. (Hemoglobin A makes up about 98% of human adult hemoglobin, and hemoglobin A_2 about 2%.) The other kind of polypeptide in embryonic hemoglobin is ζ; in both fetal and adult hemoglobin, it is α. There are yet additional complexities. Two γ genes exist (known as G_γ and A_γ), as do two α genes ($α_1$ and $α_2$). Furthermore, there are two β pseudogenes ($\psi\beta_1$ and $\psi\beta_2$) and two α pseudogenes ($\psi\alpha_1$ and $\psi\alpha_2$), as well as a ζ pseudogene. These pseudogenes are very similar in nucleotide sequence to the corresponding functional genes, but they include terminating codons and other mutations that make it impossible for them to yield functional hemoglobins. The similarity in the nucleotide sequence of the polypeptide genes, and pseudogenes, of both the α and β gene families indicates that they are all paralogous, arisen through various duplications and subsequent evolution from a gene ancestral to all.

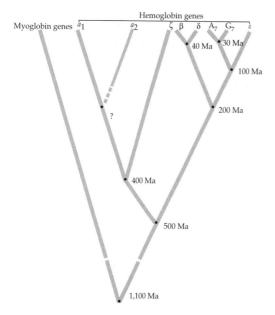

Figure 1.22 Phylogeny of the globin genes. The dots indicate points at which ancestral genes duplicated, giving rise to new gene lineages. The approximate times when these duplications occurred are indicated in million of years ago (Ma).

1.2.17 The molecular clock of evolution

In paleontology, the time sequence of fossils is determined by the age of rocks in which they are embedded as well as by other methods described in the chapters that follow. If the age of the rocks or of the fossils is determined, the evolutionary history of the organisms can be timed. Studies of molecular evolution rates have led to the proposition that DNA and proteins may serve as evolutionary clocks.

It was first observed in the 1960s that the numbers of amino acid differences between homologous proteins of any two given species seemed to be nearly proportional to the time of their divergence from a common ancestor. If the rate of evolution of a protein or gene were approximately the same in the evolutionary lineages leading to different species, proteins and DNA sequences would provide a molecular clock of evolution. The sequences could then be used to reconstruct not only the sequence of branching events of a phylogeny but also to determine the time when the various events occurred.

Consider, for example, Figure 1.21. If the substitution of nucleotides in the gene coding for cytochrome c occurred at a constant rate through time, we could determine the time elapsed along any branch of the phylogeny simply by examining the number of nucleotide substitutions along that branch. We would need only to calibrate the clock by reference to an outside source, such as the fossil record, that would provide the actual geologic time elapsed in at least one specific lineage or since one branching point. For example, if the time of divergence between insects and vertebrates is determined to have occurred 700 Ma, other times of divergence can be determined by proportion of the number of amino acid changes.

The molecular evolutionary clock is not expected to be a metronomic clock, like a watch or other timepieces that measure time exactly, but a stochastic (probabilistic) clock, like radioactive decay. In a stochastic clock the probability of a certain amount of change is constant (for example, a given quantity of atoms of radium-226 is expected, through decay, to be reduced by half in 1,620 years, its half-life), although some variation occurs in the actual amount of change. Over fairly long periods of time a stochastic clock is quite accurate. The enormous potential of the molecular evolutionary clock lies in the fact that each gene or protein is a separate clock. Each clock ticks at a different rate—the rate of evolution characteristic of a particular gene or protein—but each of the thousands and thousands of genes or proteins provides an independent measure of the same evolutionary events.

Evolutionists have found that the amount of variation observed in the evolution of DNA and proteins is greater than is expected from a stochastic clock—in other words, the clock is *overdispersed*, or somewhat erratic. The discrepancies in evolutionary rates along different lineages are not excessively large, however. So it is possible, in principle, to time phylogenetic events with considerable accuracy, but more genes or proteins (about two to four times as many) must be examined than would be required if the clock were stochastically constant in order to achieve a desired degree of accuracy. The average rates

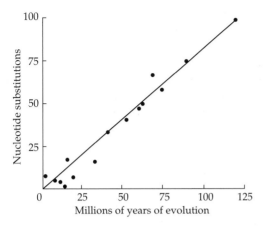

Figure 1.23 The molecular clock of evolution. The number of nucleotide substitutions for seven proteins in 17 species of mammals have been estimated for each comparison between pairs of species whose ancestors diverged at the time indicated in the abscissa. Each dot represents the sum of the number of substitutions for the seven proteins. The line has been drawn from the origin to the outermost point and corresponds to a rate of 0.41 nucleotide substitutions per million years for all seven proteins combined. The proteins are cytochrome c, fibrinopeptides A and B, hemoglobins α and β, myoglobin, and insulin c-peptide.

obtained for several proteins taken together become a fairly precise clock, particularly when many species are studied.

This conclusion is illustrated in Figure 1.23, which plots the cumulative number of nucleotide changes in seven proteins against the dates of divergence of 17 species of mammals (16 pairings) as determined from the fossil record. The overall rate of nucleotide substitution is fairly uniform. Some primate species (represented by the points below the line at the lower left of the figure) appear to have evolved at a slower rate than the average for the rest of the species. This anomaly is not unusual because the more recent the divergence of any two species, the more likely it is that the changes observed will depart from the average evolutionary rate. As the length of time increases, periods of rapid and slow evolution in any lineage tend to cancel one another out.

In the reconstruction of evolutionary history, molecular evolutionary studies have several notable advantages over paleontology, comparative anatomy, and the other classical disciplines. One is that comparisons can be made between very

different sorts of organisms. There is very little that comparative anatomy can say when, for example, organisms as diverse as yeasts, pine trees, and human beings are compared, but there are numerous DNA and protein sequences that can be compared in all three. A second advantage is multiplicity. Each organism possesses thousands of genes and proteins, which all reflect the same evolutionary history. If the investigation of one particular gene or protein does not satisfactorily resolve the evolutionary relationship of a set of species, additional genes and proteins can be investigated until the matter has been settled. Moreover, the widely different rates of evolution of different sets of genes opens up the opportunity for investigating different genes in order to achieve different degrees of resolution in the tree of evolution (see Figure 1.24). Evolutionists rely on slowly evolving genes for reconstructing remote evolutionary events, but increasingly faster-evolving genes for reconstructing the evolutionary history of more recently diverged organisms.

1.3 The classification of living beings

Taxonomy is the discipline that deals with the classification of organisms on the basis of their similarities and differences. Traditionally the traits used for this task have been primarily morphological. Aristotle (384–322 BC) and others in classical Greece had already developed a system of classification of organisms (and also inanimate objects) according to a hierarchy based on the degree of similarity. The Aristotelian classification system was further developed by Porphyry (AD 233–309) and centuries later by some medieval philosophers and naturalists, among which St. Albert the Great (c.1200–1280) stands out during the thirteenth century. The foundations of the modern system of classification of organisms were formulated in the eighteenth century by the Swede Carolus Linnaeus (1707–1778) in his book *Systema Naturae* (Linnaeus, 1735; Figure 1.25).

Linnaeus achieved his taxonomic hierarchy by grouping organisms according to degree of similarity. He established seven categories made up of groups of increasing inclusiveness: species are

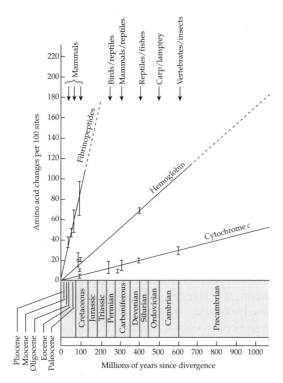

Figure 1.24 Three proteins with different rates of evolution. Cytochrome *c* evolves slowly, fibrinopeptides evolve quickly, and hemoglobin at an intermediate rate. The lines for each protein represent its average rate of evolution. The vertical lines encompass the variation observed.

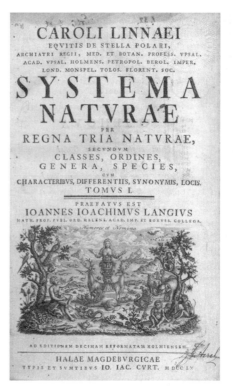

Figure 1.25 In his book *Systema Naturae* Carolus Linnaeus described his system of classification where every organism is designated with a double name written in italics: genus (with an intial capital letter) and species, as in *Homo sapiens*.

grouped into genera, genera into families, families into orders, orders into classes, classes into phyla and, finally, phyla into kingdoms. Intermediate categories were added later.

A taxon (plural: taxa) is a grouping within which organisms are classified. Thus, *Homo sapiens* is the taxon corresponding to modern humans, regarding their genus and species. But if we move up the classification hierarchy, *Homo sapiens* belongs to the tribe Hominin, the family Hominidae, the order Primata, the class Mammalia, the phylum Chordata and the kingdom Metazoa (see Table 1.4).

Linnaeus gave no scientific justification for his system of classification other than similarity. Towards the end of the eighteenth and beginning of the nineteenth century, the French biologist Jean Baptiste de Lamarck (1809) devoted much of his work to the systematic classification of

organisms, and suggested an explanation for the resemblance-based hierarchy: degree of similarity was a consequence of evolution, a gradual transition from some kinds of organisms to others. Lamarck's (1809) evolutionary theory had little influence among contemporary or later biologists, because it was metaphysical rather than biological. Lamarck's theory of evolution postulates that all organisms have an innate tendency towards improvement over time, which will continue forever and follow again and again the same path. Our ancestors of eons of time ago were worms and today's worms will have humans as their descendants eons of time hence. Although Lamarck's evolution theory was wrong, his intuitions were correct in seeking an explanation of similarity in the degree of evolutionary relationship. A scientific understanding of the similarity relations among

Table 1.4 Classification of three animals: human, lion, and a certain kind of mosquito

Category	Human	Lion	Mosquito
Kingdom	Metazoa	Metazoa	Metazoa
Phylum	Chordata	Chordata	Arthropoda
Class	Mammalia	Mammalia	Insecta
Order	Primata	Carnivora	Diptera
Family	Hominidae	Felidae	Culicidae
Genus	*Homo*	*Felis*	*Culex*
Species	*Homo sapiens*	*Felix leo*	*Culex pipiens*

Categories are used to classify organisms (species, genus, family, order, class, phylum, and kingdom). They are like drawers in which organisms are placed, so that smaller drawers are included in larger drawers. The labels that we place in each drawer are called taxa. *Homo sapiens* refers to a drawer at the species level; Hominids refers to a drawer at the family level.

Box 1.2 Tribe Hominin

For reasons we'll give in the next chapter, in this book we include modern humans and their direct and collateral ancestors, which are not the ancestors of any ape, in the tribe Hominin. Thus, we will refer to them as hominins. Tribe is a category below family, but above genus.

---

organisms came from Darwin's theory of evolution by natural selection.

Modern evolutionary theory offers a causal explanation for the similarities among living beings. Organisms evolve by means of a process of descent with modification. Changes, and thus differences, accumulate gradually over the generations. So, if the last common ancestor of two species is recent, they will have accumulated few differences. This is the same as saying that similarities in form and function reflect phylogenetic proximity. It follows that phylogenetic affinities can be inferred from the degrees of similarity. This principle currently is the scientific foundation for the reconstruction of phylogenetic relationships based on comparative analyses of living organisms through anatomical, taxonomical, embryological, molecular, and biogeographical studies.

The reconstruction of phylogeny faces several problems, in addition to occasional incompleteness of information. One important but well-understood difficulty comes from distinction between similarities that have an evolutionary origin and those that have come about independently as a result of adaptation to similar environments or ways of life, known, respectively, as *homology* and *analogy*.

The concept of homology was defined in 1843 by the biologist Richard Owen (independently of evolutionary theory) as "the same organ in different animals under every variety of form and function." (Owen, 1843). Nowadays, homology is explained in evolutionary terms. Two characters (such as human arms and dog forelimbs) are homologous when the resemblance between them reflects the presence of the same features (the various bones and muscles and their configuration and relative position) in a common ancestor from which the two current species inherited them. Analogy applies to similarities that originated independently in different lineages because they serve similar functions. For instance, the wings of bats, birds, and butterflies are analogous. These structures were not inherited from a common ancestor with wings, but evolved separately in each lineage as an adaptation to flight.

The degree of detail in the resemblance provides a practical way to distinguish between homology and analogy. Homology involves detailed similarity (as is the case with each of the bones and muscles of human arms and dog forelegs). Analogy involves similarities in the global configuration (the wings of butterflies, as those of eagles and bats, are wide, thin surfaces) but not in the details of structure and organization (the components of the wings of butterflies, birds, and mammals are very different; Figure 1.26).

Explicit principles to assess the evidence used in taxonomic classification and phylogenetic reconstruction have been formulated since 1950. In addition to traditional criteria, accumulated through the experience of evolutionists during the nineteenth and early twentieth centuries, two

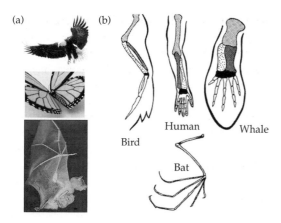

(a)   (b)

Bird

Human   Whale

Bat

Figure 1.26 Analogy and homology. (a) The wings of different animals carry out the same function of flying, but do so through different structures separately fixed in different evolutionary lineages: they are analogous. (b) The forelimb bones of mammals are very similar although some are terrestrial, others aquatic and yet others fly: they have similar bones organized in similar ways: they are homologous.

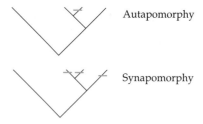

Autapomorphy

Synapomorphy

Figure 1.27 Autapomorphic and synapomorphic traits. In primates, an opposable thumb is synapomorphic (it is present in all primate lineages), whereas bipedalism is an autapomorphy of the tribe Hominini (it is not present in any other taxa). This consideration of characters is relative: the trait bipedalism is synapomorphic when considered from the perspective of the various hominin taxa (*Homo, Australopithecus*, etc.) because they all share this derived character.

new theories of classification emerged, known as *phenetics*, or numerical taxonomy, and *cladistics*.

Similarity is the basis of phenetics, but its methods seek to avoid subjectivity. Phenetics proceeds by formulating numerical algorithms, known as phenograms, in which each character can take one of two states: present or absent (they can be morphological characters, such as the thumb, or an amino acid in a particular protein, such as valine at position six in hemoglobin β). Each character receives a zero (if it is absent) or a one (if it is present) for each species (or higher-ranking categories, such as genera, families, or classes). The degree of phenetic affinity among different taxa is determined by the number of ones in the strings of zeroes and ones. This measure does not necessarily reflect evolutionary affinity: it only indicates the extent to which two organisms are similar in form. Indeed, phenetics seeks to avoid any theoretical underpinnings (such as evolution). It does not address the reason behind the resemblances.

Cladistics, on the contrary, starts from the requisite that species (or other taxa) be classified according to their phylogenetic relationship, rather than on their degree of morphological or phenetic similarity. The graphical representation

of phylogenetic relationships is a cladogram: a branching diagram where one branch splits into two whenever one species (or other taxon) splits into two species (or other taxa).

Cladistics distinguishes between primitive or ancestral characters, known as *plesiomorphic* characters, and derived or *apomorphic* characters. When an apomorphic character is present in two or more descendant taxa, it is called *synapomorphic* (meaning jointly derived); if an apomorphic trait is present in only one of the descendant taxa, the trait is *autapomorphic* (autonomously derived; see Figure 1.27). Primitive characters in any lineage are those that were already present in the ancestors. Derived characters are those that have just appeared in the lineage.

Cladistics establishes precise rules to determine phylogenetic relations. For instance, similarities based on primitive (plesiomorphic) characters are not useful for determining relationships among descendant taxa. Characters present in only one descendant taxon (autapomorphic) are also useless for determining phylogenetic relationships. Only shared derived characters (synapomorphies) are useful to determine phylogenetic relations. For example, mammary glands and hair are found in mammals, which groups them together and separates them from birds, reptiles, and fishes. From this perspective mammary glands are a synapomorphy shared by all mammals and differentiating them from other vertebrates. However, if we

EVOLUTION, GENETICS, AND SYSTEMATICS **37**

want to classify different mammals, mammary glands are plesiomorphic, a primitive character that all current mammals have inherited from the first mammals. The lack of placenta is a plesiomorphy of birds, reptiles, and fishes (that is to say, a character they inherited from a common ancestor), which does not tell us anything about the phylogenetic relations among these three groups of organisms.

1.3.1 The taxonomic concept of species

The purpose of grouping organisms in different categories is strictly taxonomic and might, at first, seem purely *nominalist*. The objective is to order the diversity of living beings by organizing them into manageable sets. We could use similar rules (whether phenetic or cladistic) to classify any other objects, such as ceramics, books, or cars. It suffices to specify the distinctive traits that are relevant to the different levels of the classification. In the case of organisms, these levels are species, genus (which includes similar species), family (which includes related genera), and so on. Linnaeus followed such a practice and defined, for example, the order Primates as animals with two pectoral breasts and four parallel superior incisors. Thereby, Linnaeus provided the distinctive traits that justified grouping certain species, genera, and families as primates (although today we do not include animals such as bats in this order, as Linnaeus did).

What about the concept of species used in human paleontology to distinguish hominin taxa that appear contemporaneously or sequentially in the fossil record? The fourth edition of the *International Code of Zoological Nomenclature* (ICZN) states that "the Code refrains from infringing upon taxonomic judgments, which must not be made subject to regulation or restraint". Scientists are entitled, if they so wish, to classify any new-found exemplar in a new species. The freedom to create species is only limited by their later acceptance or rejection by specialists in each discipline: botany, zoology, paleontology, and so on.

It seems clear that such a free-wheeling practice to name new species can lead to uncertainty and other difficulties. The differences among current conceptions of human evolution are a good example of this: the validity of many hominin taxa has been repeatedly challenged. For example, do Neanderthals belong to our own species, *Homo sapiens*? A conclusive answer to this and similar questions requires that two conditions be met. First, that we use an adequate concept of species, one which goes beyond arbitrary nominalism. Second, that we have adequate information to decide the species to which a certain exemplar belongs.

Can a nonnominalist concept of species be established? The answer is yes. That there is an objective basis to identify species is supported by the observation that common names in different languages correspond to the same organisms and coincide with scientific classifications, as noted, for example, by Mayr (1976). Margaret Mead (1966) noted that the "abominations" mentioned in *The Bible* (Leviticus) correspond to distinct species. At least since the times of Aristotle it has been commonly accepted that living beings have characteristic traits that justify grouping them in a nonarbitrary way. Tigers are different from lions and, thus, they can be separated into two sets (species), the set of lions and the set of tigers. As Aristotle pointed out, the traits that distinguish those sets maintain their integrity generation after generation, so that species are stable sets of organisms.

The most widely accepted concept of species was formulated by Dobzhansky (1935, 1937) and promoted extensively by Mayr (1942, 1963, 1970). It is based on the criterion of reproductive isolation between groups of organisms. Two populations belong to the same species if they can interbreed, and they belong to different species if they are reproductively isolated from each other. In practice, it may be difficult to verify whether two populations that are geographically separated can actually reproduce, but the principle holds. However, when applying this concept of species to populations living at different times, it is clear that reproductive compatibility or isolation cannot be determined in practice. We surely cannot verify whether two fossil specimens could interbreed.

Species, as defined by Dobzhansky and Mayr, is a key concept to understand the biology of organisms and their evolution by natural selection. Every event related to the appearance of new traits (mutation, genetic recombination, natural selection) takes place within a reproductively closed cluster of organisms: within a species. The events determining the make-up of a species cannot jump to organisms of other species, because reproductive isolation prevents it.

The process of speciation was characterized simply earlier in this chapter. Each new species can later split into new species. Evolutionary lineages, even broad encompassing ones, start as single species. The reconstruction of the lineages of species is the purpose of any phylogenetic theory, such as cladistics. Such reconstructions are not easy. For instance, modern humans are *Homo sapiens*. We belong to the genus *Homo* and, within it, to the species *sapiens*. There is no controversy about this. Consider now the specimen found on the island of Java in the late nineteenth century by Dubois (1894). It consisted of a femur, which was very similar to our own, and a very primitive skull, and was named *Pithecanthropus erectus*. The same specimen was later reclassified as *Homo erectus*. For what reason was the genus proposed by Dubois changed so that the specimen was (1) included in our own genus but (2) not in our own species?

There are no firm and generally accepted criteria to determine definitively whether a fossil has been adequately classified. Cladistics is the most widely used system seeking to reach nonarbitrary classifications, but it is not free from difficulties, which we will examine next.

1.3.2 Classification of fossils

Without evolution, the stability of living beings, as perceived by Aristotle, would seem a solid basis for an objective system of classification. Different sets for classifying lions and tigers would seem to be valid. Members of such sets that would be discovered as fossils would be classified in the same groupings as living organisms. The only problem would be the proper identification of badly conserved fossil remains.

There have been different proposals to guide the classification of organisms that lived at different times. The concept of *chronospecies* (see below) seeks to overcome the impossibility of testing empirically whether organisms that lived at different times could interbreed. Phenetics has proposed the concept of the operational taxonomic unit (OTU), defined according to the strings of zeroes and ones based on morphological comparisons, as described above, but with little success. Cladistics is the most extensively used practice in paleontological taxonomy. Cladistics was, in fact, developed by the German zoologist Willi Hennig (1950, 1966) as a useful method of classification in paleontology.

Consider a certain species which we'll call a stem species. Because it is a species, it constitutes an isolated reproductive unit. Its reproductive characteristics, according to Hennig—who called them "tokogenetic relations"—will end, and the stem species disappear, when it becomes replaced by two new descendant groups, which he called "daughter species". The set comprising the stem species and the two daughter species—sister species to one another—constitutes a clade (see Figure 1.28). The representation of the split or speciation moment is called a node.

By definition, a cladistic episode requires the presence of at least one apomorphy (derived trait) characteristic of each of the sister species appearing after the node. These species can conserve, of course, primitive traits inherited from the stem species, which will be identical in the daughter species (plesiomorphic traits). Thus, phylogenies can be inferred simply by comparing derived and primitive traits of current species with those of fossil species, or of fossil species living at different times. The identification of speciation events through time consists, by definition, of the identification of clades and nodes. A cladistic event requires the presence of at least one apomorphy, a characteristic derived trait in each of the sister species that appear at the node.

The species concepts of Hennig and Mayr consider reproductive isolation as the main trait that characterizes species. Mayr wrote that "the essence of the biological species concept is discontinuity

---

Box 1.3 Reproductive isolation

Reproductive isolation is so significant that some philosophers of science, including David Hull (1977) and Michael Ghiselin (1987), have suggested that species should be given ontological consideration as individuals: they are born, change, and disappear after a certain time, which is how individual organisms are characterized. Each species is distinguishable from others and cannot be

reduced to any of them. From this viewpoint, grouping species is not the same as grouping books or automobiles. The notion of species as individuals has been the subject of interesting, and at times acrimonious, debate. One difficulty is that populations become reproductively isolated gradually; some populations are only partially reproductively isolated.

---

due to reproductive isolation" (Mayr, 1957). Hennig's species concept refers to "reproductively isolated natural populations" (Meier and Willmann, 2000; see De Queiroz and Donoghue, 1988).

The important difference between the biological (Mayr's) and cladistic (Hennig's) species concepts is the inclusion of the time dimension in the latter. Hennig's objective was the reconstruction of phylogenies. Time is crucial in cladistics, but it is relative time, not absolute time. When a stem species S splits into two sister species, D1 and D2, absolute time is present in a trivial way, in that S must precede D1 and D2 (Figure 1.28). But the key concern is relative time: after the speciation episode, each sister species occupies an isolated temporal niche. Once the two daughter species have appeared, the evolutionary events affecting species D1 have nothing to do with those involving species D2. They live two separate *specific times*. For instance, D1 could originate a large clade, with many new speciation episodes, while D2 could remain as a single species for a long time. Mayr's species concept also allows this possibility, but the difference lies in the relevance that cladistics awards to the question of what is and is not a speciation event in any given clade.

1.3.3 The problem of phyletic lineages and reformed cladistics

In the original Hennigian formulation, a speciation event involves the appearance of two sister species and the extinction of the stem species. The stem

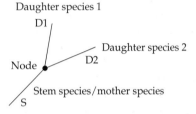

Figure 1.28 Cladistic phylogenies. Hennigian cladistics defines speciation as the split (node) of a stem or mother species (which thereby becomes extinct) into two sister species.

species cannot survive the speciation event and the ancestral lineage of the sister species cannot include two different stem species (Meier and Willmann, 2000). In other words, anagenetic speciation (the transformation of one species into another through time without the split of a stem species into two daughter species) is not allowed. Anagenetic speciation may occur in nature, but it is irrelevant for clade reconstruction and, therefore, it is ignored in cladistics. The rationale for the decision is that there is no criterion that would establish the precise boundary at which one species becomes another; indeed, the process is gradual as we know. But the split of a stem species into two descendant species (cladogenesis) can be unambiguously identified at the node.

In order to become a new species, S2 must achieve complete reproductive isolation from S1. This involves the appearance of mechanisms (ecological, genetic, or otherwise) that separate populations that live at the same time. Without such a temporal coincidence, the reproductive

Figure 1.29 Phyletic process (speciation without ramification). The node is not identifiable.

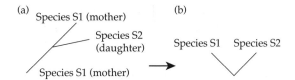

Figure 1.30 A phyletic episode with the persistence of the mother species (a) in traditional cladistics (which does not allow such speciation) and (b) in transformed cladistics.

isolation concept makes little sense. To say that 1-million-year-old *Homo erectus* was reproductively isolated from any *Australopithecus* species, which lived several million years earlier, makes no sense, cladistics says. Reproductive isolation requires that S2 becomes a new species while S1 still exists (Figure 1.29). If S1 disappears when S2 appears, as in anagenetic speciation, then we have abandoned the theoretical concept of speciation through reproductive isolation episodes. The concept of anagenetic speciation depends, rather, on operational prescriptions applied to the fossil record.

Nevertheless, it is common in paleontology to name species along a phyletic lineage. The term *chronospecies* is applied to groups of organisms living in different time periods which appear to be ancestors and descendants when these groups are morphologically as different from each other as contemporary organisms classified in different species. For instance, modern horses, *Equus*, and their 50-million-year-old ancestors, *Hyracotherium*, receive different names because, from a morphological point of view, they are at least as different from one another as either one from, say, modern zebras. The concept of chronospecies allows recognizing phyletic evolution when cladogenetic events are unknown.

The classification of ancestors and descendants into different chronospecies is appropriate when the temporal sequence of known fossils is fragmentary. The absence of transitional fossils facilitates classification into different species fossils that are quite distinct and separated by many years of evolution. If the fossil record were sufficiently complete through time, the situation would be different. If we documented small sequential changes in a long phyletic sequence, we might consider the extreme members of the sequence as different chronospecies, but there would not be a particular point in time at which one species would have become another. Of course, the fossil record is rarely sufficiently complete to display this situation.

It would be possible to identify a chronospecies S2, daughter of S1 in a phyletic lineage, if we found contemporary species. This situation could be considered exemplar of a variant of the formation of two sister species in a node, in which the mother species takes the place of one of the daughter species (Figure 1.30). However, original Hennigian cladistics does not allow the simultaneous presence of stem and daughter species.

Transformed cladistics (Platnick, 1979) tried to overcome this obstacle (Figure 1.30b) by allowing a daughter population to be considered a different species if it has at least one apomorphy (derived character) that distinguishes it from the mother species. But, how to represent the process? The solution adopted by transformed cladistics is to represent the mother species S1 and daughter species S2 as sister species in the cladogram.

An important consequence of placing the mother and daughter species as sister species is the transformation of the original sense of Hennig's stem species. Schaeffer *et al.* (1972) had already suggested, before the proposal of reformed cladistics, that all taxa, fossil or living, might be placed as terminal taxa in a cladogram. As a consequence, only hypothetical ancestors can be placed in the nodes. Once a fossil taxon is correctly identified, it must be placed as a terminal taxon. Hence, stem species disappear as parts of branches or as nodes. Not only do their representations disappear, but also the concept itself. Thus, cladograms lose their meaning as a representation of the evolution process (in the style of phylogenetic trees) and are reduced to representations of the way lineages are divided by means of sister species.

A price paid for this transformation of cladistics is the loss of the temporal dimension. Cladograms would no longer represent ancestry relations. Furthermore, speciation processes through time cannot be established by means of cladistics. According to Delson *et al*. (1977), the concept of "sister species" is a methodological instrument that must be applied even if (1) the taxa under consideration are two species that hold an ancestor–descendant relation and thus are not true "sister species" and (2) the taxa under consideration have close relatives which are as yet unknown. According to Delson *et al*. (1977), a cladogram constructed in such a fashion does not allow deciding whether the branches stemming from a node represent sister or mother/daughter species. Ancestry relations disappear as objects of scientific inquiry given that, within cladistics, the hypothesis that a taxon is the ancestor of another cannot be tested (Nelson, 1973; Cracraft, 1974; Delson *et al*., 1997). The same idea is expressed by Siddall (1998), who states that seeking to describe evolutionary relations by searching for ancestors in the fossil record is a resurgence of the cult of the golden calf.

1.3.4 Beyond species

We have elaborated that the species concept is fundamental to understand organisms and their evolution by natural selection. The Linnaean taxonomy includes other classification categories in addition to species. The category genus lies immediately above. As we mentioned, genus and species are the two categories used in the Linnaean nomenclature to identify organisms. A genus includes closely related species (although some genera may include a single species).

The species concept is not only a taxonomic category, but actually refers to groups of organisms that are importantly related to one another by relations of mating and parentage. What about the concept of genus? Is it a completely artificial construct, or does it have significance beyond its condition as a

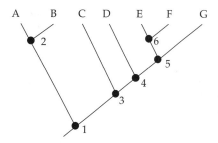

Figure 1.31 A cladogram representing seven species originated by means of six nodes.

taxonomic artifice? If the genus category is purely an artificial construct, clustering organisms in genera would be completely arbitrary, though not insignificant. Grouping beings in certain genera would make a difference, just as we might classify books by subject. But different ways to construct genera could be suggested. An alternative is to consider that the category of genus (as well as the more inclusive ranks, such as family, order, and so on) have certain distinctive traits precisely because they refer to biological attributes. If this were the case, it would be inadequate to create a taxon belonging to a genus or another higher category if it distorts the distinctive sets of attributes that characterize the taxon. Cladistics emphasizes an essential feature of organisms: they evolve, forming lineages. Figure 1.31 represents the lineages corresponding to seven species (A–G) that appeared by means of six speciation events (nodes 1–6). Each node is the source of two sister taxa. In this way, species A is B's sister group (and vice versa), whereas C + D + E + F + G is the sister group of A + B.

A genus is a set of species. If a genus is purely a taxonomic artifice, we could define a genus that would include species A, C, and E, for instance. But if taxonomy should respect evolutionary processes, then not just genera, but any category must only include taxa that constitute complete parts of the cladogram. This is the same as saying that genera (and families, and so on) are evolutionary lineages with real existence. They reflect the way in which phylogenesis occurred. This is why we include bats, lions, and dolphins in the taxon mammals. Our classification does not cluster bats, eagles, and butterflies, although they all fly. Bats, eagles, and butterflies do not constitute a lineage just because they have wings.

Monophyletic groups include whole lineages: they reflect the process of evolution. In Figure 1.31, A + B and C + D + E + F + G are monophyletic groups. The set E + F is also monophyletic because it is a group including all taxa stemming from a particular node. Paraphyletic groups are those which leave out some taxa pertaining to the lineage. In Figure 1.31, a group including C + D + E + G would be paraphyletic because it does not include taxon F. Polyphyletic groups include lineages that have not arisen from the same node, excluding intermediate ones. Thus the grouping A + B + E + F is polyphyletic.

Let's now turn to the question of rank. The rank of the genus category is superior to that of species, because a genus is a set of species. A group of genera is a family (if we do not consider intermediate categories, such as tribe), and so on. In Figure 1.31, the set A + B constitutes a genus because it is a group of species. For the same reason, E + F must be considered another genus. But now we find ourselves with a problem. Node 5 gives rise to two sister groups, E + F on the one hand and G on the other. We said that G is a species, and E + F a genus. How is it possible that the sister group (E + F) of a specific taxon (G) belongs to a higher category than the latter? A possible solution is to award sister groups the same category. Thus, although G is a species, it must be classified also as a genus, just as E + F, even if it is a genus with a single species.

Problems do not disappear with this taxonomic maneuver, however. If each new node requires elevating the rank of the categories we'll soon run into difficulties in the case of lineages with numerous branches. This is the reason why new intermediate categories are introduced (tribe, subfamily, superfamily, infraorder, and so on) but such a proliferation may become excessive. Basing taxonomy on numbers rather than names (such as phenetics does, for example) would resolve the problem, but this is not a common practice. Thus, to avoid an excessive number of categories, it is advisable not to apply the strict rank equivalence

of sister groups. We will now consider a practical case that refers to the group formed by the great apes and humans.

1.3.5 The adaptive concept of genus

As we have seen, it is not easy to come up with objective criteria to decide whether several species should be grouped into one or several genera. Hybridization between two organisms indicates that they belong to the same species, but there is not an equivalent test to verify whether they belong to the same genus. The international code of taxonomy does not provide objective classification criteria regarding categories above the species level. Authors suggest particular classifications hoping that the scientific community will accept them. But with respect to the genus category it is often helpful to follow Ernst Mayr's (1950) proposal that a genus refers to a particular way of adaptation to specific conditions. A new genus, according to Mayr's proposal, refers to a new kind of organism that adapts to its ecosystem in a different way from other organisms included in other genera.

It is not easy to determine how fossil specimens adapted to their environment. However, certain inferences can be made from morphological traits. For instance, the presence of thick molar enamel indicates a diet that included hard materials. The robusticity of the masticatory apparatus points in the same direction. The mode of locomotion can be inferred from the analyses of forelimbs and hind limbs. A large brain in relation to body size is associated with the ability to construct and carve complex tools and instruments.

The adaptive concept of genus must be used with caution. It is not a method to decide how different lineages evolved. Cladistic analyses of apomorphies are much better suited to that end. But once the distribution of lineages is known with a certain degree of confidence, it is useful to consider adaptive specializations, because they are helpful to avoid naming a new genus almost every time a new fossil is discovered, as has happened in the reconstruction of human phylogeny. In the following chapters, as we review the evolution of

Miocene and Pliocene hominins, we will see how convenient it is to avoid the excessive multiplication of genera; although a radical reduction of genera is not a good solution either. Mayr's adaptive criterion represents a step forward in the search for phylogenies and taxonomies unbiased by the classifier's preconceptions, even though there are serious difficulties in discovering adaptive strategies from the fossil and archaeological records.

CHAPTER 2

The evolution of hominoids

2.1 Hominoid taxonomy

The systematic concepts and practices reviewed in the previous chapter are applicable to any lineage, but we are particularly interested in humans and their direct and close ancestors. The evolutionary relationships between humans and our closest living relatives, the African (chimpanzees and gorillas) and Asian (orangutans) great apes, are an appropriate starting point.

With respect to phylogeny, a genus corresponds to a lineage in which different species have arisen. Figure 2.1 reflects the relations between gorillas and chimpanzees. There are two chimpanzee species, *Pan troglodytes* (common chimpanzee) and *Pan paniscus* (bonobo), included in the genus *Pan*, whose sister taxon in this cladogram is *Gorilla*.

Why do we have two genera? In terms of adaptive strategy, chimpanzees and gorillas seem to be very similar animals. Their groups are formed by a dominant male, several females and offspring. They use the same kind of locomotion (knuckle-walking,

a quadrupedalism that places the hand's knuckles and not the palm on the ground). Although gorillas are more decisively folivorous (leaf-eating) than chimpanzees, which have an omnivorous diet, they all live in tropical forests. Couldn't common chimpanzees, bonobos, and gorillas be included in a single genus, apart from others, such as *Homo*? The uncertainty increases if we consider fossil specimens about which we know little. Can we find a suitable guide to help us establish different genera?

The introduction of molecular techniques opened the way for studies that could go beyond morphological and adaptive comparisons, offering a new range of possibilities. They allowed one to address in a new way the issue of the classification of apes and humans, which together make up the superfamily Hominoidea.

According to the traditional classification, the order Primates includes several suborders (Figure 2.2). A suborder is an intermediate category between order and family. Below this suborder category we find infraorder and, one step lower, superfamily. The suborder of anthropoids (*Anthropoidea*) includes two infraorders, catarrhines (African, European, and Asian monkeys) and platyrrhines (American monkeys), which diverged after continental drift separated South America from Africa. Catarrhines are divided into two superfamilies: cercopithecoids, or Old World monkeys, and hominoids (apes and humans).

The evolutionist G.G. Simpson (1931, 1945) in his classification of mammals distinguished humans from apes at the family level: Hominidae and Pongidae, but he classified australopithecines (which nowadays are considered part of the human clade) with pongids. Except for this, Simpson's classification of primates was widely accepted for three decades.

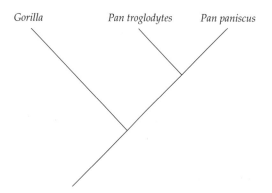

Gorilla   *Pan troglodytes*   *Pan paniscus*

Figure 2.1 Two living chimpanzee species are grouped in the same genus, which is different from that of gorillas.

Simpson's classification of hominoids as including apes and humans requires that all existing apes, their direct ancestors, and their descendants be included in a single family, Pongidae, reserving another family, Hominidae, for the human lineage. Is such a separation justified? Simpson's 1945 classification proposal was based on morphological similarities: leaving gibbons aside, it seems orangutans, gorillas, and chimpanzees are more similar among themselves than any of them is to humans. If morphological similarity would reflect evolutionary relatedness, the traditional classification implies that the hominid branch was the first to separate from the pongids, which would later split into all existing ape genera and species.

In the 1960s immunological methods contradicted such inferences (see Figure 2.3). With analyses of proteins in the blood serum of hominoids, Morris Goodman, a molecular geneticist at Wayne State University in Detroit, determined that humans, chimpanzees, and gorillas are closer to each other than any of them is to orangutans (Goodman, 1962, 1963; Goodman *et al.*, 1960). According to Goodman's results the evolution of hominoids proceeded very differently from what Simpson's taxonomy implied. The lineage leading to orangutans was the first to split, then, the lineage of gorillas split from the others and, finally, the chimpanzees and human lineages diverged from each other. Accordingly, the

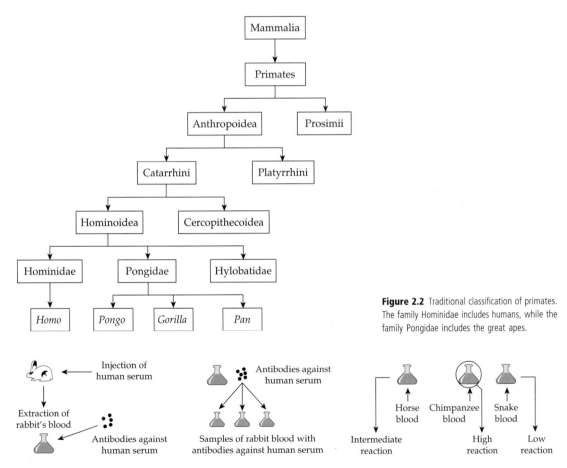

Figure 2.2 Traditional classification of primates. The family Hominidae includes humans, while the family Pongidae includes the great apes.

Figure 2.3 The activity of antibodies during the invasion of the organism by foreign proteins allows determination of the relatedness of immune systems. After an injection with human blood, a rabbit generates specific antigens against human proteins. If we then apply those antigens to the blood serum of other animals we can deduce evolutionary closeness between humans and those other animals by the strength of their immune reaction.

classification of apes in the family Pongidae and humans in a separate family, Hominidae, is not appropriate. Such a classification includes a paraphyletic group and artificially separates the taxon *Homo* from the common lineage it evolved from (Figure 2.4).

According to molecular findings, a correct classification should place orangutans in a taxon of the same category as the set gorillas + chimpanzees + humans (Figure 2.4c). The former would constitute the family Pongidae and the latter the family Hominidae, if we want to keep the division in two families. But before we discuss this issue any further, we'll turn to the age of separation of the different lineages.

2.1.1 Age of hominoid lineages

Morris Goodman's results on the sequence of separation of hominoid lineages had profound taxonomic consequences, but provided no information about the timing of the divergences. Although order of sequence and timing are related, determining the time of divergence between two lineages requires determining the rate of molecular evolution of the trait under consideration. We need something like a clock that would allow us to determine how much time had elapsed for each degree of immunological differentiation between proteins. Vincent Sarich and Alan Wilson (1967a, 1967b) argued that the immunological differentiation between

chimpanzees and humans indicated that the two lineages separated between 5 and 4 Ma. According to this calculation, no Miocene fossil could be a direct ancestor of humans, given that the two lineages diverged later. Goodman (1976, for example), on the contrary, argued that the rate of molecular evolution was slow in the hominoids and that the divergence of the hominoid lineages occurred earlier than estimated by Wilson and Sarich.

Goodman's and Sarich and Wilson's work was based on immunological methods. Greater resolution can be achieved by other studies, such as obtaining the amino acid sequence of proteins or the nucleotide sequence of the DNA, which could not be done readily at the time. An intermediate degree of resolution could be achieved by DNA–DNA hybridization. The two strands of the DNA helix are separated by heating and the rate of reannealing between strands from different sources are compared: human with human DNA, human with chimp DNA, and so on. This method was pursued, among others, by Charles Sibley and Jon Ahlquist (1984; see also Sibley *et al.*, 1990).

To estimate time of divergence on the basis of molecular differentiation, this must occur at a constant rate. This is the *molecular clock hypothesis*, which we discussed in section 1.2. As pointed out there, the molecular clock is not expected to time events precisely, but rather it would be a stochastic clock, in which events occur with a constant *probability*, such as in radioactive decay. Numerous

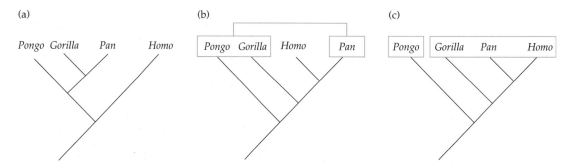

Figure 2.4 Paraphyletic and monophyletic hominoid groups. (a) Cladogram derived from traditional classification into pongids and humans. (b) Cladogram deduced from morphological and functional similarities. When grouping apes to the exclusion of humans, the former become a paraphyletic group (it leaves out a taxon of the considered lineage). (c) The correct grouping, by means of monophyletic groups, includes all the members of each lineage.

investigations have shown by now that the molecular clock is more erratic than expected from a stochastic clock. Nevertheless, because so many different genes and other DNA sequences, as well as proteins, can be studied, molecular investigations have provided very valuable information about the time of evolutionary events.

An issue that may be problematic with the molecular clock is that it needs to be calibrated by reference to some evolutionary event that has been dated with paleontological information. This calibration determines the rate at which a particular molecular clock (gene or protein) "ticks."

Some authors, such as Phillip Tobias, a paleoanthropologist at Witswatersrand University in South Africa, have pointed out that the molecular clock hypothesis involves a circular argument. We determine dates corresponding to the fossil record on the basis of molecular time rates obtained from the fossil record itself (Tobias, 1991). But this misrepresents the method, which uses fairly well-ascertained fossil dates to determine the rate of evolution of, say, a particular protein and then uses this rate to estimate the time of divergence for lineages with an uncertain fossil record. In any case, molecular evolution dates are subject to the two problems mentioned: the assumption that the rate is constant and the determination of the rate. Thus, it is not surprising that Tobias (1986) carried out a comparison among different

dates obtained for the human and chimpanzee divergence, using different kinds of calibration and observed dates that varied from 9.2 to 2.3 Ma.

Tobias (1991) has pointed out that, in addition to the issues of calibration and rate constancy, molecular investigations of phylogeny face other methodological problems. Such problems are inherent to all systems for calculating phylogenetic distances. First, there is a tacit assumption that the resulting date for the divergence of two human and chimpanzee DNA sequences, for instance, reflects how long ago the two lineages themselves diverged. This excludes the possibility of mosaic molecular evolution, which might preserve certain primitive molecular features of the molecules in one or both lineages. Thus, estimates of the age of divergence between humans and great apes may be different for different proteins or DNA sequences. Second, we are ignorant of the extent to which convergent adaptation—that is to say, the appearance of analogous traits—may be expressed at the molecular level. Both of these problems are real, but as Tobias (1991) himself has admitted, the multiplication of molecular studies carried out with different techniques is likely to yield converging estimates that approximate true time values.

Obtaining DNA sequences has now become a readily available and relatively inexpensive process. The sequencing of the human and

---

Box 2.1 Bias in the molecular clock

The molecular clock may yield erroneous estimates due to lineage-specific bias; that is, a given molecular clock (a particular gene or protein) may tick at different rates in different lineages. If the clock is calibrated using data from a certain lineage, it may yield erroneous time estimates when applied to other lineages. Some lineage-specific biases may be systematic; that is, they may occur not only with respect to a particular gene, but with respect to all genes. Thus, hominoids seem to evolve more slowly at the molecular level than, say, rodents or even other primates with shorter generation times. Some results even suggest that molecular evolution in humans may be slower than in

other hominoids. Thus Elango and collaborators (2006) have performed a large-scale analysis of lineage-specific rates of single-nucleotide substitutions among hominoids. They found that "humans indeed exhibit a significant slowdown of molecular evolution compared to chimpanzees and other hominoids. However, the amount of fixed differences between humans and chimpanzees appears extremely small, suggesting a very recent evolution of human-specific life history traits. Notably, chimpanzees also exhibit a slower rate of molecular evolution compared to gorillas and orangutans in the regions analyzed."

chimpanzee genomes has provided valuable information as a reference for phylogenetic or taxonomic investigation. The comparison of DNA sequences of genes as well as non-coding sequences has become the prevailing molecular method for systematics and phylogeny. Nevertheless, for historical completeness we'll review earlier studies.

The early immunological and blood-serum protein studies were complemented with those comparing chromosomes. The first study of this kind was carried out by the Italian anthropologist Brunetto Chiarelli (1962). Thereafter, many authors have related primate evolution with chromosomal modifications. Jean Chaline and coworkers (1991) investigated the branching sequence of gorillas, chimpanzees, and humans by identifying seven chromosomal mutations that differ between the African great apes and humans. Chaline and colleagues (1991, 1996) have combined their investigations of chromosome structure with the available evidence obtained by molecular and immunological methods. They have concluded that the gorilla, chimpanzee, and human lineages separated almost simultaneously, an event best represented by a so-called trichotomy (Figure 2.5). The difficulty of unraveling the divergence sequence of the three lineages had been pointed out earlier by Goodman (1975), as well as Bruce and Ayala (1979) and Smouse and Li (1987). Andrews (1992a) accepted the trichotomy scenario in an influential article devoted to the reinterpretation of the status of hominoids. An attempt to resolve the issue of the classification of the gorilla, chimpanzee, and human lineages was carried out by Groves and Paterson (1991) by means of the

parsimony-maximizing cladistic computer program PHYLIP. Their results pointed to a *Pan–Homo* or *Pan–Gorilla* clade, depending on the characters selected for the comparison. The trichotomy *Gorilla–Pan–Homo* has been defended more recently by Deinard and Kidd (1999), based on the study of the evolution of the intergenic region *HoxB6*. But there is opposing evidence as well. Morris Goodman's team have reinterpreted β-globin genetic sequences as favoring separate lineages for gorillas and chimpanzees + humans (Bailey *et al.*, 1992).

It is not easy to reach a general consensus based on different molecular methods, but a revision of the available data regarding DNA sequences by Ruvolo (1997) supports an initial separation between the gorilla clade and the one formed by chimpanzees and humans. On the whole, the molecular evidence favors the chimpanzees as the sister group of the lineage leading to modern humans: chimpanzees are our closest living relatives (Goodman *et al.*, 1998).

2.1.2 The *Homo/Pan* divergence

Although both hybridization and chromosomal comparison techniques provide useful information for establishing molecular proximity between two species, the direct sequencing of nucleic acids has the last word. DNA hybridization studies (Sibley and Ahlquist, 1984) had shown that chimpanzees and humans share close to 98 or 99% of their genomes' DNA. The direct sequencing of the human chromosome 21 (Hattori *et al.*, 2000) and its ortholog 22 in chimpanzees (Watanabe *et al.*, 2004)

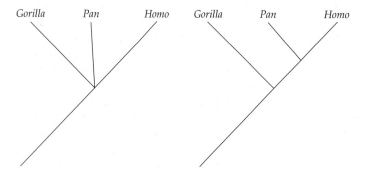

Figure 2.5 Phylogenetic relation between the genera *Gorilla*, *Pan*, and *Homo*. To the left, simultaneous appearance of the three lineages (trichotomy). To the right, initial differentiation of *Gorilla*.

allowed the detailed comparison of their genomes, confirming their genetic proximity. Excluding deletions and insertions, the differences between the two species amounted to only 1.44% of the nucleotides. It became obvious that the genomes of humans and chimpanzees are extremely similar in their DNA sequence. How similar has become known recently with the publication of the draft genome sequence of the chimpanzee and its preliminary comparison with the human genome.

The Human Genome Project was initiated in 1989, funded through two US agencies, the National Institutes of Health (NIH) and the Department of Energy (DOE), with eventual participation of scientists outside the USA. The goal set was to obtain the complete sequence of one human genome in 15 years at an approximate cost of $3,000 million, coincidentally about $1 per DNA letter. A private enterprise, Celera Genomics, started in the USA somewhat later, but joined the government-sponsored project in achieving, largely independently, similar results at about the same time. A draft of the genome sequence was completed ahead of schedule in 2001. The government-sponsored sequence was published by International Human Genome Sequencing Consortium (2001) in the journal *Nature* and the Celera sequence was published by Venter *et al.* (2001) in the journal *Science*. In 2003 the Human Genome Project was finished, but the analysis of the DNA sequences chromosome by chromosome continued over the following years. Results of these detailed analyses were published on June 1, 2006, by the Nature Publishing Group, in a special supplement entitled *Nature Collections: Human Genome*.

The draft DNA sequence of the chimpanzee genome was published on September 1, 2005, by the Chimpanzee Sequencing and Analysis Consortium in *Nature*, embedded within a series of articles and commentaries (*The Chimpanzee Genome*, Anon, 2005). The last paper in the collection presents the first fossil chimpanzee ever discovered (McBrearty and Jablonski, 2005).

In the genome regions shared by humans and chimpanzees, the two species are 99% identical. These differences may seem very small or quite large, depending on how one chooses to look at them: 1% of the total appears to be very little, but it amounts to a difference of 30 million DNA nucleotides out of the 3 billion in each genome. Twenty-nine percent of the enzymes and other proteins encoded by the genes are identical in these species. Out of the one to several hundred amino acids that make up each protein, the 71% of nonidentical proteins differ between humans and chimps by only two amino acids, on average. If one takes into account DNA stretches found in one species but not the other, the two genomes are about 96% identical, rather than nearly 99% identical as in the case of DNA sequences shared by both species. That is, a large amount of genetic material, about 3% or some 90 million DNA nucleotides, have been inserted or deleted since humans and chimps initiated their separate evolutionary ways, about 8–6 Ma. Most of this DNA does not contain genes coding for proteins, although it may include tool-kit genes and switch genes that impact developmental processes, as the rest of the noncoding DNA surely does.

Comparison of the two genomes provides insights into the rate of evolution of particular genes in the two species. One significant finding is that genes active in the brain have changed more in the human lineage than in the chimp lineage (Khaitovich *et al.*, 2005). Also significant is that the fastest-evolving human genes are those coding for *transcription factors*. These are switch proteins which control the expression of other genes; that is, they determine when other genes are turned on and off. On the whole, 585 genes have been identified as evolving faster in humans than in chimps, including genes involved in resistance to malaria and tuberculosis. (It might be mentioned that malaria is a severe disease for humans but not for chimps.) There are several regions of the human genome that contain beneficial genes that have rapidly evolved within the past 250,000 years. One region contains the *FOXP2* gene, involved in the evolution of speech.

Other regions that show higher rates of evolution in humans than in chimpanzees and other animals include 49 segments, dubbed human-accelerated regions or HARs. The greatest observed difference occurs in *HAR1F*, an RNA gene that "is expressed specifically in Cajal-Retzius neurons in the developing human neocortex from 7 to 19 gestational weeks, a crucial period for cortical neuron specification and migration." (Pollard *et al.*, 2006; see also Smith, 2006).

All this knowledge (and much more of the same kind that will be forthcoming) is of great interest, but what we so far know advances but very little our understanding of what genetic changes make us distinctively human. Extended comparisons of the human and chimpanzee genomes and experimental exploration of the functions associated with significant genes will surely advance further our understanding, over the next decade or two, of what it is that makes us distinctively human, what is it that differentiates *H. sapiens* from our closest living species, chimpanzees and bonobos, and will surely provide some light of how and when these differences may have come about during hominid evolution.

According to David Baltimore (2001) it is not clear whether "we will learn much about the origins of speech, the elaboration of the frontal lobes and the opposable thumb, the advent of upright posture, or the sources of abstract reasoning ability, from a simple genomic comparison of human and chimp." This may have been an overly pessimistic expectation, as shown by the *FOXP2* and *HAR1F* examples cited above.

2.1.3 What is *Homo* from a taxonomic point of view?

There are large functional and anatomical differences between African apes, including chimpanzees, and humans. If we are to respect phylogenetic lineages, how should the molecular similarities and phenotypic differences be reflected in the classification of hominoids?

The most commonly held point of view deduced from molecular studies proposes an evolutionary sequence that involves an initial separation of orangutans, a second separation of gorillas, and finally, the divergence between chimpanzees and humans. Goodman (1962, 1963) and Goodman *et al.* (1960) pointed out that this phylogeny brings into question Simpson's traditional classification

of the hominoids. If chimpanzees and gorillas are placed in a single genus, we would have a paraphyletic group. This could be avoided if humans were also included in the genus with the African apes.

A solution proposed by Goodman (1963) would be broadening the hominid family to include gorillas (*Gorilla*) and chimpanzees (*Pan*) in addition to the human genus (*Homo*). Goodman's proposal had considerable resonance among primatologists, but the decision to increase the scope of the family Hominidae turned out to be a slippery slope. For instance, Schwartz *et al.* (1978) and Groves (1986) also placed the genus *Pongo* (orangutans) in the family Hominidae. Szalay and Delson (1979) went a step further by also including lesser apes (like gibbons, *Hylobates*). Morris Goodman and others later agreed with this suggestion (Bailey *et al.*, 1992; Goodman *et al.*, 1994). As a consequence, the taxon including humans and their exclusive direct ancestors—that is to say, hominids—was transferred from the family category to the tribe category, *Hominini* (Schwartz *et al.*, 1978; Groves, 1986) and, later, to the genus *Homo* (Goodman *et al.*, 1994). This genus, which in human paleontology is usually used to group human ancestors that lived during the late Pliocene and Pleistocene (such as *Homo habilis*, *Homo erectus*, *Homo neanderthalensis*, and *Homo sapiens*, among others), would also include chimpanzees, according to Goodman *et al.* (1998). Thus, all humans and their numerous direct and collateral ancestors would be reduced to a subgenus. An even more extreme proposal was put forward by Watson *et al.* (2001), presented during the World Congress of Human Paleontology in Sun City, South Africa: to include gorillas as well within the genus *Homo* (*Homo gorilla*; Table 2.1).

In view of the numerous and diverse lineages that, as the following chapters will illustrate, appear in the human clade from the Miocene to the Pleistocene, it does not seem reasonable to include them all within a limited corset of a subgenus. More controversial yet (Cela-Conde, 1998) is the classification of such functionally different organisms as chimpanzees, gorillas, and humans in a single subgenus. If we accept Mayr's adaptive

Table 2.1 Changing views of the genus *Homo* and of the family Hominidae

Genus *Homo*	
Traditional view	Humans and their direct and collateral ancestors not shared with australopithecines
Goodman *et al.* (1998)	previous + chimpanzees
Watson *et al.* (2001)	previous + gorillas

Family Hominidae	
Traditional view	Humans and their direct and collateral ancestors not shared with any ape
Goodman (1963)	previous + chimpanzees + gorillas
Schwartz *et al.* (1978), Groves (1986)	previous + orangutans
Szalay and Delson (1979), Bailey *et al.* (1992), Goodman *et al.* (1994)	previous + lesser apes

criterion for characterizing a genus (see Chapter 1), it seems clear, as will become apparent in later chapters, that there are at least five different genera just in the human lineage.

It is not necessary to carry the taxonomic implications of the molecular evidence as far as it has been done by Goodman and others. A monophyletically based taxonomy can be achieved by granting the same consideration to the human clade as to the chimpanzee clade. Chimpanzees and humans are sister groups, which could be considered as subgenera. But they could also be awarded a higher-ranking category: genus, tribe, subfamily, or even family, which would be more compatible with the spirit of Simpson's traditional classification.

Goodman and collaborators (1998) argue that chimpanzees and humans should be classified in the same genus because the time elapsed since their last common ancestor is about 6 Ma. This argument, however, does not require a category as low as subgenus (Cela-Conde, 2001). Taxonomic practice shows many examples, in all sorts of organisms, where the category of genus, or even higher, has been allocated to species that diverged no more than 6 Ma; for example, Vrba (1984) classified two bovid African lineages, Alcelaphini and Aepycerotini, which also have a 6-million-year-old last common ancestor, as separate tribes. The first lineage includes 27 species and the second only two, a situation somewhat similar to that of humans and chimpanzees (numerous species and several genera are generally recognized in the human lineage, whereas only two species are known in the chimpanzee lineage).

It is a lineage's diversity, not its antiquity, which must determine its taxonomic level. The existence of numerous species may justify placing them into more than one genus. Several genera may justify different tribes, and so on for higher categories. The genetic proximity of chimpanzees and humans, and the very worthy attempt to avoid ideological biases that would consider our species as a superior category, which have led to the inclusion of African great apes in the family Hominidae, or even in the genus *Homo*, deserve to be praised. However, taxonomic decisions should be guided strictly by systematic criteria.

2.1.4 Controversies of morphological comparison

Morphological similarity should not be ignored when establishing taxonomies, but how to evaluate genetic similarity to determine taxonomic classification is a difficult issue. Andrews and Martin (1987), on the basis of all the then available morphological and molecular evidence, arrived at a cladogram of hominoid phylogenetic relations that differs little from the one resulting from exclusively molecular evidence. These authors pointed out that "the only real surprise is that shared derived traits among African apes and humans at a molecular level are not reflected firmly in morphological analyses, nor are morphological similarities reflected at the molecular level." Actually, this should not have been a surprise. Rather, it is a good example of the problems we have been tackling. There is little doubt at present regarding the phylogenetic sequence of the appearance of hominoid lineages: the orangutan clade diverging from gorillas, chimpanzees, and humans, and later gorillas separating from a common clade that includes chimpanzees and humans. But molecular similarities need not precisely translate into morphological similarities, because these also depend on factors not simply apparent by observing differences in DNA or protein sequences.

A degree of correlation between molecular proximity and morphological similarity has been pointed out in several studies. Groves (1986) included chimpanzees and human beings in the same clade on the basis of morphological similarities. A similar conclusion was reached by Gibbs and collaborators (2000) by means of cladograms based on soft-tissue traits of living hominoids. Wolpoff (1982) has also affirmed that, in morphological terms, gorillas, chimpanzees, and humans are more similar to one another than any of them is to orangutans (although he was arguing in a different context). However, Schwartz (1984), an expert in primate evolution, carried out a detailed morphological comparative study that contradicted Wolpoff's conclusion.

According to Martin (1990) the morphological similarity among the great apes is due to their slow divergence from their last common ancestor. Consequently, Martin considered the possibility of grouping the apes in the same paraphyletic taxon. These diverging proposals are a consequence of the different weight given to phylogeny relative to morphological divergence. As Tobias (1991, p. 14) has said: "there are precise definitions available . . . regarding morphological traits of hominids and apes; they are, in essence, the complex of anatomical and functional traits that *most effectively* differentiate humans from apes" (original author's emphasis).

The primatologist Russell Ciochon (1983) also opted in favor of a human clade separate from an African ape clade, based on a very complete list of morphological traits that define 11 morphotypes within hominoids. But arguments in favor of such a separation are functional as well as morphological, even though there is a correspondence between them. Bipedal gait is impossible without changes in the foot, hip, extremities, and the cranial base. In fact, paleontologists infer bipedalism based on these morphological traits. Moreover, traits do not change in isolation; usually several traits change in a coordinated fashion to achieve a

new adaptation. Le Gros Clark (1964a) and Tobias (1985a) have advanced the concept of "total morphological pattern" for taxonomically characterizing a specimen, moving away from any practice that seeks to determine adaptation and evolutionary pattern by evaluating simple traits in isolation. Nevertheless, emphasis on the relevance of a marked and relatively isolated character (such as bipedal gait) may be more reasonable than an alternative procedure that simply quantifies the number of shared (or different) traits.

The issue at hand is not only related with the weight given to molecular data. Supporters of the close taxonomic classification of gorillas, chimpanzees, and humans do not ignore that human-derived traits (from bipedalism to language) are very relevant for the adaptation of our species. What then is the base for the widespread trend among molecular primatologists to include such adaptively, morphologically, and functionally diverse beings as African apes and humans in the same family? A relevant consideration is the urgent need to avoid anthropocentrism. Often in the past an anthropocentric bias has imposed mistaken concepts, such as hierarchical relation among living beings, with humans in the role of masters of nature. Nevertheless, the discrepancies concerning classification between molecular and other primatologists are largely due to the overwhelming weight attributed by some to molecular evidence, which ignores the difference between genetic distances and the determination of phylogenetic trees.

2.1.5 A monophyletic solution that respects functional aspects

Despite arguments in favor of maintaining Simpson's (1945) taxonomy and separating the family Pongidae (great apes) from the family Hominidae (humans), the results of molecular studies are too consistent and extensive to ignore. As we have already noted, this does not necessarily lead to such a reductive classification as the one at the bottom right-hand corner of Table 2.1, with lower apes included in the Hominidae. The molecular evidence does not, by itself, determine the distribution of evolutionary lineages, but is relevant to formulate cladistic interpretations of how to incorporate new lineages. Cladistics argues that each speciation process involves the disappearance of the original taxon and the necessary appearance of two new taxa. The crucial issue here, as suggested by Martin (1990), is that cladistic principles do not impose the taxonomic categories to which the diverging clades belong. Cladistics requires that the two taxa that appear at a node—the idealized moment of their divergence—belong to the same category. Thus, if we assume that the taxon chimpanzees + humans separated from the taxon gorillas, the requirement is that we grant the taxon including chimpanzees + humans the same category as gorillas.

Here we will adopt a taxonomic classification which respects the increasing molecular evidence. We will largely follow Wood and Richmond (2000; Table 2.2). The family Hominidae embraces the set of great apes and humans. Orangutans constitute the subfamily Ponginae and gorillas the subfamily Gorillinae, while chimpanzees and humans form the subfamily Homininae. Within the latter, chimpanzees belong to the tribe Panini and humans to the tribe Hominini. The human lineage has the category of tribe: Hominini (informal name, hominins). Two subtribes are included in it. One is Australopithecina (informal name, australopiths), which encompasses four genera: *Orrorin*, *Ardipithecus*, *Australopithecus*, and *Paranthropus*. The other subtribe is Hominina (informal name, hominans), with one single genus.

2.1.6 Ape apomorphies

Humans are, technically speaking, apes. Common usage, when referring to Asian apes and African apes, incorrectly leaves out *H. sapiens*. This popular bias has an explanation: the evolution of our derived traits involved the loss of those features we associate with our closest relatives, the apomorphies that define what an ape is. Cercopithecoids, or Old World monkeys, are the sister group of apes (Figure 2.6). Cercopithecoids (African, European, and Asian monkeys) and Hominoids (apes and humans) belong to the suborder Anthropoidea, which is part of the order Primates defined by Carolus Linnaeus.

Because they are sister groups, Old World monkeys and apes share some primate primitive traits. All these traits are associated with adaptive specialization to an arboreal habitat. Given that these traits are plesiomorphies, inherited from a common ancestor, they imply that adaptation to life in the trees occurred before the appearance of apes. In the next section (2.2), we will examine when this evolutionary episode took place.

Table 2.2 The taxonomy of great apes and humans

Family	Subfamily	Tribe	Subtribe	Genus	Current species
Hominidae	Ponginae	Pongini	Pongina	*Pongo*	Orangutans
	Gorillinae	Gorillini	Gorillina	*Gorilla*	Gorillas
	Homininae	Panini	Panina	*Pan*	Common chimpanzees; bonobos
		Hominini (hominins)	Australopithecina (australopiths)	*Orrorin* *Ardipithecus* *Australopithecus* *Paranthropus*	
			Hominina (hominans)	*Homo*	Humans

Source: adapted from Wood and Richmond (2000), adding the genus *Orrorin*.

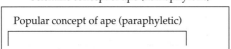

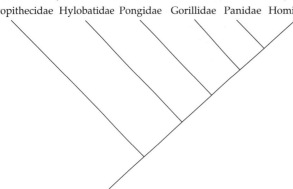

Scientific concept of ape (homophyletic)

Popular concept of ape (paraphyletic)

Cercopithecidae   Hylobatidae   Pongidae   Gorillidae   Panidae   Hominidae

Figure 2.6 Hominoids or apes and their sister group (cercopithecoids). In a technical sense, humans are apes because we belong to the same lineage as them. However, the non-technical usage of the term "ape" separates lesser apes (gibbons and siamangs), Asian great apes (orangutans) and African great apes (gorillas and chimpanzees) from humans.

2.2 Early Miocene hominoids

2.2.1 The appearance of primates

When dinosaurs disappeared 65 Ma, during the Paleocene (Table 2.3), numerous habitats were freed, among which was the floor of tropical forests. Rodents and small primate-like animals (archaic primates or plesiadapiforms; Figure 2.7) competed for the resources available on the forest floor. The latter had similar traits to rodents, such as a very pronounced diastema, great facial olfactory zones, and eyes located on each side of the head (Fleagle, 1988). Because of such traits, some primatologists assert that plesiadapiforms must not be considered true primates (Hooker, 1999). Fleagle (1999) placed them in a separate order.

After the disappearance of plesiadapiforms during the transition from the Paleocene to the Eocene (55 Ma), euprimates, or true primates, appeared in Asia (Bowen *et al.*, 2002). The families Adapidae and Omomydae, which are similar to current prosimians (Figure 2.8), are euprimates.

Phylogenetic analysis supports the hypotheses that a haplorrhine (including Anthropoidea)–strepsirrhine (Lemuriformes) dichotomy existed at least at the time of the earliest record of fossil

Table 2.3 Different geological epochs

Geological epoch	Starting time (millions of years)
Holocene	0.010
Pleistocene	1.79
Pliocene	5.32
Late Miocene	11.2
Middle Miocene	16.40
Miocene	23.80
Oligocene	33.7
Eocene	54.8
Paleocene	66

Source: adapted from Berggren *et al.* (1995) and Harland *et al.* (1990).

euprimates (earliest Eocene). Functional analysis suggests that stem haplorrhines were "small, nocturnal, arboreal, visually oriented insectivore-frugivores with a scurrying-leaping locomotion" (Kay *et al.*, 1997).

The fundamental adaptive shift occurred at the base of the lemuriform–anthropoid clade. Stem anthropoids remained small diurnal arborealists but adopted locomotor patterns with more arboreal quadrupedalism and less leaping. A shift to a more herbivorous diet occurred in several anthropoid lineages (Kay *et al.*, 1997).

Box 2.2 The size of early primates

Christophe Soligo and Robert D. Martin (2006) have challenged the widespread notion that the earliest primates were very small, which, they argue, is not supported by either the fossil record or modern species. Soligo and Martin (2006) argue, instead, "that the reduction of functional claws to nails—a primate characteristic that had up until now eluded satisfactory explanation—resulted from an increase in body mass to around 1000 g or more in the primate stem lineage."

(a)   (b)

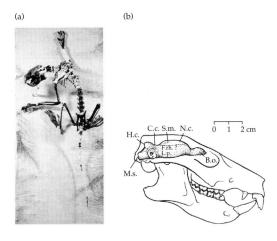

Figure 2.7 *Plesiadapis cookei*. (a) Skeleton UM 87990, as mounted in the University of Michigan Exhibit Museum; (b) cranium of *Plesiadapis*. The lateral location of the eyes, the great size of incisors, and diastema between the incisors and the rest of the teeth are traits that place plesiadapiforms close to rodents (Fleagle, 1988). Drawing from Gingerich (1976).

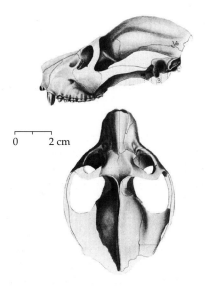

Figure 2.8 *Smilodectes gracilis*, middle-Eocene North American adapid. The limb and facial traits indicate its adaptation to arboreal life (Fleagle, 1988). Figure from the Gingerich (1981) by permission of John Wiley & Sons.

The derived traits of Eocene euprimates are primitive characters of later primates. Some of the most striking are the following (see Table 2.4 for other, less conspicuous, traits):

- fingers ending with nails but not claws, with padded fingertips,
- pentadactyl limbs, with big-toe thumb (hallux and pollex) opposable to other fingers, allowing very efficient grasping,
- separate radius and ulna in the upper limb and tibia and fibula in the lower limb,
- highly articulated limb joints,
- stereoscopic vision, with large development of both facial and brain areas related to vision,
- small or moderate facial projection and forward-facing eyes.

These traits reveal an adaptation to arboreal life. Stereoscopic vision, the mobility of the joints, the grasping capability afforded by padded fingertips

Table 2.4 Ancestral (plesiomorphic) hominoid traits

Trait
Nasal aperture higher than broad; oval-shaped*
Subnasal plane truncated
Subnasal plane stepped down to floor of nasal cavity
Orbits as broad or broader than high
Inter-orbital distance broad
Infra-orbital foramina few in number (≤ 3)
Infra-orbital foramina well removed from the zygomaticomaxillary suture*
Zygomatic bone curved and with strong posterior slope
Zygomatic foramina small
Zygomatic foramina 1–2 in number
Zygomatic foramina situated at or below the lower rim of the orbits
Glabella thickening which may occur on large individuals/species
Small, incisive foramina
Large, oval-shaped greater palatine foramina
Upper incisors lacking large size discrepancy
Thin enamel on molars

Source: Andrews and Cronin (1982).

* These traits are present in gibbons but not in African apes or humans.

and opposable thumbs, are apomorphies that facilitate climbing and jumping. Posture also began to change in adapids, showing a tendency towards an upright posture, associated with the displacement of the foramen magnum toward the inferior part of the cranium (Figure 2.9). The foramen magnum is the orifice through which the spinal cord enters into the cranium. Quadrupeds have the insertion located in the posterior part of the cranium, while in humans, bipeds, it is situated on a lower plane.

2.2.2 The puzzle of hominoid-derived traits

According to Stehlin (1909) there was a massive extinction of mammalian families—the *Grand Coupure*—during the transition between the Eocene and the Oligocene, because of major climatic shifts. As a consequence, European and Asian primates virtually disappeared, while the number of African representatives decreased notably. This is the reason why most known Oligocene primates come from a single site: Fayum, in Egypt. Up to 21 different species have been found there. They belong to such genera as *Apidium*, *Propliopithecus*, and *Aegyptopithecus*, most of them small-sized arboreal primates. According to Elwyn Simons (1965) the phyletic position of the Fayum fossils should be just before the split between cercopithecoids and hominoids (Figure 2.11).

After the Oligocene bottleneck, primates flourished during the Miocene, the epoch in which the first hominoids appeared. They did so in Africa and their morphology suggests that they were adapted to tropical forests. Members of their sister group, cercopithecoids, evolved apomorphies that allowed the exploitation of the open savanna.

Paleoecological studies have revealed that the habitats of lower- and middle-Miocene sites in Uganda, Kenya, and Tanzania (such as Napak, Songhor, Rusinga, and the Mfwangano Islands in Lake Victoria) were tropical forests with some areas of open savanna (Andrews, 1992b). Many hominoid remains have been found at those and nearby sites. It is assumed, thus, that the appearance of the superfamily Hominoidea, and its evolution during the Miocene, were associated with those ecological conditions.

That is, hominoids found a new adaptive formula that allowed them to exploit topical forests efficiently. Their large size compared to cercopithecoids exemplifies an evolutionary tendency involving the growth of the body to compete for available resources. A larger body also means a larger intestinal tract. Thus, hominoids took advantage of their ecological niche in the forest by increasing the amount of plants in their diet. This adaptation to forests abundant in edible fruits, roots, and leaves is reflected in the dental and locomotor evolutionary tendencies of fossil hominoids.

However, it is not easy to establish precisely how that evolution took place. The derived characters that separate the superfamily Hominoidea from its sister group, Cercopithecoidea, include three parallel evolutionary tendencies:

• loss of tail and broadening of the thorax, characters related with a more upright posture during feeding and traveling;

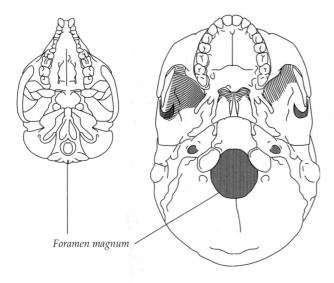

Foramen magnum

Figure 2.9 Inferior view of the crania of an Eocene primate (left) and a current human (right), showing the position of the foramen magnum.

Box 2.3 The arboreal life of euprimates

The arboreal life of euprimates was partially anticipated by some plesiadapiforms, such as those of the genus *Carpolestes* (Bloch and Gingerich, 1998). Contrary to what was previously believed, the archaic primate *Carpolestes simpsoni* already had an opposable hallux and fingers with nails (Bloch and Boyer, 2002; Figure 2.10). But, because *Carpolestes* lacked stereoscopic vision and the capacity for leaping, Bloch and Boyer (2002) considered that it

"represents the best morphological model yet known for an early stage in the ancestry of euprimates on the basis of their shared grasping capabilities and their close phylogenetic relationships." The climbing capabilities of *Carpolestes* and of later true primates could be either analogous traits or plesiomorphies inherited from a common ancestor. This issue cannot be settled by currently available evidence.

Figure 2.10 The Paleocene plesiadapiform *Carpolestes simpsoni* (*c.*55 mya; Bloch and Gingerich, 1998). Its opposable hallux and its fingertips (right) would have allowed it to climb (left). Figure adapted from Bloch and Boyer (2002).

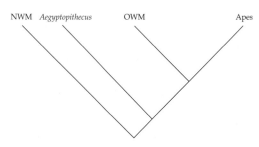

Figure 2.11 Left *Aegyptopithecus*; right, Elwyn Simons and colleagues consider *Aegyptopithecus* to be the sister group of catarrhines (Old World monkeys (OWM) and hominoids; Simons, 1965; Rossie *et al.*, 2002). NWM, New World monkeys. Modified from Rossie *et al.* (2002). Copyright, 2002, National Academy of Sciences, USA.

Box 2.4 The Oligocene bottleneck

The disappearance of European primates during the Oligocene might not have been as absolute as H.G. Stehlin believed. M. Köhler and S. Moyà-Solà (1999) have documented the presence at the Fonollosa-13 site (Barcelona, Spain) of a mandibular fragment with several teeth attributed to *Pseudoloris godinoti*, an early Oligocene primate.

• greater limb mobility, with more flexible articulations that allow improved ability to lift their arms and bestow prehensile functions to hands and feet;
• premolars with low crowns and relatively wide molars with low and rounded crowns, related to shifts in diet.

These morphological changes led to a considerable adaptive success during the Miocene, when hominoids achieved their greatest expansion throughout Europe, Asia, and Africa. But we are confronted with a paradox: the changes in mastication and locomotion that led to the traits observable in current apes seem to have appeared along two different lineages. Some Miocene specimens show apomorphies in the locomotor apparatus and retain dental plesiomorphies, while other specimens show the opposite pattern. Together with the lack of information about hominoids during the better part of the Miocene, it is not surprising to find controversies concerning hominoid evolution, plagued with doubts and contradictions.

2.2.3 Stages in hominoid evolution during the Miocene

The Miocene could be described as the golden age of hominoids. These were very quick, in geological terms, to achieve considerable evolutionary success. An important radiation had already taken place about 5 million years after the appearance of the first fossil, the East African *Proconsul*. The presence of hominoids in places far apart at that time is good evidence for this expansion.

The first discovery of a Miocene hominoid took place in the nineteenth century in Saint-Gaudens, France. It had features similar to those of living great apes, and it was given the name *Dryopithecus fontani* (Lartet, 1856). This finding was known to

Charles Darwin, who mentions it in his *Descent of Man* (1871). Since the initial discovery of *D. fontani* many more contemporary hominoids have turned up in Africa, Europe, and Asia.

The radiation of the superfamily during the Miocene is accepted unanimously. However, the subdivision of that process into stages and, especially, the relationships among the different known genera, are controversial. The available specimens span the three Miocene divisions: early Miocene (24–16 Ma), middle Miocene (16–11 Ma), and late Miocene (11–5 Ma; Table 2.3). They also appear in three different continents—Europe, Asia, and Africa—but the distribution is not uniform (Table 2.5).

There are no early Miocene specimens from Europe, nor African exemplars between 13 and 7 million years old (Figure 2.12). Hominoids are abundant in Asia during the late Miocene but are rare during the middle Miocene and inexistent during the early Miocene. This is why it is difficult to establish phylogenetic relationships among hominoids of different periods and continents and to unravel the processes of morphological change.

2.2.4 Early Miocene specimens

The oldest members of the superfamily Hominoidea belong to several genera: *Proconsul, Rangwapithecus, Nyanzapithecus, Morotopithecus,* and so on. The earliest exemplar seems to belong to *Morotopithecus* (Gebo *et al.*, 1997), found at the Moroto II site (Uganda). Its age has been estimated, using the potassium/argon

(K/Ar) method, to 20.7 million years. The best-documented genus is *Proconsul*, found in Uganda (Napak) and Kenya (Rusinga, Koru, Songhor, and Fort Ternan). The most abundant sites with *Proconsul* and *Rangwapithecus* fossils are Songhor and Koru (Kenya), which contain sediments that are 19–20 million years old (Andrews, 1996).

Proconsul (Hopwood, 1933; Figure 2.13) exhibits cranial and dental primitive traits shared with the

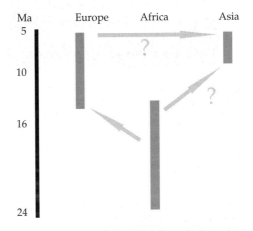

Figure 2.12 The Miocene hominoid fossil record is incomplete. We lack early Miocene specimens in Eurasia and late Miocene specimens in Africa. This lack of specimens makes it difficult to establish the phylogenesis of hominoids. How were the genera of different continents related during the Miocene? Asian late Miocene specimens could descend from European hominoids of the same period or middle-Miocene African ones.

Table 2.5 Some Miocene hominoid genera from the Old World

	Europe	Africa	Asia
Late Miocene, 11.2–5.32 Ma	*Dryopithecus, Oreopithecus, Ouranopithecus*		*Sivapithecus, Lufengpithecus, Ankarapithecus*
Middle Miocene, 16.4–11.2 Ma	*Dryopithecus*	*Afropithecus, Kenyapithecus, Equatorius, Nacholapithecus*	*Lufengpithecus*
Early Miocene, 23.8–16.4 Ma		*Proconsul, Rangwapithecus, Morotopithecus*	

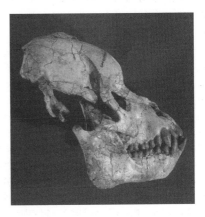

Figure 2.13 *Proconsul* skull. (c) Alan Walker.

Old World monkeys. It is considered the first member of the superfamily Hominoidea because of the presence of some features that point to the appearance of characteristically hominoid-derived traits, such as the Y 5 pattern. The Y 5 pattern (Hellman, 1928) refers to the presence in the lower molars of an occlusion surface with five cusps forming a shape resembling a Y (Figure 2.14). The Oligocene primates from Fayum (Egypt) do not show this feature, whereas *Proconsul* already does (Martin, 1990). Because all current great apes show it, the Y 5 pattern is considered a hominoid synapomorphy. There is, however, great variation in humans regarding this trait, to the point that, in some instances, one of the cusps is missing.

The morphology of *Proconsul* has been summarized by Andrews (1992a). With respect to locomotor behavior, it seems that *Proconsul* used arboreal quadrupedalism, similar to the catarrhines (Old World monkeys). However, *Proconsul* already exhibited certain derived features related to its posture and its adaptation to arboreal life. First, the posture itself is more orthograde. Second,

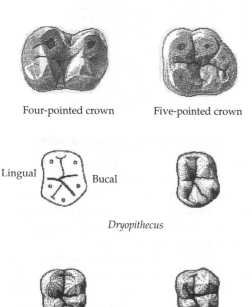

Four-pointed crown   Five-pointed crown

Lingual   Bucal

Dryopithecus

Chimpanzee   Human

Figure 2.14 The Y 5 pattern.

Proconsul possessed the capacity to lift its hands over its head and to turn its arms. The mobility of its thumbs, wrists, hips, and ankles must have been considerable. Thus, regarding mobility, *Proconsul* was quite similar to current great apes, except for having a larger thumb, and therefore, a hand more similar to that of humans. It probably lacked a tail.

The dentition of *Proconsul* shows a mosaic of primitive and derived features. The crown of the third premolar P3 is low and the molars are relatively wide, with low and rounded crowns. Lower molars show, as mentioned above, the Y 5 pattern. But its enamel is rather thin. Thus, it is usually assumed that it fed mostly on soft materials, such as fruits and tender leaves. In fact it is possible that all early Miocene hominoids shared ecologically similar habitats, and thus had similar diets to *Proconsul*, even though there are some differences. The diet of *Rangwapithecus*, for instance, is considered more folivorous (leaf-feeding) than frugivorous (fruit-feeding).

Proconsul differs from cercopithecoids in yet another trait. Its brain is larger than that of similarly sized Old World monkeys. Its encephalization quotient reached 48.8%, compared with 22–41% calculated for 11 monkey species (Walker *et al.*, 1983).

Within the genus *Proconsul* there are different species that differ, mainly, in size. The smallest, *Proconsul africanus*, as well as *Rangwapithecus gordoni*, weighed around 9–11 kg (Aiello, 1981; Walker *et al.*, 1983). The largest, *Proconsul major*, weighed close to 26–38 kg (Andrews, 1992a). However, there are a number of specimens for which the paucity of remains makes it difficult to determine their size and morphology. Based on their dental traits, Andrews (1992a) has placed them in the genus *Proconsul* (*Limnopithecus legetet*, for instance). *Proconsul africanus* is the best-known species within the genus, with remains from almost every part of the skeleton. The remains retrieved at the island of Rusinga (Kenya) have allowed a detailed description of *P. africanus*, but there are also remains from other sites.

The variation in the size of the *Proconsul* specimens from Rusinga was initially interpreted as

---

Box 2.5 Was *Proconsul* really a hominoid?

After analyzing the paranasal anatomy of *Aegyptopithecus* and comparing it with that of *Proconsul*, James B. Rossie and colleagues (2002) argued that the latter should not be considered a hominoid. Rather, *Proconsul* would be the sister group of current catarrhines, and not the first hominoid. Nevertheless, the consideration of *Proconsul* as a hominoid is widespread and is the alternative we have followed in this chapter.

Box 2.6 The encephalization quotient

If we measure the brain in absolute terms, there is no question that, generally, the larger the body of any animal, the larger the brain. To determine the adaptive significance of brain size, it is preferable to use the encephalization quotient (E.Q.), which measures the relationship between the weight of the brain and the body. When the average E.Q. is conspicuously larger for one fossil species than another, the former is usually assumed to have more advanced cognitive capacities than the latter (Tobias, 1975; Eccles, 1977). Living animals are used as a base for this attribution (Figure 2.15). Among New World monkeys, for instance, those that feed on insects have larger brains than those that feed on leaves. This difference in size is thought to be related to the greater need of insectivores to process environmental information (Jerison, 1977a).

Figure 2.15 Which has a larger brain, a dolphin or an elephant?

evidence for a marked sexual dimorphism (morphological differences between males and females; Figure 2.16). But after the specimens initially attributed to *P. africanus* were separated into two different species, *P. africanus* and *P. major*, sexual dimorphism within each of them did not seem greater than between current male and female chimpanzees (Kelley, 1992).

2.3 Middle Miocene: the migration of apes out of Africa

Hominoids appeared in African tropical forests. During the middle Miocene, their features are a mosaic of dental and locomotor primitive and derived traits. The apomorphies that would later lead to the traits observed in current apes continued to develop during the middle Miocene. But the way in which that evolution took place is intriguing and difficult to interpret.

The existence of numerous hominoid fossils from Africa, Europe, and Asia, estimated to be from between 17 Ma (early Miocene) and 12 Ma (middle Miocene), is a notable indication of an ape radiation during that period. The corresponding habitats are difficult to identify precisely, but both the analyses of paleosoils and comparison with the associated fauna suggest a further extension of

Gorilla gorilla, male

Gorilla gorilla, female

Figure 2.16 Sexual dimorphism. Natural selection has fixed different shapes and sizes in males and females in many sexually reproducing species. Sexual dimorphism is common in current apes, males being larger and, in some cases, of different color. Males exhibit great canines and sagittal crests on their crania to attach masseter muscles capable of moving their potent mandibles.

Box 2.7 *Morotopithecus*: ancestor of all apes?

The direct ancestral line of apes extended back to the beginning of the Miocene according to the interpretation by Gebo and colleagues (1997) of the postcranial remains found in 1994 and 1995 at the Moroto I and Moroto II sites in Uganda. The Moroto specimens, which include part of the left and right femora, the left shoulder and vertebrae, together with other remains found earlier, suggest, in the opinion of Gebo and colleagues, an arboreal locomotion with a certain degree of brachiation. They have proposed a new species, *Morotopithecus bishopi*, which is over 20.61 ± 0.05 million years old. This is the age of a volcanic tuff at Moroto I, dated with the $^{39}Ar/^{40}Ar$. *Morotopithecus* could either have been an

ancestor of all apes, or a predecessor of the great apes, before, in any case, the divergence of the orangutans.

A cladistic study by Young and Maclatchy (2004) revealed derived traits in *Morotopithecus*, which suggested that, if it really was an ancestor of current hominoids, it would have already fixed certain apomorphies present in current apes. These apomorphies, therefore, should be considered the primitive traits when studying the evolution of hominoid adaptive complexes, such as locomotion. Later in the book we will deal with the question of the polarity of dental and locomotor traits (that is to say, which ones should be considered primitive and which ones should be considered derived).

---

open lands with a wide range of conditions (Andrews, 1992b). This opened new possibilities for hominoids that made possible an important adaptive step: the colonization of tropical forest floors.

Chimpanzees and gorillas spend most of their time on the forest floor, although they have not lost their climbing ability. They obtain the better part of their diet on the floor, composed of harder materials than those typical of an arboreal habitat, especially in the case of gorillas. As a consequence of that return to the ground, chimpanzees and gorillas developed knuckle-walking, a distinctive locomotor habit: quadrupedalism with the forelimbs supported by the dorsal surface of the fingers. Because of their long upper limbs, typical of apes, the body is carried on a plane which is not

parallel to the ground, but slanted upright to a certain point. The tendency towards an upright posture that began in Eocene euprimates, and which is shared by many anthropoids, increased in middle-Miocene hominoids. Upright posture is not, therefore, an exclusively human feature. But it is not easy to relate posture change and locomotion.

Brachiation, typical of Asian apes, as well as the climbing and knuckle-walking of African apes, involve a certain degree of upright posture (Figure 2.17). The functional significance of the changes in the locomotor apparatus is difficult to interpret. Specifically, it is not easy to establish whether brachiation or climbing was the synapomorphic trait during the middle Miocene, and thus the primitive trait for later apes. Under the first

Figure 2.17 A female Dracma baboon. Upright posture and bipedalism are not the same thing. An upright posture is common in anthropoids. Current African apes usually adopt upright postures while feeding. Their quadrupedal locomotion also involves a more upright posture because of the greater length of their forelimbs. But no current monkey or ape is bipedal. Has there ever been a bipedal monkey or ape?

scenario, current chimpanzees and gorillas would have developed their particular locomotion later. Alternatively, the brachiation of orangutans would be the trait that appeared later. However, given that lesser apes (gibbons and siamangs) are also brachiators, suspension would be, under this second scenario, an analogous trait, appearing separately in Asian lesser and great apes.

The colonization of the floor and the adoption of a more upright posture coincided with the radiation of the superfamily throughout Eurasia. However, as mentioned earlier, the fragmentary fossil record prevents determining in detail the migration of ancestral apes out of Africa and the relationship between Eurasian fossils and their African ancestors.

The oldest known Eurasian hominoid specimen is ENG. 4/1, an upper molar fragment from the Engelswiess site in southern Germany. Its age has been estimated to be around 16.5–17.0 million years, based on magnetostratigraphy, biostratigraphy, and lithostratigraphy (Heizmann and Begun, 2001). The remains of *Lufengpithecus chiangmuanensis* from Ban Sa, northern Thailand, which include close to 20 teeth and dental fragments (which vary in size, probably due to sexual dimorphism), are between 13.5 and 10 million years old (Chaimanee *et al.*, 2003). How

are these specimens and the later Eurasian ones related to the African Miocene forms and to current apes?

In 1965 Elwyn Simons and David Pilbeam reinterpreted middle-Miocene hominoids, grouping the great variety of genera that had been suggested earlier in a single subfamily of dryopithecines. This interpretation was adopted by Andrews (1992a), who maintained the rank of subfamily for dryopithecines, but distinguished three tribes within it: Afropithecini, Kenyapithecini, and Dryopithecini. To achieve greater taxonomic coherence, Andrews (1996) modified this proposal in a later revision and placed the groups previously occupying the rank of tribe in different subfamilies: Afropithecinae, Kenyapithecinae, and Dryopithecinae.

Andrews related the differences among Afropithecinae, Kenyapithecinae, and Dryopithecinae to the adaptations of apes to the Miocene forests that we have pointed out above. According to Andrews (1996), during the middle Miocene a line represented by the subfamilies Afropithecinae and Kenyapithecinae evolved derived traits in the masticatory apparatus. These new features (thicker enamel among others) were indicative of a harder diet. Their postcranial elements did not change very much. The other evolutionary line, with the subfamily Dryopithecinae in a central position, followed the opposite adaptation: few changes in dentition (retaining the primitive trait of thin enamel, for instance) but a noteworthy variation in the locomotor apparatus. Dryopithecines moved away from primitive traits associated with quadrupedalism and developed extremities with greater mobility, in line with current apes.

Andrews' proposal shows how dentition and locomotion developed separately starting from the mosaic of early-Miocene primitive and derived features, as observed in *Proconsul*. Such a proposal faces certain difficulties. If two lineages—Afropithecinae and Kenyapithecinae—evolved changes in their dentition and a different lineage—Dryopithecinae—modified its locomotor apparatus, how did these diverse evolutionary lines lead to later hominoids, the ancestors of the different current apes? Take, for instance, the dryopithecines. If their locomotion evolved while their primitive masticatory traits did not, how could they be considered ancestors of current apes, which differ from the early *Proconsul* in postcranial and dental characters? The same question can be asked about the other subfamilies proposed by Andrews (1996), Afropithecinae and Kenyapithecinae. The ancestors of current apes must exhibit changes in both masticatory and locomotor traits.

2.3.1 Different proposals of adaptive synthesis during the middle Miocene: the role of *Kenyapithecus/Equatorius/Nacholapithecus*

The adaptive synthesis that occurred during the Middle Miocene may be interpreted in three different ways, as follows:

(a) Kenyapithecus africanus *as direct ancestor of living great apes*
One attempt to solve the evolutionary paradox of the separate appearance of traits relative to hominoid locomotion and dentition, assigns a pivotal role to one of the most interesting middle-Miocene organisms, *Kenyapithecus*. This hypothesis has been subject to disparate interpretations.

The definition of the genus *Kenyapithecus* is based on the specimens from Fort Ternan in Kenya (two maxillas and a lower molar), assigned to the species *Kenyapithecus wickeri* by Louis Leakey. Leakey (1967a) later enlarged the genus to include a new species, *Kenyapithecus africanus*, with exemplars from Songhor and Rusinga Island (Kenya). Leakey (1967a) proposed the BMNH 16649 maxilla (identified as CMH 6 in Leakey's paper) from Rusinga Island as the type specimen of the species *K. africanus*. This very specimen had been provisionally classified by Le Gros Clark and Leakey (1951) as *Sivapithecus africanus*. In 1967 Leakey considered that *Kenyapithecus*, together with '*Ramapithecus*', belonged to the family Hominidae, understood in its classical sense; that is to say, the set of humans and their exclusive ancestors. This proposal rests on the thick molar enamel of *Kenyapithecus*.

In opposition to Leakey's grouping, several authors have underscored the problems that arise from the inclusion of such different beings as *K. wickeri* and *K. africanus* in the same genus. Andrews (1996) tried to solve this problem by placing the Rusinga specimens, *K. africanus*, in the subfamily Afropithecinae and granting the Fort Ternan specimens, *K. wickeri*, the rank of subfamily Kenyapithecinae. Furthermore, he also believed the latter should include exemplars from the Turkish site of Pasalar—*Griphopithecus*—whose

---

Box 2.8 Environment, enamel, and diet

A large part of the fossil record consists of mandibles and teeth. These are hard materials, less vulnerable to the aggression of scavengers and, thus, have a better chance of fossilizing. Therefore, mandibles and teeth are often used to draw taphonomic conclusions and functional interpretations, such as those relating to the diet of different taxa.

Thick dental enamel is usually associated with the intake of hard foods, such as that of herbivores, in contrast to frugivores and carnivores. Furthermore, the analysis of microwear patterns caused by food on the enamel also provides evidence regarding the composition of the diet.

Kenyapithecus is a good example of the taphonomic conclusions that can be drawn from the analysis of teeth. Andrews (1996) attributed a tropical forest environment to

Fort Ternan. However, the studies of microwear patterns on *Kenyapithecus* teeth suggest a diet based on small and hard foodstuffs (Martin, 1985), which would be expected in a more open habitat. Kappelman (1991) suggested a possible explanation for the enamel thickness of *Kenyapithecus*, in his study of the paleoclimate at Fort Ternan, the site that yielded the *Kenyapithecus wickeri* remains. According to the analysis of the bovid fauna at Fort Ternan, which is the most abundant, Kappelman concluded that the middle-Miocene environment corresponds to open savanna. Hence, Kappelman (1991) attributed to *Kenyapithecus* a relatively more terrestrial locomotion than that of the late-Miocene *Sivapithecus*, considering it to be a primitive trait and not an apomorphic locomotor habit, which reinforces the idea that hard foods were an important part of the diet of *Kenyapithecus*.

enamel is also thick. Thereafter, other specimens from Nachola, Muruyur, and Esha (Kenya) and Candir (Turkey) were also assigned to *Kenyapithecus* (McCrossin and Benefit, 1993).

Placing *K. africanus* and *K. wickeri* in two different subfamilies (Afropithecinae and Kenyapithecinae, respectively) resolved the taxonomic problems, but not the difficulty we mentioned before. How can the dental and locomotor apomorphies that evolved separately be integrated? One possibility was suggested by a discovery in the summer of 1996 of a distal radius belonging to *K. africanus* on Maboko Island (Kenya). The study of a radioulnar joint led Thomas C. Crawford and Monte L. McCrossin to suggest at the annual meeting of the *American Association of Physical Anthropologists* (St. Louis, April 1997) the need for undertaking a reinterpretation of the adaptive path of middle Miocene hominoids.

In contrast with the separate specialization of the subfamilies Afropithecinae and Kenyapithecinae (dentition) and the subfamily Dryopithecinae (locomotion), established by Andrews, the arm bones of *K. africanus* retrieved from Maboko suggest a derived morphology. The taxon would have evolved not only dental apomorphies but also others related with locomotion. This, according to McCrossin and colleagues, would allow placement of the Maboko *Kenyapithecus* remains as the earliest direct ancestor of African great apes (Gibbons and Culotta, 1997). *Kenyapithecus* resembles these apes in its facial and dental traits and in the structure of its locomotor apparatus. Regarding the masticatory apparatus, this had been anticipated by McCrossin and Benefit (1993) in the study of the KNM-MB 20573 juvenile mandible from Maboko Island. In this interpretation, *K. africanus* is considered to be the ancestor of all great apes, including the Asian late-Miocene *Sivapithecus*.

(b) Kenyapithecus africanus *as a derived lateral lineage* (Equatorius)

The study of the KNM-TH 28860 specimen carried out by Ward and colleagues (1999) suggests a very different interpretation of *K. africanus*. This specimen was discovered in 1993 by the Baringo Paleontological Research Project at Kipsaramon, within the Muruyur formation that runs along Tugen Hills to the west of Lake Baringo in Kenya. The KNM-TH 28860 specimen was dated to 15 Ma by Ward *et al.* (1999). It is a partial skeleton consisting of:

- a nearly complete mandible with many of its teeth and some thoracic remains,
- part of the vertebral column,
- the anterior limbs, including a complete radius and most of one humerus, plus hand, wrist and finger fragments.

Ward and colleagues (1999) have suggested that the taxonomy of the specimens belonging to the genus *Kenyapithecus* should be reinterpreted by separating—as Andrews (1996) had done—the Fort Ternan more recent *K. wickeri*, dated close to 14 Ma, from the earlier Maboko *K. africanus*. According to Ward *et al.* (1999), the latter, and the KNM-TH 28860 specimen from Tugen Hills, share with *Proconsul* and *Afropithecus* the midfacial anatomical pattern and certain dental traits (such as the molar size sequence M1 < M2 < M3 and the shape of the incisors). These would be the primitive traits of *K. africanus* that later changed leading to the derived traits observed in Fort Ternan *K. wickeri*.

If indeed the Maboko (*K. africanus*) and the Fort Ternan (*K. wickeri*) specimens correspond to two separate lineages, placing them in the genus *Kenyapithecus* would render it paraphyletic. To restore the monophyly of *Kenyapithecus*, Ward and colleagues assigned KNM-TH 28860 to the taxon described by Leakey (1967a), *K. africanus*, but introducing a new genus, *Equatorius* (thus, *Equatorius africanus*; Figure 2.18). The operation is consequent with Peter Andrews' inclusion of *K. africanus* in the subfamily Afropithecinae, which separated it from the genus *Kenyapithecus*, which has its own subfamily.

Equatorius would include the material previously attributed to *K. africanus* and the new specimens from Tugen Hills. According to the members of the Baringo Paleontological Research Project, *Equatorius*' size was similar to that of an adult male baboon, with a flexible vertebral

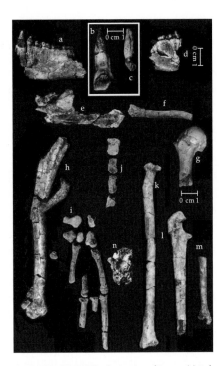

Figure 2.18 KNM-TH 28860, *Equatorius africanus*. (a) Left mandibular corpus; (b) left maxillary central incisor; (c) right maxillary lateral incisor; (d) right mandibular corpus fragment; (e) right scapula; (f) right clavicle; (g) left proximal humerus; (h) right humerus with first rib attached; (i) right hand (hamate; trapezium; trapezoid; scaphoid; pisiform; metacarpals II, III, and V; and phalanges); (j) sternum; (k) right radius; (l) right proximal ulna; (m) right distal ulna; (n) lowest thoracic vertebra. Photograph from Ward *et al.* (1999). Reprinted with permission from AAAS.

column and arms and legs of an equivalent size. *Equatorius* would divide its time between tree branches and the floor of the tropical forest (Zimmer, 1999). Thus, *Equatorius* played a major role in the descent of African apes to the ground.

Equatorius and *Kenyapithecus* were considered by Ward and colleagues as the main characters in the diversification of middle-Miocene hominoids through their adaptation to an open environment. *Equatorius* would constitute a parallel and terminal line, whereas *Kenyapithecus* (*K. wickeri*), with its derived traits, would lead to the forms that radiated throughout Europe and Asia and, ultimately, to the current ape genera (Figure 2.19).

The tectonic movements that produced the collision between the African and Eurasian plates

allowed apes to leave Africa during the middle Miocene. After the redefinition of *Kenyapithecus*, this taxon shows remarkable similarities with Eurasian hominoid specimens, such as some found at Pasalar (Turkey; Zimmer, 1999). Thus, it is possible to establish a relationship between *K. wickeri* and late-Miocene hominoids and current apes. However, David R. Begun *et al.* (2003) reject this relatively simple interpretation of the role of middle-Miocene hominoids. Begun has argued that there were several departures from Africa, which resulted in phylogenetic relationships that are currently difficult to trace. Under this scenario the evolution of hominoids towards the direct ancestors of current apes would have taken place in Eurasia, with a "return to Africa" during the late Miocene.

(c) *The diversity of Kenyapithecus: Nacholapithecus*
Considering *K. wickeri* as the direct ancestor of current apes, as Ward and colleagues (1999) suggest, does not resolve one of the most controversial questions regarding the evolution of hominoid posture during the middle Miocene. Gorillas and chimpanzees are capable of climbing trees and traveling on the ground (knuckle-walking), whereas orangutans brachiate. What appeared first: climbing or suspension?

A possible answer comes from KNM-BG 35250 (Figure 2.20), discovered in 1997 at the BG-K site (Nachola, Kenya), in the lower part of the Aka Aiteputh formation, dated to 15–14 Ma by the Joint Japan Kenya Samburu Hills–Nachola Paleoanthropological Expedition (Tatsumi and Kimura, 1991). KNM-BG 35250 includes a fairly complete skeleton, the size of a male baboon, and provides new evidence about the evolutionary changes in hominoid posture during the middle Miocene. Nasako Nakatsukasa and colleagues (1998) included the specimens in *Kenyapithecus*, while noting that such a classification might be inappropriate. Within *Kenyapithecus*, KNM-BG 35250 would be "a new species based on its gnathodental morphology". A year later, Ishida and colleagues (1999) classified the specimens in a new genus and species, *Nacholapithecus kerioi*.

What supported the initial grouping of the Nachola specimen with *Kenyapithecus*? The answer

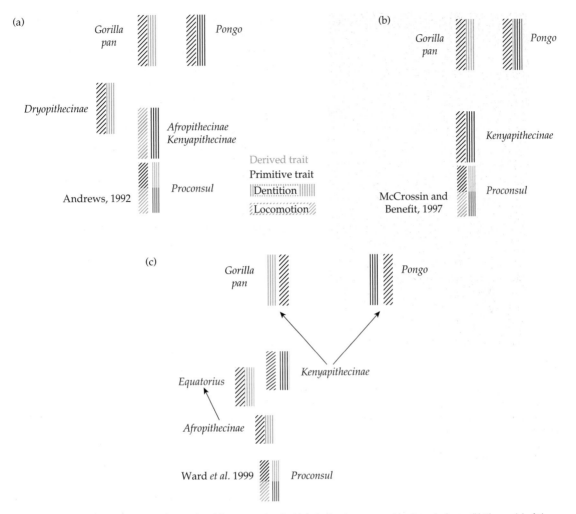

Figure 2.19 (a) The model of the evolution of middle-Miocene hominoid-derived traits suggested by Peter Andrews. (b) The model of the evolution of *Kenyapithecus*-derived traits suggested by McCrossin and Benefit. (c) Ward's model of two evolutionary lineages, a terminal one (*Equatorius* = *K. africanus*) and another related to current apes (*K. wickeri*).

might reside in Peter Andrews' observation that locomotor primitive traits are maintained in the subfamilies Afropithecinae (which include, let us not forget, *K. africanus*) and Kenyapithecinae (*K. wickeri*). The definition of the new genus was based on the identification of some dental apomorphies but, most of all, on the reconsideration of the postcranial anatomy and posture of *Nacholapithecus*.

From the viewpoint of Nakatsukasa *et al.* (1998), the specimen's primitive traits, shared with *Proconsul*, suggest a climbing locomotion and arboreal quadrupedalism, though with frequent suspension. However, these authors noted the possibility of a terrestrial locomotion—secondary to the arboreal behavior—related with foraging strategies imposed by the growing seasonality of the middle Miocene climate. Thus, the specimen was considered capable of almost any form of locomotor behavior.

In a later detailed study of the KNM-BG 35250 specimen, Ishida *et al.* (2004) noted that "the

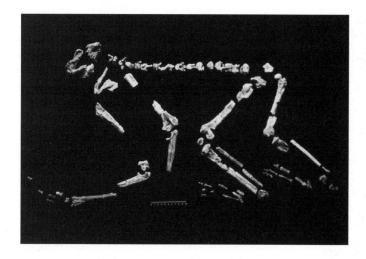

Figure 2.20 The Nachola (Kenya) KNM-BG 35250 specimen. It was initially attributed to *Kenyapithecus* (Nakatsukasa *et al.*, 1998), but it was later included in a new genus, *Nacholapithecus*. Photograph reprinted from Journal of Human Evolution. Hidemi Ishida *et al* 'Nacholapithecus skeleton from the Middle Miocene', Journal of Human Evolution, 2004, 46:1, 69–103.

postcranial specializations suggest a greater commitment to orthograde locomotor and postural modes. In this sense, *N. kerioi* assumes a modern aspect compared to *Proconsul*. However, it retains primitive features for the lumbar vertebrae and humeral trochlea." Nakatsukasa (2004) arrived at the same conclusion and interpreted the posture of *N. kerioi* as similar to that of *Proconsul* because of shared primitive traits, but with a greater specialization towards orthograde climbing: "'hoisting' and bridging, with the glenoid fossae of the scapula probably being cranially orientated, the forelimbs proportionally large, and very long toes." Nakatsukasa admitted that due to the gap in the hominoid postcranial record in Africa until 6 Ma, the last stages of locomotion evolution immediately before the divergence of *Homo* and *Pan* cannot be determined with certainty. However, his study revealed that *Nacholapithecus* had fixed climbing, and not suspension, as the form of arboreal locomotion.

If the evolutionary line that includes *Nacholapithecus* and *K. wickeri*, possible ancestors of current great apes, fixed climbing, and not suspension, this has consequences for hominoid systematics. It requires the introduction of locomotion-related homoplasies. As Ishida *et al.* (2004) have noted, "*Morotopithecus* from the early Miocene of Uganda . . . exhibits extant great ape-like climbing and/or suspensory adaptations of the lumbar vertebra."

If the climbing traits of KNM-BG 35250 are apomorphies inherited by current great apes, then brachiation must be regarded as a separately fixed homoplasy in lesser apes (*Hylobates*) and great apes (*Pongo*).

2.3.2 *Pierolapithecus*: the European link?

There is additional evidence suggesting that the brachiation of gibbons and orangutans is a homoplasy. In 2004 Salvador Moyà-Solà and colleagues announced the discovery of a partial skeleton which included the cranium's facial region (IPS-21350.1; IPS stands for Paleontologic Institute of Sabadell) from Barranc de Can Vila 1 (Barcelona, Spain). Its age is estimated by means of biostratigraphic methods (compared fauna), around 13 or 12 Ma, the middle Miocene. The importance of this discovery, in addition to its age and location, is the combination of well-preserved cranial, dental, and postcranial materials that are more recent than KNM-BG 35250 (*Nacholapithecus*). The evolutionary significance of the specimen involves interesting methodological consequences for a systematics approach. We shall examine its case in detail.

The discoverers named a new genus and species for IPS-21350, *Pierolapithecus catalaunicus* (Moyà-Solà *et al.*, 2004; Figure 2.21). The specimen reveals once again a combination of primitive and derived

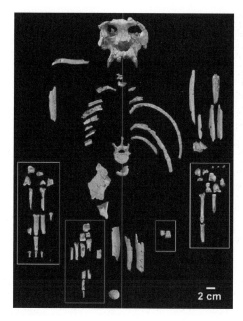

Figure 2.21 *Pierolapithecus catalaunicus*; middle Miocene from Barranc de Can Vila (Barcelona, Spain). Photograph from Moyà-Solà *et al.* (2004). Reprinted with permission from AAAS.

features in the thoracic and lumbar areas, the face, and the joints (Table 2.6). In particular:

• the thorax, lumbar region, and wrist include apomorphies that accentuate the similarities between *Pieralopithecus* and apes, such as an orthograde body design;
• the face has an overall ape-like structure, but its profile shows that the nasals form an acute angle with the palate, contrasting with the more orthognathous profile shared by Eurasian late-Miocene hominoids (*Ouranopithecus, Sivapithecus*) and extant great apes.

Moyà-Solà *et al.* (2004) suggest that *Pierolapithecus* is the common ancestor of current great apes and humans (see the cladogram in Figure 2.22), based on the increased capacity of adduction and supination at the wrist; the wide, shallow thorax; the long and chimp-like clavicle; and the stiff lumbar region. This set of traits suggests a tendency towards an orthograde locomotor and positional behavior.

Under this scenario, posture becomes the key apomorphy that establishes the relationship between *Pierolapithecus* and the diverse ape and human lineages that appeared during the late Miocene. However, the specimen's hand does not show the traits usually associated with suspensory behavior. Rather, it is similar to the short hands of monkeys. Moyà-Solà and colleagues (2004) suggest that "the primitive morphology of the *Pierolapithecus* hand, indicating little (if any) suspensory behavior, strongly suggests that the two basic components of extant ape locomotion—vertical climbing and suspension—appeared independently. Thus, modern ape-like below-branch suspensory locomotion is likely to have been acquired later and independently by the extant members of this clade".

Suspension is seen, once again, as an analogous trait—a homoplasy—and not as a current ape synapomorphy. But parsimony requires minimizing homoplasies, which would argue against Moyà-Solà and colleagues' interpretation of *Pierolapithecus*. Begun and Ward (2005) have pointed out that *Pierolapithecus* suggests a certain degree of suspension, which translates, in their parsimonious cladogram, into a clade including *Pierolapithecus* and *Dryopithecus*, which would constitute the sister group of African apes and humans (Figure 2.23).

Begun's interpretation is consistent with his proposal of a connection between European *Dryopithecus*, the Turkish specimens, and—with the return to Africa—African apes and humans, advanced after studying the dispersal patterns of Eurasian hominoids (Begun *et al.*, 2003). In their reply Moyà-Solá and colleagues (2005) stress an important consideration already mentioned: cladograms rely heavily on preliminary procedures, such as the consideration of the characters assigned to the taxa. In Moyà-Solà's opinion, Begun and Ward's argument in favor of suspensory behavior of *Pierolapithecus* is not persuasive. Gorillas and chimpanzees have limited suspensory behavior and hands which are not very large. Consequently, suspension is not, according to Moyà-Solà and colleagues, a significant behavior in evolutionary terms. The key event would have

Table 2.6 Primitive and derived traits of *Pierolapithecus*

Derived traits	Homologies	Primitive traits	Homologies

Thorax
  increased rib curvature and
    angulation (indicating a
    broad and shallow thorax)
	Oreopithecus
	Dryopithecus
	Extant great apes		

Lumbar vertebrae
  robustness of the wide and
    short pedicles
  caudally oriented spinous process
  reduced wedging
  lack of the distinct ventral keel and
    associated concave shape of the
    ventrolateral sides
	Extant great apes		

		Hand
		  middle and proximal phalanges
		    less curved and shorter than
		    those of extant apes
			Monkeys

Wrist-antebrachial joint	Extant great apes		

Face
  maxillae, nasals, and orbits on a
    same plane
  flat nasals that project anteriorly
    beneath the level of
    the lower orbital rims
  high zygomatic root
  high nasoalveolar clivus
  deep palate
  broad nasal aperture widest
    close to the base
	Extant great apes	Face
		  low face with a
		    posteriorly situated glabella
		  frontal squama forming an
		    open angle with the orbital plane
			Afropithecus

been the acquisition of an orthograde posture in association with climbing. This line of reasoning places *Pierolapithecus* as a sister taxon of African and Asian great apes and humans; that is, before the divergence of current hominoids (see Figures 2.22 and 2.23).

2.4 Late Miocene: the divergence of great apes

The process of hominoid evolution that we have reviewed up to here establishes that the superfamily Hominoidea appeared in Africa during the early Miocene, and later expanded towards Eurasia with the development of locomotor apomorphies. The increased reliance on climbing while maintaining an orthograde posture is probably the most significant evolutionary development. However, it is difficult to follow the footsteps of later hominoids due to the gap in the African fossil record after *Nacholapithecus*.

There is clear evidence supporting an important radiation of apes during the late Miocene. Middle-Miocene genera were still present and some new ones appeared, mostly in Eurasia. At no other time has there been such a diversity of hominoid forms. According to molecular data, the late Miocene is when African and Asian great-ape lineages split. But the absence of African late-Miocene hominid fossil remains hinders the study of the great-ape

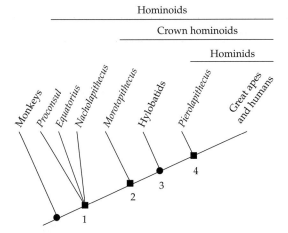

Figure 2.22 Cladogram showing early-Miocene forms (*Proconsul*, *Equatorius*, *Nacholapithecus*, *Morotopithecus*), middle-Miocene forms (*Pierolapithecus*), current lesser apes (*Hylobates*), and the set of great apes and humans. According to this proposal, *Pierolapithecus* is the ancestor of great apes and humans, but not of lesser apes. Monkeys constitute the sister group of all Hominidea (Moyà-Solà et al., 2004).

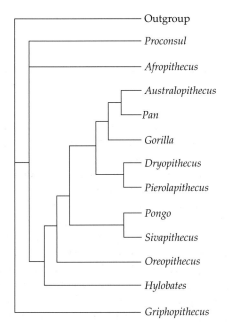

Figure 2.23 Cladogram showing the position of *Pierolapithecus*. According to Begun and Ward (2005).

diversification leading to orangutans, gorillas, and chimpanzees.

During the late Miocene we find *Dryopithecus*, *Oreopithecus*, and *Ouranopithecus* in Europe; *Ankarapithecus* in Turkey; *Sivapithecus* in Pakistan; *Lufengpithecus* in China; and *Otavipithecus* in Namibia (Africa). Most were moderately large, between the size of current mandrills and chimpanzees, except for *Gigantopithecus*, which may have been as large as a female gorilla. They all had marked sexual dimorphisms and characteristic adaptations to life in a tropical forest (Kelley, 1992).

2.4.1 *Dryopithecus* (Lartet, 1856)

Dryopithecus was the first Miocene ape to be discovered. The initial specimen, found in Saint Gaudens, France, in the mid-nineteenth century, was included by Edouard Lartet in the species *Dryopithecus fontani*. *Dryopithecus* is essentially a European genus. Most exemplars come from Hungarian sites (Rudabánya; see Kordos and Begun, 2001) and from northeast Spain. It may have also extended through the Asian continent, judging from remains found in Wudu and Keiyuan, China. *Dryopithecus* probably lived in seasonal forests, which suggests that they were predominantly arboreal primates, but their diet is unknown.

As mentioned, *Dryopithecus* retained primitive dental traits but exhibited derived postcranial traits (Andrews, 1992a, 1996). The most significant changes of *Dryopithecus* took place in the forearm and elbow, which show similarity with current great apes and favor its consideration as an ancestor of some of them. The question of which apes descend from *Dryopithecus* is debated. After examining the face of the CLL 18000 *Dryopithecus laietanus* specimen from Ca'n Llobateres (Barcelona, Spain; Figure 2.24), Begun (1992) concluded that it shares some subnasal traits with African great apes. Moyà-Solà and Köhler (1993a, 1993b), conversely, found similarities with orangutans in the zygomatic. This latter interpretation seems more reasonable if we consider that *D. laietanus* was probably a brachiator.

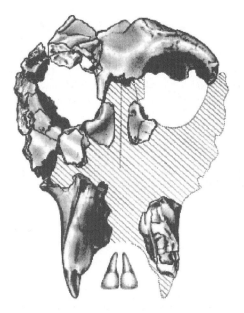

Figure 2.24 CLL 18000, *Dryopithecus laietanus*. Drawing from Moyà-Solà and Köhler (1993b). Reprinted by permission from Macmillan Publishers Ltd. *Nature* 365:6446, 543–545, 1993.

Figure 2.25 *Sivapithecus* from Potwar, Pakistan. (c) Peabody Museum of Archaeology and Ethnology, Harvard University.

Jordi Agustí (see Agustí *et al.*, 1996; Agustí, 2000) accepts the brachiation hypothesis, but suggests that European *Dryopithecus* evolved from African ancestors and colonized forested areas in middle-Miocene Europe, when the intercontinental bridge between Arabia and Eurasia was reestablished.

2.4.2 *Sivapithecus* ("*Ramapithecus*"; Lewis, 1934)

Specimens from the Siwaliks Hills, Pakistan, initially received the name of *Ramapithecus*. They are now considered members of the genus *Sivapithecus* and, for this reason, "*Ramapithecus*" usually appears in inverted commas. The genus *Sivapithecus* includes species *Sivapithecus punjabicus* (Figure 2.25), *Sivapithecus sivalensis*, and *Sivapithecus indicus* (Pilbeam *et al.*, 1977), which are between 12.5 and 7 Ma, in the middle and late Miocene. Paleoecological studies indicate that the environment at Siwaliks was heavily seasonal subtropical forest.

Because of the considerable thickness of *Sivapithecus*' molars, Kelley (1992) has argued that its diet, like that of *Ouranopithecus* and *Gigantopithecus*, consisted of hard materials. However, the shape of the teeth of *Sivapithecus*, especially the incisors, is different from that of the two other genera. Following Teaford and Walker (1984), Andrews (1992a) has proposed for *Sivapithecus* a soft, frugivorous diet. This, however, is inconsistent with the presence, at the same time, of thick molar enamel.

Andrews and Martin (1991) explained this anomaly by arguing that the trait of thick molar enamel can be considered primitive for *Sivapithecus*; that is to say, it appeared in its ancestors and, therefore, existed before the change in diet. The middle-Miocene specimens from Paçalar, Turkey, have wear patterns on their molars that suggest the mastication of hard materials. If the thick enamel of *Sivapithecus* is a plesiomorphy, a vestige inherited from previous times, it would not be useful for determining its diet.

The rest of the skeleton of *Sivapithecus* is also difficult to interpret. Two humeri from Siwaliks exhibit primitive features, such as those of *Proconsul* and *Kenyapithecus* (Pilbeam *et al.*, 1990), which may be traced back even to Old World monkeys. The primitive features of the humerus

---

Box 2.9 *Dryopithecus* species

Abundant *Dryopithecus* specimens have been found in late-Miocene sites of northeast Spain, such as El Firal in the Pyrenees, and Ca'n Llobateres, Viladecavalls, Ca'n Vila and Ca'n Ponsic in Vallés-Penedés. These specimens were grouped into a single species, *D. laietanus*, by Simons and Pilbeam (1965), but were later split into different species. In spite of the consensus regarding the inclusion of the El Firal mandible in *D. fontani*, the materials retrieved in the Vallés-Penedés region, which vary in size, have led to disparate interpretations since Crusafont (Crusafont and Hürzeler, 1961) initially identified

three species in the sample. The doubts concern mainly the Ca'n Llobateres and Ca'n Ponsic specimens. These can be viewed as belonging to two species (*D. laietanus* and *Dryopithecus crusafonti*; Begun *et al.*, 1990; Moyà-Solà *et al.*, 1990) or as a single one (*D. laietanus*; Harrison, 1991; see the analysis of the controversy by Ribot and Gibert, 1996). These are all primates with thin molar enamel and, judging from the Ca'n Llobateres skeleton, the anatomy of their arms suggests brachiation was their usual form of locomotion (Moyà-Solà and Köhler, 1996).

Box 2.10 "*Ramapithecus*" and human evolution

David Pilbeam (1978) says that by the end of the 1960s Elwyn Simons and he had concluded that pongid and hominid taxa might have already existed during the middle Miocene, although they did not look too much like the current species. *Dryopithecus* would be the ancestor of pongids and "*Ramapithecus*" the ancestor of hominids. The evolutionary model that Simons and Pilbeam had in mind in the late 1960s was very simple: "*Ramapithecus*" had evolved some 14 or 15 Ma from a species of *Dryopithecus*. Nearly 2 Ma it had led to *Australopithecus*, the Pleistocene fossil that was considered at the time the direct ancestor of modern humans.

In 1968, Pilbeam ventured to identify the ancestors of all great apes: *Dryopithecus africanus* was the ancestor (or almost) of chimpanzees, *Dryopithecus major* was the precursor of gorillas, and an Asian genus, *Sivapithecus*, the ancestor of orangutans (Pilbeam, 1968). Regarding the human ancestors, he followed the firmly held idea, which he shared with Simons (1961, 1964), that "*Ramapithecus*", a 14-Ma relative of *Dryopithecus*, was

the first hominid. According to Simons and Pilbeam most of the Pleistocene hominid dental traits were already present in "*Ramapithecus*": premolar size, canine orientation, U-shaped dental arcade, and molar shape and crown. Thus, there were reasons to believe that rather than a pongid, *Ramapithecus* was one of our direct ancestors.

The fossils of "*Ramapithecus*" did not include any postcranial remains. But Pilbeam (1966) speculated about the bipedalism of his candidate for first hominid. Because "*Ramapithecus*" had such small canines, it would have had very limited defense capabilities, and thus, in the middle Miocene, it may have already used weapons. Because the hands are needed to use weapons and prepare food, it could also be surmised that "*Ramapithecus*" was bipedal and completely terrestrial (Pilbeam, 1966). Pilbeam admitted that these conclusions were based on mere circumstantial evidence.

The most widely accepted hypothesis today is that "*Ramapithecus*" is a false taxon that includes female *Sivapithecus* specimens.

---

include the proximal end, which curves laterally and anteriorly, as in Old World monkeys, and the flattened deltoid plane. But the surface of the proximal joint—the shoulder—has, on the contrary, derived traits characteristic of great apes (Andrews, 1992a; Pilbeam *et al.*, 1990). Kelley (1992) has also noted in the forelimbs of *Sivapithecus* a combination of an upper part of the arm similar to monkeys and a more modern elbow.

2.4.3 *Ouranopithecus* (de Bonis *et al.*, 1974)

Martin and Andrews (1984) identified *Ouranopithecus macedoniensis* with *Graecopithecus freybergi*, previously described by von Koenigswald. The remains of *Ouranopithecus* come from Macedonia, Greece (Ravin de la Pluie and Xirochori sites), and have been dated, by faunal comparison, to 9–10 Ma.

Box 2.11 Dentition of *Ouranopithecus*

O. macedoniensis specimens are characterized by a high degree of dental metric variation. This variation leads to a multiple-species taxonomy, or the sample can be accommodated within one species. A study by Caitlin Schrein (2006) examined variation and sexual dimorphism in mandibular canine and postcanine dental metrics of an *Ouranopithecus* sample. The result showed that "most of the dental metrics of

Ouranopithecus were neither more variable nor more sexually dimorphic than those of *Gorilla* and *Pongo*". Therefore, it is probable that all but one *Ouranopithecus* specimens are morphologically homogeneous, and the specimens included in Schrein's study are from a single population. It is unlikely that the sample includes specimens of two sympatric large-bodied hominoid species.

The habitat at *Ouranopithecus*' sites, also inferred from the fauna, was open savanna. A specimen with a well-preserved face, XIR-1 (de Bonis *et al.*, 1990), reveals a similarity between some of its facial features and those of African great apes. Other traits, such as the supraorbital torus (the bony protuberance that protects the eyes at the height of the eyebrows) and dentition, differ from current great apes and place *Ouranopithecus* close to Pliocene robust australopithecines. This has led de Bonis and colleagues (1990) to argue that *Ouranopithecus* is the best available candidate as an ancestor of hominids. The great enamel thickness of *Ouranopithecus* (Andrews, 1990) would also support this idea.

2.4.4 *Ankarapithecus* (Alpagut *et al.*, 1996)

The discovery of an *Ankarapithecus meteai* specimen (AS95-500) in late-Miocene deposits at Sinap, Turkey, dated to around 10 Ma by geomagnetic polarity analysis (Alpagut *et al.*, 1996), forced the revision of previous opinions that included the *Ankarapithecus* remains in *Sivapithecus* (Simons and Pilbeam, 1965).

Ankarapithecus shares traits with diverse hominoid genera: *Pongo*, *Pan*, *Gorilla*, *Ouranopithecus*, *Dryopithecus*, and even *Afropithecus*. Such dispersion led Alpagut and colleagues (1996) to argue that the AS95-599 specimen requires that the European late- and middle-Miocene genera *Dryopithecus*, *Ouranopithecus*, and *Ankarapithecus* be considered lateral members of the clade of the great apes and humans. It is not possible to determine whether *Ankarapithecus* is closer to

orangutans or African apes (see Box 2.12). According to Begun *et al.* (2003) *Ankarapithecus meteai* "is a relict population of the *Sivapithecus-Pongo* clade that survived in Turkey after the initial appearance of *Sivapithecus* in the Siwaliks" (Figure 2.26).

2.4.5 Genera that are difficult to classify

A great number of late-Miocene hominoid specimens (up to 1,185) have been found at Shihuiba (Yunnan, China), in the Lufeng site. They are 8 Ma and their facial traits are very similar to those of the orangutan. The Lufeng specimens were initially classified as *Sivapithecus yuannensis*, but they were later placed in a different genus, *Lufengpithecus lufengensis* (Rukang, 1987). Their phylogenetic interpretation is uncertain. Schwartz (1990) placed the species as an ancestor of orangutans. Kelley (1992) placed it on a lateral branch, as a descendant of *Sivapithecus* and with no ancestry relationship whatsoever with current great apes. Martin (1990) and Andrews (1992a) believe its relationship with other contemporary hominoid families remains to be determined. This seems to be the most reasonable point of view, at present.

Oreopithecus, found in Italy (Baccinello and Monte Bambolini), is perhaps the most difficult to classify of all late Miocene specimens. Its dentition has been variously interpreted (Martin, 1990): sometimes this has led to its classification as a cercopithecoid and even as an ancestor of hominids (Hürzeler, 1958). Its locomotion, deduced from postcranial remains, was deemed closer to that of current great apes rather than *Sivapithecus*

Box 2.12 Shared traits between *Ankarapithecus* and other hominoids

Ankarapithecus AS95-500 (new) and MTA 2125 (previous, reinterpreted) share a multitude of traits with other hominoids (Alpagut *et al.*, 1996):

• the heteromorphism of upper incisors is similar to that of *Pongo*, *Ouranopithecus*, and *Sivapithecus*;

• the orbits are as wide as high, as in *Ouranopithecus*, *Pan*, and *Gorilla*;

• the zygomatic is flat, as in *Dryopithecus*, *Sivapithecus*, and *Pongo*;

• the interorbital region is thin, as in *Pongo*, but the frontal sinus (obtained by computerized tomography) is less similar;

• the nasal and lacrimal region is similar to that of *Afropithecus*, *Ouranopithecus*, and *Gorilla*;

• the extension of the premaxillar is similar to the orangutan's.

The general profile of the face is more similar to that of African great apes than to *Pongo* or *Sivapithecus*.

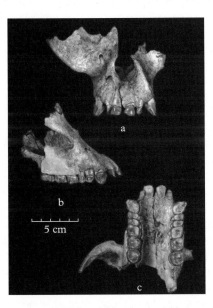

Figure 2.26 Frontal (a), lateral (b) and occlusal (c) views of the palate of a male *A. meteai*. Photograph from Begun *et al.* (2003).

or *Ouranopithecus* (Kelley, 1992). After a restoration work of almost 400 specimens from the Basel Natural History Museum carried out by Salvador Moyà-Solà, Meike Köhler, and Lorenzo Rook at the Instituto de Paleontología M. Crusafont (Sabadell, Spain), it was suggested that *Oreopithecus* might have been bipedal (Rook *et al.*, 1999; Moyà Solà, 2000). This hypothesis rests on the morphology of its hand—especially the thumb—with features suggesting that *Oreopithecus*' locomotion was similar to that of hominids, while retaining a

prehensile foot suitable for climbing up trees. The similarities between the pelvic and vertebral apomorphies observed in *Oreopithecus bamboli* and the corresponding traits of current hominoids had been noted previously by Aguirre (1996).

What is the phylogenetic position of *Oreopithecus*? We find again the problem of deciding about the primitive or derived status of several traits. The presence of curved phalanxes and the absence of a tail led Martin (1990) to move the taxon away from Old World monkeys. But the lack of cusps with five protuberances in *Oreopithecus* molars (the Y 5 pattern, common in great apes and humans) highlights its dissimilarities with hominoids. Andrews (1992a) acknowledged the ambiguities regarding the classification of *Oreopithecus*, due to its postcranial synapomorphies (shared derived traits) with current great apes, and the highly derived dentition. Moyà-Solà and other supporters of the bipedalism of *Oreopithecus* believe this trait is a homoplasy, convergent with hominid bipedalism and, thus, fixed independently in each lineage.

Otavipithecus (Conroy *et al.*, 1992) was described on the base of a mandibular fragment found in a Namibian diamond mine in southern Africa, and is the most austral of all hominoids. By faunal comparison the specimen has been dated to 13 Ma. The mandibular and dental morphology and the size of *Otavipithecus namibiensis* are similar to those of East African (*Kenyapithecus*) and Eurasian (*Sivapithecus*, *Dryopithecus*) hominoids (Conroy,

---

Box 2.13 Navicular measurements of *Oreopithecus*

Esteban Sarmiento and Leslie Marcus (2000) compared the os navicular of *Oreopithecus* with navicular measurements of Olduvai and Hadar hominids, and with a representative sample of humans and great apes; the measurements chosen for comparison were relative orientation, articular area, and curvature of the navicular facets. They identified three different groups:

- one that relates Hadar fossils (see Chapter 3) with African apes,

- one that includes OH 8 from Olduvai (see Chapter 3) and current humans,
- *Oreopithecus* as a special and particular case.

Given that the bipedalism of the Hadar specimens has been determined, as we will see later in the book, based on the function of the knee and the morphology of the pelvis, the study of Sarmiento and Marcus reveals "the fallacies inherent in constructing phylogenies on the basis of single bones and/ or fragmentary remains, and of reconstructing locomotor behaviors on the basis of localized anatomy."

Box 2.14 Reduction in tooth size in *Oreopithecus bambolii*

Its possible bipedalism is not the only striking trait of *Oreopithecus bambolii*. Alba *et al.* (2001) have noted that, even accounting for a notable sexual dimorphism, the size of the canine and postcanine dentition of *O. bambolii* underwent extra-allometric reduction. The teeth are smaller than expected for an ape of its size, similar to what is observed in bonobos (pygmy chimpanzee; *Pan paniscus*)

and hominids. The reduction in canine size is usually explained by the relaxation of selective pressures due to the use of canines as a weapon. This was the argument of Simons and Pilbeam in their interpretation of "*Ramapithecus*" as a hominid. But Alba and colleagues (2001) add the possible reduction in facial prognathism as an additional pressure towards microdontia.

1997). However, it lacks the former's thick dental enamel.

The fragmentary remains of three lower mandibles and isolated teeth served to describe *Gigantopithecus* (von Koenigswald, 1935). The first exemplar was located in a Chinese chemist's shop, where they were considered to be "dragon teeth". von Koenigswald identified one of them, a lower last molar, as belonging to a new genus and species that he named *Gigantopithecus blacki* as a tribute to the discoverer of Peking man, Davidson Black. The story of *Gigantopithecus* honors the golden age of paleontology (von Koenigswald, 1981). But its taxonomy and evolutionary significance are so uncertain that it has been interpreted as an aberrant ape, a genus reducible to *Ouranopithecus* and even as a possible direct ancestor of humans (Martin, 1990; not to mention the identification of *Gigantopithecus* with the popular legend of Big Foot). *Gigantopithecus* used

to be quoted and discussed very often, but today it has lost much of its relevance. Andrews (1992a) does not even mention it. Martin (1990) asserts that, just as in the "*Ramapithecus*" case, it is reasonable to conclude that *Gigantopithecus* represents a partial episode in the radiation of Miocene great apes with thick enamel. It is not saying much, but it avoids making a mistake.

2.4.6 Synthesis of Miocene hominoid evolution

The book *Function, Phylogeny and Fossils*, edited by David R. Begun, Carol V. Ward and Michael D. Rose, provided in 1997 an up-to-date overview of hominoid adaptive processes. In the preface, the authors wrote that the numerous fossil ape discoveries during the late twentieth century have allowed a certain consensus about their phylogeny during the Miocene. But MacLatchy (1998) pointed out in her review of the book that the claim is rather optimistic. The only consensus that can be inferred from that book concerns the enormous functional (and taxonomic) diversity of this period's hominoids, also noted throughout this chapter.

Certain issues are well established. For instance, it seems clear that apes evolved in arboreal environments, contrary to the tendency of cercopithecoids, their sister group, to colonize the savanna. A relationship has been established between this adaptive option and certain dental and locomotor changes, which had probably already appeared in middle-Miocene specimens. Thereafter, the paucity of remains from the African continent hinders any attempt to understand how the direct ancestors of current apes appeared.

The remaining problems are not only taxonomic—to determine the taxa for each lineage—but also systematic in a general sense, concerning the difficulty of explaining how the apomorphies that appeared during the middle Miocene, and possibly before, were transmitted to current Asian and African apes. The suggestions made all stumble with the inconvenience of the extreme variability of the traits. Regarding locomotion, the tendency towards an orthograde posture is constant during the Miocene. But the specific type of locomotion associated with a more upright posture is not the same in all middle-Miocene or current hominoids. Regarding dental traits, the general tendency leads to low crowns on wider molars. But dental enamel thickness is also diverse in Miocene and current apes (Table 2.7).

Taking this into account, can a relationship be established among climate, diet, and hominoid evolution?

2.4.7 The ecological script of Miocene hominoid evolution

Hominoids, as we have repeated so many times throughout this chapter, appeared at the beginning of the Miocene—or even towards the end of the Oligocene—in tropical jungle environments in Africa. Those early hominoids can be described as very similar to current Old World monkeys,

Table 2.7 Locomotion and enamel of different current and middle-Miocene hominoid genera

Genus	Period	Enamel	Locomotion
Hylobates	Current	Thick	Brachiation
Dryopithecus	Middle/late Miocene	Thin	Brachiation
Kenyapithecus (*K. wickeri*)	Middle Miocene	Thick	Climbing
Kenyapithecus (*K. africanus*)	Middle Miocene	Thick	Quadrupedalism
Pongo	Current	Thick	Brachiation
Gorilla	Current	Thin	Knuckle-walking
Pan	Current	Thin	Knuckle-walking
Homo	Current	Thick	Bipedalism

cercopithecoids (mangabeys, geladas, and baboons), and colobus. This is how they have been perceived since the 1950s, for example by Le Gros Clark and Leakey (1951). According to Andrews (1981), from a functional and ecological point of view, the first Miocene hominoids were equivalent to monkeys.

The planet's cooling around 14 Ma produced a certain degradation of East African tropical forests and the so-called protosavanna expanded. This environment consists of forests with patches of open vegetation, similar to that associated with the appearance of *Kenyapithecus* at Fort Ternan. The climatic changes led in Europe to a dry and cold period, but by the end of the middle Miocene there was an increase in temperature and recovery of European and Asian forest lands (Agustí, 2000). Hominoids (*Sivapithecus* in Pakistan, *Dryopithecus* in Europe) radiated and expanded, taking advantage of the new ecological niche afforded by the floor of those forests. Agustí (2000; Agustí *et al.*, 1996) believes that the thin enamel of *Dryopithecus* was the result of a frugivorous diet, but the phylogenetic development leading to that trait is not explained easily. It can be considered either an apomorphy, a derived trait fixed by *Dryopithecus*, or a plesiomorphy inherited from *Proconsul*. Neither option changes the fact that the African middle-Miocene ancestors of European *Dryopithecus* remain unknown.

The transition from the middle to the late Miocene coincided with a new decrease of temperature, some 11 Ma. With the accumulation of ice on the continents and the descent of sea level, new bridges appeared, favoring faunal exchanges. Although hominoids diversified considerably, the replacements that could be expected did not happen. Agustí (2000) has explained this fact, which affects mammals as a whole. A million years later, around 9.6–8 Ma, there was a profound crisis (the Vallesian crisis, which takes its name from the studies carried out by Agustí and Moyà-Solà at Vallés-Penedés) which led to a faunal impoverishment that also affected hominoids. The essential factor that triggered the Vallesian crisis was the substitution of subtropical forests for European temperate forests, with deciduous trees. This change might have led to the extinction of hominoids in Europe. The African and Asian specimens survived and led, in time, to the ape lineages we know today.

Where and how did the evolution leading to orangutans, gorillas, and chimpanzees take place? Chaline and colleagues (1996) have noted that we still ignore how the divergence process among the three lineages occurred, but they believe that it was associated with the late-Miocene climatic change, which took place between 6 and 5.3 Ma. Chaline and colleagues propose a correspondence between evolutionary processes and ecological adaptation, similar to the one put forward by Yves Coppens (1994). The determining factor may have been geographical separation:

- pre-gorillas to the north of the barrier constituted by the Zaire River,
- pre-australopithecines in the eastern area of the African Rift,
- pre-chimpanzees in the center and west of Africa.

Under this scenario, the derived traits of each branch were fixed allopatrically, separately in each region (Figure 2.27). The allopatric speciation model would explain the absence of chimpanzee ancestors in hominid sites. However, the discovery of a middle-Pleistocene fossil chimpanzee specimen within the Rift valley, in Tugen Hills (Kenya) has challenged this model of allopatric speciation, as we will see in a subsequent chapter.

The association of hominoid evolutionary diversification with climatic change at the end of the Miocene is both reasonable and interesting. The issue is whether it is sufficiently grounded on available evidence to be considered anything more than a mere hypothetical model. Isotope analyses of the paleoenvironments of the Kenyan Rift carried out by Kingston *et al.* (1994) have revealed there have not been widespread savannas in those localities for the last 15.3 Ma. Apparently, during this period there has not been any significant environmental change (besides the planet's global climatic alternations). If this is true, it would mean

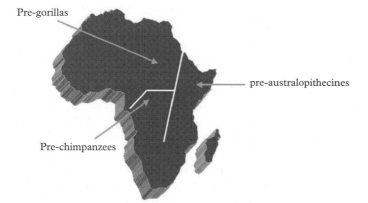

Pre-gorillas

pre-australopithecines

Pre-chimpanzees

Figure 2.27 The hypothesis of the allopatric separation of chimpanzee, gorilla, and australopithecine ancestors favored by geographical barriers (Rift Valley and Zaire River).

Box 2.15 The link between enamel thickness and bipedalism

The link between thick enamel and terrestrial quadrupedal locomotion is based on Jolly's seed-eaters hypothesis (Jolly, 1970). Jolly established a relation between the reduction of the canines, development of potent molars, feeding on seeds, bipedalism, and even development of language and social groups with a dominant male. He did not, however, mention lithic instruments as significant elements. In Wolpoff's (1982) view this was not an accident. This model

of a predominantly herivorous diet moved away from the original idea in which Darwin related the use of tools and weapons with hunting behavior. But, in any case, the relation between molar enamel and the diet of an inhabitant of tropical forest floors is not easily supported. Kay's (1981) and Gantt's (1979) studies argue against any correlation between the trait of dental enamel and terrestrial locomotion.

---

that there was no general development of open savannas during the middle and late Miocene in East Africa. Then, the situation was not ecologically very different there from that in northern and western tropical forests.

The Miocene climatic conditions probably led to very diverse habitats in all sub-Saharan Africa. The divergence of hominoids is not readily related to a certain climatic episode, unless the geographical location under consideration can be determined with great precision. In later chapters dealing with the paleoclimatology of early hominid sites we will review some of the relevant arguments. But, because we know nothing about the direct ancestors of chimpanzees and gorillas, or about the places in which they might have lived, the plausibility of an allopatric speciation process remains hypothetical.

2.4.8 Locomotion and dentition: difficulties for an integrative model

Current apes are not homogeneous in their posture or diet-related dental features (Table 2.7). They share certain general features, such as orthograde posture and a varied diet, but living apes are highly diverse. Orangutans use brachiation, chimpanzees and gorillas use knuckle-walking. Chimpanzees and gorillas have thin molar enamel, but orangutans—like humans—have thicker enamel.

Ascertaining the continuity of the apomorphies appeared during the middle Miocene is not easy. The specimens that seem better candidates for

current African ape ancestors have thick enamel. Their relationship with middle-Miocene Eurasian apes, such as *Dryopithecus*, is not clear either. Orangutans have thick molar enamel, whereas the enamel of *Dryopithecus* is thin.

The work of Andrews (1981) concerning the diets of Miocene monkeys and apes has explained the paradox of the persistence of thin dental enamel in gorillas after their change from a frugivorous to a leaf-eating diet. Andrews (1981) underlined the fact that the diet of a terrestrial quadruped is actually very varied (seeds, roots, rhizomes, bulbs, fruits, and soft leaves). Thick molar enamel is not required even for diets based on materials which are harder than fruits, as is the case with folivorous diets. In gorillas, it is the shape of the molars, with pointy cusps and long cutting surfaces, which allows their folivorous diet. Siamangs, which are also leaf-eaters, have similar molars. Andrews noted that those two species are the largest of their respective families, although he did not determine whether the large size is due to the folivorous diet or the other way around.

If we accept Andrews (1981) conclusions, African great apes could have developed terrestrial quadrupedalism while retaining the trait of thin dental enamel, generally considered primitive (Andrews and Cronin, 1982). However, some authors, such as Martin (1985) and Verhaegen (1996) argue that the primitive trait is thick enamel.

A cautious approach is to establish taxonomies and evolutionary relationships based on several morphological characters, rather than just one.

It does not make sense to draw phylogenies based solely on the trait of enamel thickness. This trait has an undisputable importance when classifying particular specimens, especially in instances in which there are only mandibular fragments available for identifying a species or even a genus. But progress in fossil extraction techniques has made more postcranial remains available. Moreover, advances in paleoecology have provided very relevant information. Three kinds of evidence have usually been taken into account to determine taxonomic and phylogenetic relationships: dentition, locomotion, and ecological context. Explanations for the appearance of a new adaptive lineage should be grounded on a comprehensive study of, at least, these three features.

CHAPTER 3

The hominin lineage

3.1 The origin of hominins

3.1.1 What is a hominin?

The term hominin is used extensively in the scientific literature, although it has not yet been adopted widely in popular writing. The tribe Hominini defines current humans and our extinct ancestors as a distinct group; it does not include our closest living relatives. As Phillip Tobias (1971) put it, a distant relative of modern humans initiated the necessary evolutionary thrust to establish a new phyletic lineage, different from that of gorillas and chimpanzees. The hominin taxon includes humans and their direct or collateral ancestors that are not also ancestors of other living hominoids.

As stated in the previous chapter, the most plausible cladogram that represents the phyletic relationships between higher apes and humans considers chimpanzees as the sister group of current humans. There are two possible phylogenetic interpretations of fossils belonging to early genera, which we will describe in the following chapters, such as *Orrorin*, *Ardipithecus*, *Kenyanthropus*, *Australopithecus*, or *Paranthropus*. The first option is to place them within the lineage of current humans (the one leading to the genus *Homo*). The alternative is to consider them as ancestors of

chimpanzees or gorillas. The first option implies that the specimens are hominins, pre-humans.

What is a pre-human? We have already given an initial answer: any of our direct or collateral exclusive ancestors (Figure 3.1). So, the lateral branches of our lineage's evolution, such as the robust australopiths (*Paranthropus*) are considered hominins. Here is where hominoid cladistics is so important.

Chimpanzees share a clade with us to the exclusion of gorillas and orangutans. Why do we not also consider them hominins? They have striking derived traits. But, as we will see in later chapters, we classify certain fossil specimens as hominins even though they exhibit derived traits which are remarkably different from our own. However, there is a critical difference between the chimpanzees and the fossils that we classify as hominins; namely, that the former are living organisms. It is not obvious how chimpanzees would be classified if they had gone extinct soon after their appearance and their remains had been found.

The interpretation of human evolution is strongly influenced by current living hominoid species. If robust australopiths were still alive, would we emphasize their derived traits? Would they take the place of the hominin sister taxon that we currently award to chimpanzees?

Box 3.1 Australopiths

As already noted, we use the term australopith for the specimens of the subfamily Australopithecina. This includes the genera *Orrorin*, *Ardipthecus*, *Australopithecus*, and *Paranthropus*. We will refer to the members of the genus *Paranthropus* as robust australopiths, because they constitute the robust branch of the hominin cladistic event reviewed further on.

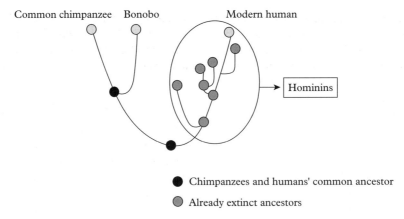

Common chimpanzee   Bonobo   Modern human

Hominins

● Chimpanzees and humans' common ancestor

◐ Already extinct ancestors

Figure 3.1 The term hominin applies to current humans and their extinct exclusive direct and collateral ancestors. Exclusive means that they are not also ancestors of our sister group, common chimpanzees and bonobos. The ancestors included in the figure are hypothetical; they are meant as an illustration. There are many known hominin ancestors, but there are no known ancestors of current chimpanzees, with the exception of two central upper incisors from the Kapthurin Formation of the Tugen Hills (Baringo, Kenya), dated to around 545,000 years ago (McBrearty and Jablonski, 2005).

Would we then reduce the hominin taxon to the genus *Homo*?

In our opinion all these questions can be answered affirmatively. Indeed, if australopiths were alive, the adaptation to the savanna initiated by *Homo habilis* might be considered more significant for our concept of what a human being is than the bipedalism of early australopiths. In this case, hominins would be those exhibiting derived traits related to a mostly carnivorous diet, enlargement of the cranium, production of culture, and gradual achievement of a high level of semantic and syntactic communication. Because there are no living australopiths, we include them, together with all *Homo* species and others, among the hominins, while leaving the chimpanzees out of the taxon.

We saw in section 2.1 that other authors, like Morris Goodman, argue against this idea. Given the genetic proximity between African apes and humans, they should all be considered hominins, and even members of the same genus, *Homo*. Such radical claims are not appropriate. We have noted that the concept of hominin takes into account the traditional classification of hominoids. But it is important to understand the existing problem. Andrews (1995) has referred to early australopiths as "ecological apes", suggesting that their adaptive role

was not part of that "push or kick" phase towards current humans mentioned by Tobias (1995). This is a reasonable position that should be taken seriously, even though it is reminiscent of some work carried out during the 1930s by Robert Broom and John Robinson, who claimed that australopiths constitute a different group from apes and humans.

To what taxonomic category should the hominin taxon be assigned? The question of whether hominin and African ape clades constitute tribes, as we support, or families, subfamilies, or even genera, is irrelevant to the present discussion. It does not alter the basic question regarding the relative distribution of lineages. The fact is that the common ancestors of current humans, gorillas and chimpanzees, lived during the middle Miocene. Although we do not know much about them, we do know that a little later, some 7 Ma, the first hominin appeared, leading, in time, to the other members of our lineage. What does this mean? To what kind of primate did that episode lead?

3.1.2 What were the first hominins like?

In his *Descent of Man* (1871) Charles Darwin noted that "Whether primeval man, when he possessed but few arts, and those of the rudest kind, and when his power of language was extremely imperfect, would have deserved to be called man, must depend on the definition we employ. In a series of forms graduating insensibly from some ape-like creature to man as he now exists, it would be impossible to fix on any definite point when the term 'man' ought to be used. But this is a matter of very little importance".

These words reflect what Darwin believed about early human evolution. Translated into cladistic terminology, we could reformulate this notion as: "we expect that specimens corresponding to hominins and their sister group will look alike soon after their divergence". The reason for this is obvious: apomorphies will have not yet developed much in each separate lineage, so they would lack clear-cut differences. Shortly after the cladistic event, the differences between diverging lineages will be small, if they can be detected at all.

The distinctive apomorphies of the tribe Hominini can be clearly identified only when comparing its single extant species, modern humans, with current chimpanzee species, the sister group: *Pan troglodytes* and *Pan paniscus*. But current phenotypic differences have their origin in the divergence between the lineages, about 7 Ma. What would the chimpanzee and human ancestors look like shortly after the cladistic event? We do not know. As we will see in the following chapter, hominin specimens have been discovered with age close to the presumed time of the divergence, but there are no equivalent specimens from the chimpanzee lineage, except for the Tugen Hills exemplar that we will describe in section 4.2. Taking into account Darwin's warning, how can we tell whether a 6- or 7-million-year-old specimen belongs to the human or chimpanzee lineage? Would they not be almost identical?

Even though the answer would be affirmative, the apomorphies separating the lineages can be identified by comparing current chimpanzees (and other great apes) with humans. It may be that some of these apomorphies appeared early enough in our own clade as synapomorphies, derived traits shared by all hominins. If one such trait existed, it could serve to identify the members of our family. Any specimen, regardless of its age, would be considered a "hominin" if it exhibited this apomorphy, but only if we could be reasonably sure that it is not an independently fixed homoplasy.

3.1.3 Human apomorphies

Paradoxically, comparison of the members of our species with our closest relatives, gorillas and chimpanzees, produces contradictory results: both are very similar to or very different from humans, depending on the chosen trait. The differences between humans and chimpanzees are now known, as we pointed out in section 2.1. Humans and chimpanzees are 99% identical in the overlapping genome regions, and 29% of the enzymes and other proteins encoded by the genes are identical in both species. The nonidentical proteins differ between humans and chimps by only two amino acids, on average. Concluding that minor protein differences and a 1% DNA sequence difference implies a virtual biological identity between chimpanzees and humans would, however, ignore the immense importance of developmental processes.

The anatomy and behavior of chimpanzees and humans are very different and some differences are quite remarkable, such as language, brain size, bipedal gait, and culture. One of the main features of human language is its dual patterning (we combine basic sounds, phonemes, to form words, and words to form sentences), absent in any ape communication system. Our brain is much larger than that of the higher apes relative to body size and much more complex. We usually walk on our two feet, whereas orangutans use brachiation to travel, and gorillas and chimpanzees are rather special quadrupeds that lean on the knuckles of their hands. The sophistication of our cultural traditions is a notable divergent trait.

There are other differences that may not quite jump out so conspicuously. We lack body hair or have very little. Our face is not so prognathus. Gestation is shorter than in the case of higher apes. Ovulation in human females is cryptic and they are continuously sexually receptive without specific periods when females enter into heat.

Tobias (1994) lists 24 morphological traits and five physiological ones that differentiate humans from higher apes. One additional feature difficult to ignore is population size. Today, orangutans, gorillas, and chimpanzees are reduced to relatively small populations living in receding tropical forests in Africa and southeastern Asia. There are more than 6 billion humans distributed across the planet. Numbers are not an unmixed blessing, but they are a measure of adaptive success.

Table 3.1 lists the apomorphies identified by Carroll (2003) in his study of the genetic basis of the physical and behavioral traits that distinguish humans from other primates. Some apomorphies listed by Carroll are functional, like language. Others that are anatomical are not shown in fossils, like hair or the brain's topology. However, body shape, brain size, relative length of limbs, and vertically placed cranium above the vertebral column are morphological traits that clearly set us apart from any ape.

Some distinctive human traits have appeared recently; if we look back in time they disappear from our lineage. Ten thousand years ago neither writing nor agriculture existed. Fifty thousand years ago there were no people in America. These are negligible time intervals relative to the

Table 3.1 Some distinctive human traits (human apomorphies)

Human apomorphies
Body shape and thorax
Cranial features (brain case and face)
Brain size
Brain morphology
Limb length
Long ontogeny and lifespan
Small canine teeth
Skull balanced upright on vertebral column
Reduced hair cover
Elongated thumb and shortened fingers
Dimensions of the pelvis
Presence of a chin
S-shaped spine
Language
Advanced tool making

Source: Carroll (2003).

7 million years that have passed by since the divergence of the evolutionary branches leading to the African great apes and humans, or the 6 million years since the fossil *Orrorin tugenensis*, described further on, was buried in the ground.

Thus, current human features are generally not very helpful to reach conclusions about our initial apomorphies. What we are looking for are derived ancient traits that can be considered synapomorphic, shared by every hominin that ever existed. These would define the earliest member of our lineage as adaptively distinct. Leaving aside the necessarily dark period surrounding the exact moment when the lineages split, are there any such traits? Can we find a trait that will allow us to determine whether a given fossil specimen is a hominin?

3.1.4 What about the brain?

If someone unfamiliar with anthropology and paleontology had to choose the most "human" apomorphy among those listed by Carroll, he or she would probably choose some feature related to the mind. It could be the size of the cranium, the brain's topology, or some functional trait, like language or complex tool making. But out of these characteristics, only the size of the cranium can be directly detected in the fossil record. Human and ape cranial volumes are very different (Figure 3.2). The average cranial capacity of modern humans is 1,350 cc., while that of chimpanzees, with a comparable body size to ours, is 450 cc. Does this mean that a large cranium evolved at the beginning of the human lineage and, thus, constitutes a hominin synapomorphy?

The notion that a large brain is an essential trait identifying the appearance of the human lineage is very old, but not enough to find it among Darwin's hypotheses about this matter. In Chapter II of his *Descent of Man* he speculated on how our first progenitors might have been and how they might have evolved. The earliest would have inherited, naturally, some traits from their own ancestors, such as large canines. In Darwin's opinion, the gradual reduction of the size of canines in the human lineage was a consequence of the appearance of culture: they became smaller because of

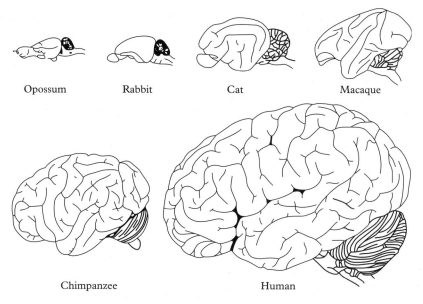

Opossum   Rabbit   Cat   Macaque

Chimpanzee   Human

Figure 3.2 Brains of diverse mammals represented at the same scale. A large brain is a very distinctive feature of our species.

their disuse in favor of tools and weapons. The manipulation of tools required a bipedal posture, or at least the former was facilitated by the latter. Because the reduction in canine size was accompanied by the reduction of the muscles that move the jaw, the cranium was able to grow, and with it, the brain and mental faculties that, of course, improved culture (Darwin, 1871, pp. 435–436).

Darwin's hypothesis about the gradual reduction of canines because of their disuse, caused by the appearance of culture and tools, must be understood within Lamarck's model of inheritance, which suggested that the function created the organ and vice versa. But a Lamarckian framework is not necessary to support that relationship; Elwyn Simons and David Pilbeam's hypothesis regarding the possible use of tools by "*Ramapithecus*" (which we encountered in section 2.4) was based on the interaction between culture and the reduction of canine size. We now know that acquired traits are not biologically inherited, but adaptive evolution achieved through cultural changes occurs at a fast pace.

The Darwinian notion of hominin evolutionary change may be interpreted as a closed feedback loop. Culture required bipedalism and, at the same

time, reinforced it. The reduction in canine size was a consequence of the use of weapons; but that reduction facilitated brain size, through the restructuring of the cranium; further, mental development allowed devising, making, and using better weapons. Brain increase improved bipedal balance and permitted the development of language. Language facilitated the transmission of culture and collective hunting using meat as food allowed further reductions in dentition size. This is a feedback model: each factor depends on the others and, at the same time, promotes them. The process involves a functional and anatomical integration in which several coordinated factors participate. One of them must have appeared before the others, serving as an initial thrust for the loop to start running. Which might it be?

3.1.5 Piltdown man

Darwin envisioned the following chain of events: descent from the trees, bipedalism, brain-size increase, language, and appearance of culture (with all its components, both intellectual and technological). Some of these elements can be traced in the fossil record, but not others.

Phenomena such as the development of moral sense, which Darwin believed was extremely important, are not associated with fossil remains. Language does not fossilize either. But the cranium and bones of the hip and lower limbs leave fossil trails that can provide firm evidence regarding whether it was our bipedal posture or our large brain that developed first.

During the early twentieth century there were defenders of two opposite hypotheses. Arthur Keith was one of the most prominent advocates for bipedalism as the initial trait, while Grafton Elliot Smith argued that a large encephalization appeared first. The swords were drawn when the Piltdown fossil specimen appeared on the scene.

The Piltdown fossil was a fraud that does not have any bearing on our understanding of hominin evolution. But it is worth examining it for two reasons. First, it was an attempt to ground the search for distinctive derived hominin traits on the fossil record. Second, the idea of very primitive high encephalization became widespread because of Piltdown man. That was a burden that hindered the understanding of the evolutionary thrust and the phylogenetic history of our species for decades.

The fossil popularly known as Piltdown man seemed an adequate specimen to establish the phyletic sequence of the traits. The specimen had a large cranium, the size of the cranium of a modern human, combined with a very primitive mandible, resembling that of an orangutan or gorilla. The different fragments that formed the specimen were found in 1912 by Charles Dawson, an amateur archaeologist in the English town of Piltdown (Figure 3.3). The fossil would have never attained great popularity if it were not for the support given by Arthur Smith-Woodward, a very prestigious scientist at the time.

The story of its discovery and the controversy it sparked has been told many times (e.g. Reader, 1981; Lewin, 1987; Spencer, 1990; Walsh, 1996; Weiner and Stringer, 2003). The article that the *London Illustrated News* devoted in September 1913 to the finding renders a very good picture of the challenges posed by the specimen's interpretations.

The fossil was reconstructed several times. The reconstruction made by Arthur Smith-Woodward—curator of the Geological Department of the British Museum—rendered a cranial volume of $1070\,cm^3$. It was diagnosed as an intermediate between apes and humans, which Woodward named *Eoanthropus dawsoni*. Alternatively, the reconstruction made by Arthur Keith—Curator of the Museum of the Royal College of Surgeons—attributed it $1500\,cm^3$, which suggested it belonged to an advanced human, with a similar cranial capacity to our own: *Homo piltdownensis* was the name given by Keith.

 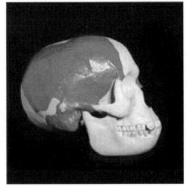

Figure 3.3 Left: Charles Dawson (on the left) and Arthur Smith-Woodward. Right: the Piltdown cranium reconstructed by Arthur Smith-Woodward. Left: photograph from www.nhm.ac.uk/nature-online/life/human-origins/piltdown-man/field_a_01.html, ©Natural History Museum; right: photograph from www.nhm.ac.uk/nature-online/life/human-origins/piltdown-man/skull_a_02.html, ©Natural History Museum.

This controversy was not trivial. If indeed it was a *missing link*—in the words of the 1913 London magazine—during the evolution of our lineage, the cranium had developed considerably at a time when the mandible and dentition were still ape-like. The Piltdown fossil exhibited some unconvincing traits, such as the awkward connection between the cranium and mandible—raising the suspicion that they belonged to different specimens. Many paleontologists were indeed suspicious. Its discoverers did not allow its examination, alluding to the fragility of the original fossil. It was necessary to use copies made with a mold. The suspicions turned out to be well founded. In 1953 (thanks to Joseph Weiner, Kenneth Oakley, and Wilfred Le Gros Clark) it was confirmed that the Piltdown fossil was a fraud. Someone had filed an orangutan's mandible and canines to reduce them and fit them, quite sloppily, to a human cranium. The main suspect of the fraud is Martin Hinton, curator of the Natural History Museum, London. Hinton was the owner of a trunk found in the museum's attic in 1996, with bones manipulated in a similar way to those constituting the Piltdown specimen (Gee, 1996).

The controversy between Keith and Elliot Smith and the consequent divergent interpretations of the Piltdown fossil is a good example of the influence of preconceived notions on the assessment of available evidence. The fact that illustrious scientists like Woodward and Elliot Smith supported the authenticity of Piltdown man, in spite of contrary evidence, merits attention. The Piltdown specimen fitted well with the insight, common at the time—and even now—that our intellect had a major role in human evolution. It demonstrated that evolution towards humanity involved acquiring a large brain—that is to say, a great mental development—very early on. The specimen suggested that the trait leading to the appearance of the first hominins was the mind, the human spirit. The specimen seemed to put an end to the controversy of what drove human evolution. The fact that the first well-evolved hominin was European and, specifically, English, rounded off a completely satisfactory perception of human

evolution. The alternative that has turned out to be true is quite different. The first hominins had similar-sized brains to chimpanzees and lived in East Africa. It is not hard to understand that the prevalent preconceived ideas during the first three decades of the twentieth century spoke in favor of Piltdown man.

3.1.6 In search of the missing link

Before the Piltdown deception, there already was evidence contrary to the early evolution of a large brain. Remains of fossil beings that were very similar to us were known since the beginning of the nineteenth century, before the controversy between evolutionists and antievolutionists reached the virulence sparked by Darwin's work. The discovery of a very famous specimen in 1856, the Neander valley cranium (Germany), which would christen the Neanderthals, occurred several years before the publication of Darwin's *Origin of Species*. But the first modern discoveries—that is to say, interpreted in terms of evolutionary ideas—were made after 1887, subsequent to the arrival of the Dutch physician Eugène Dubois in Indonesia. As a hobby, Dubois searched for fossils that could prove Darwin was right. At the Javanese site of Trinil, Dubois discovered in 1891 remains that completely transcended the realm of scientists and became universally known. The specimen includes a primitive and small cranium (with a capacity of about 850 cm^3) found beside a femur that was very similar to that of modern humans (Figure 3.4). The name given to the taxon, *Pithecanthropus erectus*,

Figure 3.4 Calotte and femur of the Trinil specimen discovered by Eugène Dubois, *Pithecanthropus erectus* (currently classified as *Homo erectus*).

means upright ape-man, conveying the idea that it was an intermediate being between humans and apes (*pithecus* for ape, and *anthropus* for human); and that it had a posture distinctively upright (*erectus*). We will return to this specimen in section 7.1.

Subsequent discoveries have required revision of Dubois' interpretations. The Trinil fossil is not an intermediate form between humans and apes, but a fairly advanced hominin. The upright posture was not a new apomorphy; rather it was already present in its ancestors. But Dubois' phylogenetic interpretation was correct: the ancestors of current humans had fixed a bipedalism similar to our own before the brain reached its current size.

It is generally accepted nowadays that bipedalism is a hominin synapomorphy—an apomorphy shared by all the members of the lineage. Any specimen close to the divergence between the chimpanzee and human lineages is attributed to the latter if it is bipedal. Most of the modifications to the trunk, limbs, hip, and the insertion of the vertebral column in the skull are related to bipedalism, which distinguishes our species from the apes. But before we analyze the morphological correlates of bipedal locomotion, we'll consider the fossil footprints from Laetoli in Tanzania which show that more than 3 Ma there were creatures that walked upright.

3.1.7 The Laetoli footprints

The first Laetoli fossil footprint was discovered on July 24, 1978 by Paul I. Abell (Leakey, 1981). The next day a footprint specialist, Louise Robbins, attributed the trail to two overlaying bovid hoof-prints. An initial excavation of a square meter allowed Mary Leakey and others to inspect the site and conclude that they were hominin footprints. This is the first direct proof of early bipedalism, more than 3.5 Ma. The footprints were described by Mary Leakey and Richard Hay (1979), and subjected to different interpretations (reviewed in White and Suwa, 1987).

The Laetoli footprints (Figure 3.5) are one of several found in the Eyasi Plateau, northeast of

Lake Eyasi. The eruptions of the Sadiman volcano, about 20 km from Laetoli, deposited successive layers of ash at Eyasi during the Pliocene, forming volcanic tuffs, known as the Laetolil Beds, which cover 1,500 km^2 and are 130 m thick at Laetoli (Leakey and Hay, 1979). Ash deposits are, generally, quite ephemeral. The conservation of fossil footprints is due to the concurrence of several fortunate, unusual circumstances:

- soft-enough texture to allow the impression of the footprints,
- adequate degree of humidity for an animal trail to be clear,
- high compression to allow large animal prints to have well-defined vertical edges,
- rapid deposit of new and dense ashes capable of cementing and protecting the prints,
- the possibility of easily removing the covering material, though in some instances it had disappeared on its own (White and Suwa, 1987).

These circumstances came together at Laetoli allowing the preservation of dozens of sets of thousands of fossilized animal tracks in ash deposits. One of the three air-fall tuffs, known as the Footprint Tuff, contains 16 outcrops—sites A–P—in which the protective upper layer has disappeared exposing the underlying fossil footprints. In 1977 Peter Jones and Mary Leakey found five very imprecise and badly conserved footprints in a 1.5-m section of site A. Mary Leakey attributed them to hominins "with a 75% certainty" (White and Suwa, 1987), judging, mostly, from the position and size of the big toe's impression. Abell's discovery a year later at site G is known in human paleontology as "the fossil footprints of Laetoli". These are two parallel series of tracks running for 25 m^2 from south to north produced by clearly bipedal animals (Figure 3.6). One of the series (G1) corresponds to the steps of a small primate that seems to have stopped at a certain point and turned around before continuing (Leakey and Hay, 1979). The other is more difficult to interpret. It was described by Leakey and Hay (1979) as prints made by a larger primate (G2), but was later described as a double trail: a larger-sized biped (G2) would have left a trail of steps in which

Figure 3.5 The 3.5-Ma old Laetoli fossil footsteps (Tanzania). Photograph by Martha Demas. © The J. Paul Getty Trust, 1995. All rights reserved.

something smaller (G3) would have stepped (an interpretation that is not universally accepted). "Large" and "small" are relative terms; if we consider the relationship between the length of the foot and height of modern humans, then the smallest primate (G1) would be about 1.20 m tall and the medium-sized one (G3), which followed the footsteps of the largest, would measure close to 1.40 m. The height of the largest primate (G2) cannot be calculated because its footsteps are partially hidden by those of the medium-sized one (Hay and Leakey, 1982).

The overlaying G2–G3 Laetoli tracks can inspire the imagination beyond reasonable limits. They have led some to envision the existence 3.5 Ma of such a human child's game as stepping in someone else's footsteps. Tracks G1 and G2 are about 25 cm apart, too close for both individuals to walk side by side without touching each other. Either one was walking in front of the other, or they were hugging while walking. This last image—the male embraces his mate while they stroll along the savanna—appears on the cover of *The Fossil Trail* (Tattersall, 1995a). Tattersall is against indulging in empty speculations, and that book is an excellent demonstration. Tattersall does not mention the G3 series, and describes G1 and G2 as the footprints of two individuals that walked beside each other, without searching for further interpretations. A question remains: what direct evidence do the Laetoli footprints provide regarding bipedalism?

Leakey and Hay (1979) hardly addressed functional aspects in their original description.

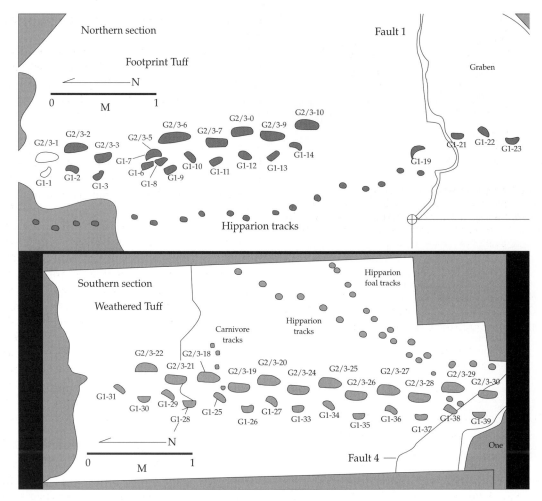

Figure 3.6 The two series of hominin footprints at Laetoli.

The tracks at sites A and G were considered "presumably hominin", that is all. The authors noted that the longitudinal arches of those feet were well developed, resembling those of modern humans, and that the big toe was parallel to the others. But these indications are enough to conclude that the way the foot was placed to cause these prints was similar to the way current humans do when walking, thus suggesting well-developed bipedalism (Figure 3.7). The conclusion that such early hominins were already capable of a functionally developed bipedalism was explicitly put forward by Day and Wickens (1980), and has been suggested by many others (e.g., White, 1980; Lovejoy, 1981; Tuttle, 1981; Robbins, 1987; White and Suwa, 1987; Tuttle *et al.*, 1991).

Tuttle *et al.* (1991) have argued that the footprints found at Laetoli site G were made by feet more similar to those of *Homo sapiens* than to those of australopiths. Regarding the bipedal footprints at site A, these authors, after studying 16 circus bears, concluded that bipedal bears leave very similar footprints to the trail found at site A, although they note that there is no specimen

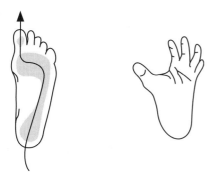

Figure 3.7 The longitudinal arch and the big toe produce the greatest differences in the footprints left by a bipedal being like a human (left) and a terrestrial quadruped like a chimpanzee (right). The first trait is very marked in humans, whereas the big toe of chimpanzees separates in an oblique direction.

attributable to bears in the limited collection of Laetoli carnivores. They settled the issue with a "pathetic poem" (their own expression):

> Fuzzy Wuzzy was a bear
> Fuzzy footprints were found there
> Was Fuzzy at Laetoli?
> Or, wasn't he?

The evidence of a clearly bipedal trail at Laetoli has often been used as irrefutable proof of the existence, 3.5 Ma, of a bipedalism very similar to our own. If this is correct, a functionally complete bipedalism would be a derived trait that characterizes all the members of the tribe Hominini, distinguishing them from our closest relatives, the African apes.

However, this is far from certain. The photograph of the Laetoli footprints in the *Cambridge Encyclopedia of Human Evolution* (Potts, 1992, p. 325) is accompanied by a categorical statement: "The discovery of the footprints confirmed that early hominins walked upright on two legs in a characteristically human fashion and that their footbones were arranged like a modern human's, with no gap between the big toe (toe 1) and the other toes". Yet the author of the accompanying article, Richard Potts (1992), as well as Bernard Wood (1992c), in the article dedicated to australopiths in the same book, caution that there are reasonable doubts about the interpretation

of the Laetoli footprints and favor partial bipedalism.

Monographs devoted to footprints also reflect a diversity of interpretations. Day and Wickens (1980, pp. 386–387) wrote: "The pattern of weight and force transference throughout the foot, well known in modern man, also seems to be very similar in the fossil footprints and indicates that even at this early stage of hominid evolution bipedalism had reached an advanced and specialized stage". But some illustrations that accompany the text reveal certain traits (narrow print of the heel, absence of the medial prominence in the base of the big toe, its orientation) that are reminiscent of the shape of a chimpanzee's print. Susman and collaborators (1984) concluded that the footprint molds are not similar to the prints left by the feet of humans. However, White and Suwa (1987), after suggesting the Laetoli footprints might have been made by *Australopithecus afarensis*, rejected Stern and Susman's (1983) idea that *A. afarensis* were intermediate between African apes and humans from a locomotor point of view, but accepted that the maker of the Laetoli footprints might differ in certain locomotor aspects from modern humans. The thorough study carried out by Deloison (1991) agrees with that of Susman *et al.* (1984). After studying the best-defined Laetoli footprint, G1/34, Deloison compares its contour from a picture taken of a cast of the print with the prints left by the right feet of a human and a chimpanzee, arriving at the conclusion that the Laetoli footprint shows an intermediate shape between the other two. In fact, the contour is closer to that of chimpanzees' feet. Deloison pointed out, moreover, that some features of the Laetoli footprints support the prehensile functionality of the foot. In her opinion the maker of the prints used that prehensile capability to keep its balance on the humid floor, curving the foot and separating its big toe.

It is difficult to arrive at an interpretation of the Laetoli footprints that would integrate the diverse points of view. Susman and colleagues (1984), who were among the first to interpret the Laetoli footprints as associated with a partial bipedalism, suggest that their scientific value is limited to the demonstration that they were produced by individuals that walked on their two lower limbs, which is not disputed by anyone. However, when the fieldwork was finished in 1979, and with the intention of conserving the footprints, the research team covered them with sand first and lava pebbles afterwards (Agnew and Demas, 1998) without realizing that there were acacia seeds mixed in the sand. These germinated, growing into trees that menaced the destruction of the trails with their roots. A program for the conservation of the Laetoli footprints was initiated in 1994. It involved removing the trees, covering the trails again with sand that had been protected with herbicide and extending a mantle of gravels and plastic materials crowned with lava pebbles (Agnew and Demas, 1998). Given the deterioration of the footprints, Susman and colleagues (1984) conclude that the bipedalism of early hominins should be determined by the direct study of available fossil specimens, without need to depend of the Laetoli footprints.

3.2 Bipedal locomotion

3.2.1 The morphology of the hip

The locomotor habit of catarrhines is quadrupedalism, that of apes is brachiation and knuckle-walking, and the human posture is completely orthograde, with locomotion performed using only the lower limbs. Bipedal locomotion involves anatomical modifications of almost the whole skeleton: head, trunk, hip, and upper and lower limbs (Figure 3.8).

If complete sequences of fossil specimens were available within the chimpanzee and hominin lineages, it would be possible to determine how and when bipedalism evolved. But this is not the case. Remains are very rare and partial, especially the oldest, and there are no informative remains of chimpanzee fossil ancestors. The best we can do is to compare the traits of different hominins with living apes and humans.

Some features, such as the insertion of the vertebral column in the cranium, the foramen magnum, provide evidence regarding bipedal habit,

but the two most conspicuous morphological traits associated with bipedal locomotion are the shape of the hip—including the femur's insertion in it—and the shape of the limbs, especially the feet.

As Poirier (1987) has pointed out, no postcranial anatomical element shows greater differences between current apes and human beings than the pelvis. He concludes that the main factor leading to the divergence between the lineages was the modification of the locomotor apparatus. Accordingly, the shape of the pelvis is a good indication of the taxonomic status of a given fossil. There are biomechanical reasons to argue, according to Deloison (1996), that the rotation of the body's axis during the transition towards an upright posture necessarily affected the structure of the pelvis. The transformations that led to bipedal locomotion very possibly began with this anatomical element.

The pelvis is a ring-shaped osseous structure constituting two lateral parts, which articulate frontally, and a posterior part, the sacrum, composed by several welded vertebrae. Each lateral section, colloquially known as a hip, is also the result of the welding of three bones: illium, ischium, and pubis. The differences between the shape of the hips of current chimpanzees and humans are related to the function performed by the pubic skeleton and the musculature necessary to carry out different kinds of locomotion. Most differences between the hips of apes and humans are located in the upper part, the illium, while the lower part is quite similar between these taxa (although the ischium is shorter in our species). These similarities and differences have led to a general agreement that the functional evolution of the hip, and consequently locomotion, took place through changes in the illium (Schultz, 1930; Napier, 1967). Accordingly, the illium is the most relevant feature for comparing the locomotor habits of different specimens.

The illium of African great apes is longer and thinner than that of humans. The widening and shortening in our species is the result of the adaptation of muscular insertions to allow hominins to keep their balance in an upright posture while using a bipedal locomotion. The earliest available fossil pelvic bones are from

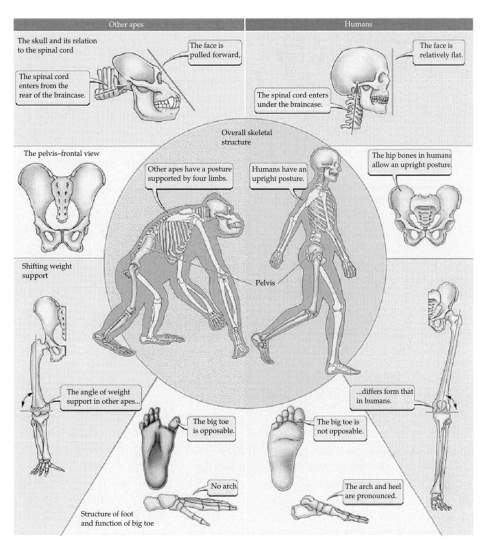

Figure 3.8 Comparison between the locomotion of a gorilla (left) and a human (right), highlighting the main modifications associated with bipedalism.

3-million-year-old australopiths from South Africa (*Australopithecus africanus*, Sterkfontein and Swartkrans) and East Africa (*A. afarensis*, Hadar). The pelvis of these australopiths is intermediate between chimpanzees and humans, but it is closer to the latter (Figure 3.9). Thus, it can be concluded that around 3 Ma there were already hominins whose locomotion was relatively similar to our own (Table 3.2).

Pelvic functions are not exclusively related to locomotion: in females these bones must allow the complete development of the fetus and its birth. Locomotion and bearing offspring call for conflicting pelvic designs. The widening of the illium, required for the acquisition of bipedal posture, tends, in females, to close the canal through which the head and body of the fetus has to pass during birth (Figure 3.10). The appearance of bipedalism and the increase

Figure 3.9 The hips of a chimpanzee (left), a modern human (center), and an australopith (*A. africanus*; right), at different scales to facilitate morphological comparison.

Table 3.2 Earliest fossil hip specimens

Specimen	Locality
A.L. 288–1	Hadar (Ethiopia)
SK 50	Swartkrans (South Africa)
Sts 14	Sterkfontein (South Africa)
Sts 65	
MLD7	Makapansgat (South Africa)
MLD252	
TM 1517	

Figure 3.10. The fetus passing through the female pelvic canal of a chimpanzee (left), australopith (center), and modern human (right). Michael Day (1992) has suggested a transversal position for the australopith fetus.

in cranial size represent a paradox: natural selection has yielded traits that make healthy births difficult. But nature has alleviated this difficulty by selecting soft fetal cranial bones and birth at a very immature stage with a long period of exterogestation. In addition, human male and female pelvic anatomy are different: the design of women's pelvic structures is less strained by our bipedal posture.

However, in the case of australopiths the conflict between a hip design apt for bipedal posture and the birth constraints caused by a large head would be less, because their cranial size is approximately that of chimpanzees (Lovejoy, 1975; Tague and Lovejoy, 1986). Furthermore, australopith hips are variable, as well as scarce.

3.2.2 Upper and lower limbs

The limbs also reveal information regarding bipedal and quadrupedal locomotion. The morphology of the upper limbs can reveal whether or not they are used for locomotion. Napier's (1980) study of the human hand has shown that, among all primates, only hominins have "true hands". The morphology of the hands is an indication of the degree of bipedalism. As Jouffroy (1991) has noted, the hands of the first hominins that developed incomplete bipedalism exhibit biomechanical features associated with their locomotor activities. Bipedalism was possibly acquired before the structure of the hand's carpal bones underwent significant modifications. The studies of Jouffroy (1991) and Susman and Stern (1979) on the OH 7 specimen from Olduvai, Tanzania, reveal that its anatomy is associated with certain locomotor functions, but the numerous authors that have compared the morphology of different fossil hands have reached disparate conclusions (Bush, 1980; Bush *et al.*, 1982; Stern, 1983; Susman *et al.*, 1984). The same features have been interpreted as similarities between australopith specimens and modern humans, and as similarities between the former and current apes. The lower limbs provide the best evidence regarding bipedal locomotor function. However, William Jungers (1994) and Tim White (1994; White *et al.*, 1993) disagree about the locomotor significance of the very robust MAK-VP-1/3 humerus, found at the Maka site in Ethiopia.

The fibula and tibia set and the footbones may provide definitive evidence for ascertaining locomotor biomechanics. Whereas climbing capacity is associated with great joint mobility, bipedalism requires solid articulations able to resist the weight distribution during upright posture. But authors differ in their interpretation of the available evidence. The OH 6 specimen from Olduvai is precisely formed by a tibia–fibula set. Davis (1964) characterized certain traits of the fibula and the distal region of the tibia as practically identical to those of modern humans, although there are some differences in the proximal area of the tibia. After comparing the OH 6 and KNM-ER-741 specimens

(Koobi Fora, Kenya), Lovejoy (1975) went even further: "the australopith tibia approximates the modern human pattern with such fidelity that no locomotor or mechanical differences are implied by the morphology of these bones". With respect to the A.L. 288–1 specimen from Hadar in Ethiopia, Latimer *et al.* (1987) and Stern and Susman (1991) reached contrary conclusions concerning the

tibia's flexibility, probably because of their different understanding of the type of locomotion of the specimen. According to Latimer and collaborators *A. afarensis* was fully bipedal, while Stern and Susman believe it was partially arboreal.

Evidence about the morphology of early hominin feet was enriched by the discovery of articulated footbones at the Sterkfontein site in South

---

Box 3.2 Bipedalism and *A. afarensis*

The different functional interpretations of the bipedalism of some *A. afarensis* specimens may be due, according to Carol Ward (2002), to two reasons: "First, there are divergent perspectives on how to interpret primitive characters. . . . Second, researchers are asking fundamentally different questions about the fossils. Some are interested in reconstructing the history of selection that shaped *A. afarensis*, while others are interested in reconstructing *A. afarensis* behavior." Ward concludes: "Evidence from features affected by individual behaviors during ontogeny shows that *A. afarensis* individuals were habitually traveling bipedally, but evidence presented for arboreal behavior so far is not conclusive". If so, bipedalism very similar to ours would be already present at least 3 Ma.

This is how Harcourt-Smith and Aiello (2004) see the matter: "The central point is that contemporary fossil taxa may well have been mosaic in their adaptations, but, critically, may have been mosaic in different ways to each other . . . Further analyses of other skeletal elements are needed to reinforce this interpretation. If correct, this would imply that there was more locomotor diversity in the fossil record than has been suggested, and raises questions over whether there was a single origin for bipedalism or not. At the very least, if bipedalism appeared only once in the hominin radiation and is therefore monophyletic, such evidence would suggest that there were multiple evolutionary pathways responding to that selection pressure."

Box 3.3 The story of the *Little Foot* specimen

The StW 573 bones, *Little Foot*, were discovered 15 years before Clarke and Tobias interpreted them as belonging to a hominin. Phillip Tobias has recounted the finding in a story that serves as an excellent testimony of the diverse demands sometimes faced by a paleontologist: "On February 28, 1980, one of our field assistants, David Molepole, extracted a very small left astragalus, or ankle, followed the same day by the navicular, the ship-shaped bone which articulates immediately after the astragalus. As usual, each bone was carefully marked with the source of the material and the date of the extraction. The third bone, the left first cuneiform, was extracted on February 29th (it was a leap year). After the following weekend, Tuesday March 4th, Molepole extracted a fourth bone, the proximal half of the left big toe's metatarsal. Probably due to the small size of the bones, the team must have assumed that they belonged to a baboon or a monkey. As I had just

been named Dean of the Medicine Faculty of the Witwatersrand University, and I had little time to work on the fossils, the four footbones were placed in a box with other small pieces of postcranial bones of primate and carnivore extremities. Alan Hughes, who was in charge of the Sterkfontein excavation from 1966 to 1990, must have included those specimens among the 63 postcranial bones of baboons, monkeys and carnivores annotated in the Annual Report of the Palaeo-anthropology Research Unit of the year 1980 (September 1979 to September 1980). The number of limb bones rose to 190 in September 1981. So the four carefully labeled footbones remained in that box for fourteen years" (Tobias, 1997a). The second part of the story, the examination of the box containing the footbones, their identification and the surprising discovery of the rest of the StW 573 skeleton, has been told by Ron Clarke (1998).

Africa. The specimen known as *Little Foot* (StW 573) provided relatively solid evidence that the locomotion of the first hominins cannot be characterized as functionally developed bipedalism, equivalent to that of current humans.

The StW 573 specimen was found in Sterkfontein Member 2, dated to between 3.5 and 3 Ma (Tobias and Clarke, 1996). However, for a long time it was stored among unclassified cercopithecoid bones, from the Silberberg Grotto at Sterkfontein, in a box stored in the Department of Anatomy of the Medical School, University of Witwatersrand, South Africa. While searching for a bovid specimen in the box, Clarke and Tobias came across the bones that received the colloquial name of *Little Foot*. They include a left talus, a left navicular that articulates with the head of the talus, a medial left cuneiform that articulates with the distal surface of the navicular, and the proximal half of the first metatarsal of the left foot's big toe, articulated with the cuneiform (Clarke and Tobias, 1995; Figure 3.11). The fact that these bones articulate together is precisely what makes possible understanding the biomechanical function performed by the foot to which they belonged.

The morphology of *Little Foot* reveals a mixture of apelike and humanlike traits. Functionally speaking, the heel suggests that the foot belonged

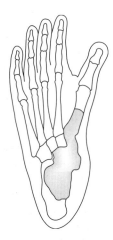

Figure 3.11 StW 573, *Little Foot* (Clarke and Tobias, 1995), from 3.5 to 3 Ma sediments at Sterkfontein in South Africa. Drawing by Ron Clark; from Tobias (1997a).

to a biped which supported the body's weight on its lower limbs. But its bipedalism was not completely developed; the anterior region of *Little Foot* retains the prehensile capacity required for arboreal activity. From back to front, the anatomy of the StW 573 articulated bones shows the intermediate condition between total bipedalism and arboreal activity. The talus is the closest bone to human morphology, though it is smaller (even accounting for the body size differences between *Little Foot* and modern humans). The navicular and the medial cuneiform exhibit intermediate morphology. The metatarsal and its articulation with the medial cuneiform evinces an apelike character. Based on the articulation's shape, it seems that the foot's big toe adopted a diagonal position and was medially separated from the foot's axis (as happens in chimpanzees), different from the parallel position in relation to the foot's axis of the big toe of modern humans.

The mobility of the big toe is an important trait (similar in significance to the freedom of movement of the hand's thumb), determining the ability to climb trees. Tobias' (1997a) conclusions are that while the talus had began the way towards the shape associated with habitual bipedalism, part of the navicular, the medial cuneiform, and the base of the first metatarsal, seem not to have done so: rather, they retained apelike traits. *Little Foot* represents an intermediate stage in the evolutionary conversion from a foot adapted to arboreal life, with a diverging and prehensile big toe, to an extremity adequate for regular bipedalism. "It seems that the astragalus and the proximal ankle joint adopted the human form quite early on, whereas the anterior part of the foot retained its primitive state for a long time" (Tobias, 1997a).

In East Africa, Hadar (Ethiopia) has provided australopith specimens which convey information about feet morphology. The 333 series, discovered in 1975 in Member DD of the site, is composed by two calcaneus bones (A.L. 333–8 and A.L. 333–55) and a cuneiform (A.L. 333–28), the proximal part of a first metatarsal (A.L. 333–54), and a partial foot (A.L. 333–115) with 13 bones, including phalanxes and metatarsals, with the head of the first metatarsal of the big toe and the first distal phalanx of

---

Box 3.4 Specimens from Hadar

Some discrepancies between authors regarding the bipedalism of Ethiopian australopiths would disappear if it were accepted that there are two species at Hadar. One would be represented by the A.L. 333 series and the other by specimens like A.L. 288–1. But this can hardly be taken as an argument in favor of an early functionally complete bipedalism. As Groves (1989) noted, the presence of a *Homo* sp. (unnamed) at Hadar, represented by A.L. 333

and, maybe, A.L. 400, is based on Coppens' (1983a) observation that certain traits in those specimens are reminiscent of *Homo*. Coppens qualified certain traits of A.L. 333 as "surprisingly modern". However, the work of Deloison and Susman and Stern suggests a "mosaic evolution" in the mobile traits of precisely the same specimens, a pattern characteristic of a bipedalism that is not completely developed.

---

the same finger, which, obviously, articulates with the previous one. The age of Member DD is between 3.22 and 3.18 Ma (Walter and Aronson, 1993). Certain traits of the 333 series suggest a foot morphology intermediate between human and chimpanzee. Deloison (1991) argued that the various position of the big toe, the convexity of the calcaneus, the flattened and long phalanxes of the fingers, and the mobile articulations are features related to prehensile ability and, thus, suggest the possibility of an arboreal behavior. Susman and colleagues (1984) reached a similar conclusion after a comparative examination of the metatarsal heads of gorillas, bonobos, A.L. 333–115, and modern humans.

The OH 8 specimen from Olduvai in Tanzania is another fossil that affords information about foot traits. It is between 1 and 2 million years old, younger than *Little Foot* and the 333 series. Hence, it reflects the direction of the evolution of loco-motion. OH 8, found in Bed I of the site, and estimated at 1.85–1.71 Ma, contains (in addition to a clavicle, part of a hand, and a partial molar) an almost complete foot. Louis Leakey, Tobias and Napier (1964) believed it possessed most of the specializations associated with modern humans' plantigrade propulsive foot. However, later the morphology of OH 8 was interpreted differently. Some authors, such as Oxnard and Lisowski (1980), argued that its function seems to have been associated mainly with arboreal behavior. Even though this individual could probably also walk upright, this kind of locomotion would be far from human bipedalism and closer to that of gorillas and chimpanzees, as suggested mainly by the

shape of the transversal arch. Other authors have described it as intermediate between complete bipedalism and arboreal life (Lewis, 1972, 1980), or as an unquestionably bipedal being but without yet reaching the posture of *Homo sapiens* (Day and Napier, 1964; Day and Wood, 1968). Some researchers that have delved deepest in the study of the evolution of bipedalism (White and Suwa, 1987; Susman and Stern, 1991; Deloison, 1996) see in the foot of OH 8 a very similar morphology to current humans. The fact that Randall Susman and Jack Stern, on one hand, and Yvette Deloison on the other, firm advocates of the gradual evolution of bipedalism, consider that the foot of *Homo habilis* is functionally modern, speaks in favor of this thesis.

3.2.3 The reasons for bipedalism

Given that bipedalism is a synapomorphy shared among different hominin genera for 7 million years, it must have had an undoubtable adaptive advantage. What was the advantage? What was it about a permanent upright posture that improved resources? As we have seen, Darwin suggested a hypothesis that related bipedalism, free hands, and tool use to the extent that their combination would amount to a single complex phenomenon with morphological and functional aspects. The "hit'em where it hurts" hypothesis, as Tuttle and colleagues (1990) called it, is undeniably attractive. Darwin's initial sug-gestions regarding this issue were preserved in later models that relate the appearance of savannas with bipedalism and tool making. The number of published articles devoted to bipedalism and its role during human evolution possibly outnumber those dedicated to any other hominin functional feature. But, as Tobias (1965) noted, bipedalism is not a requisite for making or using tools. Chimpanzees use instruments quite ably, and they do so sitting up. The essential element in the relation between posture and the use of cultural elements is upright posture, not bipedalism. But there is more. Bipedalism appeared in human evolution long before culture.

If bipedalism is not explained by the manufacture and use of instruments, what drove its appearance? Table 3.3 summarizes different hypotheses concerning the adaptive advantages of bipedalism in pre-cultural conditions, without reference to tool use. We have retained the original names used by Tuttle and colleagues (1990) to keep their casual tone. But that does not mean they should not be taken seriously. For instance, studies on the mechanics of locomotion have shown the benefits of the bipedal solution in terms of energetic economy (Kimura *et al.*, 1985; Reynolds, 1985). The results of Deloison's (1991) biometrical studies of the Hadar *A. afarensis* remains and the Laetoli footprints support a kind of locomotion that combined climbing and bipedalism.

Table 3.3 Possible explanations for the adaptive advantages of bipedalism

Hypothetical bipedal action	Possible explanation
schlepp	Food transportation, caring for offspring; involves the presence of a kind of home-base
peek-a-boo	Vigilant behavior, standing up over the savanna's long grasses
trench coat	Phallic exhibition in males to attract females
tag along	Following herds of herbivores during their migrations through the savanna
hot to trot	A way to lose heat when exposed to the solar radiation in the open savanna
two feet are better than four	Bipedalism has a favorable energetic balance for long treks

Source: Tuttle *et al.* (1990).

As Tuttle and colleagues (1990) noted, it is possible that several factors provided adaptive advantages and that some, or many, of them combined to achieve the result of bipedal behavior. There are two separate issues underlying the search for hypotheses to explain the adaptive advantage of bipedalism. First, the motives behind the appearance of the first bipedal behaviors in a tropical forest environment. The second issue concerns the benefits of bipedalism as an adaptation to the savanna. These two questions must not be confounded: bipedal behavior existed long before savannas were extensive in the Rift Valley. The two questions are often confounded by seeking a "general explanation of bipedalism".

The hypotheses summarized in Tuttle and colleagues' (1990) classification refer to the adaptive advantages of bipedalism in the savanna, not on the forest floor. Thus, they are inadequate to explain the reason for an upright locomotion, unless this evolution is considered to have taken place only during the last 2.5 Ma. We will later specify some of the functional traits that allowed the adaptation to open savannas, such as the dietary patterns of 2.5-million-year-old hominins. It seems clear that the explanations listed in Table 3.3, except maybe the trench-coat hypothesis, were irrelevant at a time when our ancestors were creatures with a precarious bipedalism that lived in tropical forests.

Coppens (1983a, 1983b) suggested the progressive reduction of the tropical forest thickness as a possible explanation for the gradual evolution of bipedalism. If the distance among the trees gradually increased, it would become necessary to travel longer distances on the ground to go from one to another. At the same time it would be imperative to retain the locomotor means for climbing. Distinct functional responses appeared in the different lineages leading to current primates: knuckle-walking bipedalism in the ancestors of gorillas and chimpanzees and an incipient bipedalism in the first hominins (Coppens, 1983a, 1983b, 1991; Senut, 1991).

The gradual substitution of forests for open savanna spaces would be an increasing selective pressure towards more complete bipedalism, functionally speaking. The final result of this process was two evolutionary lineages of bipedal primates based on different adaptive strategies, close to 3.5 Ma. One million years later this divergence would increase with the decrease in temperatures and the appearance of extremely robust australopiths and the genus *Homo*.

The explanation given by Coppens and Senut has a considerable advantage: simplicity. Brigitte Senut noted that the locomotor hypothesis of the origin of bipedalism has been among the least favored. This hypothesis suggests that hominins had become bipedal for reasons strictly associated with locomotion itself (Senut, 1991); that is to say, the need for traveling on the ground of open forests. Senut explored eight hypothetical ways in which bipedalism could have originated from the locomotion of other primates, but ended up developing with greater detail the explanation favored by Coppens (1983a, 1983b).

3.2.4 The origin of bipedalism

There are many different kinds of bipedalism. For a quadruped to adopt and maintain a bipedal posture it must solve the problems of balance and lifting the body (Kummer, 1991). This can be done in two ways, fast and slow. Getting up by means of the thrust of acceleration requires no specific anatomical prerequisite, just having enough muscular strength. However, standing up slowly requires keeping the center of gravity within the support area, which generally are by the feet soles or these and the hind limbs. Balance can be achieved, as chimpanzees do, by means of very long upper limbs and a pronounced angulation of the lower limb's articulations. In a bipedal posture, the center of gravity is located in a clearly ventral point and, thus, the feet must also be placed in that position. Chimpanzees and other animals that adopt an upright posture in this way keep their balance owing to intense action of ventral and dorsal muscles. This mechanism consumes great amounts of energy and does not allow the bipedal posture to be maintained for very long.

An upright posture can also be achieved starting from a sitting position, with the main supporting area constituted by the lower limbs and the pelvis, as

many small mammals do. The energetic consumption is very low, but it does not allow traveling. Great apes adopt this kind of "bipedalism" for activities that take quite a time, such as eating or sleeping.

Birds exhibit an authentic and permanent bipedalism. The weight is balanced in front and behind the vertical axis of the legs, assuring that the center of gravity falls within the legs. This form of bipedalism evolved from early reptilian tripedalism, which involved leaning the tail on the ground; it has led to a vertebral column with a pronounced S shape. The case of humans is completely different. The vertebral column is almost completely straight, and the vertical axis of the center of gravity, which practically coincides with it, passes through the articulations of the lower limbs. The skeletal and muscular anatomical modifications required by human bipedal posture are quite conspicuous. In addition to the shape and position of the vertebral column, the foramen magnum is displaced towards the inferior part of the head, the bones of the lower limbs are elongated and those of the upper limbs are shortened, and there are changes in the shape of the hip, the structure of the foot and in flexor and extensor muscles. The modifications also affect the shape and mobility of the articulations. During walking, for instance, human bipedalism turns into a successive monopedalism that requires placing each foot on the center of gravity's vertical and leaning the distal part of the femur inwards. This does not happen in quadrupeds, which can always walk on two extremities at a time.

This overview of the different ways of placing the body in an upright position allows a better understanding of how human bipedalism could have evolved. The second upright posture we mentioned, the one related with sitting, is common among all primates—arboreal or terrestrial—and appeared very early in their evolution. It is a first stage in the evolution of bipedalism (Tobias, 1982a, 1982b). From there on, primate locomotion diverged into many different solutions, from climbers to leapers to arboreal quadrupeds and those that use a more or less complete brachiation (Napier, 1963; Napier and Walker, 1967). The anatomical modifications that take place during phylogenesis reflect, of course, the kind of loco-motion of a given species (see Fleagle, 1992).

From which previous kind of locomotion did hominin and African apes locomotion evolve? As we saw in previous chapters, it is not easy to deter-mine the phylogenetic lineage of hominoid posture. When reviewing Miocene hominoid evolution we concluded that it is difficult to trace the evolutionary sequence back to that period. We have no definitive evidence whatsoever about the phyletic relation-ships between Miocene hominoids and current great apes and hominins. There are no postcranial remains of *Ouranopithecus*, a good candidate to be the common ancestor of the clade formed by African apes and humans. The remains of *Dryopithecus* are very informative in regard to locomotion, but the phyletic relations between these beings and gorillas, chimpanzees, and humans are not clear.

There is a hypothesis, which can be traced back to Arthur Keith (1903), suggesting that brachiation required the development of postcranial traits capable of leading, through several evolutionary stages, to bipedal locomotion. Brigitte Senut (1991) has warned about the risk of referring to "brachiation" in the case of orangutans. Gibbons, which only use their arms when moving from branch to branch, are the true brachiators. Oran-gutans, and to a lesser degree, chimpanzees and gorillas, use suspension and not brachiation. In any event, as Tuttle and Basmajian (1974) suggest, traveling along branches—climbing trees—could be the kind of locomotion that finally led to bipedalism. Alternatively, other authors, such as Napier (1967) and Washburn (1967) relate the evolution of bipedalism to knuckle-walking, which would be, under this scenario, an obligate stage towards human bipedalism shared by gorillas, chimpanzees, and pre-hominins (Figure 3.12).

Richmond and Strait's (2000) study of aus-tralopith specimens has revealed that early homi-nins retained certain morphological traits in their wrists related to knuckle-walking. It is worth not-ing that the question of whether knuckle-walking is a derived or a primitive trait is crucial to understanding hominoid locomotor evolution. If it is a plesiomorphy, as suggested by Richmond and Strait (2000)—in line with Napier (1967) and

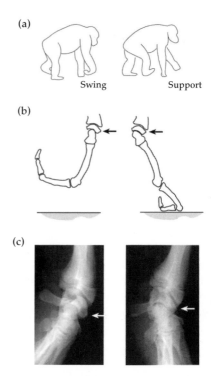

Figure 3.12 The wrist joint during the swing phase (left-hand panels) and support phase (right-hand panels) of knuckle-walking. Reprinted by permission from Macmillan Publishers Ltd. *Nature*, 404: 6776, 382–385, 2000.

Washburn (1967)—then it is feasible that the common ancestors of chimpanzees and humans already used this locomotor habit, from which bipedalism necessarily would have evolved. But if knuckle-walking is a homoplasy, a convergent adaptation that appeared separately along the gorilla and chimpanzee lineages (Dainton and Macho, 1999), the primitive trait must have been another one, and hominin bipedalism would derive from a different locomotor habit to that of current African great apes. As Collard and Aiello (2000) have commented about Richmond and Strait's study, the consideration of knuckle-walking as a primitive trait for gorillas, chimpan-zees, and the first hominins is consistent with the cladogram suggested by molecular geneticists such as Morris Goodman (Figure 3.13).

Senut (1991) has argued that the comparative examination of fossil specimens and current great

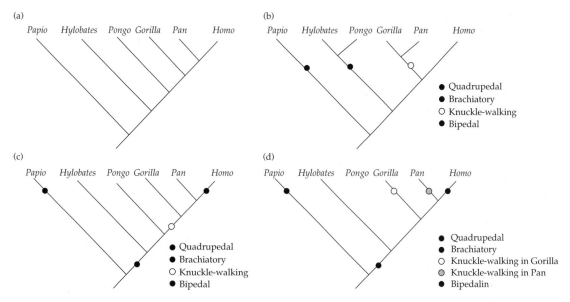

Figure 3.13 (a) Most-parsimonious cladogram according to hominoid genetic distances. (b) Most-parsimonious cladogram if knuckle-walking is a derived trait characteristic of chimpanzees and gorillas. (c) Most-parsimonious cladogram if knuckle-walking is a primitive trait of African apes, modified in hominins. (d) If the cladogram established by molecular distances has to be made compatible with knuckle-walking as a derived trait of chimpanzees and gorillas, then this form of locomotion must be a homoplasy fixed separately in *Gorilla* and *Pan*.

apes does not lead to definitive conclusions. The combination of traits observed in fossil hominin species suggests that their locomotor habit would be very different from the very specialized locomotion of current great apes. If so, the only clues regarding the evolution of bipedalism are provided by fossil hip and limb postcranial remains of Miocene and early Pliocene members of our tribe, australopiths. The answers concerning the evolution of bipedalism depend on the analysis of these morphological traits in early and current specimens of our lineage. If they are similar but not identical, what do the differences mean phylogenetically?

3.2.5 "Partial" and "complete" bipedalism

Hominin bipedalism is currently widely considered as a homologous trait, shared by the whole lineage. It is thought to have developed in several stages from the incipient bipedalism of early australopiths to the complete bipedalism of the specimen found in Java by Dubois, *Homo erectus*. But this is not the only possible interpretation. There are authors who reject the idea that there were

different stages in the evolution of bipedalism along the hominin lineage. For instance, the comparative examination of the tibia of australopith specimens from Olduvai (Tanzania), Koobi Fora (Kenya), and Hadar (Ethiopia) led—as we pointed out above—Owen Lovejoy, renowned specialist in hominin locomotor patterns, to the conclusion that the bipedal locomotion of early hominins was as developed as our own (Lovejoy, 1975; Latimer *et al.*, 1987). The study of australopith specimens from South Africa also indicated, according to Lovejoy (1975), that there is no morphological reason to consider that their locomotion was "intermediate" between that of African apes and modern humans. The morphology of the pelvis of those early hominins is very similar to that of living current humans, according to this author. Their illium is equivalent to human beings (this, by the way, had already been noted since the discovery of the first exemplars—Dart, 1949a—and generally admitted since then). The differences observed in their ischium probably have no functional consequences. And the pubis, in any case, has little bearing on the question of locomotion.

Box 3.5 Lovejoy's view

According to Owen Lovejoy, despite what some isolated traits might suggest, the overall biomechanical pattern of australopith postcranial anatomy supports the notion that the only difference between the bipedal locomotion of

Australopithecus and *Homo sapiens* is advantageous to the former. All the necessary adaptations for bipedal locomotion, in Lovejoy's opinion, were already present in those early hominins, although in a different way in males and females.

The idea that the very wide pelvis of australopiths would have been favorable for bipedal locomotion has been rejected, however, by Berge (1991) after the examination of the A.L. 288–1 specimen from Hadar in Ethiopia. At the level of the iliac crests and the pelvic cavity, the pelvis of A.L. 288–1 is much wider than that of modern humans. In Berge's biomechanical reconstruction the long neck of the femur, acting as a lever arm, does not constitute an advantage, as Lovejoy surmised; rather, it introduces balancing problems. The vertical of the center of gravity would fall, in the case of *A. afarensis*, far from the knee articulation when leaning on one foot while traveling, leading to a greater instability of the lower limb (Berge, 1991). As a consequence, the kind of bipedal locomotion exhibited by *A. afarensis* would have required a higher degree of hip rotation to place the leaning knee within the body's vertical axis. In her morphometric study of the mobility of the hip of *A. afarensis*, and to obviate the difference in height between *Lucy* and current humans, Berge carried out the comparison with the pelvis of a pigmy woman 137 cm tall.

Regarding the possible reconstruction of the insertion of the gluteus in the hip of A.L. 288–1, Berge (1991) pointed out a noteworthy circumstance. Not much is known about that insertion, but the two possible alternatives are the "human" way, with the *gluteus maximus* inserted in the illium, and the "ape" way, in which the muscle would be inserted for the most part in the ischium. When Berge reconstructed the internal rotation movements of the thigh, she argued that the hip's morphology, together with the "human" reconstruction of the *gluteus maximus* insertion, would not allow A.L. 288–1 to perform the necessary movements for bipedalism. These could only be performed with an "ape" insertion of the gluteus. This point is especially important, given that the role of the *gluteus maximus* in the evolution of bipedalism had been considered in a different way by Washburn, who believed that the transition from quadrupedalism to bipedalism began precisely with "human" changes to the *gluteus*, and Napier (1967), who believed that this change did not take place until later stages in the evolution of bipedalism and carried out functions related only with balance while running or going up slopes, but not walking. Berge's (1991) study supported Napier's point of view and concluded that the hip of A.L. 288–1 suggests that its locomotion included partially arboreal behavior (Figure 3.14).

Latimer and colleagues (Latimer, 1991; Latimer *et al.*, 1987) have put forward an argument against the notion of australopith "partial" bipedalism, which, by the way, can be applied to any evolutionary process. They argue that the earliest hominins were bipedal, although they preserved some climbing traits. This claim is based on the fact that, within a Darwinian scenario, the persistence of primitive traits is not significant.

In Latimer's (1991) opinion, the functional value of primitive and derived traits is not the same. No arboreal primitive traits are retained by late-Pliocene African great apes (Latimer and Lovejoy, 1989). This means, according to Latimer, that if australopiths are considered arboreal, they should be so based on certain derived traits that reveal the specific way in which they had adapted to their particular arboreal life. But all australopith derived traits are related to bipedalism, not arboreality. Therefore, Latimer (1991) concluded that there can be no talk of "intermediate degrees" of bipedalism. Locomotion is determined by the new derived bipedal traits, while the presence of primitive

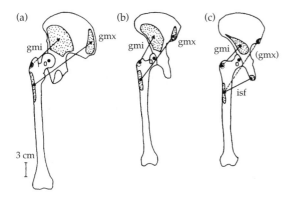

Figure 3.14 Different ways of interpreting the insertion of the muscles that attach hip and femur in A.L. 288–1, *A. afarensis*. (a) *H. sapiens* pelvis and femur. (b) "Human" insertion in A.L. 288–1. (c) "Ape-like" insertion in A.L. 288–1. gmi, *gluteus minumum*; gmx, *gluteus maximus*; isf, *ischio-femoralis*. Picture from Berge (1991).

characters must be understood as atavisms. Latimer (1991) believed this is supported by the morphology of the lower limbs, very evolved towards bipedalism, while primitive traits present in the upper limbs have little evolutionary significance.

3.2.6 The two adaptive strategies of bipedalism

Rejecting Lovejoy's notion of an advantage of australopith bipedalism has often led to the opposite conclusion. The bipedalism of early hominins is considered to be partial, something like a stage prior to development of complete bipedalism—which is believed to begin with *H. erectus*, such as the Trinil specimen. A partial bipedalism—like that described by Tuttle and Basmajian (1974)—is well adjusted to the morphology of australopith hands and feet and would be an adequate way to respond to the environmental demands of a tropical forest in which a considerable amount of traveling, but not all—and not even most of it—was done on the ground. But, is such locomotion a first step towards complete bipedalism?

Susman and Stern (1991) argue that the evolution of bipedalism was gradual and venture how it occurred differently in the gracile and robust australopith species. The earliest exemplars of our genus, *Homo habilis*, exhibit bipedal features in

their feet (OH 8 from Olduvai, Tanzania) and their fingers (OH 7, OH 62), which are functionally apt for climbing. However, Spoor *et al.* (1994) have suggested a new way of studying the evolution of bipedalism: examination of the vestibular apparatus, the inner structure of the ear that is part of the system that controls movement. Using high-resolution computerized-tomography techniques, these authors analyzed the morphology of the bones of the semicircular canals of the ear in 31 current primate species, including our own and several higher apes, as well as different fossil specimens. Spoor and colleagues concluded that *H. erectus* was the first species to exhibit an undoubtable modern human morphology. The dimensions of the australopith and paranthropine semicircular canal are similar to those of current higher apes. According to Spoor and colleagues (1994) *H. erectus* would necessarily be completely bipedal, while *A. africanus* had locomotor habits including optional bipedalism and arboreal climbing. Among australopiths, bipedalism would be a matter of posture and would not allow them to perform more complex movements, such as running or jumping.

The studies by Susman and Stern (1991) and Spoor and colleagues (1994) suggest that posture and balance differ between australopiths and the genus *Homo*. The study by Kramer and Eck (2000) of the energetic balance of bipedalism puts the finger on the central question in the evolution of hominin locomotion. Can the same criterion be applied to calibrate the efficiency of different forms of hominin locomotion? From the point of view of our current locomotion, the way in which australopiths walked can seem inefficient. But the energetic balance of early bipedalism should be seen as an optimization to a different ecological niche. Thus, there would be two different adaptive strategies related with our lineage's bipedalism:

1 slow bipedalism, characteristic of australopiths, with an excellent energetic balance in foraging tasks at low velocity, but inefficient for running at higher speeds;
2 fast bipedalism, apt for running, with high energetic efficiency when great distances have to

---

Box 3.6 Primitive derived traits

George Gaylord Simpson (1953) presented evolution as a directional vector in which certain traits (the primitive ones) are retained and others (the derived ones) characterize a new form of adaptation. The identification of primitive and derived traits is not easy. Depending on the direction of a vector, the same characters can be considered primitive or derived. However, if there is agreement in considering bipedalism as the characteristic hominin vector, then it is clear which traits are primitive and which are derived.

Box 3.7 Doubts about bipedalism

Not all authors are categorical about the kind of bipedalism of early hominins. Carol Ward's analysis of the posture and locomotion of the Hadar specimens (Ward, 2002) raises doubts regarding the extent of their similarities and differences with the genus *Homo*. Ward, consequently, suggested that the characterization of the polarity of primitive and derived traits needs to be improved before arriving at any definitive conclusions about australopith locomotion.

---

be traveled; this is the characteristic locomotion of *H. erectus* and later *Homo* taxa.

Accepting that the bipedalism of australopiths was different from that of modern humans does not imply that it was an incipient stage in human locomotion. This is theoretically robust, because intermediate stages do not make much evolutionary sense. Each taxon has evolved its own distinctive adaptations, which are, in this sense, final, rather than intermediate. A given species does not evolve a partial organ as an intermediate step towards later complete versions of it.

Within such a scheme, slow australopith bipedalism is not a transitory stage towards more developed locomotion processes. Morphological and functional indications suggest that their locomotion was apt for individuals that lived in tropical forests and traveled short distances in their foraging activities. Sarmiento (1998; Sarmiento and Marcus, 2000) has proposed that Hadar australopiths would adopt quadrupedalism when they needed to move fast or travel long distances.

The importance of fast bipedalism in the evolution of the genus *Homo* has been brought to light in Bramble and Lieberman's (2004) study of the role of running. It is evident that current humans are not among the fastest animals in the savanna; nor were our hominin ancestors. However, running is related not only to speed itself. After comparing the metabolic costs of running and walking Bramble and Lieberman (2004) conclude that several anatomical traits of the genus *Homo*—including narrow pelvis, long legs, short neck of the femur and big toe—improved the energetic balance of fast bipedalism, running, because of enhanced features of fast marching: balance, thermoregulation, shock absorption, stress reduction, stabilization of the head and trunk, energy storage, and so on. The most important characteristic of running would be related with energy balance factors and not pure speed. This kind of locomotion would have been efficient for hunting and scavenging in open savannas when long distances had to be covered.

3.2.7 An environment for the first steps

Since Darwin's time, it has been assumed that bipedalism is closely related with the environment to which an organism is adapted. Higher apes, tropical forest inhabitants, exhibit diverse locomotor habits (from brachiation to terrestrial quadrupedalism), which were substituted, along the hominin lineage, for bipedalism as an

adaptation to more open ground. This is the ecological basis of the theory of hominization based
on an early bipedalism. Whether or not the first
hominins used bipedalism is a question that can,
thus, be replaced by whether or not they were
adapted to open savannas.

What kind of environment characterized the
Transvaal (South Africa) and Rift regions at the end
of the Miocene and during the early Pliocene? Some
problems inherent in the study of South African
caves, which we will examine in a later chapter,
make matters difficult. Rayner and colleagues (1993)
suggest Makapansgat (South Africa) during the
early Pliocene was forested. The Rift Valley represents a kind of climatic frontier that confers East
Africa an ecological particularity. Towards the end
of the Miocene the rains were abundant enough to
sustain the tropical forest to the west of the Rift
Valley. But, the east of the valley and the Rift itself
saw a drier climate that caused the expansion of
open lands, savannas with low vegetation, which
replaced jungles and forests (Roberts, 1992). This has
often been taken as an ecological argument in favor
of the very early presence of a developed bipedalism. But the Rift turning into a savanna corresponds
to current conditions, not to the early Pliocene
paleoenvironment.

The rates between the two isotopes ($^{18}O/^{16}O$)
present in calcareous microfossils found on the
bottom of deep tropical seas reveals information
about paleoclimatic changes. When the climate
cools down, the ratio increases and vice versa: a
greater relative amount of ^{18}O means the presence
of a cold period, possibly a glacial period. By
means of this method, Prentice and Denton (1988)
have identified a sudden climatic change 14 Ma,
during the middle Miocene. During this period
there was gradual but pronounced average cooling, although there were several oscillations (glaciations and interglacial periods; Table 3.4).

At the end of the Miocene and beginning of the
Pliocene, between 6 and 4.3 Ma, the oscillations
provided periods of intermediate coolness on the

Table 3.4 Climate alternations since the middle Miocene

Period	Conditions
0.7 Ma onwards	"Ice age"
0.9–0.7 Ma	Great glaciation; new ice level maximum
2.1–0.9 Ma	Fluctuations every 40,000 years
2.4–2.1 Ma	Gradual decrease of the ice, reaching an intermediate level
2.5–2.4 Ma	Great glaciation; ice level maximum
4.3–2.8 Ma	Fluctuations with low ice level
6–4.3 Ma	Fluctuations with medium ice level
14 Ma	Fluctuations with medium ice level

Source: Prentice and Denton (1988).

Earth. Between 4.3 and 2.8 Ma the frozen mass
decreased so that the climate was warmer than
before and after. This time precisely corresponds
to the earliest known hominins (and the Laetoli
footsteps). This conclusion has been confirmed as
follows. The Rift paleoclimate has been studied
with great detail at certain sites that have yielded
very early hominins. The work carried out by
Kingston and colleagues (1994) on the Rift's
paleoclimate, WoldeGabriel and colleagues (1994)
at Aramis in Ethiopia, and Rayner *et al.* (1993) at
Makapansgat in South Africa leaves no room for
doubt that the habitat at all those places corresponded, during the early Pliocene, to tropical
forests and not savannas.

Thus, the basic panorama is fairly complete. The
hominin lineage appeared about 7 Ma in the tropical
forests of the Rift depression, associated with an
essential apomorphy: bipedal locomotion. In time,
that lineage diversified and dispersed, colonizing
the whole planet. In a certain sense, even the distinctive hominin synapomorphy, bipedalism,
changed its function to adapt to running in open
savannas. The different clades of the lineage gradually developed adaptive specializations, and one
of them, the genus *Homo*, managed to live until the
present. In the following chapters we will examine
in detail the steps of that evolutionary trail.

CHAPTER 4

Miocene and Pliocene genera and species

4.1 The first hominins

4.1.1 Hominini genera and species

The identification of genera and species within the evolutionary lineage leading to *Homo sapiens* is crucial for the understanding of human evolution. How can we determine whether known fossil specimens are correctly classified within a certain genus and species?

The international code of taxonomy does not provide objective criteria regarding the classification of genera. Moreover, the biological definition of species—a group that is reproductively isolated from other close groups—is not applicable to the fossil record, because it is impossible to know whether two specimens of extinct groups might have been able to produce fertile offspring (section 1.3). This state of affairs requires a pragmatic approach: based on reasonable grounds, fossil specimens are included in certain genera and species hoping that the scientific community will agree with the decision. This is also the procedure followed in previous chapters with taxa that include the earliest members of the tribe Hominini, the Miocene and Pliocene genera and species.

The taxonomic freedom granted by the international code may lead to conflicting and sometimes seemingly arbitrary situations. There is a great number of taxa included in Hominini in the specialized literature. Neanderthals, for instance, have been classified as a species on their own, *Homo neanderthalensis*, as a subspecies of *H. sapiens*, and in at least four more ways. Given that such dispersion is, obviously, undesirable, it becomes appropriate to

search for reasonable criteria to guide our decision. A solution, suggested in section 1.3, is to follow Ernst Mayr's idea of a genus that includes a set of closely related organisms that are adapted to particular ways of life and environmental conditions. According to this criterion, a new genus should only be proposed for a truly new type of hominin: one that does something different, irreducible to other known genera. This leads to the definition of only five unique kinds of hominin, corresponding to the following five genera.

1 The first organisms that diverged from their apelike relatives adopting a bipedal posture when moving on the ground. They retained the better part of primitive traits in the masticatory system: genus *Orrorin*.
2 The hominins that took advantage of bipedalism to gradually occupy open savannahs as the African climate got colder and forests receded, adapting their dentition to the new conditions: *Australopithecus*.
3 Bipedal organisms, therefore hominins, whose adaptive patterns and dental enamel thickness are similar to those of African apes: *Ardipithecus*.
4 Hominins that developed large masticatory apparatus around 3.5–2.5 Ma, with a diet specialized on hard savannah vegetation: *Paranthropus*.
5 Hominins that retained relatively gracile masticatory apparatus around 3.5–2.5 Ma. This genus also includes the descendants within the lineage that, in time, developed large brains and constructed tools: *Homo*.

Other recent taxonomical proposals include different genera, such as *Kenyanthropus* and

Sahelanthropus, whose distinctive adaptive traits are not clear (Table 4.1).

Could the number of hominin genera be reduced further? It might be argued that *Orrorin* specimens are reducible to *Australopithecus*, as we have suggested elsewhere (Cela-Conde and Ayala, 2003). *Paranthropus* could also be included in that genus, as many scholars have recommended. But the five aforementioned genera fit the results of different cladistic episodes well. The first of these divergences separated hominins from chimpanzees. The second separated ardipithecines from australopithecines. The third one separated *Paranthropus* from *Homo*. The corresponding nodes are usually placed around 7, 4.5, and 3.5–2.5 Ma (Figure 4.1).

Table 4.1 Genera and species belonging to the tribe Hominini suggested by different authors

Genus	Species	Age (million years)
Sahelanthropus	*Sahelanthropus tchadensis*	7
Orrorin	*Orrorin tugenensis*	6
Ardipithecus	*Ardipithecus kadabba*	5.8
	Ardipithecus ramidus	4.4
Australopithecus	*Australopithecus anamensis*	4
	Australopithecus africanus	3.5
	Australopithecus bahrelgazhali	3.5?
	Australopithecus garhi	2.5
Paranthropus	*Paranthropus africanus*	3.5?
	Paranthropus aethiopicus	2.5
	Paranthropus robustus	2.0
	Paranthropus boisei	1.7
Kenyanthropus	*Kenyanthropus platyops*	3.5
	Kenyanthropus rudolfensis	2.5
Homo	*Homo habilis*	2.5
	Homo ergaster	1.8
	Homo georgicus	1.8
	Homo erectus	1.6?
	Homo floresiensis	?
	Homo antecessor	0.8
	Homo heidelbergensis	0.4
	Homo neanderthalensis	0.3
	Homo sapiens	0.2

The table does not include all species that have been named.

The different genera of the tribe Hominin include many species, but there is no consensus regarding their number and appropriateness. Table 4.1 is the result of adopting a conservative criterion. To include all the species that have ever been named is undesirable and endless. But a very restrictive criterion would reduce the list to very few species that would include seemingly quite diverse hominins. This would happen if we were to follow a strict cladistic methodology, so that only species originated through cladistic episodes were considered valid. Chronospecies (different species belonging to a single phyletic lineage, without ramifications) would be eliminated. Such a criterion would eliminate some of the *Homo* species listed in Table 4.1, such as *Homo heidelbergensis* or *Homo antecessor*, at least on the basis of current knowledge. We will discuss the pertinence of maintaining the taxa included in the table later in this book.

4.1.2 The last common ancestor of apes and humans

The appearance of the tribe Hominin took place, as we saw in the previous chapter, at the end of the Miocene. According to what we mentioned there, and constrained by the limited knowledge of late-Miocene hominoids, the divergence episode could be summarized as follows:

- around 10–9 Ma there was a genus, *Ouranopithecus*, documented in Macedonian sites, which was

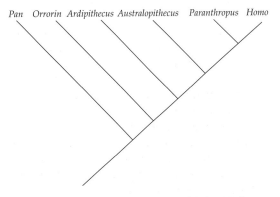

Figure 4.1 Cladogram showing the genera of the hominin lineage.

possibly the common ancestor of all African great apes and humans (Andrews 1996);

• about 8–7 Ma, or maybe slightly less, the evolutionary branches that would eventually lead to gorillas, chimpanzees, and modern humans split;

• chimpanzees are the sister group of hominins.

Scientific studies of apes and humans have not agreed unanimously on the estimate of 7 Ma for this last cladistic event. Indirect evidence allows for a very wide range of discrepancies, from close to 4 Ma—suggested by some molecular geneticists, to close to 12 Ma—implied by the consideration of "*Ramapithecus*" as a hominin. At present there is direct proof—fossils—of the existence of bipedal hominoids—that is to say, hominins—close to 7 Ma. This seems a valid estimate for the time elapsed since the divergence of apes and humans.

We lack good African fossils located immediately before the separation of the chimpanzee and human lineages. One possibility is a right mandibular fragment with three molars, of which only one retains the crown (KNM-LT 329), found at Lothagam Hill in Kenya (Patterson *et al.*, 1970), but its interpretation is difficult. The K/Ar method places the fauna at the Lothagam I bed between two basalt intrusions 8.5 ± 0.2 and 3.8 ± 0.4 million years old, respectively (Brown *et al.*, 1985a). Biostratigraphic analyses have revealed that the mandible is older than 5.6 Ma (Hill *et al.*, 1992). Given its humanlike traits, it seems that it belonged to a very early hominin, rather than to an ape ancestor.

The KNM-SH 8531 maxilla found at Samburu Hills, Kenya (Ishida *et al.*, 1984), and thought to be 9 million years old, shares certain traits with current gorillas (large size, prognathism, shape of the nasal aperture), but not others (thick enamel, low and rounded cusps; Figure 4.2). Some traits suggest a resemblance with gorillas, others with chimpanzees, others with humans, while the rest are primitive traits. It was classified by Ishida *et al.* (1984) as *Samburupithecus kiptalami*.

Although the Samburu Hills specimen shows no derived traits from pongids (the family including orangutans and their direct ancestors), some features are distinctive of the gorilla + chimpanzee + human set. But, the specimen lacks specific derived traits of each of these three branches. Thus, Groves (1989) concluded that this individual lived before the separation of gorillas, chimpanzees, and hominins. In that case, the phylogenetic role of the Samburu Hills specimen would be similar to the one of *Ouranopithecus*, as a common ancestor of African great apes and hominins.

A scene of emptiness in the fossil record prior to the appearance of our own tribe is the starting point for the presentation of the earliest hominin specimens. They are African and have all been found in East Africa (the Rift Valley) and South Africa, with one important exception, *Sahelanthropus tchadensis*, from central Africa (see below). The temporal order in which the discoveries were made does not coincide with the age of the fossil finds. In fact, because research has been directed towards progressively older terrains, the situation has turned out to be rather the opposite. Thus, the findings may be presented in a historical sequence, starting by those which were found first, or following the real chronological order. We have opted for this second possibility, and thus, the

---

Box 4.1 The Lothagam specimen

The Lothagam specimen was initially classified by Patterson and colleagues (1970) as *Australopithecus* cf. *africanus*, but Eckhardt (1977) saw in it a possible pongid. Studies by Kramer (1986) and Hill *et al.* (1992) reveal traits resembling *Australopithecus afarensis*, such as molar width. McHenry and Corruccini (1980) agree that it shows hominin-derived traits, but due to the lack of some notable *A. afarensis* traits, they recommend its classification as Hominini indet. White (1986) classified the specimen within *A. afarensis*. Wood and Richmond (2000) have pointed out its affinities both with *A. afarensis* and *Australopithecus ramidus*.

Box 4.2 The void between Samburu Hills and current great apes

There is a notable void between the Samburu Hills remains and current African great apes. We know of none of their remote or close ancestors, with the exception of the Tugen Hills exemplar, classified as a member of the chimpanzee lineage, which we will discuss later (McBrearty and Jablonski, 2005). Conversely, there are abundant fossil hominin specimens, and as years go by, their number increases. There are some authors, such as Leonard

Greenfield (1983) and Russell Ciochon (1983) who, agreeing with Darwin, maintain that we will never identify the fossils corresponding to organisms that lived immediately before the separation of African apes and hominins. To put it another way, we would not recognize them if we had them in front of us. Their traits would not have differentiated enough to allow their identification as members of the chimpanzee, gorilla, or human lineages.

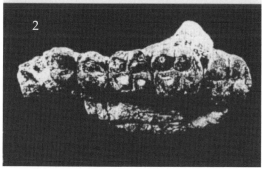

Figure 4.2 The KNM-SH 8531 maxilla, *Samburupithecus*. Photograph from Ishida and Pickford (1997).

western region of Kenya will be the first place we will visit in search of our earliest ancestors.

4.1.3 The time and place of the appearance of hominins

There is little doubt that the tribe Hominin appeared in Africa. According to currently available

evidence, East Africa's Rift Valley is, in all probability, the place where the first hominins evolved. Moving thousands of kilometers south, South Africa is another key area in the interpretation of the beginnings of hominin evolution. South African sites are not as old as East African ones, but because they were excavated before, some of the specimens shaped the early ideas about our tribe's evolution. It is helpful to compare South African and Rift specimens. Chad, in central Africa, is the third place where early hominins appear. However, only two taxa come from there: an australopithecine (Brunet *et al.*, 1995) and the specimens which have been named *S. tchadensis* (Brunet *et al.*, 2002; Figure 4.3).

The Rift Valley is a long and thin fracture depression, with numerous volcanoes, which extends from the south of Turkey to East Africa and Mozambique. It formed as a consequence of the movement of the African and Arabian tectonic plates. The East African Rift is a discontinuous succession of valleys that run 3,000 km southwards from the Afar region (Ethiopia) to southern Malawi. The crucial sites in the early history of hominization are found in Ethiopia, Kenya, and Tanzania: Olduvai, Omo, Hadar, Laetoli, Koobi Fora, West Turkana, Tugen, Aramis, and so on. These names will come up again as we go over the main specimens found in those sites.

The great volcanic activity that took place in the Rift during the Pliocene is a valuable asset for paleoanthropologists, because the numerous ashes and tuffs can be analyzed by means of the K/Ar method to determine their age. Few interesting

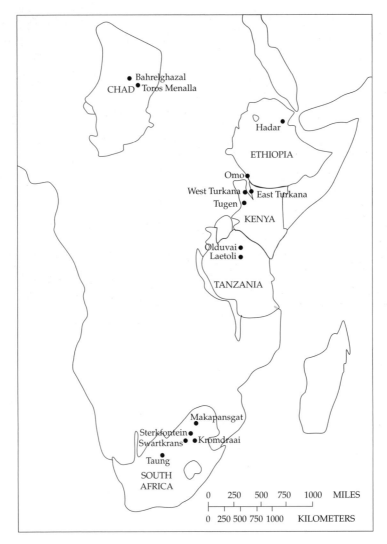

Figure 4.3 The main African areas where early hominins have appeared.

places for human paleontology have been dated with such precision as the sites located in this extensive fault zone.

4.1.4 Tugen Hills: *Orrorin tugenensis*

The Rift's earliest relevant exemplars of our tribe's genera and species come from the Lukeino formation of Tugen Hills (Figure 4.4), in the Lake Baringo district (Kenya). During the months of

October and November of 2000, the Kenya Paleontology Expedition (KPE), organized by the Collège de France (Paris) and the Community Museums of Kenya (Nairobi), found up to 12 hominin mandibular, dental, and postcranial fragments at four sites belonging to the Lukeino formation (Cheboit, Kapsomin, Kapcheberek, and Aragai). An inferior molar (KNM LU 335), found by Martin Pickford at the Cheboit site in 1974, was included together with those new fragments.

Figure 4.4 Upper strata of the Lukeino formation at Tugen Hills (Kenya). Photograph by C.J. Cela-Conde.

Box 4.3 Chronostratigraphy of the Tugen Hills

The chronostratigraphy of the Tugen Hills in which the Lukeino mandible was found has been confirmed by Deino and colleagues (2002). The techniques used were ^{39}Ar/^{40}Ar single-crystal laser-fusion dating, K/Ar dating, and paleomagnetic reversal stratigraphy. The Lukeino, Kaparaina Basalt, and Chemeron formations constitute a sequence with ages ranging from 6.56 to 3.8 Ma. The upper Lukeino Formation at Kapcheberek is constrained to the interval 5.88–5.72 Ma. The combined ^{39}Ar/^{40}Ar and paleomagnetic data constrain the age of the Chemeron formation at Tabarin to 4.63–3.837 Ma see Box 5.15.

The Lukeino formation's geological conditions have been described by Martin Pickford (1975), and Hill and colleagues (Hill *et al.*, 1985; Hill, 1999, 2002). Its inferior limits are marked by the Kabarnet Trachyte formation, with an age of between 6.7 ± 0.3 and 7.2 ± 0.3 Ma calculated with the K/Ar method (Pickford, 1975). Hill (2002) attributed to the Lukeino formation an age of between 6.2 and 5.6 Ma. According to the discoverers of *Orrorin*, it is 6 Ma. This age places the KPE's findings close to the appearance of the tribe Hominin, the earliest known hominins, with the exception of *Sahelanthropus*, which we will review later.

Brigitte Senut and Martin Pickford referred to the Lukeino specimens as Millennium Man in the announcement of the discovery. There was a dispute regarding which research group possessed the authorized excavation permits to carry out research at Baringo (Butler, 2001). But, be that as it may, Millenium Man was a very important discovery, because of the specimen's age and because of the presence of dental and postcranial remains that allow definition of our tribe's primitive traits. Senut and colleagues (2001) noted that its thick dental enamel, its small dentition relative to body size, and the shape of the femur indicate that they are hominins. These hominins are different from *Ardipithecus* (with thin enamel) and from *Australopithecus*, with larger dentition and femora less *Homo*-like than *Orrorin*'s (see Box 4.4). Senut *et al.* (2001) named the new genus and species *Orrorin tugenensis*. The BAR 1000 00 fragmentary mandible in two pieces constitutes the holotype and the remaining specimens, including the KNM LU 335 molar, found in 1974, are paratypes (Figure 4.5). Orrorin means "original man" in the Tugen language, while the species name honors the toponym of the hills in which the fossils were found.

Orrorin shows a mixture of primitive and derived traits. Apelike traits can be seen in canines, incisors and premolars. The anatomy of the humerus and phalanx are similar to those of climbing primates. Derived traits include those pertaining to the femur, which indicate that it is a bipedal organism, and its molars, which are relatively small and have thick enamel.

Haile-Selassie (2001) has questioned the hominin condition of *Orrorin* on the basis of the observation of a primitive trait: the low crowns on the upper canines. Regarding locomotion, Halie-Selassie suggested that "[it] remains uncertain at this time

---

Box 4.4 *Orrorin*

The issue of the shape and thickness of the *Orrorin* femur, and how they relate to the specimen's hip motion, remains controversial. Based on tomography scans of the neck-shaft junction of BAR 1002 00, Galik *et al.* (2004) conclude that "the cortex is markedly thinner superiorly than inferiorly, differing from the approximately equal cortical thicknesses observed in extant African apes." Accordingly, Brigitte Senut suggested, at a symposium on Prehistoric Climates, Cultures, and Societies (Paris, France, September 13–16, 2004), that "*Orrorin*'s gait was more humanlike than that of the 2- to 4-million-year-old australopithecines" (see Gibbons, 2004).

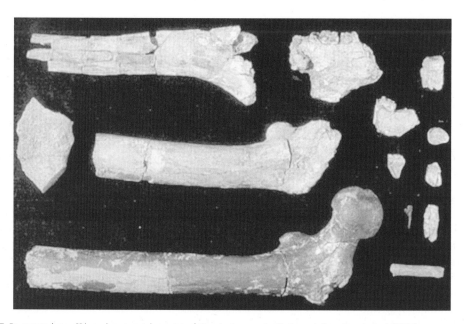

Figure 4.5 Fragmented mandible and postcranial remains of *Orrorin tugenensis*. Photograph from Senut *et al.* (2001).

because its description lacked comment on characters directly diagnostic of bipedality, such as the presence of an obturator externus groove or an asymmetrical distribution of cortex in the femoral neck." Consequently, "there is nothing to preclude *Orrorin* from representing the last common ancestor, and thereby antedating the cladogenesis of hominins…or an exclusive precursor of chimpanzees, gorillas or humans." However, the question of whether *Orrorin* was an ancestor of chimpanzees, gorillas, and humans, or, alternatively, a member of Hominini, cannot be approached based on primitive traits, such as those related with canines. This issue requires the consideration of derived traits. In our view, the femur's proximal end is indication enough to regard *Orrorin tugenensis* as bipedal and, consequently, a hominin. Haile-Selassie and colleagues (2004) later accepted this interpretation, as we will see when we review *Ardipithecus ramidus*.

An interesting aspect brought to light by the KPE's research team is the size of *Orrorin*'s femur and humerus. It is 1.5 times larger than that of A.L. 288–1, an *Australopithecus afarensis* found at Hadar, which we will review below. According to Senut *et al.* (2001) this fact contradicts the widespread idea that our first ancestors were small in size. Moreover, A.L. 288–1 corresponds, as we will see, to a female, which raises the important question of sexual dimorphism.

4.1.5 Ardipithecines: hominins or apes?

With the exception of *Sahelanthropus, Orrorin* is the oldest known hominin. But that does not mean that it represents the first member of a direct lineage that leads to us. In its beginnings, and until very recently,

human evolution produced a wide range of different adaptive options, each of them with their own particular features. A very different organism, almost coinciding in time with *Orrorin*, also left a trace of its existence in the Rift Valley. In 1994, Tim White, Gen Suwa, and Berhane Asfaw published the results of the research campaigns of the two previous years at the Aramis site, in the Middle Awash region of Ethiopia (Figure 4.6). The findings included 17 possible hominin specimens from sites 1, 6, and 7, which at the time were the oldest documented hominin remains (White *et al.*, 1994). Sixteen of them, and a great number of fossils of other vertebrates (more than 600), were found in strata comprised between two markers, the complex of vitreous tuffs known as the Gàala Tuff Complex, GATC (*Gàala* means "dromedary" in the Afar language), and the basalt tuff Daam Aatu Basaltic Tuff, DABT (*Daam Aatu* means "monkey" in the same language), with an average 4 m of sediments between them (WoldeGabriel *et al.*, 1994).

The hominin specimens of Aramis include, among other remains, three fragmentary bones of a left arm (ARA-VP-7/2), found half a meter above DABT, associated dentition (ARA-VP-1/128; the holotype ARA-VP-6/1), a deciduous molar (ARA-VP-1/129; Figure 4.7), a complete right humerus (ARA-VP-1/4), and cranial fragments (ARA-VP-1/125 and -1/500; White *et al.*, 1994). White and colleagues initially classified all those specimens in the same genus as *A. afarensis*, but in a different species, suggesting *Australopithecus ramidus* for the Aramis exemplars. However, 1 year later they decided to elevate the differences to the rank of genus, adding the taxon *Ardipithecus ramidus* (*ardi* means "ground" or "floor" in the Afar language) for the Aramis specimens (White, *et al.*, 1995).

Box 4.5 Locomotion in *Orrorin*

Senut and colleagues (2001) emphasize that the appearance of the locomotor apparatus of *Orrorin* is more modern than that of australopithecines. The justification of the new genus *Orrorin* rests on locomotion and not just on dentition. However, Aiello and Collard (2001) question

such an early divergence between two different kinds of locomotion. As we saw in the previous chapter, the bipedal locomotion of the first hominins is a particular form of adaptation that contrasts with current bipedalism, acquired by the genus *Homo*.

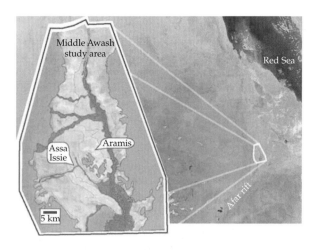

Figure 4.6 Location of the Aramis and Asa Issie sites in Ethiopia. Reprinted by permission from Macmillan Publishers Ltd. *Nature*, 440: 7086, 883–889, 2006.

Box 4.6 The age of Aramis specimens

By means of the ^{39}Ar/^{40}Ar laser fusion method, the GATC tuff was estimated to be 4.387 ± 0.031 Ma, setting the maximum age for the hominin remains at the site (WoldeGabriel *et al.*, 1994). The DABT tuff could not be dated due to the high contamination of the Miocene soils. However, WoldeGabriel and colleagues dated the strata containing hominins between 4.48 and 4.29 Ma by biochronology and paleomagnetism. The rounded estimate usually attributed to the Aramis specimens is 4.4 Ma.

Figure 4.7 ARA-VP-17129, deciduous molar. Found by A. Asfaw in 1992 in Aramis (Ethiopia), *Ardipithecus ramidus*. Photograph from White *et al.* (1994).

Aramis hominins share a wide range of characters with *A. afarensis*, also found in Ethiopia. But they are different in some features, mainly those relative to dentition. The first deciduous teeth of ARA-VP-1/129 is closer to chimpanzees than to any hominins. Regarding adult dentition, some traits (the area of the ARA-VP-6/1 canine crown, for instance) are also similar to chimpanzees, but others are not. Incisors do not have the great width of current chimpanzees (the relationship between the incisors, molars, and premolars of Aramis specimens is typical of Miocene hominins and gorillas) and the morphology of canines is different from apes. The position of the foramen magnum, indicative of posture and, thus, of the possibility of bipedal locomotion, is close to that observed in the rest of hominins and distant from chimpanzees (White *et al.*, 1994). Senut and colleagues (2001) expressed an opposite view, arguing that the bipedalism of *Ardipithecus* cannot be demonstrated on the basis of the described specimens.

Box 4.7 *Ardipithecus*

The proposal of *Ardipithecus* as a genus to fit the Aramis specimens was done almost telegraphically. The new *Ardipithecus* characteristic traits were defined on the basis of a mandible discovered towards the end of 1994. Michael Day (1995) protested the addition of a new

genus in the tribe Hominin without offering a full argument. He also added, somewhat ironically, that the hurry to name the new genus before someone else did, seemed directed to achieve priority rather than scientific clarity.

Box 4.8 Enamel thickness in *Ardipithecus*

The most controversial trait of *Ardipithecus* is the thinness of its enamel. This trait is often used in the discussion regarding hominoid phylogenesis and classification as a criterion to identify hominins. "*Ramapithecus*" was previously considered as a hominin precisely because of its thick molar enamel (Pilbeam, 1978). The appearance of such a very early being as *Ardipithecus ramidus* and its thin molar enamel raised doubts. Other hominins have thick enamel while the enamel of chimpanzees and gorillas is thin, and that of orangutans is of an intermediate thickness. Thus, *Ardipithecus* led to a new edition of the discussions concerning the value of enamel thickness for the determination of lineages. Peter

Andrews, for instance, noted that thick dental enamel seems to be shared by hominins and by other 10-million-year-old fossil apes, which suggests that the chimpanzee thin enamel is a derived trait (cited by Fischman, 1994).

White and colleagues (1994) and Fischman (1994) have warned of the need for accounting for global features of teeth shape and other masticatory aspects when classifying specimens. The specific trait of enamel would, then, lose much of its significance. However, as Ramirez Rozzi (1998) has shown by means of the study of its microstructure, enamel retains relevance when specimens found in the same or close sites are compared.

Henry Gee (1995) advanced a pair of prophecies regarding the *ramidus* specimens. First, by the year 2000, *A. ramidus*, as they were known at the time, would have been placed into another genus. It was not necessary to wait so long: a few months later White and colleagues (1995) introduced the taxon *Ardipithecus*. Indeed, ardipithecines are such special hominins that it is reasonable to classify them in a separate group. Andrews described them as "ecological apes" (Andrews, 1995), meaning that their enamel must be related with an adaptation to the tropical forest, as is the case with chimpanzees and gorillas. But, were they bipedal? Should the genus *Ardipithecus* be included among other hominins or in the chimpanzee or gorilla lineages?

Gee's (1995) second prophecy forecast that, again around the year 2000, *ramidus* would be considered a member of the "ramidopithecines", the common ancestors of chimpanzees and humans. Senut and colleagues (2001) considered

them in such a way, but, in 2001 Haile-Selassie reported the finding of 11 *Ardipithecus* specimens whose interpretation is contrary to such a hypothesis. The new specimens are dated between 5.2 and 5.8 Ma and come from five localities of the Ethiopian part of Middle Awash (Saitune Dora, Alíala, Asa Koma, and Digiba Dora on the western margin of the Middle Awash, and Amba East from the Kuserale Member of the Sagantole formation of the Central Awash Complex; Haile-Selassie, 2001).

The Middle Awash specimens include postcranial fragments—like a manual phalanx (ALA-VP-2/11), a pedal phalanx (AME-VP-1/71), and arm bones—that lend support to the notion that *Ardipithecus* was bipedal (Figure 4.8). In particular, the dorsal orientation of the AME-VP-1/71 phalanx indicates, according to Haile-Selassie, a similar pedal morphology to that of *A. afarensis*—in mosaic, with traits shared with apes—and indicative of a similar locomotion to that of *A. afarensis* and *Ar. ramidus*. Dental

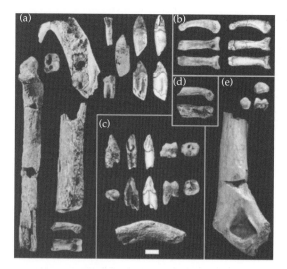

Figure 4.8 *Ardipithecus ramidus kadabba* from the Late Miocene Middle Awash deposits (Haile-Selassie, 2001). (a) ALAVP-2/10, mandible and all associated teeth; ALA-VP-2/120, ulna and humerus shaft; ALA-VP-2/11, hand phalanx. (b) AME-VP-1/71, lateral, plantar and dorsal views of foot phalanx. (c) STD-VP-2, teeth and partial clavicle. (d) DID-VP-1/80, hand phalanx. (e) ASKVP-3/160, occlusal, mesial and buccal views; ASK-VP-3/78, posterior view. All images are at the same scale. Scale bar, 1 cm. Illustration from Haile-Selassie (2001). Reprinted by permission from Macmillan Publishers Ltd. *Nature*, 412: 6843, 178–181, 2001.

remains, composed by a right mandible with associated teeth (ALAVP-2/10) and other dental material, show a mixture of primitive traits, shared with apes, and hominin features—lower canines with developed distal tubercles and expressed mesial marginal ridges. The primitive traits that separate them from the Aramis *Ar. ramidus* led Haile-Selassie (2001) to attribute the Middle Awash exemplars to a new subspecies: *Ardipithecus ramidus kadabba*.

Regarding the phylogenetic position of *Ar. ramidus*, Haile-Selassie defined the taxon as a hominin close to the divergence from chimpanzees. Because the ages attributed to Middle Awash *Ardipithecus* and *O. tugenensis* places both of them very close to the cladogenesis that originated the tribe Hominin, it is striking that they are so different. However, it must be recalled that Haile-Selassie (2001) believes that it has not been demonstrated that *Orrorin* is a hominin. Alternatively, the discoverers of *Orrorin* believe *Ardipithecus* is an ape.

The question of the relative positions of *O. tugenensis* and *Ar. ramidus* changed notably after the finding in 2002 of six new *Ardipithecus* teeth in the Asa Koma locality, with an age of between 5.6 and 5.8 Ma (Haile-Selassie *et al.*, 2004). In the article that christened the taxon *Ar. ramidus kadabba*, Haile-Selassie (2001) pointed out the "possible absence of a fully functional honing canine/premolar complex in *Ardipithecus*." The Asa Koma sample also included a lower canine. After comparing available canines with those of chimpanzees and australopithecines (Figure 4.9), Haile-Selassie and colleagues (2004) stated: "the projecting, interlocking upper and lower canines, and the asymmetric lower P3 with buccal wear facet imply that its last common ancestor with chimpanzees and bonobos retained a functioning C/P3 complex. But wear on the upper and lower canines of *Sahelanthropus* and the lower canine of *A. kadabba* from Alayla suggest a lack of consistently expressed functional honing in these earliest hominins." Haile-Selassie *et al* (2004) suggested that the scarce but meaningful dental derived traits confirm the hominin condition of the Alayla and Asa Koma samples. However, the new species *Ardipithecus kadabba* was defined based on primitive characters.

After the Asa Koma discoveries, Haile-Selassie and colleagues (2004) believed the early hominin phylogenetic sequence was as follows:

• *Sahelanthropus*—a taxon we will discuss further on—*Orrorin*, and *Ar. kadabba* provide important outgroup comparisons to younger *Ar. ramidus* and *Australopithecus anamensis*;
• metric and morphological variation within the available small samples of late-Miocene teeth attributed to *Ar. kadabba*, *O. tugenensis*, and *S. tchadensis* is no greater in degree than that seen within extant ape genera;
• the interpretation that these taxa represent three separate genera or even lineages can be questioned;
• it is possible that all these remains represent specific or subspecific variation within a single genus.

This is the same systematic interpretation that we have followed. But following taxonomic rules, the

Box 4.9 Variability of enamel thickness

Haile-Selassie (2001) thought that the controversy regarding the thin enamel of *Ar. ramidus* was not very important, given the great variability of the trait, even within a single species. Nevertheless, he stated that "studies of enamel thickness are underway, but the available broken and littleworn teeth suggest that molar enamel thicknesses . . . were comparable to, or slightly greater than, those of the younger Aramis samples of *A. ramidus*."

Pan troglodytes   Ar. Kadabba

Figure 4.9 Lateral views of a female common chimpanzee (left) and *Ar. kadabba* (right) upper and lower canines and premolars (upper canine ASK-VP-3/400, lower canine STD-VP-2/61, upper premolar ASK-VP-3/160 reversed, lower premolar ASK-VP-3/403 reversed). Photographs from Haile-Selassie *et al.* (2004). Reprinted with permission from AAAS.

Table 4.2 Reorganization of very early hominin genera according to the systematic criterion used to interpret the specimens suggested by Haile-Selassie *et al.* (2004). The designation of *Orrorin* for the species included in this genus is a proposal by the authors of this book, not by Haile-Selassie *et al.*

Genus	Species
Orrorin	*Orrorin tugenensis*
	Orrorin tchadensis
	Orrorin kadabba
Ardipithecus	*Ardipithecus ramidus*
Australopithecus	*Australopithecus anamensis*
	Australopithecus afarensis
	Australopithecus bahrelgazhali
	Australopithecus garhi

single genus of early hominins should be the one suggested first, that is to say, *Orrorin*. It is reasonable to include in a single genus all hominins close to the cladistic separation regarding chimpanzees. The specialization of *Ar. ramidus* as an "ecological ape", with its thin dental enamel, should be considered a derived trait and not a plesiomorphy inherited from ancestors they shared with chimpanzees. From a taxonomical point of view, this would require separating *Ar. kadabba* and *Ar. ramidus* in two different genera, placing the first within *Orrorin* (Table 4.2).

New specimens, corresponding to nine hominins, were discovered in the As Duma site (Gona, Ethiopia) in 2004. They were found in soils dated between 4.5 and 4.3 Ma—estimation obtained by means of paleomagnetism and ^{39}Ar/^{40}Ar. The exemplars include a partial right mandible (GWM3/P1), a left mandibular fragment (GWM5sw/P56), and other dental and postcranial

fragments (Figure 4.10). The dentition allows these specimens to be included in *Ar. ramidus*, according to Sileshi Semaw *et al.* (2005). The authors infer its bipedal character from the dorsal orientation in the transversely broad oval proximal facet of GWM-10/P1, a quite complete manual left proximal phalanx.

The As Duma specimens support the view that *Ardipithecus* are a group of peculiar hominins that combine bipedalism with ape masticatory traits. Unfortunately, the site's Pliocene climatology corresponds to a mosaic of environments, which prevents considering those ardipithecines as organisms adapted to a precise habitat.

4.2 *Australopithecus*

4.2.1 What can *Australopithecus* include?

We have argued that *Australopithecus* is a genus corresponding to a particular type of hominin that developed certain locomotion patterns and had a

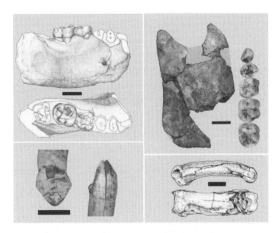

Figure 4.10 Early-Pliocene hominid fossils from As Duma, Gona Western Margin. Top left, GWM3/P1, lateral and occlusal view of right mandibular corpus. Top right, GWM5sw/P56, mesial view of left mandibular ramus and occlusal views of right P3–M3. Gray shaded area of ramus is an area of reconstruction. Bottom left, GWM9n/P51, labial view of maxillary left canine, and GWM9n/P50, labial view of mandibular right canine. Bottom right, GWM10/P1, lateral and palmar view of manual proximal phalanx. Scale bar, 1 cm. Drawings by L. Gudz, from Semaw et al. (2005). Reprinted by permission from Macmillan Publishers Ltd. *Nature*, 433: 7023, 301–305, 2005.

derived dentition when compared to chimpanzees and earlier hominins. Several taxa fall within such a characterization. Strait *et al.* (1997) have pointed out that the taxon *Australopithecus*, conventionally defined as the set of all hominins prior to *Homo*— except *Ardipithecus*—that were known at the time, constitutes a paraphyletic group. That is, *Australopithecus* was at that time a hodgepodge in which to place all early hominins that did not fit elsewhere. Strait and colleagues (1997) took into account the following taxa in their study: *A. afarensis*, *A. africanus*, *A. aethiopicus*, *A. robustus*, and *A. boisei*. Table 4.1 includes other species that were not known at the time, such as *A. anamensis* and *Australopithecus garhi*, but the issues raised by Strait *et al.* do not change by this inclusion. The consequence of defining the genus *Australopithecus* in such a broad way is that specimens that cannot be grouped in a single evolutionary lineage are ultimately included together.

One way to resolve the problem, so that genera can be defined that constitute true lineages, is to separate "robust" australopithecines from the other taxa.

The distinction between robust and gracile hominins was set up as a consequence of the discovery of fossils that exhibited very different features, but were found in close South African sites during the first half of the twentieth century. It is necessary to introduce a historical note here. The cranial traits of the Taung specimen, used by Raymond Dart to define, in 1925, the genus *Australopithecus* (Dart, 1925) and those found later at Sterkfontein and Makapansgat did not appear massive, lacking a sagittal crest. Robert Broom (1938) later discovered much more robust specimens at Kromdraai, similar to those found later at Swartkrans. Although Broom suggested the genus and species *Paranthropus robustus* for them, many authors just distinguished between gracile *A. africanus*, and robust *Australopithecus robustus*. However, later findings at Olduvai (Tanzania) required the revision of this relative concept of robusticity and gracility. Early *Homo* from Olduvai were more gracile than *A. africanus*. Moreover, gracile *Australopithecus* are much older than the robust specimens and than Olduvai *Homo habilis*. These two last types of hominin are approximately the same age. So, the distinction between gracile and robust forms should be used to refer to two lineages, one specialized in the intake of hard vegetables, and the other a carnivorous lineage.

If we group robust australopithecines in *Paranthropus*, a separate genus of their own, as Broom suggested, the sets are more coherent with the evolutionary lineages. However, this does not solve all the difficulties. In the cladistic analysis performed by Strait and colleagues, *A. africanus* could be considered both as the sister clade of *Paranthropus* + *Homo* or the sister clade of *Paranthropus*. There is a third option, which stems from the evolutionary interpretation of early hominin taxa. This is to consider that *A. africanus* is, in actual fact, an ancestor of the later "robust" hominins, but had not yet developed the adaptive features of robusticity (Cela-Conde and Altaba, 2002). This leads to the reduction of the genus *Australopithecus* to the taxa listed in Table 4.1: *A. anamensis*, *A. afarensis*, *A. bahrelgazhali* and *A. garhi*, ordered by decreasing age. They will be the starting point of our analysis of the genus. We will deal with *A. africanus* and other South African forms later.

Box 4.10 Use of *Praeanthropus africanus*

In this book we follow the usual designation of the genus *Australopithecus*, which includes *A. afarensis* and other related taxa found in East Africa and Chad. However, authors like Strait, Grine and Moniz (1997), Wood and Collard (1999b), and Cela-Conde and Ayala (2003) prefer to rescue the original designation of *Praeanthropus africanus* that Weinert (1950) gave to certain exemplars later included in *A. afarensis*. To use the genus *Praeanthropus* to refer to australopithecines before the division of hominids in gracile and robust has three advantages. First, it complies with the rules of the International Code of Zoological Nomenclature, which requires using the first name suggested for any taxon, except when there are well-founded reasons not to do so.

Second, it allows reserving the genus *Australopithecus* for the robust australopithecines, which Raymond Dart gave to the first discovery of *Australopithecus africanus*. Finally, if the taxon *Australopithecus* is reserved for *A. africanus* it is possible to give the genus *Paranthropus* up and place all its members in *Australopithecus*. *A. africanus* would be, as we have suggested, the species that leads to the forms which, after the climate change that occurred 2.5 Ma, evolved into robust australopithecines. Despite so many reasons in favor of such taxonomy, the extended use of the names *Australopithecus afarensis*, *A. bahrelghazali*, *A. anamensis*, and *A. garhi* has deterred us from following the change suggested by Strait *et al.* (1997), while defending its appropriateness.

4.2.2 Hadar: *Australopithecus afarensis*

In a map of hominin sites in the Rift Valley, Laetoli (Tanzania) and Hadar (Ethiopia) represent almost the southern and northern extremes. Both sites have been linked to certain Pliocene hominins that are among the best known from a morphological perspective. The Hadar formation is in the Afar triangle (Figure 4.11), an extensive desert area around the River Awash, some 300 km northeast of Adis Abeba, the capital of Ethiopia. The sites are located in an area more than 60 km². That is where the International Afar Research Expedition, organized by Maurice Taieb, carried out intense paleontological research between 1972 and 1977. This research found many Pliocene mammal remains (elephants, pigs, cercopithecine monkeys; up to 6.000 specimens belonging to 73 different species) in an excellent conservation state (Johanson and Taieb, 1976; Johanson and White, 1979; Taieb *et al.*, 1976). The dating of the remains was somewhat problematic due to the geological history of the soils. Hadar is a region with numerous crisscrossing gullies, faults, and folds, which make it difficult correlating strata. We will only review the chronology of the three Hadar members above the sterile basal member in which hominin remains have been found. These are Kada Hadar (KH), Denen Dora (DD), and Sidi

Hakoma (SH), ordered by increasing antiquity (Figure 4.12).

The set of Kada Hadar, Denen Dora, and Sidi Hakoma has a thickness of around 180–280 m of sedimentary deposits, depending on the areas. The sediments are intermixed with volcanic tuffs, and their origins are lacustrine, riverine—from the edge of the lake—and fluvial (Taieb *et al.*, 1976). The hominin remains are concentrated in three main groups: one in the lower part of the Sidi Hakoma Member, another in the transition between Denen Dora and Kada Hadar, and the third in the upper part of this last member.

The first discovery of fossil hominin remains at Hadar took place on October 30, 1973. It included four associated fragments of lower limb bones (left femur, A.L. 128-1, and right tibia, A.L. 129-1), which permitted reconstruction of the knee of an individual that, judging from this articulation's morphology, was bipedal (Johanson and Taieb, 1976; Johanson and Coppens, 1976; Figure 4.13). The specimens were found in the lower part of the Sidi Hakoma Member, just above the SHT tuff, in soils dated around 3 Ma. The following year, 1974, 10 additional specimens offered a much broader vision of the Hadar hominins. Among them was the famous almost complete A.L. 288-1 skeleton, Lucy. The 1974 specimens

Figure 4.11 Location and view of the Hadar site (Ethiopia).

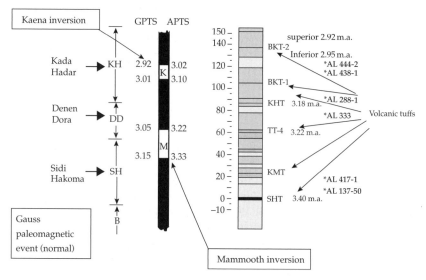

Figure 4.12 Kada Hadar (KH), Denen Dora (DD), and Sidi Hakoma (SH) in the Hadar formation. APTS, astronomical polarity time scale; GPTS, geomagnetic polarity time scale. Picture modified from Walter and Aronson (1993).

include, in addition, a complete palate with all its teeth, A.L. 200-1a, and the right half of a maxilla, A.L. 199-1, together with other remains: mandibles and teeth (A.L. 666-1, A.L. 188-1, A.L. 277-1, A.L. 198-1, A.L. 198-18, A.L. 198-17a) and femora (A.L. 211-1, A.L. 228-1).

The retrieval of fossils at Hadar was initially done on the surface, taking advantage of the cleaning of the ground by infrequent but torrential rains. The remains of A.L. 288-1 were collected later in 3 weeks of work that led to the retrieval of the well conserved partial skeleton of a single

---

Box 4.11 Age estimates for the Hadar formation

The age estimates proposed by Aronson and colleagues in 1977 for the members of the Hadar formation are between 2.6 and 3.3 Ma. They based this estimate on geochronological, paleomagnetic, and biostratigraphical evidence. Walter and Aronson (1982) have revised the dates, but, as they state, the conventional K/Ar method can be used on the BKT-2 deposits, in the upper half of Kada Hadar, stratigraphically above all the levels containing *Australopithecus* remains, and aged between

2.8 and 3.1 Ma. Most of the other volcanic soils pose problems for age estimation. However, Walter and Aronson (1993) were later able to directly establish the age of the Sidi Hakoma (SHT) tuff for the first time by means of the laser-fusion ^{39}Ar/^{40}Ar technique: 3.4 Ma, an estimate confirmed by Kimbel *et al.* (1996), who, in addition, have provided a very complete stratigraphy of the three members containing hominins. The Hadar hominins can now be dated quite precisely.

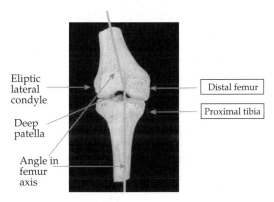

Eliptic lateral condyle

Deep patella

Angle in femur axis

Distal femur

Proximal tibia

Figure 4.13 Knee joint A.L. 129 from Sidi Hakoma, Hadar, *c.* 3 Ma. Photograph of the specimen from http://www.mnh.si.edu/anthro/humanorigins/ha/al129.htm; details added.

individual, an adult female, from sediments in the inferior part of the Kada Hadar Member, just above the KHT tuff. This is the most complete Pliocene hominin skeleton known to date: up to 80% if we include lateral symmetry. The preservation of fossils at Kada Hadar is excellent. Even fossilized tortoise and crocodile eggs have been found (Johanson and Taieb, 1976). A.L. 288-1 thus provided an exceptional opportunity to study the morphology of very early hominins. According to Walter and Aronson's (1982) estimate, Lucy would be 3.5 Ma. A later revision estimated the KHT tuff age to be 3.18 Ma, which makes A.L. 288-1 about 3.1 Ma.

In 1975 the discoveries at Hadar were complemented with up to 13 individuals of different

ages and sexes, A.L. 333. They are fragmentary and incomplete specimens compared to *Lucy*, but conserved well enough to allow certain determinations relative to dimorphisms and to juvenile and adult forms (Johanson, 1976; Johanson and White, 1979). The series 333 was found under the KHT tuff and, thus, would be about 3.2 Ma.

4.2.3 AL 288: morphology and classification

The excellent conservation of the A.L. 288-1 specimen allows us to get a clear picture of the morphology of this 3-million-year-old ancestor (Figure 4.14). It was a small individual, between 1.10 and 1.30 m high. This height is confirmed by other Hadar remains, such as A.L. 128 and 129, whereas specimens belonging to the 333 series indicate a larger size. We'll comment on these differences later. The relation between the length of its humerus and femur, which gives an idea of how long the arms are in comparison to the legs, is greater than in current humans. The hand bones of A.L. 288-1 and the 333 series, as well as the feet of the latter, are different from current morphology. The structure of Lucy's hip suggests a bipedal posture. A significant element of the morphology of A.L. 288-1 is its cranium.

One of the most notable missing pieces of A.L. 288-1 is the face. The specimen includes only a few cranial fragments. The absence of crania is a common circumstance in all Hadar discoveries of the 1970s. This is why the reconstruction of the sample's cranium was done by grouping fragments from

Box 4.12 Lucy's discovery

The circumstances of Lucy's discovery are well known. On November 30, 1974, while exploring Hadar Locality 162, Donald Johanson and Tom Gray, came across numerous bone fragments that at first sight seemed to belong to a singe individual (Johanson and Edey, 1981). The finding was registered as A.L. 288–1 and named Lucy. That same night, during the celebration of the discovery there were drinks, song, and dance at the Hadar campsite. The magnetophone played the Beatles' song *Lucy in the Sky with Diamonds* over and over. No one remembers when or who suggested it, but the skeleton was baptized with the popular name it has been referred to since (Johanson and Edey, 1981).

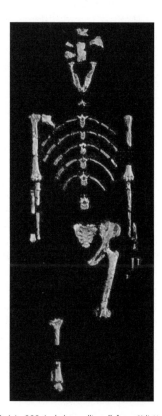

Figure 4.14 A.L. 288-1 skeleton, "Lucy" from Sidi Hakoma, Hadar, c. 3.1 Ma. Pʰ   ʰph from Johanson and Edgar (1996).

ᵗnder such conditions the
ᵗapacity is not very precise
ᵗs that these individuals
ᵗ Donald Johanson says
ᵗhan three and a half
ᵗ yet walked erect"
ᵗ).

The Hadar organisms are an excellent sample of mosaic evolution. Some traits, such as hip shape, indicate a notable proximity to later hominins. Others, such as the dental arcade (V-shaped, unlike our own, which has a parabolic form) are primitive characters, as are the length of the anterior limbs or the small and robust cranium (Johanson and White, 1979). It was, thus, "an ape-brained little creature with a pelvis and leg bones almost identical in function with those of modern humans" (Johanson and Edey, 1981, p. 181).

How to classify Lucy? The book by Johanson and Edey (1981) on the Hadar discoveries devotes three chapters (13–15) to answering this question. Johanson recounts interesting details of his discussions with Tim White regarding the species in which the Hadar remains should be placed. The alternatives were clear: either Lucy belonged to one of the known species (and, in that case, which one?) or to a new one. The problems involved in placing the Hadar hominins in a known species, leaving aside the fact of their great age, were that their molars were as small as those of *Homo habilis* but their other features had little to do with the genus *Homo*. Thus, Johanson and White decided, in the first place, that it was an australopithecine. That decision, which may now seem obvious, required rejecting some of the taxonomic alternatives put forward for the Hadar remains. Johanson and Taieb (1976) did not suggest any formal classification for the specimens in their initial description; they only pointed out affinities and similarities: "On the basis of the present hominin collection from Hadar it is tentatively suggested that some specimens show

---

Box 4.13 Lucy's bipedalism

Johanson's emphasis on Lucy's bipedalism must be toned down, in favor of a different style from modern human locomotion. The Hadar discoveries proved the existence of bipedalism with chimpanzee-like cranial size. During the 1970s, when the main Hadar discoveries were made, no

paleontological authority accepted the Piltdown man fraud. But if the influence of such a trick had lasted until then, Lucy would have provided the irrefutable proof that human evolution involved an early appearance of bipedal locomotion and a much later increase of cranial capacities.

---

affinities with *A. robustus*, some with *A. africanus* (*sensu stricto*), and others with fossils previously referred to *Homo*". This diagnosis covered all imaginable possibilities except a new species. It was a hasty statement that Johanson admits to have regretted very soon (Johanson and Edey, 1981). As he asserted, a more careful examination determined that all the specimens found at Hadar corresponded to a single taxon (Johanson *et al.*, 1978). Johanson suggested the taxon *A. afarensis* for the specimens found at Hadar at a Nobel symposium held in May 1978, in Stockholm. Johanson, White, and Coppens (1978) published a detailed description of *A. afarensis* that same year. However, the suggested type specimen was not any of the Hadar exemplars, but Laetoli specimen L.H.-4, described by Mary Leakey and colleagues (1976).

The Laetoli site, in northern Tanzania, is located some 40 km south of Olduvai (see section 3.1 about the fossil footprints). Since the first discovery made by Kohl-Larsen in 1938–1939, Laetoli has yielded some very early hominin specimens. The K/Ar method yields an age estimation of 3.8–3.5 Ma (Harris, 1985) for the sediments that contain hominin remains, the same age as the fossil footprints (see Section 3.1.7). Up to 30 hominin specimens, including mandibles, maxillas, isolated teeth, and a partial juvenile skeleton were discovered between 1938 and 1979 (Day, 1986). The research team led by Mary Leakey found 13 of them, consisting of teeth and mandibles, during the 1974 and 1975 campaigns. Among them is the L.H.-4 specimen, constituting a relatively undistorted mandible without ramus and partial adult dentition (Leakey *et al.*, 1976). Mary Leakey and colleagues pointed out the similarity between this and the other specimens, on one side,

and gracile australopithecines, specifically early *Homo* (considered by the authors as australopiths), on the other; they also noted differences regarding robust australopithecines. They did not assign the remains to any species or genus, but underlined their "strong resemblance" with the East African *Homo* specimens. However, in a description of the specimens found between 1976 and 1979 Tim White included a brief note suggesting that they should be ascribed to *A. afarensis* (White, 1980b).

We face difficulties related to the characterization of *A. afarensis*. By designating a specimen from Laetoli as the holotype, *A. afarensis* becomes bound to that exemplar, and, therefore, the Hadar specimens would be paratypes. To what extent can such geographically distant specimens be regarded as members of the same species? Phillip Tobias (1980) noted in his criticism of the *A. afarensis* proposal that between the Hadar and Laetoli specimens there is a distance of 1,600 km and a time gap of 800,000 years, in addition to their morphological differences, which strengthens the case for alternative proposals. After comparatively examining the morphology of the Hadar, Laetoli, and Transvaal (Sterkfontein and Makapansgat) samples, Tobias (1980) maintained that all these specimens belonged to the species described as *A. africanus* by Dart, while remaining open to the possibility that they may constitute subspecies. Thus, Tobias suggested that the species *A. africanus* would include, in addition to *A. africanus transvaalensis*, the subspecies *A. africanus afarensis*, for the Laetoli specimens, given that Johanson, White, and Coppens designated the paratype of *A. afarensis* on the basis of one of them, and *A. africanus aethiopicus* for the Hadar specimens (Tobias, 1980). However, most of the initial problems regarding the

---

Box 4.14 The type specimen of *A. afarensis*

The proposal of a hominin from Laetoli, far from Hadar as the type specimen of *A. afarensis* would surprise those who are not familiar with the ins and outs of human paleontology. Lewin's (1987) narration of the episode includes the reasons behind this decision: Mary Leakey's annoyance; the letters exchanged on the subject of the new species name; the inclusion of

Mary Leakey as co-author of the *Kirtlandia* article in which the new species *A. afarensis* was proposed and her demands for her name to be removed, even if the number was already printed. There occurred a frontal collision between Donald Johanson and Mary Leakey, motivated by various reasons, regarding the Laetoli hominin remains.

Box 4.15 The debate around L.H.-4

It is prudent to distinguish two different questions related with the debate surrounding the description of *A. afarensis* with L.H.-4 as holotype. The first is whether paleontological data from other authors can be used without their permission. The second refers to the procedures to be followed when defining a new species. The problems involved in the description and characterization of *A. afarensis* did not end with the initial confrontation between Mary Leakey and Johanson. The

controversy also reached the taxonomic level, given that the species *A. afarensis* was suggested in an irregular but valid way by a reporter without reference to a holotype. Thus, and in accordance with the rigorous rules of taxonomic procedures, L.H.-4 could not be used later as holotype (Day, 1986). As we mentioned before, some authors like Strait and colleagues (1997) and Wood and Collard (1999b) have suggested that *A. afarensis* should, in actual fact, be called *Praeanthropus africanus*.

---

interpretation of *A. afarensis* have mitigated, and the criticisms of Tobias and the Leakeys directed at the new species have not had much success. *A. afarensis* is at present widely accepted as the species that includes the Hadar specimens.

A different question is whether the *Australopithecus* specimens found in Hadar constitute a homogeneous group; that is, whether they belong to a single species. It might be appropriate to divide them into two or more species. According to Poirier (1987), the consideration of the Hadar sample as one or several species depends on two factors. First is the amount of morphological variability that must be allowed within a single species. Second is the amount of variability attributable to sexual dimorphism. The most striking size and weight differences could be explained by sexual dimorphisms, which are very common among the australopithecines. In view of the morphological similarity between large and small Hadar exemplars, Johanson and White

(1979) argued that the variation owed only to sexual dimorphisms. Large and small palates, mandibles and femoral distal fragments would be morphologically identical copies, but at different scales (Johanson and White, 1979). In line with that idea, Kimbel and White carried out a reconstruction of a "complete" cranium of *A. afarensis* by combining cranial and facial fragments from different individuals. This sparked the accusation that they had mixed remains pertaining to more than one species (Shreeve, 1994).

Todd Olson's (1985) comparative study of the cranial morphology (the base of the cranium and the nasal region) of Hadar and South African specimens showed that these areas are affected by selective pressures related mainly with bipedalism, and, to a lesser degree, with dietary specializations. The large and small Hadar specimens, according to Olson, would not be explained by sexual dimorphism, but by those selective pressures. Some, A.L. 333-45 and A.L. 333-105 for instance, would have developed specializations that would even justify their classification within the *Paranthropus* clade; that is, as robust australopithecines. Accordingly, the idea that *Paranthropus* and *Homo* separated 2.5 Ma would be destroyed for the simple reason that the robust specimens present in Hadar would be at least half a million years older. This is coherent with the notion that we will examine in the next chapter regarding an early separation of the gracile and robust branches.

The analysis of the pelvis of A.L. 288-1 by Häusler and Schmid (1995) introduced another controversial element: the possibility that Lucy may have been a male. If this was true, the morphological differences in the Hadar sample could not be attributed only to sexual dimorphisms. However, Häussler and Schmid's hypothesis is based on a speculative argument: the attribution to the large Hadar specimens of a cranial size such that at birth they would not have been able to go through the pelvic canal of A.L. 288-1. Lovejoy and Johanson have criticized those calculations (Shreeve, 1995).

The debate regarding the number of species in the Hadar sample, grouped under the label of *A. afarensis*, received new light after the discovery of the first cranium that conserved the face, the specimen A.L. 444-2, described in 1994 by Kimbel, Johanson, and Rak (1994). It is the most recent specimen of all *A. afarensis*, found in the intermediate part of the Kada Hadar Member, approximately 3.0 Ma. Because it was an adult male it soon received the popular name of Lucy's child. The A.L. 444-2 cranium (Figure 4.15) is very broad; the broadest in all the Hadar samples, with a wide mandibular body, but less robust than the average. Additionally, the facial projection is considerable.

Overall, A.L. 444-2 retains typical morphological features of other Hadar *A. afarensis* specimens, which fact led Kimbel and colleagues (1994) to argue that A.L. 444-2 refuted the notion that the reconstructions carried out with different specimens of the site involved the superposition of two different kinds of contemporary hominin. Lucy's child certainly supports the hypothesis of a single

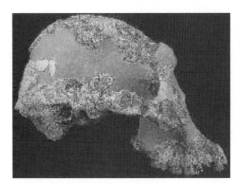

Figure 4.15 Lucy's child, A.L. 444-2 from Hadar; *A. afarensis.* Photograph from Johanson and Edgar (1996).

species, although highly variable, at Hadar. Other findings speak in favor of large sexual dimorphisms. For instance, a maxilla associated with a partial mandible and cranial-base fragments (A.L. 417-1d) from the middle of the Sidi Hakoma Member (close to 3.25 Ma), which probably belonged to a female (Kimbel *et al.*, 1994), shows smaller canines and less prognathism than A.L. 444-2.

Kimbel *et al.* (1994) have also described an ulna (A.L. 438-1; from the Kada Hadar Member, a bit older than the A.L. 444-2 cranium) and the A.L. 137-50 very robust humerus (from the lower part of the Sidi Hakoma Member). In their opinion, the combination of the humerus and the ulna indicates, just as Lucy's (A.L. 288-1) did, that the length of the anterior limbs of *A. afarensis* was closer to that of chimpanzees than humans. Aiello (1994) agreed with the idea of a highly variable single species at Hadar, but pointed out that the ulna A.L. 438-1 does not show traits related with knuckle-walking. It contains a mosaic of traits that, together with the robust form of the humerus, would be ideal for a creature that walked erect on the floor but also climbed trees. Regarding the cranium A.L. 444-2, Aiello (1994) suggested a certain link with the upper-Miocene *Ouranopithecus macedonensis*.

The most recent discovery of *A. afarensis* has been found in Dikika, in the Afar region of Ethiopia. A team under the leadership of Zeresenay Alemseged found there, towards the end of 2000, a skull sticking out of the sand. According to Alemseged, "the sandstone preserved the face and a cast of its skull cavity—allowing researchers a glimpse of her appearance and a measure of brain volume. That first foray also revealed shoulder blades, collar bones, ribs and spinal column, hinting that there was more to unearth" (Alemseged et al., 2006).

Once the specimen was extracted, the outcome has been described as "the intact skull and a partial skeleton of a 3-year-old *A. afarensis* girl: the most complete, earliest specimen ever found" (Alemseged et al., 2006). It was named DIK-1-1, but with the popular name of Selam,

or "peace" in the Afar language (Dalton, 2006; Wood, 2006).

The DIK-1-1 locality belongs to lower Sidi Hakoma Member of the Hadar formation (Figure 4.16). Wynn et al. (2006) have established its age as 3.31 to 3.35 Ma, on the basis of stratigraphic scaling and known chronostratigraphy.

DIK-1-1 includes a nearly complete cranium with the articulated mandible incorporated (Figure 4.17). Also found are numerous elements of the articulated axial skeleton (both scapulae and clavicles, the cervical, thoracic and the first two lumbar vertebrae, and many ribs) recovered in a

---

Box 4.16 Ongoing work on the DIK-1-1 specimen

Alemseged spent most of his summer days for 5 years describing the DIK-1-1 fossil and cleaning it "under a microscope with dental instruments, because I decided not to use acid treatments that could destroy it." He asserts that fully exposing and isolating the many postcranial

elements is a complex task that will take several more years to complete (Alemseged et al., 2006).

DIK-1-1 is so well preserved that it includes the hyoid bone, one of only three fossil hyoids found, the others being from Atapuerca (Spain) and Kebara (Israel; see section 9.2).

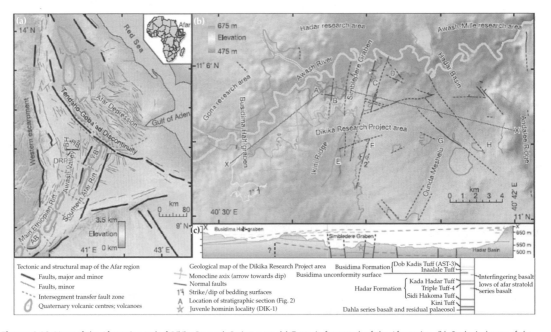

Figure 4.16 Maps of the Afar region and of Dikika Research Project area. (a) Tectonic framework of the Afar region. (b) Geological map of the eastern DRP area. (c) Cross-section of the DRP area (labeled in a). From Wynn et al. (2006). Reprinted by permission from Macmillan Publishers Ltd. *Nature*, 443: 7109, 332–336, 2006.

block of sandstone matrix. Additional postcranial parts were found separately.

The morphological interpretation of DIK-1-1 faces similar difficulties as those of any juvenile exemplar. Alemseged *et al.* (2006) conclude that in the detailed morphology of the face, DIK-1-1 resembles *A. afarensis* and differs from *A. africanus*, including the Taung exemplar, which is also a juvenile. The mandible and the nasal bones are also similar to those of A.L. 333–43, *A. afarensis*. In addition, "most bipedal features seen in *A. afarensis* specimens are observed on the lower limb and foot of DIK-1-1" (Alemseged *et al.*, 2006). Thus, it is not surprising that Alemseged and collaborators attribute DIK-1-1 to *A. afarensis* unambiguously "because the diagnostic facial morphology of this species is evident even at this juvenile stage."

Although the reconstruction is not yet complete, the postcranial remains of DIK-1-1 favor some climbing ability as a retained primitive feature. As we saw earlier (section 3.1), the locomotion of early hominins and the extent of their bipedalism are unsettled. According to Alemseged *et al.* (2006), "the scapula morphology, together with forelimb features, such as the long and curved manual phalanges of DIK-1-1, will raise new questions about the importance of arboreal behavior in *A. afarensis*."

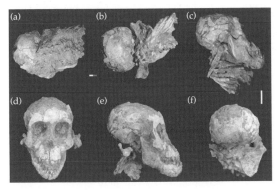

Figure 4.17 The DIK-1-1 juvenile skull and partial skeleton. (a) Dorsal and inferior view as discovered and (b) after partial preparation. (c) Lateral view after partial preparation showing the scapula and many ribs. (d–f) Anterior (d), lateral (e), and posterior (f) views. Scale bars, 2 cm (a and b; c–f). From Alemseged *et al.* (2006). Reprinted by permission from Macmillan Publishers Ltd. *Nature*, 443: 7109, 296–301, 2006.

Other Kenyan localities, such as Lothagam Hill and South Turkwel, are among the places with possible presence of *A. afarensis*. We mentioned Lothagam when, in the section devoted to the last common ancestors of African apes and hominins, we presented the KNM-LT 329 specimen, consisting of a partial maxilla. South Turkwel was the place where the KNM-WT 22936 juvenile mandibular fragment and other associated fossils, including postcranial remains (KNM-WT 22944), were found. Its age, obtained by means of geological correlations and faunal analysis, could be close to 3.5 Ma. With regards to taxonomy, Carol Ward and colleagues (1999) only noted that the specimens morphology is reminiscent of *A. afarensis* and *A. africanus*. One specimen, BEL-VP-1/1, found in 1981 in Belohdelie, on the Ethiopian side of the Middle Awash area, consisting of a partial frontal with a small fragment of the left parietal, had been estimated as being 3.9 Ma. But Asfaw's (1987) study of the specimen's morphology by specular imaging revealed that it was a very generalized hominin, close to the divergence between hominins and African apes. He did not suggest it belonged to any particular species, although he did indicate that it shared some traits with *A. afarensis*.

The Maka site, in Ethiopia, has also provided specimens attributed to *A. afarensis*. Among them, there is an almost complete mandible (MAK-VP-1/12; Figure 4.18), other mandibular and dental materials, a partial (proximal) femur, an ulna and an almost complete humerus, with an age of 3.4 Ma (White *et al.*, 1993). The study by White *et al.* (2000) of the mandibles dealt with the evolutionary significance of the C/P3 complex. As we saw, the C/P3 is an apomorphy in *Ar. ramidus*. Thus, it is one of the dental derived traits that first evolved from characters shared with the great apes. The presence of large and small specimens in the Maka site supports Johanson and White's (1979) idea of a single highly variable *A. afarensis* taxon.

4.2.4 Kanapoi and Allia Bay: the earliest *Australopithecus*

Although *A. afarensis* may be the most representative taxon of the australopithecines, it is not the

Box 4.17 DIK-1-1 in relation to bipedalism in *A. afarensis*

Bernard Wood (2006) has highlighted the significance of DIK-1-1 in order to ascertain the bipedalism of *A. afarensis*: "If its mode of locomotion was exclusively on two legs, one would expect that the limb bones and the organs that help it to balance would be more similar to those of the only living bipedal higher primate (that is, us) than to those of chimpanzees and gorillas. These primates walk on two feet only rarely, if at all." But the scapula and phalanges of the exemplar is similar to those of African apes; and, moreover, the semicircular system—part of the inner ear—in DIK-1-1 "is similar to that of African apes and *A. africanus*, and this has been associated with limited head decoupling and absence of fast and agile bipedal gaits" (Alemseged *et al.*, 2006). Ward, like Tim White (cited by Gibbons, 2006), argues that it will be necessary to have available the full components of DIK-1-1 to reach definitive conclusions about its locomotion.

(a)

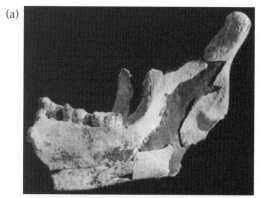

(b)

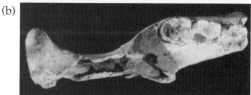

(c)

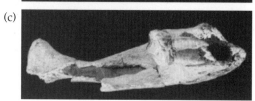

Figure 4.18 The MAK-VP-1/12 mandible; lateral (a), superior (b), and anterior views (c). Photographs from White *et al.* (2000).

oldest. Allia Bay (on the eastern shore of Lake Turkana) and Kanapoi (on the western shore and somewhat south of the same lake) are two Kenyan sites that have provided the oldest *Australopithecus* specimens known to date. Together with

O. tugenensis and *Ar. ramidus*, they complete the landscape of the earliest hominins. The nine specimens from Kanapoi and the 12 from Allia Bay described by Mary Leakey and colleagues (1995) have an age that ranges from 4.1 Ma for the former to 3.9 Ma for the latter; they push back the existence of australopithecines by half a million years.

The sedimentary sequence at Kanapoi includes an interval spanning from 4.17 Ma to around 3.4 Ma, and the hominin specimens are found in the oldest layer of the paleosoils. The age estimation of the soils where the specimens were found could not be carried out during the 1994 campaign; it was based on the application of the ^{39}Ar/^{40}Ar method to nearby volcanic tuffs. The KNM-KP 29281 mandible (Figure 4.19), and the KNM-KP 29283 maxilla come from the inferior stratigraphic level (between 4.17 and 4.12 Ma), while the NM-KP 29285 tibia, the KNM-KP 271 humerus, two mandibular fragments (KNM-KP 29281), and a large mandible (KNM-KP 29287), presumably male and found in the higher level, are between 4.1 and 3.5 Ma (M.G. Leakey *et al.*, 1995).

The Kanapoi and Allia Bay specimens fill the wide temporal range of 1 million years between *Ar. ramidus* and *A. afarensis*. Meave Leakey described the morphology of the new specimens as a mixture of primitive and derived traits, confirming the view, already quite established, that evolution within the tribe Hominini took place in a mosaic fashion during the Pliocene. For instance, the study of the KNM-KP 29285 tibia indicated a bipedal posture. Also, the KNM-KP 271 humerus includes, according to Meave Leakey *et al.* (M.G. Leakey *et al.*, 1995), many hominin-derived traits. However, the persistence of some plesiomorphies in the tibia, which are shared with African apes and *A. afarensis*, show that bipedal locomotion was at a different stage of that of *Homo*. All these features had already been detected in *A. afarensis*. But together with the traits that place Kanapoi and Allia Bay specimens close to *A. afarensis*, there are other traits that move them apart, especially Hadar. These differences led Meave Leakey and colleagues to reject the idea of a long stasis in *A. afarensis*, the presence of the species for an extensive lapse of time with little variation. Rather, M.G. Leakey *et al.* (1995) suggested a new

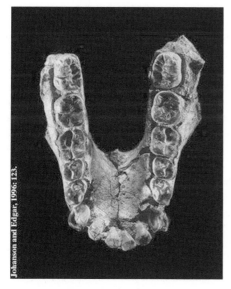

Figure 4.19 The KNM-KP 29281 mandible, holotype of *A. anamensis*. Photograph from Johanson and Edgar (1996).

species for the Kanapoi and Allia Bay findings: *Australopithecus anamensis* (*anam* means "lake" in the Turkana language). The type specimen (holotype) is the KNM-KP 29281 mandible, which retains all its teeth, found by Peter Nzube in 1994 (Figure 4.19). The paratype is constituted by the remaining specimens of both sites.

New findings during the 1995–1997 campaigns confirmed, in the eyes of Meave Leakey and colleagues (M.G. Leakey *et al.*, 1998), the age attributed to *A. anamensis* and the presence of a single species in the Kanapoi and Allia Bay samples. The authors reported the $^{39}Ar/^{40}Ar$ measurement of the Kanapoi tuff, located in the higher part of the most recent lacustrine sedimentary sequence, rendering an age of 4.07 ± 0.023 Ma. All the exemplars described in 1995, including the KNM-P 29285 tibia, would have been recovered from paleosoils located in a very tight interval: between 4.17 and 4.07 Ma. Only the KNM-KP 29287 mandible is slightly younger. These results support, thus, the hypothesis of a single species for all the Kanapoi and Allia Bay samples that is more than 4 million years old. The 1995–1997 campaigns also provided a juvenile which includes teeth and cranial fragments (KNM-KP 34725), pieces that complete the KNM-KP 29287 anterior mandible, a manual proximal phalanx, and a maxilla. The comparative analyses of this extended sample confirm, in the opinion of Meave Leakey and colleagues (1998), the presence of a single species.

Is this a distinctive species, different from other australopithecines? According to Meave Leakey *et al.* (M.G. Leakey *et al.*, 1995, 1998), *A. anamensis* can be distinguished from *A. afarensis* in certain

---

Box 4.18 Andrews' critique

Andrews (1995) criticized the taxonomic solution of classifying the Kanapoi and Allia Bay hominins in a single species. The stratigraphic sequence of Kanapoi and Allia Bay separates early fluvial sediments from later ones originated in the ancestral Lake Lonyumun. Andrews pointed out that the more recent specimens from the lake sediments could belong to a different and

more derived species, which would explain the relative robusticity of the KNMP-KP 29287 mandible. But, if this was the case, the locomotor characteristics of *A. anamensis*—inferred from the KNM-KP 29285 tibia, belonging to the same lithic interval as the cited mandible—would not be applicable to the earlier specimens.

dental traits, such as the longer canines and robust roots. But they are also different form *Ar. ramidus* in that the enamel of *A. anamensis* is much thicker and similar to that of *A. afarensis*.

The discovery of additional exemplars of *A. anamensis* in the region of Asa Issie, Middle Awash (Ethiopia) has provided new insight about the phylogenetic relationships between that taxon and *A. afarensis*. The new specimens, dated at 4.12 Ma, have been described by Tim White *et al.* (2006) as follows.

• Specimen ARA-VP-14/1 is a left maxilla with fragmentary teeth, and adjacent palatal and lateral maxillary surface. The palate is very shallow anteriorly on the left. Its roof is distorted superiorly on the right. The canine jugum would have formed the margin of the pyriform aperture. The specimen is slightly smaller but anatomically similar in preserved parts to the *A. anamensis* paratype.
• Associated dental rows ASI-VP-2/2 and ASI-VP-2/334 are from separate individuals. They definitively place the Asa Issie sample within expected ranges of *A. anamensis* variation. Molar crown dimensions are at or slightly above (ASI-VP-2/334) the upper end of the known *A. anamensis* range. Combined with the slightly smaller ARA-VP-14/1 dentition, these Middle Awash postcanine teeth

are distinctly larger than *Ar. ramidus* but broadly equivalent to both *A. anamensis* and *A. afarensis* counterparts (White *et al.*, 2006).
• The ASI-VP-2 and ASI-VP-5 postcrania include a metatarsal shaft without ends, an eroded distal foot phalanx, and an intact intermediate hand phalanx. According to White *et al.* (2006), the last specimen "is morphologically similar to those from Hadar, but is longer relative to its breadth. Four vertebral fragments include an atlas larger than its single Hadar homologue and a thoracic arch larger than any in the Hadar A.L. 288-1 specimen."

4.2.5 Bouri and *A. garhi*

The Hata Member of the Bouri formation in Middle Awash (Ethiopia) has yielded hominin remains since 1990. The most complete specimens were described in Asfaw *et al.* (1999), with the proposal of yet another australopithecine taxon: *Australopithecus garhi*. The holotype of the species is the specimen BOU-VP-12/130, a set of cranial fragments including the frontal, parietals, and maxilla with dentition, found in 1997 by Haile-Selassie (Figure 4.20). The volume of an endocranial cast made by Ralph Holloway was 450 cm^3. The age obtained by means of the ^{39}Ar/^{40}Ar method, paleomagnetism, and associated fauna, is about 2.5 Ma (2.496 ± 0.008 Ma,

---

Box 4.19 The relationship between *Ar. ramidus* and *A. anamensis*

Ar. ramidus and *A. anamensis* represent early stages of the tribe Hominini, which suggests the evolutionary sequence: *Ar. ramidus* → *A. anamensis* → *A. afarensis*. However, morphological studies do not support such a lineage. The conclusion reached by Meave Leakey *et al.* (M.G. Leakey *et al.*, 1995) is that *A. anamensis* represents a new 4-Ma ancestor leading to *Homo*. But this lineage cannot consist of the descendants of *Ar. ramidus*. It is probable, according to these authors, that *Ar. ramidus* formed a lateral branch, a sister species of the Kanapoi and Allia Bay specimens and all subsequent hominins. In their 1998 article, Meave Leakey *et al.* proposed the following alternative: either all hominins between 4.4 Ma and close to 3 Ma consist of an evolving single species, or there are three separate species

(*Ar. ramidus*, *A. anamensis*, *A. afarensis*) whose phylogenetic relations are imprecise.

According to White *et al.* (2006), two alternative hypotheses are possible concerning the three taxa: (a) the first hypothesis derives *A. anamensis* phyletically from *Ar. ramidus* within a 200,000-year interval; (b) the second involves cladogenesis of *A. anamensis* from an ancestor (presumably *Ardipithecus* or some close relative) even deeper in the Pliocene or late Miocene. Under the latter hypothesis, *Ar. ramidus* would represent a relict species in an ecological refugium.

According to Meave Leakey, co-author of the paper where *A. anamensis* was named, its relationship to *Ar. ramidus* remains uncertain (see Gibbons, 2006).

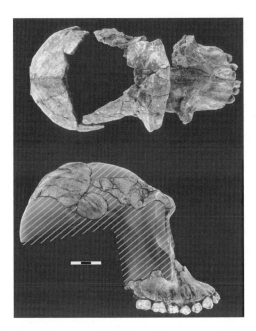

Figure 4.20 Superior and lateral view of the BOU-VP-12/130, *Australopithecus ghari*. The scale indicates centimeters. Photos ©David L. Brill 1999/Atlanta.

from radioisotopic studies; de Heinzelin *et al.*, 1999). Other postcranial remains, including a left femur and right humerus, radius, and ulna, were found during the 1996–1998 campaigns.

The small cranial capacity, together with wide premolars and molars, led Asfaw and colleagues (1999) to argue that the Bouri remains belonged to an australopithecine posterior to *A. afarensis*. The differences between the specimen and *A. africanus*, or *Paranthropus aethiopicus* (*Australopithecus aethiopicus*, according to the authors), justify, in the view of Asfaw *et al.* (1999), the creation of a new species. In phylogenetic terms, this suggestion is conservative in a certain way: Asfaw and colleagues noted the unresolved cladistic event that must account for the phylogenetic relationships between *A. africanus*, the genus *Paranthropus*, and now *A. garhi*. However, they pointed out that the Bouri specimens are found in the right place and time to be the ancestors of the early *Homo*. David Strait and Frederick Grine (1999) rejected this possible ancestral relation, arguing that *A. garhi* lacks any specific *Homo* synapomorphies. Strait and Grine

noted that *A. garhi* is further removed from *Homo* than the earlier *A. africanus*; *A. garhi* would be no more than another lateral branch of the divergence process of early hominin forms.

A related controversy concerning *A. garhi* arises from the discovery in Gona, 96 km north of Bouri, of an abundant sample of 2.6-million-year-old Oldowan lithic artifacts (Semaw *et al.*, 1997). Jean de Heinzelin *et al.* (1999) attributed to *A. garhi* their manufacture and use on the grounds that this is the only hominin present at Bouri. These authors acknowledged that the lack of tools at Hata is inconvenient for the hypothesis of *A. garhi* as a stone carver. But they noted that "rare, isolated, widely scattered cores and flakes of Mode I technology appearing to have eroded from the Hata bed have been encountered during our surveys." The bruise and cut marks on bovid bones found in the Hata formation where *A. garhi* was found speak in favor of this hypothesis. This is the earliest evidence of cut marks made by hominins. We will return to the issue of the relationship between species and cultures in section 8.2.

4.2.6 Out of the Rift: early hominins from Chad

All Miocene and Pliocene hominins we have described up to here come from the Rift Valley. The data accumulated during the golden age of human paleontology pointed towards East Africa as the birthplace of hominins. The discovery of a very early australopithecine specimen, more than 3 million years old, in Chad, a considerable distance from all previously known deposits, represented a challenge. The authors of the article in which the new specimen was announced (Brunet *et al.*, 1995; see Morell, 1995) argued that this finding completely changed the scene regarding humanity's birthplace. The need to take into account the Chad discoveries became pressing when another specimen, estimated to be between 6 and 7 million years old, was later discovered.

In January 1995 Michel Brunet found a mandibular fragment of an adult hominin (KT 12/H1) at the locality KT 12 of the Bahr el Ghazal region ("river of gazelles" in classic Arab), in the Djourab

Desert, 45 km east of Koro Toro in Chad (Figure 4.21). The fragment includes an incisor, two canines, and two premolars of both sides. The morphological examination carried out by Brunet *et al.* (1995) indicated that the specimen was similar to *A. afarensis*, even though some features differentiated it from that species and other australopithecines. The provisional classification of KT 12/H1 was *Australopithecus* aff. *afarensis* (Brunet *et al.*, 1995). One year later, and after the comparison with *A. afarensis* specimens kept at the National Museum of Adis Abeba (Ethiopia), that assignation was revised. Some derived traits of KT 12/H1, such as those related with the morphology of the mandibular symphysis, indicate a less prognate face than *A. afarensis*. The decision was made to name the taxon *Australopithecus bahrelghazali*, with KT 12/H1 as the holotype, and a premolar discovered in 1996 (KT 12/H2) as the only paratype for the moment (Brunet *et al.*, 1996).

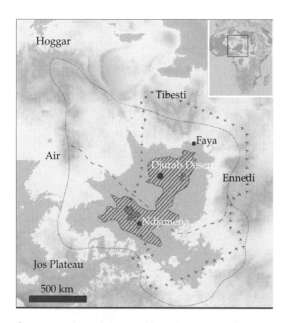

Figure 4.21 The Djourab Desert (Vignaud *et al.*, 2002). The crosses represent the Chad border. The boundary of the ancient Lake Chad basin is shown as a dotted line (with northern and southern sub-basins divided by a dashed line). Shaded circle, the Toros-Menalla hominin site (TM 266). Illustration from Vígnaud *et al.* (2002). Reprinted by permission from Macmillan Publishers Ltd. *Nature*, 418: 6894, 152–155, 2002.

As a homage to a late colleague, Abel Brillanceau, Brunet and his team informally baptized KT 12/H1 as Abel. No radiometrically measurable soils were found, but the fauna associated with the specimen show a close similarity with that of Hadar soils between 3.4 and 3.0 Ma (Brunet *et al.*, 1995). Paleontological and sedimentological studies poin ted towards a lakeshore habitat, with a mosaic vegetation: gallery forests and brushy savannas with open grasslands; that is to say, a similar habitat to Hadar at the time. This coincides, certainly, with the ecological model usually related with the first hominins. But the notion that *A. bahrelghazali* was contemporary to Hadar *A. afarensis* and, thus, represents one of the first hominins, stumbles with the diversification model of hominoids put forward by Yves Coppens, one of the authors of the articles that introduced the new species. The geological transformation of East Africa as a consequence of the separation of the African and Arabian tectonic plates, which produced the Rift Valley chain of faults, led to important changes in the ecosystem, which could have been a source of selective pressures for African late-Miocene hominoid populations, leading to the separation of the different lineages we know today.

Coppens (1994) and Jean Chaline *et al.* (1996) have explained such a separation by means of an allopatric model (see Section 2.4.7). The geographic separation between the different populations would be: pre-gorillas to the north of the barrier constituted by the River Zaire, pre-australopithecines in the Levant region of the Rift, and pre-chimpanzees in central and Western Africa. A separate evolution in each geographical location would fix specific derived traits (autapomorphies) in each of the branches. This allopatric speciation model would explain why there are no remains of chimpanzee ancestors in hominin sites.

However, the finding of KT 12/H1 suggests an alternative because it indicates that the presumed continuity of thick forests to the west of the Rift did not exist during the Pliocene. Brunet *et al.* (1995, 1996) believe that the forests and savannas, capable of housing the first hominins, extended from the Atlantic Ocean, through the Sahel, along East Africa and up to the Cape of Good Hope. The reason why no specimens have been found in intermediate zones is the lack of Pliocene and lower-Pleistocene soils. The fast appearance and expansion of hominins would render speculations about their birthplace useless.

This is an attractive idea, and a recent discovery speaks in favor of it. Sally McBrearty and Nina Jablonski found three teeth, KNM-TH 45519, KNM-TH 45520, and KNM-TH 45521, which represent the first known chimpanzee fossil specimens (McBrearty and Jablonski, 2005), in the Kapthurin formation (Tugen, district of Lake Baringo, Kenya; Figure 4.22). The Kapthurin formation has been dated using the ^{39}Ar/^{40}Ar method (Deino and McBrearty, 2002), attributing to the chimpanzee fossils an age of $545,000 \pm 3,000$ years. (Notice that the Kapthurin formation is placed within the "pre-chimpanzee" region proposed by Coppeus—see Figure 2.27.)

A. bahrelghazali is not the only hominin taxon from Chad. A year after Senut and Pickford's proposal of *Orrorin*, an international team led by Michel Brunet reported the discovery of another fossil very close to the time of the divergence between chimpanzees and hominins (Brunet *et al.*, 2002). It was found at the locality TM 266, in the Toros-Menalla fossiliferous zone, not far from Bahr el Ghazal. Although the estimation of the

Toros-Menalla specimen's age is a little imprecise, biochronological studies estimate it to be between 6 and 7 Ma (Vignaud *et al.*, 2002). Very close, also, to the age of *O. tugenensis*.

The Chad discoveries included an almost complete cranium (TM 266-01-60-1; Figure 4.23), which lacked a good portion of the occiput, and six mandibular fragments. The cranium received the colloquial name of Toumai ("hope for life" in the Goran language). Its morphology shows, once again, a mixture of primitive and derived traits. The very low cranial capacity, the great supraorbital arch, and a very developed mandibular ramus are reminiscent of African apes. The small canines (if it is a male, as the discoverers proposed), the reduced subnasal prognathism, and the molar enamel thickness— intermediate between that of gorillas + chimpanzees and *Australopithecus*—move the specimen towards hominins. The authors describing the Toros-Menalla specimens suggested, consequently, a new genus hominin, baptizing the corresponding species as *Sahelanthropus tchadensis* (Brunet *et al.*, 2002).

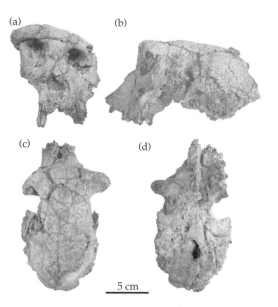

Figure 4.23 TM 266-01-60-1, *S. tchadensis* (Brunet *et al.*, 2002). Top left, facial view. Top right, lateral view. Bottom left, dorsal view. Bottom right, basal view. Photograph from Brunet *et al.* (2002). Reprinted by permission from Macmillan Publishers Ltd. *Nature*, 418: 6894, 145–151, 2002.

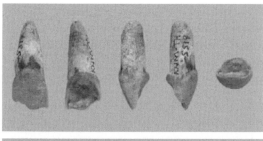

Figure 4.22 Top: KNM-TH 45519 (labial, lingual, mesial, distal). Bottom: enlargement of the incisal edge of KNM-TH 45519 (left) and KNM-TH 45521 (right). Photograph from McBrearty and Jablonski (2005). Reprinted by permission from Macmillan Publishers Ltd. *Nature*, 437: 7055, 105–108, 2005.

Determining the phylogenetic significance of *S. tchadensis* is difficult. If it is a hominin, it contradicts the widely held notion that the first exemplars of our tribe appeared in the Rift. This issue, as we have seen, was already raised by *Australopithecus bahrelgazhali*. But the age of this latter taxon would allow considering it as an immigrant coming from the Rift. However, *S. tchadensis* is another story. Even granting a wide margin of doubt, this organism is very close to the separation of the ape and hominin lineages. The problem is not only the hominin place of origin. The similar age of *O. tugenensis* and *S. tchadensis* requires positing in very far apart locations the almost simultaneous appearance of beings that could be the ancestors of australopithecines.

Which of these two very early specimens is closest to later hominins? As Bernard Wood (2002) noted, *S. tchadensis* displays a strange morphological combination: the posterior part of its cranium resembles that of an ape, while its face, with a great supraorbital arch and moderate subnasal facial projection, is similar to the hominins, although not to the australopithecines, as might have been expected if *Sahelanthropus* was an ancestor of *Australopithecus*, but to the much later *Homo erectus*. The facial morphology of *S. tchadensis* could not have been inherited directly by *H. erectus* without the same traits appearing in the different intermediate species between *Australopithecus* and *Homo*.

Milford Wolpoff and colleagues (2002) suggested that *S. tchadensis* could be an ancestor of African apes, possibly gorillas, rather than a hominin. Its most gracile traits could be explained if the specimen were a female. If the smaller dentition is due to sexual dimorphism, then the small

canines of *Sahelanthropus* lose their evolutionary relevance; a female ancestor of gorillas would be expected to have canines like those. But Michel Brunet's team believed the Toros-Menalla specimen is a male, which endorses its central role in hominin phylogeny.

The Toros-Menalla localities TM 247, TM 266, and TM 292 later provided three new *S. tchadensis* specimens: two mandibular fragments (TM 292-02-01 and TM 247-01-02) and an upper right premolar P3 (TM 266-01v462) (Brunet *et al.*, 2005). Brunet and colleagues described the presence of some derived traits in these materials, such as a non-honing C/P3 complex and radial enamel thickness intermediate between chimpanzees and australopithecines. These features also support the hominin character of the taxon.

Thus, we are faced with the difficulty of understanding how the evolutionary history of early hominins could have taken place in places as far apart as Kenya and Chad. This difficulty and the doubts that it generates may probably be inevitable. According Darwin, as we have cited frequently in this book, a mixture of primitive characters is foreseeable in any specimen that is close to the divergence process that separated the evolutionary lineages of hominins and panids. Thus, the presence of certain primitive traits in *Sahelanthropus*, *Orrorin*, and *Ar. kadabba* should be expected. The mixture of plesiomorphies and apomorphies that appear incipiently complicates the task of establishing precise phylogenetic relations in the context of the process of our tribe's appearance. We, therefore, favor the suggestion made by Haile-Selassie and colleagues (2004) of grouping for now all early specimens in a single genus, *Orrorin*.

CHAPTER 5

Pliocene hominins: South Africa and the Rift Valley

5.1 South African sites

5.1.1 Taung

Several fossiliferous areas from South Africa provided in the first decades of the twentieth century the earliest reliable evidence of the evolution of the tribe Hominini during the Pliocene and early Pleistocene. These include, among other sites, Taung, Sterkfontein, Makapansgat, Kromdraai, Swartkrans, Gladysvale (see Figure 5.1), and Drimolen. Some fossils of other human ancestors were already known: Neanderthals, *Pithecanthropus erectus*, and the Piltdown fraud. None of

Figure 5.1 South African sites in which Pliocene and early Pleistocene hominins have appeared.

them was African, which seemed contrary to any notion that our earliest ancestors might come from Africa. The notion that hominins originated in Africa took long to become widely accepted. We will examine the way in which this came to be established. We will also see how indications of two very different kinds of hominin—robust and gracile—appeared in South Africa and East Africa.

The early Pleistocene site of Taung is the starting point for the modern interpretation of the early ancestors of current humans. The so-called Taung Child, the specimen described by Raymond Dart in 1925 (Figure 5.2), was found there. The story of the discovery has been told by Dart (see Box 5.1): the surprise of finding himself before a fossil baboon, the coincidence of Professor Young's trip to Taung, the fortune of coming across fitting pieces, the difficulties involved in cleaning the molds and, at last, the prize of an astonishing discovery.

Phillip Tobias—successor to Raymond Dart's chair at the University of Witwatersrand—has

Figure 5.2 Portrait of Raymond Dart (1893–1988).

135

Box 5.1 The discovery of the Taung Child

"Towards the close of 1924, Miss Josephine Salmons, student demonstrator of anatomy in the University of Witwatersrand, brought to me the fossilized skull of a cercopithecid monkey which, through her instrumentality, was very generously loaned to the Department for description by its owner, Mr. E.G. Izod, of the Rand Mines Limited. I learned that this valuable fossil had been blasted out of the limestone cliff formation—at a vertical depth of 50 feet and a horizontal depth of 200 feet—at Taungs, which lies 80 miles north of Kimberley on the main line to Rhodesia, in Bechuanaland.... I immediately consulted Dr. R. B. Young, professor of geology in the University of Witwatersrand, about the discovery, and he, by a fortunate coincidence, was called down to Taungs almost synchronously to investigate geologically the lime deposits of an adjacent farm. During his visit to Taungs, Prof. Young was enabled, through the courtesy of Mr. A. F. Campbell, general manager of the Northern Lime Company, to inspect the site of the discovery and to select further samples of fossil material for me from the same formation. These included a natural cercopithecid endocranial cast, a second and larger cast, and some rock fragments disclosing portions of bone.... In manipulating the pieces of rock brought back by Prof. Young, I found that the larger natural endocast articulated exactly by its fractured frontal extremity with another piece of rock in which the broken lower and posterior margin of the left side of a mandible was visible. After cleaning the rock mass, the outline of the hinder and lower part of the facial skeleton came into view. Careful development of the solid limestone in which it was embedded finally revealed the almost entire face depicted in the accompanying photographs". (Dart, 1925, p. 195)

added useful insights regarding the fossil, its appearance, and South African fossil baboons known at the time of the Taung discovery, as well as a detailed chronology of the whole episode (Tobias, 1990). Dart's anatomical expertise allowed him immediately to realize that the pieces provided by Young did not correspond to any Old World monkey (cercopithecid). Dart noticed that it was the cranium of an organism intermediate between anthropoids and current humans (Figure 5.3). Its humanoid traits were evident: the mandible, the dentition (it was a juvenile fossil, with a dental development equivalent to that of a 6-year-old modern human), the position of the foramen magnum (which suggested bipedal locomotion), and the brain's organization. The brain's features led Dart to believe that it corresponded to an "ultra-simian". But some brain features, such as its volume and the lack of an expansion of the temporal area, suggested the Taung Child was "pre-human". Thus, Dart believed it was an intermediate between apes and humans. Dart named the species *Australopithecus africanus*. The Taung specimen was the holotype and the only available specimen at the time.

Figure 5.3 The Taung Child, *Australopithecus africanus* (Dart, 1925). Photograph from Johanson and Edgar (1996).

5.1.2 The controversy over the significance of *A. africanus*

Dart's detailed and firm arguments regarding ultra-simian features of *Australopithecus africanus* were compelling. But the suggestion that it was a creature intermediate between current pongids and humans sparked numerous criticisms. Not even the name itself was spared from snide remarks. An editorial in the same journal that had published the Taung discovery accused Dart of

Box 5.2 Dart's proposal of the Homo-simiadae family

The consideration of the Taung fossil as "ultra-simian" and "pre-human" led Raymond Dart to suggest a new Homo-simiadae family that would include the species *Australopithecus africanus*, to which the specimen belonged. He conceived *A. africanus* as a biped half way between great apes and humans, with a similar cranial volume as chimpanzees, which would be the first of the gracile hominins. The proposal of the Homo-simiadae family was not successful, however, in contrast with the general acceptance of the genus *Australopithecus*.

incompetence in etymological matters. After publishing Raymond Dart's article, *Nature* commissioned review commentaries of the *A. africanus* proposal from several distinguished paleontologists. They were published in the next issue, February 14, 1925. Arthur Keith, Elliot Smith, Arthur Smith Woodward, and W.L.H. Duckworth (Duckworth, 1925; Elliott Smith, 1925; Keith, 1925a; Woodward, 1925) expressed their hostility to the definition of a new genus placed between anthropoids and humans. The cranial capacity of the Taung Child was within expected values for a 4-year-old chimpanzee or gorilla. Thus, it was described by the experts as an "anthropoid ape", though "a very remarkable one" (Keith); an "unmistakable anthropoid ape" (Smith); an "extinct anthropoid ape" (Woodward); and "an African, not an Asiatic form of anthropoid ape" (Duckworth). Additionally, Woodward lamented that Dart had "chosen for it so barbarous [Latin–Greek] a name as *Australopithecus*".

Five months after the publication of those reports, and having had the chance to examine a cast of the Taung fossil at the South African pavilion of the British Empire Exhibition held at Wembley in 1925, Arthur Keith (Figure 5.4) wrote: "The skull is that of a young anthropoid ape—one which was in the fourth year of growth, a child, and showing so many points of affinity with the two living African anthropoids, the gorilla and chimpanzee—that there cannot be a moment's hesitation in placing the fossil form in this living group" (Keith, 1925b). Keith added: "In every essential respect the Taungs skull is that of a young anthropoid ape, possessing a brain which, in point of size, is actually smaller than that of a gorilla of a corresponding age.

Figure 5.4 Arthur Keith (1866–1955). Detail of a pencil drawing by William Rothenstein, 1928; in the National Portrait Gallery, London.

Only in the lesser development of teeth, jaws, and bony structures connected with mastication can it claim a greater degree of humanity than the gorilla".

Arthur Keith's categorical assertions are but a sample among other notable opinions that spurned *A. africanus*. There are several reasons behind this rejection of the australopith as a human ancestor. One of the most important was the widespread notion during the early twentieth century that the missing link would turn up in Asia, and in a much earlier epoch (Oligocene, or Miocene at the very most) than the Taung remains (Lewin, 1987). The existence of the Piltdown fraud, with completely opposite features to those observed in the Taung specimen (mainly the large cranium), also counted against the australopith. That the Taung child was a young specimen was a contributing factor, because it is not easy to distinguish the morphologies of immature specimens belonging to different primate species.

It took a long time to reach today's generalized consensus that South African australopiths are hominins. Wilfred Le Gros Clark played a crucial role in the reestablishment of the taxonomic justice and Raymond Dart's scientific standing. In 1947 Le Gros Clark traveled to South Africa. The australopith specimens he was able to examine there, including the robust ones discovered by Robert Broom (see below), led him to the firm conclusion that australopiths were human ancestors (Le Gros Clark, 1947). In honor of Keith's intellectual integrity, it must be said that shortly after Le Gros Clark's interpretation he sent a letter to *Nature* admitting that Dart was right and that it was him, Keith, who had erred. But even admitting his mistake, Keith was still against the name given to the Taung fossil (Keith, 1947; see Box 5.3).

5.1.3 After Taung: new South African specimens

During the 20 years between the finding of the Taung fossil and its final acceptance, it was Robert Broom (Figure 5.5), from the Transvaal Museum of Pretoria, South Africa, who continued the task of searching for new fossils and fitting them into our family's evolutionary scheme. In the summer of 1936, while visiting the South African site of Sterkfontein, the quarry's foreman showed Broom a partial cranium that his workmen had found after a detonation. An intensive search produced other fragments of the cranium. After its reconstruction, Broom obtained what he believed to be an adult specimen, TM 1512 (constituting a fragmented

cranium, part of the face, and the mandible and some teeth), belonging to a different species but same genus as Taung. The best-preserved specimen from Sterkfontein is Mrs Ples (Sts 5; Figure 5.6), an almost complete cranium, though without teeth or lower mandible. It is an adult specimen, with moderately marked brow ridge and glabella and a weak occipital torus (Broom *et al.*, 1950). Both specimens are now considered to be the same species as Taung, *A. africanus* (see Box 5.4).

Another site, Kromdraai, yielded new australopith remains 2 years after Robert Broom's initial discovery at Sterkfontein. The finding made by a schoolboy, G. Terblanche, and identified by Broom, turned out to be very important for our family's history. It was a craniofacial fragment with five teeth and other associated skeletal fragments (TM 1517) with a very massive appearance (Broom, 1938), in contrast with previous specimens. Whereas some traits (small cranium, incisors, and canine orifices) highlighted its resemblance with previously found australopithecines, others (wide molars and large and projected face) suggested that it was a different kind of individual. Broom (1938) proposed a new genus and species for the specimen, *Paranthropus robustus*. Thus, a new branch was added to our family. Taken together, the evidence from Taung, Sterkfontein, and Kromdraai revealed that there was not just one very early kind of hominin in South Africa, but at least two types: gracile and robust.

Two other sites were soon added to the list of South African sites yielding very early hominins: Swartkrans and Makapansgat. In 1938 Broom found at Swartkrans an incomplete cranium with

---

Box 5.3 Keith's apology

Letter sent by Arthur Keith to *Nature* in 1947: "Like Prof. Le Gros Clark, I am now convinced, on the evidence submitted by Dr. Robert Broom, that Prof. Dart was right and that I was wrong; the Australopithecinae are in or near the line which culminated in the human form. My only complaint now is the length of the name which the extinct anthropoid of South Africa must for ever bear. Seeing that Prof. Dart not only discovered them but also rightly

perceived their true nature, I have ventured, when writing of the Australopithecinae to give them the colloquial name of "Dartians," thereby saving much expenditure of ink and of print. The Dartians are ground-living anthropoids, human in posture, gait and dentition, but still anthropoid in facial physiognomy and in size of brain. It is much easier to say there was a 'Dartian' phase in man's evolution than to speak of one which was 'australopithecine'."

Figure 5.5 Robert Broom (1866–1951).

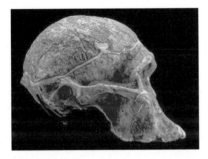

Figure 5.6 Sts 5, Mrs Ples, *A. africanus* from Sterkfontein (South Africa). Photograph from Johanson and Edgar (1996).

part of the face and the upper mandible (SK 46), which was classified as *Paranthropus crassidens* (Broom, 1939). Its robustness was reminiscent of the Kromdraai specimen. In 1950 another nearly complete cranium, with almost the whole face (SK 48), was found in the same site (Broom and Robinson, 1952). SK 48 revealed some previously unknown traits of *Paranthropus* (Figure 5.7). Although the specimen broke during the excavation works, the reconstruction showed a pronounced brow ridge and glabella. The absence of sagittal crest was attributed to sexual dimorphism: SK 48 was thought to correspond to a female. (Incidentally, it is still one of the most complete *P. robustus* craniums available.) The discoveries at Swartkrans documented dental dimorphisms. A complete mandible (SK 23), which was very robust, with a high and solid ramus, wide molars and premolars, and small canines and incisors, was attributed to a female. Two

mandibles, which were even more robust, were classified as male specimens (Day, 1986).

The findings at Makapansgat were the fruit of Dart's renewed interest for paleoanthropology after his work became universally recognized. Such interest moved him to lead the research at the Anatomy Department of the University of Witwatersrand. In 1948 Dart suggested the name *Australopithecus prometheus* for the Makapansgat hominins (Dart, 1948). Nevertheless, the partial cranium found in 1962 in two parts (MLD 37 and MLD 38), lacking the frontal region and the face (Dart, 1962), turned out to be very similar to the Sts 5 specimen from Sterkfontein. It was believed to be a female because of the lack of occipital torus.

Taung, Sterkfontein, Swartkrans, Kromdraai, and Makapansgat are the South African sites with the earliest known hominin specimens, but not the only ones. For instance, Cooper's B site has been excavated since 1938. At the end of the twentieth century others were added: Gladysvale, Drimolen, and Gondolin, in addition to the hominin sites corresponding to the late Pleistocene, such as Duinefontein, Border Cave, Klasies River Mouth, Equus Cave, Witkrans Cave, and Kelders Cave, which we will discuss in subsequent chapters. For instance, Gladysvale has yielded two australopith teeth (GVH-1 and GVH-2), associated with abundant fauna, which are difficult to classify (Berger *et al.*, 1993). Drimolen has provided a very complete cranium of *P. robustus*, DNH 7, attributed to a female, known as Eurydice. A mandible with its dentition almost complete appeared together with the cranium. It was more robust and has been attributed to a male (DNH 8, Orpheus; Keyser, 2000; see Figure 5.17, below). *Paranthropus* teeth (GDA 1 and GDA 2) have been found at Gondolin, a site excavated by Elizaberth Vrba in 1979 (Menter *et al.*, 1999).

Altogether, South Africa provides a broad record of Pliocene and early Pleistocene hominin remains, the most abundant in the world. Classifying and dating those exemplars has been far from straightforward. To proceed in an orderly way, we will first deal with the structure and age of the South African sites. Thereafter, we will examine the different hominin species they contain.

Box 5.4 *Australopithecus transvaalensis*

Broom included the TM 1512 specimen in *Australopithecus transvaalensis* (Broom, 1936), although 2 years later he suggested changing it to *Plesianthropus transvaalensis* (Broom, 1938). Robert Broom, John T. Robinson, and Girrit Willem H. Schepers (1950) later included the Sts 5 specimen in *Plesianthropus transvalensis*. But

the Sterkfontein discoveries were finally included in the taxon *A. africanus* as specimens belonging to the same species as the Taung Child. Thus, they were a firm support for Dart's thesis and resolved many of the questions raised by the Taung specimen's infantile morphology.

Box 5.5 Postcranial remains from Makapansgat

Makapansgat has also yielded postcranial remains, such as the iliac fragments of a juvenile male (MLD 7) and a juvenile female ischium (MLD 25) (Dart, 1949a). The

Makapansgat hips have been used persistently since the time of Raymond Dart to argue in favor of the modernity of the australopithecine pelvis.

 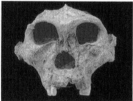

Figure 5.7 SK 48, *P. robustus* from Swartkrans (South Africa). Lateral and frontal views. Photographs from http://www.msu.edu/~heslipst/contents/ANP440/robustus.htm.

5.1.4 Structure and dating of the South African sites

South African australopith sites are located in limestone soils with dolomitic caves formed during the Precambrian and subsequently filled by limestone breccias. The dolomitic caves are excavated by the action of water, which dissolves calcareous soils and forms stalactites and stalagmites inside the caves. Usually the caves also fill with a conglomerate of rocky materials, debris, bones, and carbonated compounds, forming the breccias in which fossils are embedded. The original structures, very old, are generally deteriorated, and found in different stages of destruction (Figure 5.8).

In some instances, such as Kromdraai, the deposits were found on the surface of a hill,

because the cave had completely collapsed. Excavation works were required to identify the original geologic structure of that site. Swartkrans shows a completely eroded outer cave, while the roof is preserved in the interior. The structure of Sterkfontein is probably the most complex. This site consists of an outer cave open to the surface and successive deeper rooms that contain different kinds of breccia. In regard to Taung, the only discovery (besides some much later tools and bones), the Child, took place in the deepest and farthest part of a cave contained in a small hill. The mining works completely destroyed the original cave, whose structure is unknown. A brief but clarifying summary of the tormented geology of South African sites has been provided by Day (1986).

The complex geological structure casts many shadows on the ages assigned to South African sites. The use of paleomagnetism has recently provided a new way to obtain direct datings. However, paleomagnetic results, with the exception of Makapansgat, are not very consistent (Partridge, 1982). The use of fission track techniques in South African sites has not been fruitful either. Thermoluminescence or electronic-spin-resonance techniques do not reach the ages of the Members containing australopith remains. The absence of

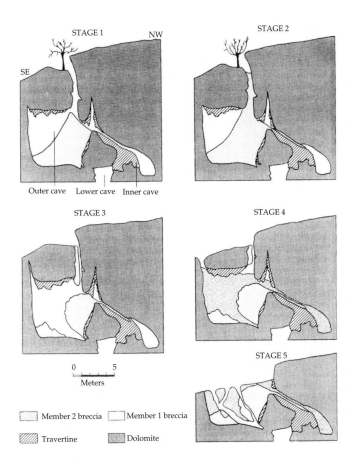

STAGE 1

SE   NW

Outer cave   Lower cave   Inner cave

STAGE 2

STAGE 3

0   5
Meters

STAGE 4

STAGE 5

Member 2 breccia   Member 1 breccia

Travertine   Dolomite

Figure 5.8 Phases in the destruction of a dolomitic cave.

materials that allow the use of radiometric methods (volcanic tuffs), in addition to the difficulties entailed in stratigraphic comparisons, have required researchers to be guided by studies of associated fauna and their correlation with other sites which are easy to date, such as East African ones.

Sterkfontein
The stratigraphy of the Sterkfontein cave (Figures 5.9 and 5.10) is the most thoroughly studied of all South African sites. It serves as a comparative reference to calibrate the datings at other sites, and as an example of the difficulties involved in the description and dating of a dolomitic cave. Thus, we will discuss it as an example. The stratigraphic sequence of Sterkfontein has been described profusely by Timothy Partridge (Partridge, 1975, 1978, 1982, 2000; Partridge and Watt, 1991; Partridge

et al., 1999, 2003). Six Members have been identified. Members 1, 2, and 3 are the deepest. Members 4, 5, and 6 are on the open surface of the cave, or close to it, and were initially identified by the color of their breccias. The different Members are not only vertically distributed, on top of each other, but they are also distributed horizontally, extending through the different chambers. As a consequence, it is sometimes difficult to identify the different infills within the deposits.

The Sterkfontein Members that contain hominin remains are: Member 2, which yielded the StW 573 specimen, Little Foot, discussed in section 3.2; Member 4, with many *A. africanus* specimens; and Member 5, with *P. robustus* specimens as well as exemplars attributed to *Homo*. The ages of these Members is controversial. Let us begin with Member 2, the oldest. There are those who argue that it is very old. The discoverers of StW 573,

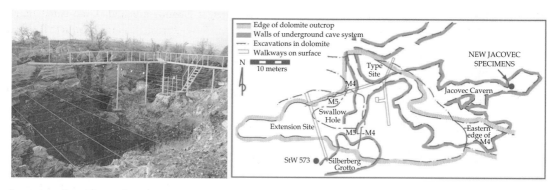

Figure 5.9 Left: aerial view of the Sterkfontein site at present. Right: diagram of the underlying caves. From Partridge *et al.* (2003). Reprinted with permission from AAAS.

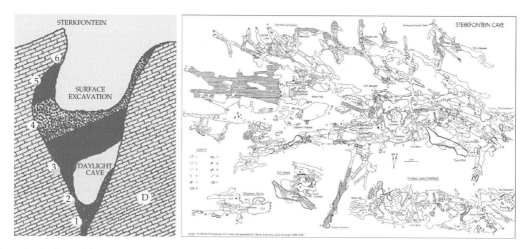

Figure 5.10 Left:, Sterkfontein stratigraphy, according to Timothy Partridge. Right: speleological map of the cave: (1) cavity outline; (2) cavity outline under single overlay; (3) cavity outline under double overlay; (4) scarp and edge of pit in cave; (5) scarp and edge of pit at surface (entrance); (6) blocks and scree; (7) walls of breccia and boulder blockages; (8) *Australopithecus* skeleton; (9) masonry wall; (10) gates and fences; (11) flight of stairs. Picture from Martini *et al.* (2003).

Little Foot, date it to about 3.5 Ma (Clarke and Tobias, 1995). Partridge and colleagues (1999) provided a more accurate estimation by means of paleomagnetic studies; they identified five inversions (Figure 5.11). The StW 573 skeleton belongs to an interval between the end of the Mammoth subchron (3.22 Ma) and the Gauss–Gilbert boundary (3.58 Ma); the skeleton's age was constrained to 3.30–3.33 Ma by assuming a constant sedimentation rate during this interval (Partridge *et al.*, 1999). Kuman and Clarke (2000) have accepted this estimation of 3.3 Ma (although the focus of Kuman and Clarke is Member 5).

McKee (1996) has expressed doubts regarding the age attributed by Clarke and Tobias (1995). On the basis of faunal comparisons, he attributed to the StW 573 specimen an age of 2.6–2.5 Ma. This makes the specimen younger than Makapansgat Member 3, which is 3.0 Ma. In their response Tobias and Clarke (1996) have rejected the faunal comparison on the grounds of the great persistence of carnivores (present since 3.5 Ma). Berger and colleagues (2002) have argued against the age of Member 2 suggested by Partridge *et al.* (1999). Their interpretation of the fauna, the archeometric results, and the magnetostratigraphy of Sterkfontein

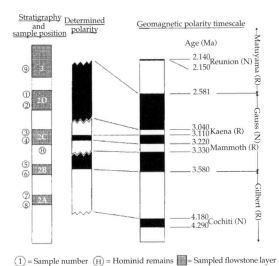

Stratigraphy and sample position / Determined polarity / Geomagnetic polarity timescale

Age (Ma)

2.140 Reunion (N)
2.150

2.581

3.040
3.110 Kaena (R)
3.220
3.330 Mammoth (R)

3.580

4.180
4.290 Cochiti (N)

Matuyama (R) | Gauss (N) | Gilbert (R)

① = Sample number   Ⓗ = Hominid remains   ▨ = Sampled flowstone layer

Figure 5.11 Paleomagnetic column of Sterkfontein (Partridge *et al.*, 1999). The column on the left shows the position of the samples taken. The column on the right shows the geomagnetic polarity timescale (Cande and Kent, 1995). The column in the middle shows the correlation of the paleomagnetism at Sterkfontein with the geomagnetic timescale, although the broken lines indicate magnetic inversions whose positions cannot be specified because they fall outside the studied sample. From Partridge *et al.* (1999).

indicate that "it is unlikely that any Members yet described from Sterkfontein are in excess of 3.04 Ma in age" (Berger *et al.*, 2002). Berger and colleagues estimate the age of Sterkfontein Member 2 to under 3.0 Ma. According to Partridge *et al.* (2003), "although the paleomagnetic measurements in Member 2 are apparently precise, proper dating relies on the correct identification of magnetic reversals in a sequence that may not have preserved all of these changes. The paleomagnetic age assignment is also subject to uncertainties in both the faunal correlations and sedimentation rates, and so requires independent verification." In order to carry out such verification, Partridge *et al.* (2003) conducted a new study based on burial dating by the radioactive decay of cosmogenic ^{26}Al and ^{10}Be in quartz (see Box 5.6). According to their results, the breccia that contained StW 573 is approximately 4.0 million years old. Partridge (1978) estimated other ages of the Sterkfontein Members containing hominins: Member 4, between 2.8 and 2.6 Ma. This Member shares with

Makapansgat Members 3 and 4 the same fauna found in East Africa, where it is well dated between 3 and 2.9 Ma. But Berger *et al.* (2002) have estimated the age of Sterkfontein Member 4 as being 2.5–1.5 million years old.

An interesting specimen from Bed B of Member 4 is the StW 431 partial skeleton described by Toussaint and colleagues (2003). Although the exemplar is lacking the cranium, it includes a total of 18 postcranial bones, with the better part of the vertebral column. The specimen shows a mosaic of primitive traits—elbow, small size of the lumbar vertebrae and sacro—and derived traits—a pelvis broadly similar to that of modern humans. Toussaint *et al.* (2003) assigned the specimens to *A. africanus*, though they note morphological similarities with *A. afarensis*, which support the proximity of these two taxa. Toussaint *et al.* (2003) believe StW 431 was not an obligate biped; that is to say, its locomotion was different from that of modern humans.

Sterkfontein Member 5 includes the Oldowan, east and west breccias, ordered from oldest to youngest probable age. The Oldowan breccia is 2–1.7 million years old (Clarke, 1994), while the set of tools from the west breccia—the appearance of which places them between advanced Oldowan and Acheulean—suggests it is younger, from about 1.7 to 1.4 Ma. Regarding Sterkfontein Member 6, it could be between 0.2 and 0.1 Ma (Kuman and Clarke, 2000).

Makapansgat

Faunal comparisons led Basil Cooke (1964) to estimate the age of Makapansgat deposits 3 and 4 as 3.0–2.5 Ma. Through an extensive review of available studies, Partridge (1982), estimated Makapansgat Member 3 at 3 Ma, which is in agreement with paleomagnetic results. The comparison of the Makapansgat suid (pig) series with their East African equivalents pointed in another direction (White *et al.*, 1981). Tim White and colleagues believed that the age of Makapansgat Member 3 is only 2.6 Ma. But in Partridge's (1982) opinion, both the paleomagnetic results and the evidence showing that *A. africanus* from Member 3 are older cannot be overlooked. Regarding Member 4 of that site, its fauna resembles that of

---

Box 5.6 Dating using ^{26}Al and ^{10}Be

"Burial dating is based on the radioactive decay of ^{26}Al (radioactive mean life $\tau_{26} = 1.02 \pm 0.02$ Ma) and ^{10}Be ($\tau_{10} = 1.93 \pm 0.10$ Ma) in quartz. These two cosmogenic radionuclides are produced in a known ratio by secondary cosmic-ray nucleons and muons near the ground surface. Quartz grains near the surface accumulate an inventory of these radionuclides, whose concentrations depend on the mineral's exposure time to cosmic rays, which in turn depends on the erosion rate of the host rock. If quartz from the surface is suddenly buried—for example, by deposition in a cave as at Sterkfontein—then production of ^{26}Al and ^{10}Be drastically slows or ceases. Because ^{26}Al decays more rapidly than ^{10}Be, the ^{26}Al/^{10}Be ratio decreases exponentially with burial time, offering a means to date the sediment burial. The sediment's age is calculated by solving simultaneously for burial time and preburial ^{26}Al and ^{10}Be concentration." (Partridge et al., 2003).

Box 5.7 Features of Sterkfontein Member 4

Kuman and Clarke (2000) carried out a detailed study of the upper Sterkfontein Members. They identified the following significant features in Member 4: (1) absence of lithic tools, (2) vegetation abundant with lianas, and (3) fauna in which animals such as *Equus*, *Pedetes*, and *Struthio* are absent, whereas *Colobus* monkeys are present. Taken together, these observations suggest a forest and humid climate, which distinguishes Member 4 from Member 5. There is evidence suggesting that the latter corresponds to a cooler and drier climate, with animals typical of open spaces. Consequently, the frontier between both Members would coincide with the great African climatic change that took place about 2.5 Ma (2.6 for Kuman and Clarke), which would imply that Member 4 spans from 3 to 2.6 Ma. However, Kuman and Clarke (2000) noted that Member 4 from Sterkfontein might include, in its southern part, younger deposits corresponding to a cooler and drier time. The StW 53 infill, attributed by Partridge (1982) to Member 5, is considered by Kuman and Clarke as a transition estimated to date from between 2.6 and 2 Ma.

Sterkfontein Member 4 (Partridge, 1982). Both Makapansgat Members 3 and 4 have yielded *A. africanus* specimens.

Swartkrans, Kromdraai, and Taung
Swartkrans and Kromdraai can be dated, by comparison with Sterkfontein, using adequate faunal correspondences. Partridge (1982) believed that Swartkrans Member 1 and Kromdraai B Member 3 belong to the interval between 2 and 1 Ma. Swartkrans Member 2 would be around half a million years old. Brain and Sillen (1988) pushed the age of Swartkrans to 1.5–1.0 Ma, while Delson (1988; see below) favored the more restrictive estimate of 1.8–1.6 Ma. We will accept a range between 2 and 1.5 Ma, which is almost the average (Table 5.1).

In regard to Taung, the destruction of the cave caused by quarry works forced Partridge (1982) to base his estimation on indirect evidence, which suggests a similar age to Kromdraai B Member 3, between 2 and 1 Ma. Contrary to Partridge's opinion, Delson (1988) argues that the fauna at Taung is older than 2 Ma and younger than 2.5 Ma, with the most probable estimate for the site's age at 2.3 Ma. Delson's (1988) work consisted in broadening and detailing the sample of cercopithecoid monkeys, correlating them with the same East African species. In order to do so, Delson took into account the estimations of the chronology of the Turkana basin (Koobi Fora, West Turkana, and Shungura formation) provided by Brown et al. (1985a). Turning to Swartkrans and Kromdraai, Delson estimates both sites (on the grounds of Swartkrans Members 1–3 and the original fauna collected by Broom at Kromdraai) to be between 1.8 and 1.6 million years old.

Table 5.1 Estimates by different authors for the ages of South African sites

Site	Taxa	Age (Ma)									
		A	B	C	D	E	F	G	H	I	J
Sterkfontein, Mbr 2	StW 573					3.5	2.6–2.5	3.3		<3.0	>4.0
Sterkfontein, Mbr 4	A. africanus	2.8–2.6								2.5–1.5	
Sterkfontein, Mbr 5	Australopithecus/ A. africanuso/ H. habilis (?)								2.0–1.4		
Sterkfontein, Mbr 6									0.2–0.1		
Makapansgat Mbr 3	A. africanus	3–2.5	2.6								
Makapansgat Mbr 4	A. africanus	3–2.5									
Swartkrans 1	P. robustus/ H. habilis/ H. erectus (?)	2–1		1.5–1	1.8–1.6						
Kromdraai B	P. robustus	2–1			1.8–1.6						

Sources: A, Partridge (1978; 1982), stratigraphy, fauna, paleomagnetism; B, White *et al* (1981), fauna; C, Brain and Sillen (1988); D, Delson (1988); E, Clarke and Tobias (1995); F, McKee (1996), fauna; G, Partridge *et al.* (1999), paleomagnetism; H, Kuman and Clarke (2000), stratigraphy and fauna; I, Berger *et al.* (2002), fauna, archeometry, magnetostratigraphy; J, Partridge *et al.* (2003), [26]Al and [10]Be decay.

5.2 South African hominins of the Pliocene

South African hominin specimens have been classified in a variety of different genera and species. The diversity of forms has eventually been reduced to three main types: (1) a very early gracile form; (2) a later robust form; and (3) another gracile form, but different from (1) and contemporaneous with (2). These three forms are usually classified as *A. africanus*, *P. robustus*, and *Homo habilis*, respectively. The case of Little Foot deserves a separate consideration because of its age and its anatomical features.

5.2.1 Little Foot

We have already mentioned the fossil footbones, found at Sterkfontein Member 2, known technically as StW 573, and colloquially as Little Foot. Certain details of its discovery suggested that additional remains belonging to the same individual could have been preserved and were still contained within the rock. The search for them is due to Ron Clarke's initiative (see Box 5.8).

The result was the discovery of quite a complete skeleton embedded in the breccia. The work to retrieve the remains is still underway. The materials recovered to date show that it is a fossil in an excellent state of conservation, the most complete of all hominins dated to around 3.5–3 Ma.

In addition to the footbones, the StW 573 remains described to date include the cranium (Clarke, 1998; Figure 5.12) and a complete left arm and hand (Clarke, 1999; Figure 5.13). Given that these fossils are still embedded in the rock, Ron Clarke warned about the provisionality of the studies. With regard to the cranium, he noted that it corresponds to a mature adult with a zygomatic arch which is much more massive than that of *A. africanus*. Additionally, it has a small sagittal crest on the parietals, and the nuchal plane is very muscular with a pronounced, pointed inion. Clarke (1998) argued that in no way does it conform to the morphology of *A. africanus* specimens from Sterkfontein Member 4.

Clarke's (1999) preliminary observations of the hand and the arm reveal a mosaic of traits. Some of them (such as the heads of both radius and ulna) more closely resemble those of apes than they do

Box 5.8 The discovery of Little Foot

Ron J. Clarke recounts how the first bones of Little Foot were identified: "In a bag labelled as bovid tibiae, I found a shaft, with the distal end intact, of what was clearly a hominid tibia, and it fitted perfectly with the left talus of StW 573. Then I realised that the other distal tibia fragment was from the right leg of the same individual. In this light, I checked again the damaged, supposed cuneiform that I had noted earlier and found that it was indeed a mirror image of the left lateral cuneiform that I had just found. I now had part of the right foot of the same individual. I also found in a bag labelled 'Dump 20, Bovid Humeri' at Sterkfontein a heavily damaged chunk of bone that I identified as part of a hominid calcaneum. When I checked it against the previous fragment that I had thought was calcaneum, they fitted together. . . . I now had a total of 12 foot and lower leg bones of one ape–man individual—the left tibia and fibula, which joined to an articulated set of eight foot and ankle bones, and the distal fragment of a right tibia and right lateral cuneiform . . . —the implication was stunning: I stated my conviction that the rest of the skeleton was still encased in the cave breccia of the Silberberg Grotto. . . .

I gave a cast of the distal fragment of the right tibia to two of the Sterkfontein fossil preparators, Nkwane Molefe and Stephen Motsumi, and asked them to search the exposed breccia surfaces in the entire Silberberg Grotto (except for the area we had recently blasted) to find a matching cross section of bone for which this would provide an exact fit .The task I had set them was like looking for a needle in a haystack as the grotto is an enormous, deep, dark cavern with breccia exposed on the walls, floor and ceiling. After two days of searching with the aid of hand-held lamps, they found it on 3 July 1997, near the bottom of the Member 2 talus slope at the western end of the grotto. This was at the opposite end to where we had previously excavated. The fit was perfect, despite the bone having been blasted apart by lime workers 65 or more years previously. To the left of the exposed end of the right tibia could be seen the section of the broken–off shaft of the left tibia, to which the lower end of the left tibia with foot bones could be joined. To the left of that could be seen the broken–off shaft of the left fibula. From their positions with the lower limbs in correct anatomical relationship, it seemed that the whole skeleton had to be there, lying face downwards." (Clarke, 1998)

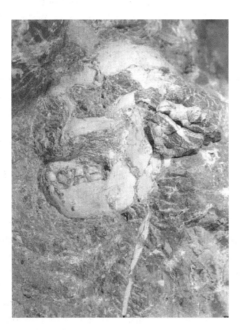

Figure 5.12 Left side of the StW 573 skull partially excavated from the breccia. Photograph from Clarke (1998).

those of modern humans. The size and length of radius, ulna, and humerus are within the range of average modern humans and of chimpanzees and are not elongated like those of orangutans, although some features observed on the distal humerus and proximal ulna resemble those of orangutans rather than those of chimpanzees and humans. Metacarpal length is similar to that of *Australopithecus* and *Homo*. The proximal phalanges of the thumb and forefinger display a curvature like that of the phalanges of *A. afarensis* specimens from Hadar (Ethiopia). The first metacarpal and its proximal phalanx indicate a thumb similar to that of modern humans. Conversely, the trapezium of the wrist is different from those of modern humans, chimpanzees, and orangutans, showing unique features. Clarke (1999) has noted that the significance of the unusual thumb joint on the StW 573 specimen "has still to be determined".

How can the taxonomy of such a specimen be resolved? Clarke's (1988) answer: "I prefer to reserve judgment on the fossil's exact taxonomic

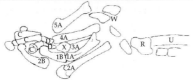

Figure 5.13 Handbones of StW 573. Legend for interpretive sketch: 1A, first metacarpal (thumb); 2A, second metacarpal (index finger); 3A, third metacarpal (middle finger); 4A, fourth metacarpal (ring finger); 5A, fifth metacarpal (little finger); 1B, proximal phalanx of thumb; 2B, proximal phalanx of index finger; C, middle phalanges; D, terminal phalanges; W, wrist bones; R, damaged radius; U, damaged ulna; X, unknown. The trapezium is at the lower right-hand edge of picture. Pictures from Clarke (1999).

affinities, although it does appear to be a form of *Australopithecus*." Preliminary descriptions suggest StW 573 could be regarded as a very early species, older than *A. africanus*, and closer, in morphological terms, to contemporary Ethiopian australopiths. As Clarke notes, the hand/arm set make possible to study an *Australopithecus* forelimb as a complete unit for the first time, allowing interpretation of its function. The strong opposable thumb, the curved phalanges, and orangutan-like elbow joint are suggestive of arboreality, as the footbones that inspired the name Little Foot did earlier. Clarke (1999) agrees with Sabater Pi and colleagues (1997) that australopiths might have nested in trees. Clarke added that they might also have spent part of the day feeding in trees, as orangutans and chimpanzees do.

5.2.2 *Australopithecus africanus*

Remains of *A. africanus* have been found at Makapansgat (Members 3 and 4), Sterkfontein

(Member 4) and, with less certainty, at Sterkfontein Member 5. The cranium found at Taung, the famous Child, is the most remarkable South African fossil hominin. Furthermore, it is the type specimen proposed by Raymond Dart for the species *A. africanus*. However, subsequent discoveries have cast doubt on the true significance of the Taung specimen. The great amount of *A. africanus* specimens from Sterkfontein Member 4 provide a hypodigm whose traits differ somewhat from Taung's. Thus, the true species to which the latter belongs is an open question.

Despite such abundance of cranial and postcranial remains, the character of *A. africanus* has been highly debated, maybe due to the difficulty of tracing phyletic relations between this species and later *P. robustus* and *H. habilis*. In any case, the cranium of *A. africanus* exhibits a modest brow ridge, a rather pronounced glabella, a weak occipital crest, and a low nuchal plane. The foramen magnum is placed in the middle of the cranium and the mastoid region is small. The cranial case is narrow, with a capacity close to 485 cm^3. The mandible (Sts 52b for instance) has a tall ascending ramus and lacks a chin. The dentition is characterized by wide and small incisors, rather projected canines and relatively large premolars, with the size of the teeth increasing from front to back. In section 3.2 we mentioned that the hips and lower limbs were indicative of a clear bipedalism combined with climbing tendencies. This suggests that its locomotion was different from that of later *Homo erectus*.

The most complete *A. africanus* cranium is StW 505 (Figure 5.14), found in Bed B of Sterkfontein Member 4. It includes most of the face, the left endocranium, the cranial anterior fossa, and part of the right middle cranial fossa. A partial right temporal bone, initially believed to belong to another specimen (StW 504), was later reinterpreted as part of StW 505, receiving the name of StW 505b. This specimen was studied in detail 10 years later (Lockwood and Tobias, 1999). Lockwood and Tobias concluded that, despite its large size and robust complexion, the StW 505 specimen's morphological features are characteristic of *A. africanus*. Nevertheless, this specimen shows

Box 5.9 The mystery of the radius associated to StW 573

"There is one other bone present that, while appearing to be undoubtedly part of the *Australopithecus* skeleton, has three uncharacteristic features. It is a left radius that lies next to the left femur in a position that might be expected for a minimally disturbed skeleton. As there are no other animal fossils apart from an occasional small fragment within a wide vicinity of the skeleton, there was no reason to think that it was not part of the hominid. Furthermore, apart from slight crushing at the distal end, this slender bone is virtually intact and unbroken, despite being adjacent to and downslope of a large rock. Thus, it seems unlikely to have rolled down the slope from elsewhere. . . . As the skeleton could not have had two left radii, it was necessary for me then to expose more of this previous left radius that was still largely encased in breccia next to the femurs. As soon as I had uncovered the proximal end of the shaft, it became clear that it was morphologically like a large cercopithecoid, although the distal end did not quite match any of our fossil or modern cercopithecoids. It appears an extraordinary coincidence that this is the only virtually complete animal bone in the vicinity of the hominid skeleton. It is also in the correct position relative to the femurs to have belonged to the hominid if the latter's arm had been resting by its side. Although it might simply be coincidence, one can at least consider the possibility that the *Australopithecus* may have been carrying that radius when it died." (Clarke, 1999)

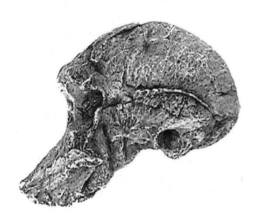

Figure 5.14 StW 505, *A. africanus*, reconstructed by Charles Lockwood *et al*. Picture from Howks and Wolpoff, Science 283: 96, 1999. Reprinted with permission from AAAS.

affinities with *A. afarensis* (the weak sagittal crest mentioned earlier, for instance, which is similar to that of A.L. 444-2, interpreted as a plesiomorphy) and with *H. habilis* (a similar brow ridge to the one observed in the KNM-ER 1470 specimen, for example).

The computer-generated three-dimensional reconstruction of StW 505 yields a cranial capacity around 515 cm^3, the highest of all *A. africanus* (Conroy *et al*., 1998). After a comparison between StW 505 and other hominin specimens from South Africa (Sts 71) and East Africa (OH 24,

KNM-ER 1813, and KNM-ER 732; see below), Conroy and colleagues concluded that the brain volumes of these specimens had been overestimated.

StW 505 is not the only specimen from Sterkfontein Member 4 exhibiting robust features. The diversity of this Member's sample supports the notion that there are different species at the site. Hence, specimens StW 183 and StW 255 show derived traits of robust australopiths, such as *P. robustus* from South Africa, but also *Paranthropus boisei* and *Paranthropus aethiopicus* from East Africa, which we will review later (Lockwood and Tobias, 2002). Even StW 252 could be included in this robust set from Member 4 on the grounds of its dental features (Clarke, 1988). The specimen showing the most robust features is StW 183. But its comparison with other South African forms is difficult because it is a juvenile specimen. Lockwood and Tobias (2002) note some derived traits of *P. robustus* in StW 183, such as the rounded lateral portion of the inferior orbital margin, found in no other specimen from Sterkfontein Member 4. In any case, Lockwood and Tobias (2002) maintain that there was a single species at Sterkfontein Member 4, *A. africanus*. Ron Clarke (1988) rather favors that the cranial sample includes a robust species (StW 252, Sts 71) and a gracile species (Sts 5, Sts 17, Sts 52).

5.2.3 Plio-Pleistocene *Homo* in South Africa

The *A. africanus* remains from Sterkfontein and Makapansgat are not the only gracile South African Plio-Pleistocene specimens. In 1949, John Robinson, Robert Broom's assistant at the time, found a mandibular fragment, SK 15, at Swartkrans. Broom and Robinson (1949) classified the exemplar as *Telanthropus capensis*, believing it to represent a completely new kind of hominin. Broom and Robinson wrote in their article that its morphology was very advanced, corresponding to an immediate ancestor of modern humans (and they even compared it with Heidelberg Man; see section 8.3). SK 15 was embedded in a dark colored breccia (Member 2), which contrasted with the pink-colored breccia at the same site (Member 1) in which robust australopithecines had been found. As we saw before, associated faunal studies had estimated the average age of Swartkrans to around 1.5 Ma. Broom and Robinson argued that, while *Telanthropus* and *P. robustus* were a similar age, the former was much more modern than the latter from a morphological point of view. This suggested it could represent an intermediate stage between australopiths and modern human beings.

A new mandibular fragment was found at Swartkrans the same year, SK 45. Its anatomy was quite similar to that of SK 15, even more modern-looking, closer to that of current humans (Broom and Robinson, 1950). But SK 45 was embedded in the same pink-colored breccia containing robust australopith specimens; it is older than SK 15. After some initial doubts, Broom and Robinson (1952) decided to assign SK 45 to the genus *Telanthropus*, while noting that it corresponded to an early hominin with similar features to those of *Homo*.

The story of *Telanthropus* has been rigorously and clearly narrated by Phillip Tobias (1978). Setting aside the taxonomic vicissitudes of the genus, currently rejected, the question remains of the presence in South African sediments of specimens which are anatomically similar to *Homo*. How can this be explained? Multiple answers have been suggested. In 1961 Robinson suggested including *Telanthropus* specimens in *H. erectus*. This was accepted, with varying degrees of

enthusiasm, by such respected authors as Tobias, von Koenigswald, B. Campbell, and Day, among others (Tobias, 1978). However, the proposal of the new species *Homo habilis* from Olduvai in 1964, which we will describe further on, opened a new perspective: that *Telanthropus*, mainly SK 45, belonged to *H. habilis*. Thereafter, advanced specimens from Swartkrans have generally been included in the latter species.

The fossil remains from Swartkrans were reinterpreted taking into account the early presence of *Homo* at the site. For instance, Clarke *et al.* (1970) assigned the reconstruction of the SK 847 cranium (Figure 5.15), previously believed to be a robust australopith, to *Homo* sp. The detailed study carried out later by Clarke, in his doctoral dissertation, could not tell whether the specimen belonged to *H. habilis* or *H. erectus*. As we will soon see, this is not an easy problem to solve.

Randall Susman *et al.* (2001) identified hominin specimens in the Sterkfontein faunal collection, increasing the *Paranthropus* and *Homo* samples from Swartkrans. The SK 1896 distal femur and other bones attributed to *Homo* cf. *erectus* are of particular interest. According to Susman *et al.* (2001) these remains indicate that male *Homo* were larger than *Paranthropus* at Swartkrans.

Sterkfontein also yielded early specimens that have been assigned to the genus *Homo*. The 1957 and 1958 campaigns retrieved several hominin remains from Member 5, estimated by associated fauna to be between 2.0 and 1.4 Ma. Most of them were mandibular fragments and teeth. They were assigned to

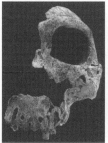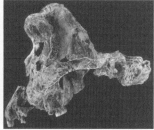

Figure 5.15 Frontal and lateral views of the SK 847 specimen from Swartkrans, *Homo* sp. (Clarke *et al.*, 1970). Photographs from http://www.msu.edu/~heslipst/contents/ANP440/ergaster.htm.

different species: *Australopithecus, Telanthropus,* and *Homo.* In 1976 A.R. Hughes found a partial cranium in the StW 53 infill. It included part of the skull, the face and palate, several teeth, and a right mandibular ramus, and was catalogued as StW 53 (Hughes and Tobias, 1977; Figure 5.16).

Even though the reconstruction of StW 53 did not allow calculation of its cranial capacity, Tobias (1978) argued that it was moderately large for such a small cranium. Given that its morphology was reminiscent of some *H. habilis* specimens from Olduvai, Hughes and Tobias (1977) classified StW 53 as *H. habilis* or *H. aff. habilis* in the first published study of the specimen. Tobias (1978) linked it with two similar-aged partial crania from Sterkfontein (StW 53) and Swartkrans (SK 847). This supported earlier decisions to assign them to the genus *Homo,* although it was not specified whether they belonged to the same species. Their faces were very similar, but regarding the supraorbital region, StW 53 resembled *H. habilis,*

whereas SK 847 resembled *H. erectus.* This is a typical problem related to the difficulty of assigning specimens to one species or another when they are close to their cladistic separation. Ronald J. Clarke (1985), author of the reconstruction of both StW 53 and SK 847, classified the former as *Homo habilis* and the latter as *Homo erectus.* However, Kuman and Clarke's (2000) reinterpretation returned StW 53 to *Australopithecus,* on the grounds of numerous traits, but thought it was a late exemplar of the genus and did not assign it to any particular species.

Curnoe and Tobias (2006) have provided a detailed description and morphological comparison of StW 53, after its reconstruction. They conclude that it shares many cranial features with *H. habilis,* and thus favor keeping StW 53 within this species. Their comparison of StW 53 with SK 847 favors assigning the latter specimen to *H. habilis.*

Are there any *H. habilis* specimens in South Africa? Or is it better to consider the late

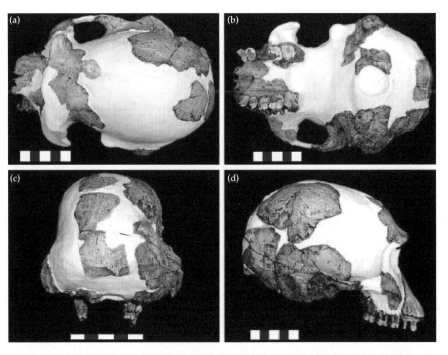

Figure 5.16 New reconstruction of StW 53 from Sterkfontein, *H. habilis* (Hughes and Tobias, 1977) or an unspecified species of *Australopithecus* (Kuman and Clarke, 2000). (a) Superior view. (b) Inferior view. (c) Occipital. (d) Right lateral. Scale bars, 5 cm. Photographs from Curnoe and Tobias (2006). Reprinted from *Journal of Human Evolution.* Vol 50: 1, Curnoe and Tobias, 'Description, new reconstruction' 36–77, 2006 with permission from Elsevier.

gracile-looking specimens from Sterkfontein and Swartkrans as *H. erectus*? We believe the latter is the most parsimonious option.

5.2.4 *Paranthropus robustus*

Many of the different genera and species used by Robert Broom and other authors to classify South African findings have eventually been abandoned. But it would be unfair to criticize the multiplication of genera and species of the past as if it were senseless. It would even be worse to believe we are the first ones to realize the problem. As different South African hominins were discovered, some researchers advocated considering them as belonging to a single genus, and even a single species. For example, Broom (1950) quoted S.H. Haughton and H.B.S. Cooke, who reintroduced an argument used against synapomorphies in many instances: if two very similar species were present at the same time and the same or similar places, then they must be regarded as a single species. But Broom (1950) gave sound arguments in favor of recognizing different South African genera. Taung corresponds to a juvenile specimen, and establishing it as the holotype of *A. africanus* was, undoubtedly, an inconvenience. Hence, Broom's comparisons between the Taung specimen and other fossils available at the time had to be based on deciduous teeth.

Despite such problems, Broom (1950) reached an important taxonomic conclusion after examining South African hominin genera and species. He suggested there were two kinds of early hominin in South Africa. He placed them at the subfamily level: Australopithecinae, including the genera *Australopithecus* and *Plesianthropus*;

and Paranthropinae, including the genus *Paranthropus*. Broom added a doubtful subfamily (Archanthropinae, encompassing *Australopithecus prometheus*).

Broom's splitting tendency has been beneficial to a certain extent. Lowering the differences from the subfamily to the genus level, and forgetting the arcanthropines, a doubtful concept in Broom's opinion, South African sites have allowed documenting an essential difference among late-Pliocene hominins: the aforementioned gracile and robust forms. This dichotomy is usually recognized at the genus level, with gracile specimens included in *Homo* and robust ones in *Paranthropus*. Some authors, such as Tobias, prefer to minimize the difference among genera, considering robust hominins as a subgenus, *Australopithecus* (*Paranthropus*). But few authors followed the steps of those reducing all late-Pliocene South African hominins to a single species.

Clarke (1985) used 20 derived cranial traits to characterize *Paranthropus* (Figure 5.17). Some of the most conspicuous features were their great molar and premolar widening, anterior teeth relatively small compared to molars, very massive mandibles, and a large masticatory musculature (inferred from the muscle insertions on the cranium), in addition to other minor dental traits. We will soon turn to another distinctive feature of *Paranthropus*: enamel thickness.

5.2.5 Robust versus gracile: dental enamel and diet

What sets gracile and robust hominins apart? The name is explicit enough: *A. africanus* appears

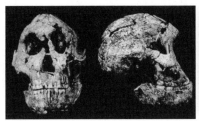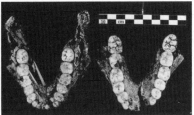

Figure 5.17 DNH 7 (left) and DNH 8 (right), *Paranthropus robustus.* Photographs from Keyser (2000).

lighter than *P. robustus*. As Grine (1988) said, Broom's choice was prophetic. But the concept of robusticity is almost phenomenological. Can this concept be described in greater detail?

In 1938 Broom chose "robust" as a designation for the hominin he had discovered at Kromdraai based mainly on the teeth and the structure of the face (Table 5.2). These features were especially noticeable in comparison with the specimens found at Sterkfontein not long before. Broom described the Swartkrans hominin later, in 1949. Its appearance was close enough to that notion of robusticity to place it in the genus *Paranthropus*. In fact, the teeth of the Swartkrans robust hominin were even larger that those of Kromdraai

specimens. The size of the teeth played, hence, an important role in determining the degree of robusticity, a task undertaken by John Robinson (1954, 1968). Together with other dental traits, the size was also an indication that paranthropines were vegetarian. This thesis, advocated by Robinson, will come up again later when we examine the diet of different hominins.

Dental enamel is another differential trait. The robust australopith lineage has very thick molar enamel compared to the thick enamel—but not extremely thick—of *A. africanus*. Grine and Martin (1988) have shown that this trait raises important phylogenetic and adaptive questions.

Table 5.2 Differences between gracile and robust australopihtecines

	'Gracile'	'Robust'
Cranial		
1. Overall shape	Narrow, with 'unmistakable' forehead; higher value for supraorbital height index (Le Gros Clark, 1950)	Broad across the ears; lacking a forehead; low supraorbital height index
2. Sagittal crest	Normally absent	Normally present
3. Face	Weak supraorbital torus; variable degree of prognathism, sometimes as little as 'robust' form	Supraorbital torus well developed medially to form a flattened 'platform' at glabella; face flat and broad, with little prognathism
4. Floor of nasal cavity	More marked transition from the facial surface of the maxilla into the floor of the pyriform aperture; sloping posterior border to the anterior nasal spine and lower insertion of the vomer	Smooth transition from facial surface of maxilla into the floor of the pyriform aperture; small anterior nasal spine that articulates at its tip with the vomer
5. Shape of the dental arcade and palate	Rounded anteriorly and even in depth	Straight line between canines, deeper posteriorly
6. Pterygoid region	Slender lateral pterygoid plate	Robust lateral pterygoid plate
Dental		
7. Relative size of teeth	Anterior and posterior teeth in 'proportion'	Anterior teeth proportionally small; posterior teeth proportionally large
8. dm_1	Small, with relatively larger mesial cusps. Lingually situated anterior fovea; large Protoconid with long, sloping buccal surface	Large, molariform, with deeply incised buccal groove and relatively large distal cusps
9. P^2-roots	Single buccal root	Double buccal root
10. $\underline{c}^1$	Large, robust and symmetric crown with slender marginal ridges and parallel lingual grooves	Small, *Homo*-like, with thick marginal ridges and lingual grooves converging on the gingival eminence
11. $\bar{c}^1$	Asymmetric crown with marked cusplet on the distal marginal ridge and marked central ridge on the lingual surface	More symmetric crown with parallel lingual grooves, weak lingual ridge and featureless distal enamel ridge

Source: Wood and Richmond (2000).

Talking about dental enamel is talking about dietary tendencies. Initial interpretations of the South African remains suggested that *Paranthropus* and *Australopithecus* specialized in different kinds of food. Dart conceived *A. africanus* as a hunter and, thus, a carnivore. He pictured *A. africanus* as using stone weapons to survive in the savannah in competition with other predators (Dart, 1925, 1949b, 1953, 1957). After a comparative study of the dentition of South African australopiths, Robinson (1954) put forward his dietary hypothesis as an interpretation of the differences between gracile and robust hominins. According to Robinson the diet of *Paranthropus* was vegetarian, whereas that of *A. africanus* was omnivorous, and included an important amount of meat. Robinson believed the structure of the face and cranium reflected this dietary difference.

The studies of Frederick Grine on the microwear patterns left by food on teeth (Grine, 1981, 1987; Kay and Grine, 1985; Grine and Martin, 1988; Ungar and Grine, 1991) provided the first direct evidence about dietary habits of robust and gracile australopiths. The microwear patterns vary with the hardness of the materials that were chewed. Grine and colleagues examined the microwear patterns of deciduous and permanent molars under the microscope, and compared them with those of current primates. They concluded that *Paranthopus* and *Australopithecus* had different dietary habits. According to Grine's model, the diet of *A. africanus* was mainly folivorous. But the impossibility of comparing their microwear patterns with those of current frugivorous primates that do not also chew hard objects led Grine to suggest fruits could have also been part of the diet of *A. africanus*. Conversely, *P. robustus* would have chewed smaller and harder objects, such as seeds. The microwear studies on incisors revealed that the diet of *A. africanus* was more varied than that of *P. robustus*, which was more specialized (Ungar and Grine, 1991; Figure 5.18).

The analysis of the strontium/calcium (Sr/Ca) ratios in tooth enamel is a second source of direct evidence of the effects food has on teeth. Andrew Sillen and colleagues (Sillen, 1992; Sillen *et al.*, 1995, 1998) have applied this technique to Swartkrans fossil mammals. They found that Sr/Ca ratios correlated with feeding habits. Leaves and grass contain less strontium than fruits and seeds. To a first approximation, a high Sr/Ca ratio would suggest an herbivorous diet (but not only leaves), or carnivorous. Because the amount of strontium varies locally, migration patterns had to be taken into account. In any case, the studies showed that for *P. robustus*, Sr/Ca ratios were higher than for carnivores, but lower than for baboons, whose diet is omnivorous (Sillen, 1992). Gracile specimens from Swartkrans—*H. habilis* and *H. erectus*—showed a similar Sr/Ca ratio to baboons. Thus, it seems their diet was more varied than that of robust specimens.

The third direct source of dietary evidence is analysis of stable carbon isotopes. The ratio of carbon isotopes $^{13}C/^{12}C$ is lower in trees and shrubs than in grasses, reflecting the way they are incorporated into the trophic chain. Herbivores that feed on grass and their carnivore predators will have higher $^{13}C/^{12}C$ ratios (Sponheimer and Lee-Thorp, 1999). Julia Lee-Thorp and colleagues

---

Box 5.10 The diet of gracile and robust forms

Not everyone agreed with the dietary distinction between *Australopithecus* and *Paranthropus* grounded on dental traits. Milford Wolpoff (1971b), for instance, denied there was a great morphological disparity between the gracile and robust forms. All South African australopiths, according to Wolpoff, are much more similar than is often maintained. Furthermore, their dentition indicates that they shared a hard vegetarian diet. As we'll see below, Wood and Strait (2004) reached the same conclusion in their study of 11 traits related with diet, habitat preference, population density, and dispersion, among others. According to Wood and Strait, "*Paranthropus* and early *Homo* were both likely to have been ecological generalists."

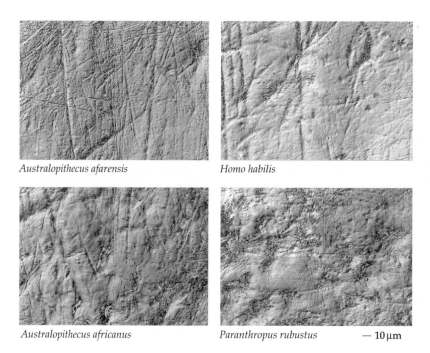

Australopithecus afarensis *Homo habilis*

Australopithecus africanus *Paranthropus rubustus* — 10 μm

Figure 5.18 Microwear texture images. Some molars have a more heavily pitted surface than others, reflecting the different diets, much softer in *H. habilis* than in *P. robustus*, for example. Photographs by and copyright owned by Peter S. Ungar.

Box 5.11 The correlation between microwear and diet

Peter Ungar and colleagues (2006), who include Frederick Grine, have examined the correlation between microwear and diet. They studied 18 cheek teeth attributed to the genus *Homo* with preserved antemortem microwear from Ethiopia, Kenya, Tanzania, Malawi, and South Africa. Microwear features were measured and compared for these specimens, five extant primate species (*Cebus apella,*

Gorilla gorilla, Lophocebus albigena, Pan troglodytes, and *Papio ursinus*), and two human foraging groups (Aleut and Arikara). Ungar *et al.* (2006) concluded that "dental microwear reflects diet, such that hard-object specialists tend to have more large microwear pits, whereas tough food eaters usually have more striations and smaller microwear features".

(Lee-Thorp and van der Merwe, 1993; Lee-Thorp *et al.*, 1994) applied this analysis to Swartkrans *P. robustus* specimens. Their results suggested that the better part of their diet was food with a relatively low $^{13}C/^{12}C$ ratio, specifically about three-quarters of their total intake. *P. robustus* either fed mostly on fruits and leaves, or on herbivores with this diet. Sponheimer and Lee-Thorp (1999) also applied stable-carbon-isotope analysis to *A. africanus* specimens from Makapansgat, obtaining higher ratios than those expected for a frugivore. According to these authors, either *A.*

africanus ate not only fruits and leaves but also large quantities of foods such as grasses and sedges, or they fed on animals that ate these plants. In the first case, *A. africanus* from Makapansgat would have exploited the resources afforded by an open savanna, searching for food in woodlands or grasslands. The second scenario implies that a carnivorous diet had appeared before the genus *Homo* and Oldowan tools (Sponheimer and Thorp, 1999).

Taken together, the $^{13}C/^{12}C$ ratio studies show that it cannot be assumed that the diet of *P. robustus* was specialized compared to the more varied one of *A. africanus*. Wood and Strait (2004) concluded, in their review of the different sources of evidence related with dentition and diet, that it is incorrect to associate stenophagy—reduced diet—with South African robust hominins (*P. robustus*) and euryophagy—broad diet—with gracile ones (*A. africanus*; *H. habilis*/*H. erectus*). Contrary to Robinson's dietary hypothesis, the diets of South African early hominins must have been more intricate and complex. This is also argued by Matt Sponheimer *et al.* (2006) based on the analysis of the enamel of four permanent teeth of *Paranthropus robustus* from Swartkrans (South Africa) using the technique of stable isotope laser ablation. According of Sponheimer *et al.* their data suggest that "Paranthropus was not a dietary specialist and that by about 1.8 million years ago, savanna-based foods such as grasses or sedges or animals eating these foods made up an important but highly variable part of its diet."

Interestingly, the work carried out by van der Merwe and colleagues (2003) on the carbon-isotope ratios of 10 *A. africanus* specimens from Sterkfontein Member 4 concluded that their diet was highly diverse. It could have included grasses and sedges and/or the insects and vertebrates that eat these plants. But this variation existed among the sample's individuals, and it was "more pronounced than for any other early hominin or non-human primate species on record" (van der Merwe *et al.*, 2003). To put it another way, it might be necessary to study differences among individuals to advance our current knowledge of the diet of *A. africanus*.

5.2.6 Phyletic relationships between *A. africanus* and *P. robustus*

As we have seen along this chapter, at least two taxa are usually identified among early South African hominin specimens: *A. africanus* and *P. robustus*. What is the phyletic relationship between these two kinds of beings that occupied very close spaces? The question is not answered easily. Schwartz and colleagues (1998) studied the molar size, enamel thickness, and cusps of *A. africanus* and *P. robustus*. They noted differences between the dentition of those two taxa, but they

reached no functional conclusions regarding them. They drew no phylogenetic implications either; on the contrary, Schwartz *et al.* (1998) argued that the traits of the cusps examined in their study (mainly Carabelli cusps) do not provide information regarding the evolutionary relationships between *A. africanus* and *P. robustus*.

Moreover, the sharp separation between gracile and robust forms in the South African early-hominin fossil record has not been accepted unanimously. During the discussion that took place in the 1950s and 1960s regarding South African australopith taxonomy, some researchers expressed their belief that there was a much greater homogeneity among them. As Emiliano Aguirre (1970) noted, there are certain common traits: marked prognathism, wide premolars and molars along the bucolingual axis, and small incisors and canines (Figure 5.19). Such similarities sometimes make the task of distinguishing gracile and robust specimens very difficult. Aguirre (1970) concluded that there might be more than one early hominin species at Makapansgat, and maybe also at Kromdraai and Sterkfontein.

These difficulties, together with the parallel problem of the robusticity of the *A. africanus* specimens found at certain locations, such as the youngest Sterkfontein Members, have interesting systematic implications. The paradoxes that have arisen could be better understood if the taxon *A. africanus* is an ancestor of the robust clade, as suggested by Rak (1983). Under this scenario, the most robust *A. africanus* specimens would appear at younger sites, as an expression of the tendency towards the apomorphies of the robust clade (Cela-Conde and Altaba, 2002). The youngest specimens from Sterkfontein seem indeed to exhibit such a tendency. We'll return in section 6.2 to the relationship between *A africanus* and the robust clade, when we examine the robust lineage from East Africa.

The age difference between *A. africanus* and *P. robustus* speaks in favor to this evolutionary relationship. In regard to the age of the former, and despite problems in the dating of South African sites, the overall picture clearly suggests that it appeared before *Paranthropus*. Skelton and coleagues (1986) examined the studies of a number of

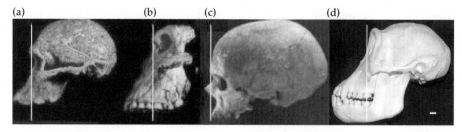

Figure 5.19 Prognathism in (a) *A. africanus* and (b) *P. robustus* in comparison with (c) a modern human and (d) a chimpanzee. Photographs of *A. africanus* and *P. robustus* faces from http://www.mc.maricopa.edu/dept/d10/asb/anthro2003/origins/hominid_journey/africanus3.html; chimpanzee skull (scale bar, 1 cm), ©John Kappelman; modern human skull from http://www.msu.edu/~heslipst/contents/ANP440/images/ Cromagnon_1_rside.jpg.

authors, and estimated the set of *A. africanus* remains to between 3.3 and 2.3 Ma. But this remarkable age raises a problem for the classification of the Taung Child. Partridge (1982) estimated the Taung site to 2.0–1.0 Ma. Can the presence of *A. africanus* at such a late time be accepted? In Tobias' (1988) opinion, from a morphological viewpoint, the Taung specimen supports the most parsimonious thesis among the alternatives presented by Skelton and colleagues (1986). Taung would be a variant of Makapansgat and Sterkfontein *A. africanus*, a derived *A. africanus*. Tobias (1988) did not seem to believe that doubts concerning the age of Taung required dismissing its morphology as the closest to that *A. africanus* variant, although he did admit that the question of the specimen's age is still open. The phylogenetic significance and even the taxon to which it belongs also seem to be unresolved issues.

Accepting the hypothesis of *A. africanus* as ancestral to *P. robustus*, the relation between these taxa and the more recent *H. erectus* (or *H. habilis*) from South Africa, remains to be explained. We will propose a phylogenetic explanation of the gracile lineage in the chapters that follow, after we analyze the East African Plio-Pleistocene specimens.

The robust lineage has traditionally been less difficult than the gracile one. But, as we shall see, the delineation of the latter's phylogenetic relations is not free from problems. South African robust australopiths exhibit some very conspicuous derived traits. Such traits also appear in some specimens discovered in East Africa. Thus,

we can ask ourselves whether they are homologous traits—synapomorphies—that would involve a common robust lineage for South African species (*P. robustus*, *P. crassidens*) and East African ones (*P. boisei*, *P. aethiopicus*). Alternatively, those traits could be analogous and might have developed separately in each species—homoplasies—that would speak against the existence of a robust clade. After describing the robust specimens from Olduvai and West Turkana we will return to the question of whether robust australopiths constitute a clade or not.

5.3 The "gracile" and "robust" alternatives in East Africa

5.3.1 Olduvai makes history: OH 5, *Paranthropus boisei*

South African sites were the first to be systematically excavated, giving birth to modern paleoanthropology, but they are not the oldest. Their ages, barely 4 Ma in the best of cases, contrast with the close to 6 Ma of some Rift Valley sites. We now return to the latter to follow the history of the discoveries of Plio-Pleistocene hominin specimens.

We will begin with the site that became a true emblem of human evolution: Olduvai (Tanzania). This site was discovered by the German scientist Wilhelm Kattwinkel in 1911. Many of its features make it remarkable. It was the first place where radiometric dating techniques—the K/Ar method—were applied. The study of the history of our tribe, Hominini, in the Rift Valley, began at this site. The first counterparts of South African gracile and robust hominins were found there. *Homo habilis* appeared there. Olduvai yielded evidence of the earliest lithic culture, called Oldowan for this reason. And Olduvai is linked forever to the Leakey family, Louis and Mary Leakey. Born Kenyan and son of a missionary couple, Louis Leakey excavated at Olduvai—together with his wife, Mary Nichols (Mary Leakey; Figure 5.20)—since 1932 and, continuously, since 1951, with Mary Leakey in charge of the works from the mid-1960s.

The Olduvai site is in the Serengeti plain, close to the Ngorongoro extinct volcano, some 200 km east of Arusha (Figure 5.21). It is a gorge that reaches a depth of up to 100 m in some places and extends, approximately, 40 km from east to west. The canyon was excavated in sedimentary terrains containing lacustrine, fluvial, and eolian deposits and volcanic tuffs from Ngorongoro by a river which is now dry. The sediments at Olduvai were deposited on a basalt layer, the IB tuff, visible only near the VEK–FLK localities (see Box 5.12 for an explanation), and dated to 1.84 ± 0.03 Ma. The beds are numbered correlatively as Bed I through IV, followed by the Masek, Ndutu, and Naisiusiu Beds.

The first dating of the volcanic tuff at the base of Olduvai Bed I, by means of the K/Ar method, yielded an estimate of 1.7 Ma (Leakey *et al.*, 1961). Many dating methods have been subsequently applied, such as sedimentation rate, geomagnetic polarity, fission track, and amino acid racemization. Mary Leakey and Richard Hay (1982) documented the available information on the stratigraphic sequences and ages of Olduvai. Following their work, Michael Day (1986) gave the following ages for Olduvai Beds: Bed I, 2.1–1.7 Ma; Bed II, 1.7–1.15 Ma; Bed III, 1.15–0.8 Ma; Bed IV, 0.8–0.6 Ma; Masek, 0.6–0.4 Ma; Ndutu, 400,000–32,000 years; Naisiusiu, 22,000–15,000 years.

The stratigraphic scope of Olduvai reaches 2 million years, but only two beds, I and II, belong to the Plio-Pleistocene. Hominins have been found at both of them, but the boundary between both beds is arbitrary. They are separated by a layer of slabs identified by Hans Reck and Louis Leakey in 1931, and established as the frontier under the assumption that geological and faunal changes coincided with the presence of this fringe. Today we know they did not. From the perspective of an appropriate reconstruction, the lower part of Bed II forms a unit with Bed I. However, we will refer to "Bed I" and "Bed II" as if their separation corresponded with real change. Bed I (from the base to 1.71 Ma) contains deposits corresponding to a salty lake. Bed II spans from about 1.7 Ma to a little more than 1.2 Ma, and includes tuff IIA, which allows calculating its lower limit. This bed's deposits underwent important geological and ecological changes, such as the salinization and reduction of the lake and its subsequent replacement by open savannas. The fauna discovered at Olduvai fits geological data well.

Figure 5.20 Dr Louis Leakey, left, and his wife Mary dig at Tanzania's Olduvai Gorge. Photo: Robert F. Sisson ©National Geographic Society.

Figure 5.21 Olduvai Gorge. Photograph by C.J. Cela-Conde.

Box 5.12 Names of the Olduvai Gorge localities

VEK stands for Vivien Evelyn (Fuchs) Korongo, and FLK for Frida Leakey Korongo. The names of the Olduvai Gorge localities usually include the letter K, which stands for korongo, meaning "gully" in Swahili. The complete list of the names is in Mary Leakey's report included in L. Leakey (1967b).

Numerous micro- and macromammals, including primates, bovids, equids, and carnivores among the latter, are present in Bed I, but there are also amphibians and fishes that indicate a lacustrine environment (Andrews, 1983; Leakey, 1967b). Bed II contains the typical Villafranchian fauna: large herbivores and carnivores, but also primates and hippopotamus (Leakey, 1967b). This is indicative of a diverse habitat in which the extension of open savannas was an important factor.

Olduvai yielded hominin remains as early as 1935, specifically belonging to *H. erectus* (or *Homo ergaster*), discovered by Mary Leakey. But, as Yves Coppens (1983) recalls, almost nobody took notice of those first specimens. Two teeth dated to 1.5 Ma found by Louis Leakey in 1955 in Bed II (Leakey, 1958) did not have much repercussion either. The discovery made the following year suddenly placed Olduvai, and all of East Africa with it, at the center of attention of paleoanthropologists around the world.

Louis Leakey (1959) has recounted the finding in detail. On July 17, Mary Leakey found a hominin cranium, partially exposed by the terrain's natural erosion, at the FLK site and about 7 m below the upper limit of Olduvai Bed I. The excavation works began the following day, and on August 6 an almost complete cranium was recovered. It was associated with animal bone fragments and tools belonging to a very primitive culture, which was named Oldowan. The specimen, one of the most famous pieces in the history of anthropology, is known technically as OH 5 (Olduvai Hominid number 5), and colloquially by the name Dear Boy (Figure 5.22).

The OH 5 cranium was found fragmented in small pieces, but the fragments had been preserved together, including some very fragile bits, like the nasal bones. The fossils of other animals associated with OH 5 were also fragmented, but they were shattered in a different way: they seemed to have been purposely broken to get to the bone marrow. Louis Leakey concluded that hominins such as Dear Boy lived there close to 2 Ma, and had used stone tools to break open animal bones. Leakey rejected the idea that Dear Boy itself had also been a victim of predators, other hominins, or cannibalism (Leakey, 1959). "Had we found only fragments of skull, or fragments of jaw, we should not have taken such a positive view of this. It therefore seems that we have, in this skull, an actual representative of the type of 'man' who made the Oldowan pre-Chelles-Acheul culture" (Leakey, 1959, p. 491).

As soon as Tobias (1967) reconstructed the cranium, it became apparent that Dear Boy was a robust exemplar; in fact, very robust. Leakey himself admitted from the start that it was an australopith resembling, in certain aspects, South African *Paranthropus*: sagittal crest; reduction of canine and incisor size, placed in a straight line in front of the palate; and the general structure of the cranium. But according to Leakey (1959) other traits observed in OH 5 were reminiscent of *A. africanus*: the size of the third upper molar, smaller than the second one, for instance. While admitting the inconvenience of multiplying taxa, Leakey chose a new genus of the subfamily Australopithecinae, giving OH 5 the name of *Zinjanthropus boisei*; "Zinj", for the ancient name for East Africa, and "boisei" in honor of Charles Boise, who contributed to financing the excavations at Olduvai. Owing to Tobias' (1967) detailed study of the specimen, it was later included in the same genus as South African robust australopiths, but in a different species, namely, *Paranthropus boisei* or *Australopithecus boisei* for authors who, like Tobias, do not accept the genus *Paranthropus*.

Box 5.13 Villafranchian fauna

Although the description of the Villafranchian fauna was based on European animal assemblages, it is also present in other continents. It persisted through the late Upper Pliocene and Lower Pleistocene eras, and indicates, hence, the transition from one era to another. The

Villafranchian fauna includes mammal genera such as *Ursus* (bears), *Elephas* (elephants) *Hypperion* (horses), the great savanna ungulates and their predators, such as *Crocuta* (hyenas) and *Smilodon* or *Homotherium* (saber-tooth cats).

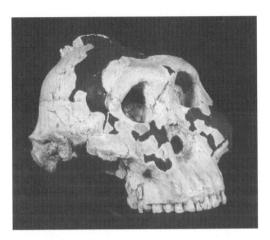

Figure 5.22 OH 5, Dear Boy, *P. boisei*. Photograph from http://www.anthrophoto.com/cgi-bin/ImageFolio31//imageFolio.cgi?action=view&link=Paleontology&image=APCOO_1541.jpg&img=&tt=.

OH 5 was described as a specimen with certain *Paranthropus* and *Australopithecus* traits, but it is not an intermediate form between South African gracile and robust australopiths. Quite the opposite, it is even more robust than the Kromdraai and Swartkrans specimens, and this was the most important consideration in favor of the initial proposal of a new genus, in addition to the great distance between Olduvai and South Africa. The description of the cranium highlighted, among other distinctive traits, a continuous sagittal crest through the occipital bone in males, a very massive supraorbital torus, and a crest on the frontal bone with a very prominent anterior margin. This margin, which is keel-shaped, had not been documented even in the most muscular male specimens of South African *Paranthropus*. *Zinjanthropus* gave

an overall impression of a much more muscular masticatory apparatus than robust australopiths known at the time when the discovery was made.

A relevant aspect for the phylogenetic characterization of Dear Boy was added in 1961. The age of OH 5 was established using a method which was new at the time: K/Ar. Jack Evernden and Garniss Curtiss (1965), from the Department of Geology of the University of California at Berkeley, established the first absolute chronological series of a hominin site: Olduvai Bed I. As Bernard Wood (1997) said, in Mary Leakey's obituary, Olduvai Gorge soon became the yardstick used to calibrate the ages of other sites. The dating of Bed I was established by studying 10 tuffs, the average age of which was estimated by Leakey *et al.* (1961) to be 1.75 Ma. This was also the age of the KA 437 tuff, separated from where hominin remains were found, but whose horizon could be correlated with it.

5.3.2 The first taxon of our genus: *Homo habilis*

The discovery of *Z. boisei* would have been enough to award Olduvai a remarkable place in human paleontology. In addition to the morphological evidence it provided, it inaugurated radiometric dating. But hominins from Olduvai include more than robust specimens similar to OH 5. Another Olduvai specimen, *H. habilis*, also deserves a privileged consideration. *H. habilis* changed our understanding of our tribe's evolution. However, it was not easy for its discoverers to gather support for a point of view which was, at the time, so new and different from the official stance. Tobias has referred to the discovery of *H. habilis* as an instance

---

Box 5.14 *Paranthropus* versus *Australopithecus*

Phillip Tobias is among the authors who prefer discarding the genus *Paranthropus*, considering robust australopiths as a subgenus of *Australopithecus*. In this case, OH 5 would be classified as *Australopithecus boisei*.

Box 5.15 The potassium/argon method

The K/Ar method can be used on volcanic rocks. These contain the radioactive, and thus unstable, potassium isotope ^{40}K which transforms into argon gas ^{40}Ar. The half-life of ^{40}K is 1,300 Ma; that is, it takes that length of time, on average, for half of a given amount of ^{40}K to change into ^{40}Ar. If the amount of original ^{40}K still remaining in a sample can be determined, then its age can be calculated. This is done by heating the sample until its complete fission and measuring the amount of argon released in the process. However, it is necessary that the examined sample has not been contaminated by current argon present in the atmosphere. The K/Ar method has very broad applications. The age of our planet has been estimated to be 4,500 Ma using the ^{40}K/^{40}Ar method. The limits of the technique are around 1 Ma.

A variant of this radiometric method is Ar/Ar dating, ^{39}Ar/^{40}Ar. It involves heating several samples to different temperatures specified before hand, releasing a certain amount of ^{40}Ar. Each of the samples is thereafter subjected to a neutron bombardment in the reactor, causing the conversion of ^{39}K into ^{39}Ar. ^{39}K is the most common potassium isotope, and it is not radioactive. By comparing the ^{40}Ar liberated with the ^{39}Ar obtained in the nuclear reactor, it is possible to estimate the apparent age corresponding to each of the samples heated to different temperatures: the so-called age spectrum. The analysis of the age spectrum allows calculation of the age of the sample.

---

of premature finding, like when in 1797 John Frere came across the first stone tools in Hoxne, Suffolk, England, or the proposal of Mendel's laws, or Fleming's discovery of penicillin or, in the field of human paleontology, the discovery of the Taung Child (Tobias, 1992). Tobias played a direct and fundamental role in the study of *H. habilis*.

Between 1960 and 1964 Louis and Mary Leakey's team found a series of specimens in Olduvai Beds I and II. Their interpretation was immediately very controversial. One of them, OH 7—Jonny's child—(Leakey, 1961a), included a mandible, a parietal, and handbones of a juvenile individual from the FLKNN I site (NN stands for Ndutu and Naisiusiu; see above and Box 5.12) at Bed I, slightly older than the sediments in which the first important specimen from Olduvai, *Z. boisei* (L. Leakey, 1959), had appeared a couple of years earlier. OH 8 (Leakey, 1961b) was found at the same FLKNN I site. This specimen included

two phalanxes, a molar fragment, and a set of footbones. Other discoveries, such as OH 4, OH 6, OH 13 (*Cinderella*; Figure 5.23), and OH 16, provided additional cranial and postcranial evidence that suggested, based on the specimens age and morphology, that they belonged to a very gracile Olduvai australopith.

However, Leakey, Tobias, and Napier (1964) suggested including all those findings in the genus *Homo*, defining the new species *H. habilis*. OH 7 (an immature specimen, unfortunately) constituted the type specimen; OH 13, OH 16, OH 6, OH 8, and OH 4 were paratypes. The description of *Homo habilis* included Leakey's analysis of OH 7 (Leakey, 1961a), together with Tobias' calculations of cranial capacity (Tobias, 1964), and Napier's study of handbones (Napier, 1962). Overall, *H. habilis* exhibits certain features which represent changes in the cranium and dentition compared with *Australopithecus*. Its face is less prognathic and its

Figure 5.23 OH 13, Cinderella, *H. habilis.* Picture from http://www.msu.edu/~heslipst/contents/ANP440/habilis.htm.

cranial capacity is larger. Its masticatory apparatus is smaller, especially molars and premolars, and dental enamel is slightly thinner. The shape of its dental arcade is parabolic, like later *Homo* specimens.

The proposal of the taxon *H. habilis* was provocative at the time. Tobias (1992) explained later on that the accepted doctrine during the mid-twentieth century maintained that the "morphological distance" between *A. africanus* and the typical middle Pleistocene hominin, *H. erectus*, was not enough to accommodate any other gracile taxon, because they are very similar. Moreover, the genus *Homo* was considered characteristic of the middle Pleistocene and thought to have a larger endocranial capacity than that of the Olduvai specimens. These two kinds of reason against the *H. habilis* proposal were, in actual fact, incompatible: on the one hand, it was too similar to *Homo erectus* and, on the other, it was not similar enough to known *Homo* specimens.

All of the morphological features of *Homo habilis* were cited with severity by some illustrious contemporary paleontologists during the discussion following its proposal, including Le Gros Clark, F. C. Howell, B. Campbell, D. Pilbeam, E. Simons, and Robinson, among others. The letter sent by Le Gros Clark (1964a) to the editor of *Discovery* shortly after the proposal of *H. habilis* summarizes these doubts. He believed the specimens discovered by Leakey at Olduvai were too similar to

Australopithecus and too different from *Homo* so that without doubt they belonged to the former genus. An examination of the available evidence, including cladistic studies, has led Wood and Collard to suggest the same idea 35 years later (Wood and Collard, 1999b).

According to Tobias (1992), the most devastating attack on *H. habilis* was written by Loring Brace and colleagues (1973). After criticizing the dental measures of the type specimen and the paratypes, Brace *et al.* concluded that, since the taxon *H. habilis* lacked a type specimen and paratypes, it constituted an inadequately proposed empty taxon and, therefore, deserved to be formally eliminated. Tobias (1992) has noted that in the 15 months following the description of the new species, the *H. habilis* specimens were reclassified by different authors as *Australopithecus africanus habilis*, *Australopithecus habilis*, *Homo erectus habilis*, or *Homo erectus* (unspecified subspecies). That is to say, in any of the possible ways that avoided admitting the new taxon as defined by Leakey, Tobias, and Napier.

The criticisms against *H. habilis* were, in a way, the best argument in favor of the proposal, premature though this was. How to explain that some renowned authorities considered the Olduvai findings attributed to *H. habilis* to be an australopith, while other authorities thought it was *H. erectus*? The new specimens exhibited an intermediate morphology, and their inclusion in one or another side depended on the emphasis placed on similarities and differences. At the time, taxonomy sharply distinguished between Pliocene fossil hominins (*Australopithecus*, including gracile and robust lineages) and Pleistocene forms (*H. erectus*). But the specimens from Olduvai could not unequivocally be considered either *Australopithecus* or *H. erectus*. In this respect, the proposal by Leakey and colleagues was firmly grounded. A new species was required, but why in the genus *Homo*?

To classify specimens, taxonomy usually takes into account their morphology, above any other consideration. This is why morphological descriptions of the type specimen and paratypes were used by Leakey *et al.* (1964) to propose

---

Box 5.16 Age at death of *H. habilis*

Surprised by the great amount of remains corresponding to immature *H. habilis* individuals found at Olduvai, Tobias (1991a) studied the relationship between the age at death, demographic patterns, and environmental conditions and concluded that only 59% of *H. habilis* at Olduvai lived to become adults. This is similar to the figure calculated for *A. robustus* (56–57%), but it is much less than for

A. africanus (75–81%). Given that child survival rate depends on environmental conditions, Tobias argued that those conditions were harder in the times of *H. habilis* and paranthropines, when Africa had undergone a cooling process, than during the more temperate period of *A. africanus*.

H. habilis. However, the genus *Homo* has become associated with features other than morphological traits, namely, the production of tools used for scavenging and hunting. This behavior requires a big enough brain to carry out the complex cognitive operations involved in those tasks. The proponents of the new taxon suggested in their 1964 article that *H. habilis* was the true author of the Oldowan culture, the lithic industry at Olduvai, while *Zinjanthropus*—the earlier candidate—was a mere intruder. Thus, *Homo* would be the genus that introduced the adaptive strategy of stone tool making, and *H. habilis* its first representative. Following Dart's suggestion, the new species was christened *H. habilis* mainly for this reason. "Habilis" means able, handy, mentally skillful, vigorous, as noted by the authors in 1964.

In October 1968, P. Nzube discovered a partial, fractured and squashed cranium at locality DKE (Douglas Korongo East), corresponding to the lower part of Bed I of Olduvai. The specimen received the technical designation of OH 24 (M.D. Leakey, 1969; Figure 5.24). The cranial capacity of OH 24 is small, around 600 cm³, which raised doubts of whether it belonged to the taxon *H. habilis*. However, as Tobias (1991b) noted in his meticulous study of the specimen, OH 24 is one of the most remarkable *H. habilis* exemplars. Its profile is similar to that of *A. africanus*, such as Sts 5, StW 13, and MLD 6, with a marked prognathism. But the palate, which forms a parabolic arcade, is different from both *A. africanus* and *P. robustus*, and although the size of its molars (measured as the sum of their surfaces) is greater than the *H. habilis* average, it is below that of

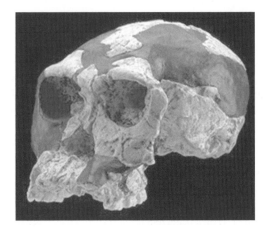

Figure 5.24 OH 24, *H. habilis*. Picture from http://www.msu.edu/~heslipst/contents/ANP440/habilis.htm.

A. africanus and *P. robustus*, as well as *P. boisei* (Tobias, 1991b). OH 24 seemed decisively to confirm the taxon *H. habilis*. However, its problems had just started.

5.3.3 OH 62 and the question of the number of species within *Homo habilis*

A specimen discovered by Donald Johanson *et al.* (1987) raised new doubts concerning the taxon *H. habilis*. The specimen was found at Dik Dik Hill, Olduvai Bed I, close to the FLK site of *Zinjanthropus*. It included up to 300 fragmented remains of an individual's face, palate, cranium, and jaw, together with a complete humerus and fragments of the radius, tibia, fibula, and femur. The assemblage was designated OH 62. Johanson and

---

Box 5.17 OH 24, Twiggy

The OH 24 specimen is popularly known as Twiggy, because of a humorous comment by Phillip Tobias when he saw the fossil: "only Twiggy has been that flat" (Tobias, 1991c; pp. 44–45). Twiggy was the name of a 1960s British model, famous for her extreme thinness. Mary

Leakey (1969) associated OH 24 with *H. habilis* since the first communication of the discovery. In their 1971 description, Mary and Louis Leakey and Ronald Clarke formally assigned it to a female *H. habilis* (Leakey *et al.*, 1971).

---

colleagues believed OH 62 belonged to the species *H. habilis*, but this ascription is not unproblematic. For a start, it is a very short adult individual, one of the smallest among all hominins. Given that previous *H. habilis* specimens hardly allowed associating cranial and postcranial remains, OH 62 afforded a great opportunity for integrating those disperse traits in a single individual. But the result yielded more shadows than lights. Due to its modest size, OH 62 fell within the body size of the much earlier Ethiopian australopiths, *A. afarensis*. In addition, the OH 62 humerus measured by Johanson *et al.* (1987) was 264 mm long. Hence, it was longer than the humeri of *A. afarensis* such as AL 288-1, or Lucy.

This is a problem for one of the central assumptions in the study of hominin evolution. The evolutionary tendency towards the acquisition of bipedalism is considered a process in which long upper limbs, a relatively opposable toe, and great joint mobility are the primitive traits. They allowed australopiths to combine bipedalism with an ability for climbing up trees. But these plesiomorphies are not present in the characteristic bipedalism of *H. erectus*. The original description of *H. habilis* was based on available morphological traits, which included few postcranial remains, preventing their integration in an ecological model. Not surprisingly, thus, the proposing authors defended the species on the grounds of dental and cranial traits (for instance, Tobias and Clarke, 1996). However, when postcranial remains appeared, it seemed that some *H. habilis* had arms longer than those of early australopiths. At the time, this did not fit the concept of the genus *Homo*. The remarkable length of the humerus seemed to indicate the presence of an arboreal

organism, like much earlier australopiths. The comparison of OH 62 with the morphology of current great apes had led some authors (Hartwig-Scherer and Martin, 1991; Aiello, 1992) to argue it belonged to an "ape-style" hominin; that is to say, something very different from what could be expected for a creature belonging to the genus *Homo*.

Should *H. habilis* be described, in light of OH 62, as a partially arboreal being? A positive answer would be easily accepted were it not for another Olduvai specimen, OH 8, included in the *H. habilis* paratype by L. Leakey *et al.* (1964). OH 8 suggests *H. habilis* possessed a similar locomotion to modern humans. *H. habilis* did not appear to be arboreal. Nevertheless, the length of the OH 62 humerus suggested that the taxon exhibited some climbing ability.

It is unfortunate that OH 62 does not include footbones, the study of which could indicate whether *H. habilis* had reached a bipedalism that was functionally similar to our own, while the upper limbs were still those of a climber. The set of OH 7, OH 8, and OH 62 supports this hypothesis, but there is an alternative interpretation of this samples: that they belonged to different species.

The problem of the morphological differences between two distinct kinds of *H. habilis* from Olduvai, with OH 13 and OH 24 on one side and OH 7 and OH 16 on the other, had already been noted by Mary Leakey and colleagues (1971), subsequently leading to diverse interpretations of the sample assigned to the taxon *H. habilis*. The first possibility is that there were two different species. The larger specimens, OH 7 and OH 16, should be considered true *H. habilis*, as suggested by M. Leakey *et al.* (1971). The rest, smaller in size

Box 5.18 Difficulties wth OH 65

In their description of OH 65, Blumenschine and colleagues (2003) have argued that "On the basis of facial morphology, parietal size and shape, and anterior mandibular morphology, the smaller brained, small-toothed hominids that have been placed in *H. habilis* (including OH 13, OH 24, OH 62, and ER 1813) do not appear to belong to that species. Phenetically they may be thought of as a gracile form of australopithecine, although cladistically they may be assigned to a primitive form of *Homo*. They lack, however, the larger neurocranium combined with raised nasal bones and everted lateral nasal margins characteristic of *Homo*."

The distinction drawn by these authors between the diagnosis obtained by morphological comparison and cladistic methods is interesting, because it shows the difficulties involved in establishing an adequate taxonomy for early hominins. As Tobias (2003) has affirmed in his commentary to the OH 65 discovery, the question concerning the species related to *H. habilis* and its equivalents at Koobi Fora (Kenya), *Homo rudolfensis*, "is by no means settled." See next chapter, particularly Box 6.5.

and cranial capacity (OH 13, OH 24), might represent late survivors of *Australopithecus* contemporaneous with *H. habilis*, according to Richard Leakey and Alan Walker (1980). The notion of two species was supported by Chris Stringer (1986). However, White and colleagues (1981), Howell (1978), and Tobias (1985c) advocated a different alternative. In their opinion, the *H. habilis* specimens from Olduvai represent a highly variable single species, with important sexual dimorphisms and geographic variants. Tobias' (1991b) monographic study of the *H. habilis* remains from Olduvai argues that there is no reason whatsoever to alter the 1964 proposal of one species. Colin Groves (1989) also rejected the possibility of two *H. habilis* species after his analysis. In a study concerning the meaning of the genus *Homo*, Wood and Collard (1999b) have noted that most authors admit that the Olduvai remains correspond to a highly variable *H. habilis* species.

A maxilla complete with its dentition and part of the face appeared in 1995, many years after the initial discoveries. Designated as OH 65, it has been described by Robert Blumenschine and colleagues in 2003 (Figure 5.25). The specimen came from Trench 57 excavated by the authors. These sediments were initially attributed to the lower part of Bed II. However, stratigraphic analyses and later dating with the ^{39}Ar/^{40}Ar method placed OH 65 in Bed I, with an age of 1.85 ± 0.002 Ma.

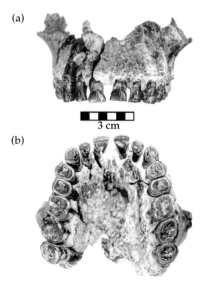

(a)

3 cm

(b)

Figure 5.25 Frontal (a) and palatal (b) views of OH 65, *H. habilis*. Photographs from Blumenschine *et al.* (2003).

The authors of the description note that the orthognathic profile is reminiscent of *Paranthropus*. But the lower nasal region and, specifically, the naso-alveolar clivus, is clearly different from that of robust australopiths. Blumenschine *et al.* (2003) believed that OH 65 is closer to the Koobi Fora (Kenya) specimens resembling *H. habilis*, attributed by some authors to the taxon *Homo rudolfensis*, which we will review in the next chapter.

In addition to *H. habilis* (OH 13) and robust australopith (OH 3, OH 38) specimens, Olduvai Bed II has yielded certain specimens (OH 9, OH 12, OH 28) resembling Asian *H. erectus*, associated with tools belonging to a more advanced industry than the Oldowan assemblage. This set shows a transition towards the middle Pleistocene *H. erectus* specimens, similar to the one we pointed out in South Africa. We will return to this transition in the next chapter.

CHAPTER 6

The Pliocene cladogenesis: *Paranthropus* versus *Homo*

6.1 Evidence from Koobi Fora

As we have seen, the different Pliocene hominins from South Africa and the Rift are assigned to two distinct kinds of hominin: gracile and robust. Regarding South African exemplars, those two kinds correspond to the two australopithecine species named during the first third of the twentieth century: *Australopithecus africanus* and *Paranthropus robustus*. But the findings at Olduvai suggest that the robust/gracile distinction fits *Paranthropus* and early *Homo* adaptive alternatives better. South African *A. africanus* and *P. robustus* correspond to quite different periods, while *Paranthropus boisei* and *Homo habilis* were sympatric in certain regions of East Africa, occupying the same kind of ecosystem in a similar temporal period. This is one of the best indications of a cladogenesis in the evolution of the human lineage.

Several authors, including Vrba (1980) and Tobias (1985b, 1991), have argued that Pliocene environmental change triggered the cladistic episode in the hominin lineage. This notion is grounded on the theoretical postulate that all evolutionary change is driven by environmental changes. Many mammals living in the African savanna that appeared at the Miocene–Pliocene boundary would have diverged as a consequence of the appearance of great extensions of open lands with a distinctive vegetation. Hominins were among these mammals. Hence, Vrba (1985) considered them as "founding members" of the African savanna biota and endemic to it for most of its history.

Data from South Africa have allowed the narrowing down of the moment of the climatic transition to the interval ranging from 2.5 to 2.0 million years, the ages of Sterkfontein Members 4 and 5. There is also record of a vegetation change, indicated by palynology (the study of pollen) in the East African site of Omo, part of the Shungura Formation. The interval has been determined there with much more precision, between 2.52 and 2.4 Ma (Brown *et al.*, 1985a). It is therefore reasonable to believe that about 2.5 Ma there was a change in vegetation that led to the appearance of extensive open savannas in Africa. Furthermore, Vrba (1980, 1985) has shown correlation between vegetation and hominins in South African sites. *A. africanus*, the gracile type, appear

---

Box 6.1 Determining ancestor/descendant relations

It is impossible to determine ancestor/descendant relations with certainty with the techniques currently used for fossil retrieval and study. Hypotheses regarding ancestry relations of close species must rely on their morphology, their temporal sequence, and the proximity of the sites where the specimens were found. As we saw in section 1.3, evolution along a lineage is called anagenesis, and the ancestral and descendant species are called chronospecies. Speciation episodes that involve the split of one species into two descendant ones are called cladogenesis.

Box 6.2 Bovids and early hominins in the savanna

Elizabeth Vrba has suggested that the evolution and adaptation of bovids (family Bovidae) throws light on the evolution of early hominins in the savanna (Vrba, 1974). Bovids are the most numerous large mammals among the fossils discovered in African Miocene, Pliocene, and Pleistocene sites, so their presence in time is very well documented. Additionally, bovids and hominins share a series of marked traits—large bodies, mobility, and herbivorous diet (at least to a great extent)—and they are endemic to the savanna. These similarities lead hominin rates of speciation, morphological change, and extinction to parallel those of antelopes (Vrba, 1984). This is not just a simple statistical correlation, but the result of a broad ecological analogy. Therefore, the study of bovid evolutionary trends (easier to carry out, given the abundance of fossils) can be useful to generate hypotheses about hominin evolutionary trends.

A study of bovid distribution in 16 different sub-Saharan African sites, carried out by correspondence statistical analysis, allowed Vrba to determine some adaptive patterns by grouping bovid taxa according to their presence in biotas. The results showed a consistent association of the set of two bovid tribes, Antilopini and Alcelaphini, with open savannas (Vrba, 1985; the original study is from 1980). Based on this demonstrated association, the fossil record allows determination of the kind of vegetation present at a certain place and time just by verifying the presence of the set Antilopini + Alcelaphini. Vrba's next step was to check whether this set was present in the hominin sites of Makapansgat, Sterkfontein, Taung, Kromdraai, and Swartkrans (Vrba, 1982). Leaving aside the Kromdraai site, with a low presence of bovids, the results point to a notable difference between Makapansgat Member 3 and Sterkfontein Member 4, with a low presence of Antilopini + Alcelaphini, and Sterkfontein Member 5 and Swartkrans Member 1, where they abound. If Vrba's theory is correct, the vegetation of the first two sites was typical of tropical forests, while the vegetation of the two latter sites would have been open savanna.

Box 6.3 Climate change 2.5 Ma

The climatic change episode that took place 2.5 Ma affected the whole planet. In the Rift, that episode coincides with the tectonic events that shaped the existing chain of faults. The mountain range that closes the Rift in the west acts as a barrier for the depressions that come, loaded with clouds, from the Atlantic Ocean, marking a frontier between the tropical forest and the Eastern drier lands. As a consequence of the two phenomena, the lower temperatures and climatic shield, the Rift became covered with savannas that now exist. The modification of oceanic currents could explain the speed of the climatic change, consequence of a glaciation, in areas far from the zone covered by ice. This has been suggested by Andrey Ganopolski and Stephan Rahmstorf (2001), on the basis of a computer-simulation model.

in sites with few Antilopini + Alcelaphini fossils, while *P. robustus* and early *Homo* appear in sites abundant with Antilopini + Alcelaphini remains (Table 6.1).

Thus, many authors, led by Tobias and Vrba, argue that an environmental change in Africa 2.5 Ma was responsible for the separation of the two hominin lineages. This coincides with the estimation, by means of the oxygen-isotope technique mentioned in Chapter 3, of the time of the advance of ice: around 2.5–2.4 Ma (Prentice and Denton, 1988). The date of 2.5 Ma for the cladogenesis of hominins seemed to run into a problem. The gracile specimens discovered at Olduvai by Louis Leakey and colleagues were considered, at that time, as the earliest members of the genus *Homo*, close to 1.8 Ma. But the cladistic episode occurred almost a million years before. Shouldn't there be *Homo* specimens that old? Such specimens exist, as we will see

Table 6.1 Paleoecological reconstructions of relevant early hominin fossil assemblages preserving early *Homo* and *Paranthropus*

Site (member)	Age (Ma)	Paleoecological reconstructions	Species
Southern Africa:			
Sterkfontein (Member 4)	≈ 2.6–2.4	Medium density woodland (Vrba, 1974, 1975)	*A. africanus, H. habilis?, Paranthropus* sp. indent.?
		Moderately open savanna (Vrba, 1985a)	
		Dry, open habitat (Shipman and Harris, 1988)	
		Open woodland to forest (McKee, 1991)	
		Open savanna (Benefit and McCrossin, 1990)	
		Open woodland with bushland and thicket (Reed, 1997)	
Sterkfontein (Member 5)	≈ 1.8	Open savanna (Vrba, 1974, 1975, 1985a; Shipman and Harris, 1988; McKee, 1991; Reed, 1997)	*H. habilis*
Kromdraai (B East)	≈ 1.8–1.7	Open savanna (Vrba, 1975)	*P. robustus*
		Dense woodland along river margin (Vrba, 1981)	
		Open grassland with patches of riparian woodland (Reed, 1997)	
Swartkrans (Member 1)	≈ 1.8–1.7	Open savanna (Vrba, 1975; Shipman and Harris, 1988)	*P. robustus, H. ergaster?*
		Mesic, closed woodland (Benefit and McCrossin, 1990)	
		Savanna woodland with riparian woodland and edaphic grassland (Waston, 1993; Reed, 1997)	
Swartkrans (Member 2)	≈ 1.7 or ≈ 1.5 or ≈ 1.1	Moderately open savanna (Vrba, 1975)	*P. robustus, H. ergaster?*
		Open, dry habitat (Shipman and Harris, 1988)	
		Wooded grassland with wetlands (Reed, 1997)	
Swartkrans (Member 3)	≈ 1.65 or ≈ 0.85 or ≈ 0.7	Open grassland, with river supporting edaphic grassland (Reed, 1997)	*P. robustus*
East Africa, north of Turkana region:			
Hadar (Kada Hadar, BKT-3)	2.33	Open, dry habitat (Kimbel *et al.*, 1997)	*H. habilis*
Konso (between KRT and TBT)	1.43–1.41	Dry grassland (Suwa *et al.*, 1995)	*P. boisei*
East Africa, Turkana region:			
Omo (Shungura C)	< 2.95–2.52	Wooded savanna with riverine woodland (Bonnefille, 1976; Bonnefille and Deschamps, 1983)	*Australopithecus sp. indet., P. aethiopicus*
		Riverine forest and savanna (de Heinzelin, 1983)	
		Closed, dry habitat (Shipman and Harris, 1988)	
		Mesic woodlands and dense thickets, with some forest and savanna (Wesselman. 1995)	
		Bushland/woodland riverine forest and edaphic grassland (Read, 1997)	

Site	Date (Ma)	Environment	Hominins
Omo (Shungura D)	2.52– ≈ 2.45	Mesic plant communities, with large forest galleries and some woodland savanna (Bonnefile and Deschamps, 1983; Wesselaman, 1995) Riverine forest (de Heinzelin, 1983) Woodland (Bonnefille, 1984) Closed, dry habitat (Shipman and Harris, 1988) Woodland/bushland with riverine forest and edaphic grassland (Reed,1997)	*P. aethiopicus*
Omo (Shungura E)	≈ 2.45–2.34	Grassland (Bonnefille, 1976, 1984; Bonnefille and Deschamps, 1983; de Heinzelin, 1983; Wesselman, 1995) Closed, dry habitat (Shipman and Harris, 1988) Well-watered woodland/bushland with riparian forest or woodland (Read, 1997)	*P. aethiopicus, Homo rudolfensis?*
Omo (Shungura F)	2.34–2.32	Dry savanna, open savanna/woodland, steppe, with few mesic woodlands (Jaeger and Wesselman, 1976; Wesselman, 1995) Open savanna (Boaz, 1977) Dry savanna with riverine forest galleries, steppe (Bonnefille and Deschamps, 1983; de Heinzelin, 1983) Grassland (Bonnefille, 1984) Desertic steppe (Bonnefille, 1985) Closed, dry habitat (Shipman and Harris, 1988) Open woodland/bushland with few edaphic grasslands (Reed, 1987)	*P. aethiopicus, Homo rudolfensis?*
Omo (Shungura G)	2.32– ≈ 1.9	Savanna, reverine forest (Bonnefille and Deschamps,1983; de Heinzelin, 1983) Closed, wet habitat (Shipman and Harris, 1988) Arid *Acacia* grassland (Wesselman, 1995) Open woodland, edaphic grassland (Reed, 1997)	*P. boisei, P. aethiopicus?, H. habilis s.s., Homo rudolfensis?*
Koobi Fora (Upper Burgi)	≈ 2.0–1.88	Grassland and desertic steppe (Bonnefille, 1985) Closed, wet habitat (Shipman and Harris, 1988) Mosaic of bushland, savanna, grassland and some gallery forest (Feibel *et al.,* 1991) Open woodland with edaphic grassland and riparian woodland (Reed, 1997)	*P. boisei, H. habilis s.s., H. rudolfensis, H. ergaster* *P. boisei, H. habilis s.s., H. rudolfensis, H. ergaster*
Koobi Fora (KBS)	1.88– ≈ 1.6	Closed, wet habitat (Shipman and Harris, 1988) Wet and dry grasslands, semi-arid savanna, some woodlands (Feibel et al., 1991)	*P. boisei, H. habilis s.s., H. rudolfensis, H. ergaster*

Continued

Table 6.1 (*Cont.*)

Site (member)	Age (Ma)	Paleoecological reconstructions	Species
Koobi Fora (Okote)	≈ 1.6–1.39	Scrub woodland, arid shrubland (Reed, 1997) Closed, wet habitat (Shipman and Harris, 1988) Wet grassland with dry grassland, woodland, scrub and some gallery forest (Feibel et al., 1991)	P. boisei, H. rudolfensis, H. ergaster
West Turkana (upper Lomekwi = Shungura C9)	> 2.52	Wetlands and edaphic grasslands (Reed, 1997) Grassland and marsh (Walker et al., 1986) Open woodland with bushland thickets, edaphic grasslands and wetlands, and riparian woodland or forest (Reed, 1997)	P. aethiopicus
West Turkana (Lokalalei)	2.52–2.34	Closed, wet habitat (Shipman and Harris, 1988)	P. aethiopicus
West Turkana (Kaitio)	1.86–≈ 1.6	Closed, wet habitat (Shipman and Harris, 1988)	P. boisei
West Turkana (Natoo)	≈ 1.6–1.33	Closed, wet habitat (Shipman and Harris, 1988) Woodland and edaphic grassland with marsh (Reed, 1997)	H. ergaster
East Africa, south of Turkana region:			
Tugen Hills (upper Chemeron)	≈ 3–≈ 1.6	Mosaic of C_3 and C_4 plants (Hill, 1995)	Homo s.l. sp. indet.
Lake Malawi (Chiwondo 3A)	≈ 2.5–2.3	Open environment at Maleman (Kullmer et al., 1999) Wooded savanna and open grassland with more closed vegetation near permanent water. At the hominin-bearing Uraha site, closed thicket to dry woodland with nearby = grassland (Schrenk et al., 1995)	P. aethiopicus or P. boisei?, H. rudolfensis
Olduvai (Bed I)	1.97–1.74	Lake-margin woodland and forest changing to open grassland higher in the Bed (Hay, 1973; Kappelman, 1984) Open, arid and closed, wet habitats (Shipman and Harris, 1988) C_4 plants and a gallery woodland in the vicinity of a stream (Blumenschine et al., 2003)	P. boisei, H. habilis s.s., H. rudolfensis?
Olduvai (Bed II)	1.71–1.33 or 1.1	Lake-margin woodland changing to open grassland higher in the Bed (Hay, 1971; Kappelman, 1984; Cerling and Hay, 1986) Open, arid and closed, wet habitats (Shipman and Harris, 1988)	P. boisei, H. habilis s.s., H. erectus
Peninj (Humbu)	1.7–1.3?	Deltaic environment (Dominguez-Rodrigo et al., 2001)	P. boisei
Chesowanja (Chemoigut)	≈ 1.4	Bushed grassland habitat, with riverine and lacustrine elements (Bishop et al., 1978)	P. boisei

Source: Wood and Strait (2004).

immediately, but to find them we need to move away from Olduvai.

6.1.1 The Lake Turkana sites: Koobi Fora

The proposal of *H. habilis*, based on specimens found at Olduvai, was received with considerable skepticism, as we saw in the previous chapter. It did not become generally accepted until similar exemplars appeared in other East African sites (Tobias, 1992). The Koobi Fora site, on the eastern shore of Lake Turkana, in Kenya, has yielded a great number of hominins, comparable to the Tanzanian gracile and robust specimens. No other place in East Africa has provided so many hominins as Koobi Fora. Up to 5,000 exemplars, among which there are some quite well preserved crania, were discovered there during the 1970s. Such paleontological treasure was linked from the beginning to Richard Leakey, son of Mary and Louis. Moreover, the *H. habilis* specimens found at Koobi Fora are older than those found at Olduvai.

The discovery of the paleontological sites around Lake Turkana (known as Lake Rudolph during the British colonial period) was, according to Richard Leakey's (1981) account, a matter of chance. When flying over the area during a trip to Nairobi that had been diverted because of a storm, Leakey noticed the presence of sandstones susceptible to containing fossils, in a place thought to have only volcanic soils. With *National Geographic* funds, and under the auspices of the National Museum of Kenya, Leakey began expeditions to the area in 1968. The discoveries multiplied very soon. The Koobi Fora peninsula, on the eastern shore of the lake (Figure 6.1), turned out to be especially fertile. There were, nonetheless, certain inconveniences. The severity of the climate, typical of a desert and with sporadic and tumultuous rains, exposed the Koobi Fora fossils to the open air, and they disappeared very fast if they were not collected immediately.

The Koobi Fora sedimentary sequence consists of eight members that take their name from the volcanic tuff below them. We are especially interested in the Okote, KBS, Burgi, and Tulu Bor

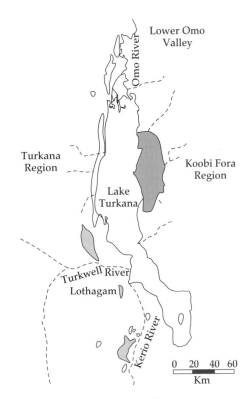

Figure 6.1 Location of the Koobi Fora site. Picture from http://www.mc.maricopa.edu/dept/d10/asb/anthro2003/origins/koobi/koobi2.html.

Members, in which hominin exemplars have been found (Figure 6.2). The Koobi Fora Research Project began its work in 1969. The most important finding during that first campaign was a robust australopithecine cranium, registered as FS-158 (later KNM-ER 406; the acronym means Kenya National Museum, East Rudolph; Figure 6.3). It was a nearly complete cranium, but without dentition, and very similar to OH 5 discovered by Mary Leakey in Olduvai. Richard Leakey (1970) classified the specimen as *Australopithecus boisei*, estimated to be 2.61 ± 0.26 Ma, by means of the $^{40}K/^{40}Ar$ method. Stone tools were found in the same campaign, as well as other mandibular fragments, and a second, very fragmentary, cranium (FS-210), with only the parietals and the baso-occipital region. Leakey (1970) indicated that these specimens belonged to a gracile hominin, although he did not precise the species.

Box 6.4 Stratigraphy at Koobi Fora

The discovery of fossils on the surface of Koobi Fora, extracted from the sediments by the water currents, raise some doubts regarding where the deposits came from. There are no reliable stratigraphic references. However, the sediment layout in the area is almost horizontal, and the erosion gorges in which the fossils appear are not very deep. Thus, even granting that a given fossil may come from a different place to where it was found, dating errors due to this circumstance are small, because close soils usually belong to the same sedimentary stratum (Walker, 1981).

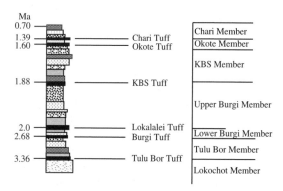

Figure 6.2 Koobi Fora stratigraphy. Picture from http://www.mc.maricopa.edu/dept/d10/asb/anthro2003/origins/koobi/koobi3.html.

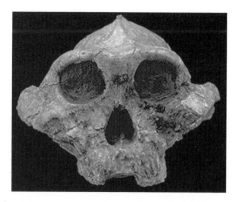

Figure 6.3 KNM-ER 406, *Paranthropus boisei* (R. Leakey, 1970). Photograph from Johanson and Edgar (1996).

These gracile hominins allowed a better characterization of *H. habilis*. In relation to this taxon, a series of interesting remains were described in 1973. These included a cranium, KNM-ER 1470, a right femur, KNM-ER 1472, and a proximal fragment of a second right femur, KNM-ER 1475 (Leakey, 1973a, 1973b, 1974; Leakey and Wood, 1973). These specimens were found in the Upper Burgi Member sediments, slightly below the KBS tuff (Figure 6.2). We will discuss the age of the KBS tuff in detail in the next section, but in any case it is older than 1.8 Ma.

The KNM-ER 1470 cranium (Figure 6.4) is the most popular of all the mentioned specimens. It was discovered by Bernard Ngeneo, Richard Leakey's assistant and a consummate expert in the location of hominin remains, who noticed a great number of fragments on the slope of an eroding gorge. Up to 150 fragments were found in an area of $20 \times 20\,\text{m}^2$ (R. Leakey, 1973b). After a laborious reconstruction, a cranium was obtained that lacked most of the base, the dental crowns, and part of the face. But the remains were sufficient to conclude that it had features that contradicted what would be expected for a hominin its age.

Its cranial capacity was large, between 770 and $775\,\text{cm}^3$, its superciliary arches not very protruding, and the face long and flat, with hardly any subnasal prognathism (Walker, 1981; Day, 1986). All these traits were considered advanced in relation to gracile australopithecines and even to Olduvai *H. habilis*. Although the general appearance of the cranium showed none of the derived robust structures found in paranthropines, the alveoli were suggestive of very wide incisors and canines. That is, the KNM-ER 1470 cranium belonged to a very early hominin with a mosaic of traits, plesiomorphies and apomorphies, the latter quite removed from the robust lineage.

Richard Leakey immediately discarded the possibility that KNM-ER 1470 was an

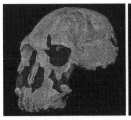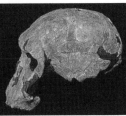

Figure 6.4 KNM-ER 1470, *Homo* sp. indet. (R. Leakey, 1973b). Left: frontal view; right: lateral view. Pictures from http://www.mnh. si.edu/anthro/humanorigins/ha/ER1470.html.

australopithecine. Morphologically, it did not fit *H. erectus* either, which included specimens that were much younger than the age attributed to KNM-ER 1470. Leakey also disregarded Olduvai *H. habilis* as the possible species in which to include his specimen, because of their younger age and lower cranial capacity. Hence, he decided to classify KNM-ER 1470 as *Homo* sp. indet., that is to say, as belonging to the genus *Homo*, without further precisions (Leakey, 1973b).

Leakey was accused of not daring to take the logical step of assigning KNM-ER 1470 to the species *H. habilis* or, alternatively, defining a new species. It could be thought that the cause was the ongoing controversy regarding *H. habilis*. However, Richard Leakey has said in an interview (Lewin, 1987), that he had not paid attention to the discussions about the meaning of *H. habilis* and that, for this reason, the controversy could not have affected his decision very much. It seems that it was lack of precision in the description of the *H. habilis* holotype and paratypes that led Richard Leakey to suggest the classification of KNM-ER 1470 as *Homo* sp. indet. The specimen ended up with the default assignation of *H. habilis*.

Although the description of KNM-ER 1470 was rather incomplete, the Russian anthropologist Valerii Alexeev (1986) suggested a new species for the specimen, *Homo rudolfensis*. The proposal gained significant, albeit not general, recognition (for instance, Groves, 1986; Collard and Wood, 1999; Wood and Collard, 1999a). Alexeev believed KNM-ER 1470 was a male and that KNM-ER 1813 (see below) was a female of the same species, *H. rudolfensis*. However, as we will see further on, the KNM-ER 1813 specimen is usually attributed to *H. habilis*. We will also review the question of the possible relationships between the taxa *Australopithecus*, *H. habilis*, and *H. rudolfensis* later.

The absence of postcranial remains associated with the holotype, KNM-ER 1470, prevents a comprehensive characterization of *H. rudolfensis*. Although the inclusion of the femora KNM-ER 1472 and KNM-ER 1481 in the *H. rudolfensis* hypodigm has been suggested (Wood, 1992a), their association with the available cranial materials is doubtful.

6.1.2 How old is KNM-ER 1470?

Another unsettled issue related with *H. rudolfensis* is age, specifically that of KNM-ER 1470. In his description of the fossil, Leakey (1973b) argued that, using the $^{40}K/^{40}Ar$ method, Fitch and Miller (1970) had firmly estimated the age of the KBS tuff, above the specimen's location, to be 2.6 Ma. Leakey assigned the cranium itself the probable age of 2.9 Ma. If the first age estimate of 2.9 Ma for KNM-ER 1470 was correct, the thesis that *Australopithecus* cannot be the ancestor of the genus *Homo*, often supported by the Leakeys, would be notably reinforced. The Hadar specimens, only half a million years older than the Koobi Fora specimen, could not have undergone the morphological changes necessary for *A. afarensis* to become like KNM-ER 1470. So many changes could not have accumulated in such a short period of time.

But that estimation is open to question because of difficulties entailed by the dating of the volcanic tuff. In 1969 Kay Behrensmeyer discovered a mantle of volcanic ashes with embedded stone tools and hippopotamus fossilized bones, which was named Kay Behrensmeyer Site (KBS). The taphonomic interpretation suggested that a group of hominins had butchered the animal *in situ*. Given that neither the tools nor the fossils could be part of the ashes, it was clear that the mantle contained different intermixed volcanic and sedimentary deposits. Thus, it is not surprising that there were many difficulties in the dating of the soils. (MacRae, 1998–2004, provides an

---

Box 6.5 *Homo rudolfensis*

Is the taxon *Homo rudolfensis* well defined? Some authors have suggested that it is an invalid taxon due to Alexeev's informal way of proposing it (Kennedy, 1999). Wood and Richmond (2000) arrived at the following conclusion in their taxonomic study of the human lineage: "In a presentation of the fossil evidence for human evolution, published in English in 1986, the Russian anthropologist Valery Alexeev suggested that the differences between the cranium KNM-ER 1470 and the fossils from Olduvai Gorge allocated to *Homo habilis* justified referring the former to a new species, *Pithecanthropus rudolfensis*, within a genus others had long ago sunk into *Homo*. Some workers have claimed that Alexeev either violated, or ignored, the rules laid down within The International Code of Zoological Nomenclature. However, there are no grounds for concluding that Alexeev's proposal did not comply with the rules of the Code, even if he did not follow all of its recommendations. Thus, if *Homo habilis sensu lato* does subsume more variability than is consistent with it being a single species, and if KNM-ER 1470 is judged to belong to a different species group than the type specimen of *Homo habilis sensu stricto*, then *Homo rudolfensis* would be available as the name of a second early *Homo* taxon."

Box 6.6 Two species at Olduvai and Koobi Fora?

The taxonomic problem of the early gracile specimens found at Koobi Fora is not resolved. G. Philip Rightmire (1993) carried out a comparative study of the Olduvai and Koobi Fora crania attributed to *H. habilis*, concluding that there are two different taxa in the sample. The first taxon would include the KNM-ER 1470 specimen, together with other Koobi Fora specimens and, probably, OH 7. According to Rightmire, this taxon can be called *H. habilis*. If the KNM-ER 1481 (femora) and KNM-ER 3228 (hip) specimens are included in such a group, then the species has a postcranial anatomy that is close to *H. erectus*. A second group, with KNM-ER 1813 and, probably, OH 13, cannot be classified within *H. habilis*. Rightmire points out that, on the grounds of the available materials, it is impossible to decide the taxon they belong to, although they seem to belong to *Homo* rather than *Australopithecus*. If OH 62 is included in this second taxon, the group would have postcranial proportions different from *H. erectus*.

---

excellent synthesis of the difficulties.) The first dating of the KBS sample rendered an estimate of 200 million years, a clear indication of contamination with ancient sediments. A second dating using the ^{39}Ar/^{40}Ar method yielded a result of 2.61 ± 0.26 Ma (Fitch and Miller, 1970). This estimate, later reduced to 2.42 Ma (Fitch *et al.*, 1976), became the "official" age of the KBS tuff, although up to 41 later measurements offered diverse results, ranging from 2.23 to 0.91 Ma (Lewin, 1987).

The "official" age of the KBS tuff created difficulties for the interpretation of the Koobi Fora fossil sequence. Its comparison with similar sequences from other sites revealed that the one at Koobi Fora was almost half a million years older (Maglio, 1972). After the comparative study of pig fossils in both sites, Cooke (1976), a worldwide expert in fossil suids, recommended revision of the estimated age of the KBS tuff, tentatively attributing it an age of 2.0 Ma at most. By fission-track dating, Hurford *et al.* (1976) estimated the age of KBS to be 2.44 ± 0.08 Ma, very close to the official age. However, a careful analysis carried out later and intended to eliminate possible errors in the fission-track method reduced that age to 1.87 ± 0.04 Ma (Gleadow, 1980). A very similar age had been obtained by Curtis *et al.* (1975) using the ^{39}Ar/^{40}Ar method. Regarding the two units identified in the KBS tuff, this study estimates an age of 1.60 ± 0.05 Ma for areas 10 and 105 and 1.82 ± 0.04 Ma for area 131.

Box 6.7 Fission-track dating

Fission-track dating analyzes the tracks left in volcanic crystals—or in ceramics—by the fission of ^{248}U atoms, radioactive materials with a half-life of 4,500 million years. The energy liberated by the fission of a nucleus leaves a track on the crystal. The number of tracks on the crystal is a function of the time elapsed since its formation. But if the crystal is heated over 60°C, the tracks begin to disappear and they do so completely at 120°C. Thus, the method allows determination of the age of the crystal containing ^{248}U and the moment it was heated, as in the production of ceramics.

The fission-track technique is often used to determine the reliability of the ages obtained by other radiometric methods. Hurford *et al.*'s (1976) fission-track study of the KBS tuff was the first to date zircon crystals with such a young age. The reanalysis by Gleadow (1980) detected certain technical problems and led to the development of a new methodology for dealing with zircons with low track densities.

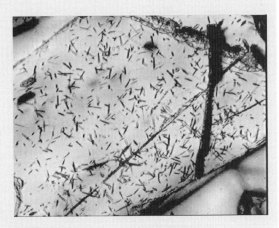

Uranium fission tracks on apatite. Photograph from the University of Cádiz (http://www.uca.es/uca-investiga/marzo-abril/fissiontrack.htm).

The ambiguities about the radiometric results were clarified to a certain extent when a correlation was established, by means of the study of trace elements, between KBS and the H2 tuff of the Shungura Formation, reliably dated at 1.8 Ma (Cerling *et al.*, 1979). Independent studies carried out by Ian McDougall *et al.* (1980) finally resolved the controversy, yielding an age of 1.89 ± 0.01 using the K/Ar method and 1.88 ± 0.02 by means of the ^{39}Ar/^{40}Ar method (McDougall, 1981, 1985).

6.1.3 Other early specimens of the genus *Homo*

The Koobi Fora discoveries had two contrary effects. On the one hand, they supported the pertinence of the species *H. habilis* described on the basis of the Olduvai specimens. There is, for instance, a great similarity between the dentition of KNM-ER 1813 and OH 13 in virtually all the common conserved characters (Walker, 1981). On the other hand, the diversity of the Koobi Fora sample contributed to complicate the scene of the first members of the genus *Homo*. This is the case with the KNM-ER 1813 specimen (Figure 6.5),

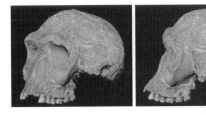

Figure 6.5 KNM-ER 1813, *H. habilis* (R. Leakey, 1974). Pictures from http://www.mnh.si.edu/anthro/humanorigins/ha/ER1813.html.

found in 1973 by Kamoya Kimeu in the Upper Burgi Member (Leakey, 1974). KNM-ER 1813 has the same cranial shape as KNM-ER 1470 and it is slightly younger—between 1.8 and 1.9 Ma—but its cranium is small despite corresponding to an adult with a developed third molar (Leakey, 1974). This is the reason why it has sometimes been considered an *A. africanus* (Walker and Leakey, 1978, for instance).

Based on the cranial shape, the small orbits, the low position of the cheek bones and the nasal bones, KNM-ER 1813 is usually included in *H. habilis*. But some of its traits are even more advanced. For instance, the arched and rounded

---

Box 6.8 Trace elements

Trace elements, or minor elements, in a geological context are defined as all elements except the eight abundant rock-forming elements: oxygen, silicon, aluminum, iron, calcium, sodium, potassium, and magnesium (Thrush, 1968). The concentration of trace elements is specific for each sediment and contributes to the reconstruction of its geological history.

---

supraorbital torus and the indications of a transverse torus bring the specimen close to the later *H. erectus*. However, the cranial capacity of KNM-ER 1813, close to 510 cm^3—or even less, if the caution of Conroy and colleagues (1998) is correct—is smaller than any *H. habilis* from Olduvai and only just above the range of *A. afarensis*. The contrast with KNM-ER 1470 is striking but, what is the explanation for these differences: a sexual dimorphism, a polymorphic sample, or the presence of different species?

The KNM-ER 1805 cranium and mandible (Figure 6.6) from Koobi Fora raises similar questions. It was discovered by Paul Abell, also in 1973, in the Upper Burgi Member and, therefore, has a similar age as KNM-ER 1813. It is an adult exemplar, with developed third molar. The cranium has an indication of a nuchal crest that brings it close to *Paranthropus*, but its dentition is very small for it to be considered a robust australopithecine. Its cranial capacity is close to 600 cm^3. KNM-ER 1805 has been assigned to different species. It was included as a paratype of *Homo ergaster*, but its cranial capacity, prognathism, and the presence of the crest challenges this classification. In Wood's (1991) detailed study of the cranial sample of Koobi Fora, it was assigned to *H. habilis*.

Other East African sites have also contributed to broaden the hypodigm of *H. habilis* (or *H. rudolfensis*), but not without doubts. This is the case with the L.894-1 fragmentary cranium from the Shungura Formation in Omo (Ethiopia), from a locality situated between tuffs estimated to be, with the K/Ar method, 1.93 ± 0.10 and 1.84 ± 0.09 Ma. The specimen was classified by Boaz and Howell (1977) as *H. habilis* or *Homo modjokertensis*, because of its similarities with the OH 24, OH 13, and Sangiran 4 specimens. The UR

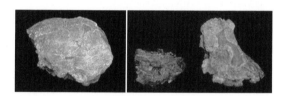

Figure 6.6 KNM-ER 1805, *H. habilis*? Left: cranium; right: mandible. Pictures from http://www.mnh.si.edu/anthro/humanorigins/ha/ER1805.html.

501 partial mandible (Figure 6.7) discovered in Malawi (Schrenk *et al.*, 1993) extended the presence of early hominins towards the Chiwondo corridor, where Oldowan tools had been retrieved since 1963 (Clark, 1995). With an age between 2.5 and 2.3 Ma, Bromage *et al.* (1995) classified the specimen as *H. rudolfensis*, asserting that the antiquity and morphology of UR 501 speak in favor of a cladistic event around 2.5 Ma that led to *Homo*. The age of UR 501 was estimated by faunal comparison. But there are other specimens that have also been attributed to early *Homo* and have similar ages, which provide very firm evidence regarding the birth of our genus. The Ethiopian specimen A. L. 666-1 has features that can be related with *Homo*, and is associated with lithic artifacts.

Ali Yesuf and Maumin Allahendu, members of the research team led by William Kimbel, discovered an almost complete upper maxilla (A.L. 666-1) on November 2, 1994, while examining an unexplored area of the upper portion of the Kada Hadar member. The specimen was fragmented in two pieces along the intermaxillary suture, and lacked some teeth, but conserved the subnasal zone. Additionally, 20 Oldowan flaked stone tools appeared in the same horizon. A.L. 666-1 was located in an outcrop immediately below the

Figure 6.7 UR 501, *H. rudolfensis*. Photograph from http://www.willighp.de/evo/funde/hr_ur501.php? PHPSESSID = 15581789a388cafad08d2f0bb2cdd7aa. Photograph: Thomas Ernsting

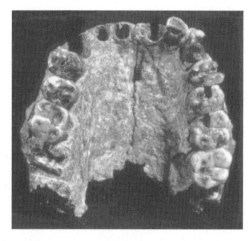

Figure 6.8 A.L. 666-1, *H. habilis*. Photograph from Kimbel *et al.* (1997).

BKT-3 ash layer, estimated by the ^{40}Ar/^{39}Ar method to be 2.33 ± 0.07 Ma (Kimbel *et al.*, 1996). This is the age attributed by Kimbel's team to the specimen and the associated tools.

The morphology of A.L. 666-1 (Figure 6.8) is different, according to Kimbel *et al.* (1996), from early Hadar *A. afarensis* and from any other *Australopithecus*. Rather, it seems closer to the genus *Homo*, on the grounds of the moderate subnasal prognathism, a relatively wide palate, the square profile of the anterior maxilla, the narrow dental crown of the first molar, and the second molar's rhomboid shape. The issue of the specific *Homo* species in which to allocate A.L. 666-1 is difficult. The traits that move the specimen close to *Homo* are derived, but at the same time they are shared by different species of the genus (they are synapomorphic traits). Kimbel *et al.* (1996) noted that the specimen's morphology was reminiscent of Olduvai *H. habilis*, so the adequate description would be *Homo* aff. *habilis*.

The question concerning the species to which A.L. 666-1 should be assigned is not as important as the finding of an almost 2.3-million-year-old member of the genus *Homo* associated with contemporary tools. Both A.L. 666-1 and UR 501 from Malawi argue in favor of the appearance of the genus as a result of a cladogenesis 2.5 Ma.

6.1.4 How many hominin species are there at Koobi Fora?

During the decade between the start of the Koobi Fora campaigns and 1978, the sample of hominins

at the site increased notably. It also turned out to be very diverse. Up to four different types of hominin appeared at Koobi Fora (Walker, 1981). The first type was constituted by clearly robust specimens (KNM-ER 406, KNM-ER 732, KNM-ER 733) found above the Okote tuff. The second and third groups, from the KBS Member or even from strata located under this tuff, included early gracile specimens. Some were specimens like KNM-ER 1813, which, as we pointed out before, have a small cranial capacity. Other exemplars had a much larger cranium (KNM-ER 1479, KNM-ER 1590). Finally, a fourth younger group (KNM-ER 3733, KNM-ER 3883) included specimens resembling *H. erectus* from Java and China, which we will discuss in the next chapter (Figure 6.9).

There are many possible ways to classify the diverse Koobi Fora sample. Alan Walker (1981) noted five alternatives:

1 All Koobi Fora hominins belong to a single species: the single-species hypothesis (Brace, 1965; Mayr, 1950; Wolpoff, 1971a). Its acceptance requires admitting that the level of variation was enormous within this single species.

2 There are three different species at Koobi Fora, one is hyper-robust and contemporary with the Olduvai, Swartkrans, and Kromdraai (South Africa) robust specimens. This interpretation is indifferent to the classification of the robust

Box 6.9 The earliest presence of *Homo* at Turkana

The earliest presence of *Homo* at Turkana is estimated to be around 2.34 Ma and corresponds to the KNM-WT 42718 specimen, a juvenile molar found in 2002 in the LA1 site (Nachukui Formation, West

Turkana; Prat *et al.*, 2005). The age was obtained by correlation with the F-1 tuff of the Shungura Formation (see section 6.2.3, dedicated to the River Omo).

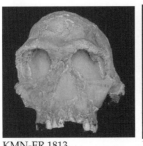
KMN-ER 1813

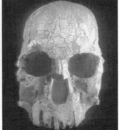
KMN-ER 1470

KMN-ER 406

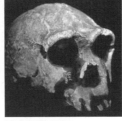
KMN-ER 3733

Figure 6.9 The four types of hominin from Koobi Fora. Top left, KNM-ER 1813, *H. habilis*; top right, KNM-ER 1470, *H. rudolfensis*; bottom left, KNM-ER 406, *P. boisei*; bottom right, KNM-ER 3733, *H. ergaster*. Pictures are not on the same scale. Sources: KNM-ER 1813, from http://www.msu.edu/~heslipst/contents/ANP440/images/ KNM_ER_1813_front.jpg; other three photographs from http://www. kfrp.com/fossils_of_koobi_fora.htm.

species as either *P. robustus* or *P. boisei*. The specimens with small cranium and dentition constitute a second species, and those resembling *H. erectus* a third one.

3 Two different species coincide at the site. The specimens defined as robust and those considered gracile but different from *H. erectus* are a single species, with the robust ones as male exemplars and the gracile ones as females. The specimens resembling *H. erectus* belong to another species. From the point of view of the cranial capacity this

is a plausible solution. However, the differences in dentition require postulating large sexual dimorphism.

4 The robust specimens and those that are similar to *H. erectus* belong to a highly variable species. The ones with small crania and teeth (the gracile ones different from *H. erectus*) belong to another, less variable, species.

5 The robust specimens from Koobi Fora constitute a species, while all the specimens with small teeth constitute another, highly variable, species, including those that are similar to *H. erectus* and those that are not (Walker, 1981).

The situation can be complicated even further. Walker (1981) did not differentiate, within the gracile early specimens, those that have a relatively large cranial capacity, such as KNM-ER 1470 or the KNM-ER 1590 specimen, and those with smaller cranial capacity, such as the KNM-ER 1813 exemplar. We have seen that this is an important difference. After a comparative morphological analysis of KNMR-1470, KNM-ER 1813, the set of 3733/3883 exemplars, and the hypodigm of *A. africanus*, Wood (1985) concluded that they are all very different. Consequently, Wood reckoned that it is possible that there were three "non-australopithecine" taxa at Koobi Fora, in addition to the taxon corresponding to the robust forms. However, Wood (1985) did not suggest any taxon for these non-australopithecine species.

Within the limitations imposed by the difficulties inherent to such a broad and diverse sample as that at Koobi Fora, it seemed reasonable to distinguish at least the four "types" of hominin indicated in Figure 6.9. Two of them, the younger ones, are clear. The robust ones can be attributed to *P. boisei* (Figure 6.9). The gracile ones (Figure 6.9)

Box 6.10 The single-species hypothesis

The hypothesis of the single species is based on a theoretical consideration: the competitive exclusion principle of Gauss (1934), which asserts that two sympatric species cannot coexist if they compete for the same resources because one will ultimately displace the other. If hominins, generally speaking, share similar ecologies, there can only be one species at any given time in each place. Wolpoff (1971a) used the single-species hypothesis to group gracile and robust australopiths in a single taxon. That same principle was later applied to include early *Homo* specimens in *Homo sapiens* (Wolpoff *et al.*, 1994; Eckhardt, 2000).

---

would be *H. erectus*, leaving aside the issue of *H. ergaster*, which we will review in the next chapter. But, what about the early gracile specimens, some with a larger cranial capacity but more primitive traits than the other? Do they constitute a single species, *H. habilis*, or two, the second being *H. rudolfensis*?

6.1.5 Taxa of the early members of the genus *Homo*

If we keep *H. habilis* strictu sensu and *H. rudolfensis* separate, we need to detail the morphological differences between the two taxa. The combination of primitive and derived traits in KNM-ER 1470 led Wood (1992a) and Wood and Richmond (2000) to establish certain criteria that any specimen must meet to be attributed to *H. rudolfensis*; these criteria have been used widely. Based on them, the KNM-ER 1590, KNM-ER 3732, KNM-ER 1801, and KNM-ER 1802 cranial remains, all from Koobi Fora, have been grouped under *H. rudolfensis*. The Malawi UR 501 mandible, and the Omo 75-14 maxilla and cranial fragments from Omo (Ethiopia) have also been attributed to *H. rudolfensis*, although Wood and Lieberman (2001) consider the last specimen to be *P. boisei*. Table 6.2 shows the available specimens generally attributed to *H. habilis* and *H. rudolfensis* from Koobi Fora and other sites.

The attribution of the specimens listed in Table 6.2 to particular species is controversial. It is clear that the classification of relatively uninformative fragments is difficult. But the underlying question, whether or not *H. rudolfensis* should be considered a different taxon from *H. habilis*, is still open. The comparison of the specimens in Table 6.2 does not lead to any conclusive result. Other ways of resolving the taxonomic situation of those early gracile hominins have been suggested. Chamberlain (1987) has argued that the Koobi Fora specimens (leaving aside the robust ones and those resembling *H. erectus*) belong to a single species, but not *H. habilis* strictu sensu. After a detailed study of the apomorphies of the genus *Homo*, Wood and Collard (1999a) argued that both *H. habilis* and *H. rudolfensis* should be considered australopithecines and moved to this genus. Such operation would clearly turn *Australopithecus* into a paraphyletic clade. Cela-Conde and Altaba (2002) tried to avoid the proliferation of new taxa and paraphyly of lineages by rescuing one of Darwin's ideas. The father of modern evolutionary theory noted that the first specimens of any lineage would be difficult to identify. This is because they typically would retain many plesiomorphies and their derived traits would not have yet completely developed. Consequently, Cela-Conde and Altaba (2002) suggested not taking into account those specimens in the construction of cladograms, and assigning them to a particular taxonomic category: the *species germinalis* of any new genus. Rather than solving the taxonomic problem, this solution avoids it altogether. It allows the inclusion of early gracile specimens in a single *germinalis* taxon instead of multiplying taxa.

There is yet one more possible way of resolving the status of *H. rudolfensis*, provided by an early specimen found on the western shore of Lake Turkana, which has been described as the new genus *Kenyanthropus*. We will examine this alternative in section 6.2.

Table 6.2 The *H. habilis* and *H. rudolfensis* hypodigm

Homo habilis	Homo rudolfensis	Specimen
East Turkana (mostly Koobi Fora)		
KNM-ER 164		Fragment of left parietal
KNM-ER 731		Mandible fragment
KNM-ER 739		Right humerus
KNM-ER 817		Mandible fragments
KNM-ER 992 (a–d)		Mandible
KNM-ER 1472		Right femur
KNM- ER 1474		Cranial fragment
KNM-ER 1481 (a–d)		Lower limbs
KNM-ER 1500		Fragments of a small skeleton
KNM-ER 1501		Mandible fragment
KNM-ER 1502		Mandible fragment
KNM-ER 1503		Proximal end of right humerus
KNM-ER 1504		Distal fragment of right humerus
KNM-ER 1506		Mandible fragment
KNM-ER 1507		Mandible fragment
KNM-ER 1805		Partial cranium, maxilla, and mandible
KNM-ER 1808		Skeleton fragments
KNM-ER 1813		Cranium
KNM-ER 3228		Hip
KNM-ER 3735		Fragmentary adult skeleton
KNM-ER 3950		Mandibular fragment
KNM-ER 3954		Mandible
KNM-ER 5429		Mandibular fragments
KNM-ER 5877		Mandibular fragment
KNM-ER 7330		Maxilla and cranial fragments
KNM-WT 42718		Juvenile molar
	KNM-ER 819	Mandible
	KNM-ER 1470	Cranium
	KNM-ER 1482	Mandible
	KNM-ER 1483 (a–e)	Mandible
	KNM-ER 1590 (a–q)	Fragmentary juvenile cranium
	KNM-ER 1801	Partial mandible
	KNM-ER 1802	Mandible
	KNM-ER 3732	Cranium
	KNM-ER 3891	Cranial fragments
Olduvai		
OH 7		Fragmentary skull and juvenile mandible
OH 8		Partial foot
OH 13		Partial skull with upper and lower dentition
OH 16		Calvaria
OH 24		Cranium
OH 62		Bone fragments
Omo		
AL 666-1		Maxilla
L.984-1		Fragmentary cranium with teeth
Malawi		
	UR 501	Mandible
Swartkrans		
SK 847		Partial cranium
StW 53		Partial cranium

6.2 Alternatives in the evolution of robust and gracile clades

6.2.1 *Kenyanthropus*: the first gracile hominin?

The fieldwork carried out at the Lomekwi basin, belonging to the Nachukui Formation, West Turkana (Kenya), during 1998 and 1999 yielded important discoveries. The research team led by Meave Leakey discovered a very complete but deformed cranium (KNM-WT 40000; Figure 6.10), a temporal bone, two fragmentary maxillas, and some isolated teeth that rendered a distinctive gracile hominin morphological picture (M. Leakey *et al.*, 2001). However, its age was much greater than any other specimen belonging to the gracile lineage. KNM-WT 40000 appeared between two volcanic tuffs, the 3.57-Ma Lokochot Tuff and the 3.40-Ma Tulu Bor Tuff. The dating was profusely discussed and correlated by Leakey and colleagues (2001). The age assigned to the fossil was 3.5 Ma, a million years older than the first *Homo* specimens, regarded as the main specimens in the cladistic event separating *Homo* and *Paranthropus*. After an anatomical comparison with the rest of Pliocene hominin species, Leakey *et al.* (2001) introduced a new genus and species, *Kenyanthropus platyops*, emphasizing the most remarkable feature of KNM-WT 40000, its flat

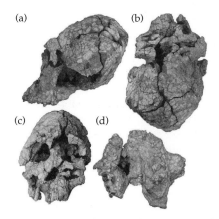

(a) (b) (c) (d)

Figure 6.10 KNM-WT 40000, *Kenyanthropus platyops*.
(a) Lateral view; (b) superior view; (c) frontal view; (d) occlusal view of palate. Photographs from M. Leakey *et al.* (2001). Reprinted by permission from Macmillan Publishers Ltd. *Nature*, 410: 6827, 433–440, 2001.

face. This specimen was considered as the species holotype, and KNM-WT 38350—a partial left maxilla found by B. Onyango in 1998—as the paratype. The remaining specimens found during the Lomekwi campaigns were left without classifying due to insufficient evidence. Nevertheless, Leakey and colleagues (2001) pointed out the affinities existing between *K. platyops* and previously found fossils, such as the KNM-WT 8556 partial mandible, attributed to *A. afarensis*.

K. platyops shows a mosaic of primitive and derived traits. The former, which draw it close to an early australopithecine, such as *A. anamensis*, include the small size of the external auditory pore; its small cranium is also typical of *Australopithecus*; and its premolar and molar enamel thickness is similar to that of *A. afarensis*. A striking derived trait of *K. platyops* is the very vertical plane below the nasal orifice, which gives it the flat-faced look its name refers to.

What is *K. platyops*? Lieberman (2001) argued that the discovery casts more shadows on the phylogenetic interpretation of Pliocene hominins. This may be an excessively pessimistic opinion. The proximity of the facial anatomy of *K. platyops* and KNM-ER 1470 was already noted by M.G. Leakey *et al.* (2001). This fact led Aiello and Collard (2001) to assign the *H. rudolfensis* specimens to the new genus, *Kenyanthropus rudolfensis*, which would have evolved from *K. platyops*.

The phylogenetic connection between KNM-ER 1470 and the genus *Kenyanthropus* suggests that *H. habilis* was the species that initiated the genus *Homo* and that *K. rudolfensis* independently developed a large cranial size, a trait that would be a homoplasy. However, it is difficult to defend placing such similar taxa as *H. habilis* and *K. rudolfensis* in different species, let alone in different genera. The authors of this book suggested a more parsimonious taxonomical interpretation (Cela-Conde and Ayala, 2003) proposing that the genus *Homo* appeared with *K. platyops*, which would be *Homo platyops*, accordingly. The *Homo* and *Paranthropus* cladogenesis would be pushed back to 3.5 Ma. This is consistent with considering *A. africanus* as the first paranthropine: *Paranthropus africanus*. The question whether or not *H. habilis* and

H. rudolfensis should remain as two separate taxa remains unsettled.

Figure 6.17 (below) suggests a possible way in which the different genera and species could be related in a phylogenetic tree, detailing the succession of species along the gracile lineage. Before we go into this clade, let us turn to the robust clade.

6.2.2 The robust clade: West Turkana and KNM-WT 17000

The controversy surrounding the number of species present at Lake Turkana is yet another episode of confrontation between paleontologists who accept only a single hominin evolutionary lineage (lumpers) and those who, on the contrary, believe there were several parallel lineages (splitters). The Koobi Fora discoveries provided solid arguments supporting several simultaneous lineages, even some coexisting in one same geographical location. Thus, in the early 1980s virtually all authors admitted at least the two lineages, gracile and robust. The latter was generally understood as a lateral branch, a late specialization in our family. Once a relation was established between robust mandibles, sagittal crest, great molars, and the intake of a hard vegetable diet, all the pieces seemed to fall into place. *A. africanus* would lead, through a specialization of its masticatory apparatus, to *P. robustus* (Rak, 1985).

The more robust forms, such as the *P. boisei* exemplars from Olduvai, Turkana, and Omo, would represent a very specialized and late version of the robust clade, based to a great extent on

the massive development of masticatory structures (Tobias, 1967; Grine, 1985; Suwa, 1988). Some authors, such as Yoel Rak (1983), suggested that *P. boisei* could be a direct descendant of *P. robustus*, drawing, thus, the evolutionary lineage *A. africanus → P. robustus → P. boisei*. Greater robusticity would tend to be associated with a later appearance in the fossil record.

Some findings on the western shore of Lake Turkana in the mid-1980s rocked this simple scheme. In 1985, Alan Walker and his colleagues discovered a cranium, registered as KNM-WT 17000, in the Lomekwi basin. A mandible (KNM-WT 16005) was found a little to the south on the same western shore, at Kangatukuseo. Both specimens were presumably attributed to the male sex. By correlation of two volcanic tuffs located above and below the place where the cranium was located with the Lokalalei basin tuffs, which had been related with the CP submember of the Shungura Formation, KNM-WT 17000 was estimated to be 2.50 ± 0.05 Ma. The mandible would be slightly younger, around 2.45 ± 0.05 Ma (Walker et al., 1986). (See below for the role of the River Omo sites in the correlation between Ethiopian and Kenyan sediments.)

KNM-WT 17000 is a nearly complete cranium, which includes most of the anterior teeth, a molar, and part of the face (Figure 6.11). It was named black skull because of its color due to the manganese of the soils. Walker and colleagues (1986) described it as a massive cranium, with a very broad face. The palate and the base of the cranium, also very broad, are similar to those of OH 5, found at Olduvai (*P. boisei*). The cranial capacity is

---

Box 6.11 Was *A. africanus* the first member of a robust clade?

Rak's notion that *A. africanus* is the first member of a robust clade that would later lead to *P. robustus* and *P. boisei* was not generally accepted. Phillip Tobias, for instance, was famously against it. Tobias considered *A. africanus* as the ancestor of both clades, the specialized (robust) and the gracile leading to *Homo*. This led him to reject *A. afarensis* as a distinct species and, additionally,

ancestral to all the rest (Tobias, 1980). In his 1980 article, Tobias argued that the specimens attributed to *A. afarensis* should be classified as a subspecies of *A. africanus*, namely *A. africanus aethiopicus*. Tobias later included *A. afarensis* as a separate species in his writings, but placing *A. africanus* as the common ancestor of all hominins (see Tobias, 1992, for instance).

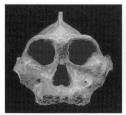

Figure 6.11 KNM-WT 17000, Black Skull, *Paranthropus aethiopicus*. Photographs from http://www.mnh.si.edu/anthro/humanorigins/ha/ WT17k.html.

low, close to 410 cm^3, which represents the lowest cranial capacity for any adult fossil hominin, except maybe A.L. 162-28 from Hadar. The sagittal crest is huge, the largest among all hominins. The cranium corresponds, thus, to a hyper-robust specimen.

Some features of KNM-WT 17000 differ from other robust australopithecines. The specimen has a very marked prognathism (a forward facial projection, beyond the vertical plane that passes through the ocular orbits). Its dentition is also different from that of *P. boisei*. The conserved molar, the upper right P3, is, in comparison with that of OH 5, larger in the longitudinal (mesodistal) direction and smaller in the transversal (bucolingual) direction. Furthermore, its robusticity is comparable to the most extreme instances of *P. boisei*.

What kind of hominin is KNM-WT 17000? When Walker and colleagues (1986) classified the West Turkana discovery they emphasized its distinctive features, seeking to establish that it might be a species different from *P. boisei*. Twenty years earlier Arambourg and Coppens had classified a robust mandible, L7A-125, found at the Omo site of the Shungura Formation, as *Paraustralopithecus aethiopicus* (Arambourg and Coppens, 1968). Consequently, Walker's team assigned the West Turkana specimens to the same species. However, they did not believe that the proposal of a different genus of robust australopithecines was justified.

It is common to classify KNM-WT 17000 as *P. aethiopicus*, although the identification of this specimen with the mandible found at Omo is not unambiguous. The L7A-125 mandible is attributed to a female and, thus, is somewhat different from

KNM-WT 17000 and KNM-WT 16005 (attributed to males). But it would seem unreasonable to assume that there were two quite similar species living at the same time at Omo and West Turkana, two very close sites.

In any case, Walker and colleagues attributed KNM-WT 17000 to *P. aethiopicus* tentatively, only if it could be convincingly shown that it was not a *P. boisei*. When Walker and colleagues compared the black skull's morphological traits with the list of traits established for *P. boisei*, they noted that there were similarities between both taxa in most of the traits. Walker and Richard Leakey (1988) argued that the specimens form West Turkana (KNM-WT 17000, KNM-WT 16005, KNM-WT 17400), Koobi Fora (like KNM-ER 406, KNM-ER 732, KNM-ER 1590, KNM-ER 3230, and KNM-ER 13750, among others), and the Omo mandible belonged to the same species as OH 5 from Olduvai. If so, *P. boisei* would be a highly variable taxon; extremely variable in some traits, such as the sagittal and nuchal crests. This taxon's sexual dimorphisms would be comparable to those of current gorillas.

Walker and Leakey's (1988) main reason for not distinguishing two species was the danger of giving different names to a segment of short duration within one evolutionary lineage, which is a risk when the exemplars are so scarce. The set of robust australopithecine specimens from East Africa includes those found at Olduvai and Peninj (Tanzania), at Chesowanja, on both shores of Lake Turkana (Kenya), and at Omo: 60 individuals, in addition to isolated dental remains. However, there are only 16 facial skeletons and the same number of crania that are informative enough to

---

Box 6.12 Clarification of L7A-125 and KNM-WT 17000

Two terminological clarifications need to be made. The L7A-125 Omo mandible initially received the identification number 18–1967–18. We have opted to refer to this exemplar by its current catalogue number, despite the one that figures in the papers mentioned.

Alan Walker and colleagues classified KNM WT-17000 as *Australopithecus aethiopicus*, and not *Paranthropus aethiopicus*. However, to avoid confusion we have kept the genus *Paranthropus* for the robust australopithecines.

---

document a taxon ranging for more than a million years (Walker and Leakey, 1988). Therefore, it is difficult to decide whether the West Turkana and Omo specimens are within the variation range of *P. boisei* or whether they should be classified as *Paranthropus aethiopicus*. Faced with this state of affairs, it is more parsimonious, according to Walker and Leakey (1988), not to multiply species.

William Kimbel and colleagues (1988) favored a contrary point of view. These authors re-analyzed the features of KNM-WT 17000 highlighted in Walker *et al.*'s (1986) original article. They made a comparative study of: (a) the apomorphies (derived traits) the specimen shares with (1) *A. africanus*, *P. robustus*, and *P. boisei*, (2) *P. robustus* and *P. boisei*, and (3) only *P. boisei*; and (b) plesiomorphies (primitive traits) shared with *A. afarensis* (Kimbel *et al.*, 1988). Out of 32 traits, 12 were, according to Kimbel and colleagues, primitive characters shared with *A. afarensis*; six were derived and shared with *A. africanus*, *P. robustus*, and *P. boisei*; 12 were derived traits shared with *P. robustus* and *P. boisei*; and they found only two derived characters shared exclusively with *P. boisei*. Kimbel *et al.* (1988) argued that their study supported the classification of KNM-WT 17000 as a separate species, *P. aethiopicus* (Table 6.3).

The existence of hyper-robust specimens almost 2.5 Ma contradicts the existing notion that robusticity is a late trait. The sagittal and nuchal crests are much more developed, robust, traits in KNM-WT 17000 than in *P. robustus* or *P. boisei*. The discovery of the black skull, thus, required reinterpreting the evolutionary history of robust hominins and, hence, of our whole family. The same year the specimen was discovered, 1986, saw the publication of several articles proposing

new phylogenetic alternatives to accommodate the new discovery (see Figure 6.12, and Cela-Conde, 1989).

According to Walker *et al.* (1986), KNM-WT 17000 helps us to understand the common traits shared by South African and East African robust specimens. Some of these traits, such as the very thick dental enamel, suggest that robust australopithecines constitute a clade; that is to say, they have a common robust ancestor (Robinson, 1963; Tobias, 1967; Rak, 1983). As we saw above, *A. africanus* could be considered the ancestor of the robust clade. However, the evolutionary sequence *A. africanus* → *P. robustus* → *P. boisei* was rejected by Walker *et al.* (1986) on the grounds of the evidence provided by KNM-WT 17000. These authors argue that the West Turkana robust specimen is closer to *A. afarensis* than to any *A. africanus*. Furthermore, its age would add plausibility to a *P. aethiopicus*–*P. boisei* common clade, but separate from the evolutionary line leading to *P. robustus*. Hence, Walker *et al.* (1986) consider the common features shared by South African and East African robust exemplars as homoplasies: analogous traits independently fixed in both lineages. This would mean that rather than a robust clade there are two: one including the sequence *P. aethiopicus* → *P. boisei* and another with *P. robustus*. If this were the case, then the different species would belong to two separate lineages and could not be grouped in a single genus *Paranthropus*.

In his study regarding the monophyly or polyphyly of *Paranthropus*, Wood (1988) noted that the evolutionary convergence of traits from complex functional structures—such as masticatory ones—can confound conclusions regarding phylogenetic proximity. Wood believes the evidence concerning

Table 6.3 Affinities of KNM-WT 17000 with different taxa

A. Primitive features shared with *A. afarensis*
  Strong upper facial prognathism*
  Flat cranial base*
  Posterior–anterior temporalis large
  Temporomandibular joint flat, open anteriorly
  Postglenoid process anterior to tympanic plate*
  Extensive temporal squama pneumatization
  Strongly flared parietal mastoid angle
    (asterionic notch?)
  Large horizontal distance between molars and
    temporomandibular joint
  Absolutely large anterior tooth row*
  Maxillary dental arch convergent posteriorly
  Flat, shallow palate
  Nasion coincident with high glabella

B. Derived features shared with all post-*A. afarensis* species
  Short cranial base*
  Vertically inclined tympanic plate inferosuperiorly
    concave*
  Reduced medial inflection of mastoid process
  Nasoalveolar contour protects weakly anterior
    to bicanine line

C. Derived features shared with *A. africanus*, *A. robustus*, and
  A. boisei
  Maxillary lateral incisor roots medial to nasal
    aperture margins
  Zygomaticoalveolar crest weakly arched in facial view

D. Derived features shared with *A. robustus* and *A. boisei*
  "Dished" midface*
  Zygomatic process forward relative to palate length
  Guttered nasoalveolar clivus grades into nasal cavity floor
  Anterior vomer insertion coincident with
    anterior nasal spine
  Nasals widest superiorly
  Supraorbitals in form of "costa supraorbitalis"
  Receding frontal squama with "trigonum frontale"
  Relatively enlarged postcanine toothrow
  Incisors in bicanine line
  Petrous inclined coronally*†
  Tympanic vertically deep, with strong vaginal process†
  Mastoid bulbous, inflated beyond supramastoid crest

E. Derived features shared exclusively with *A. boisei*
  Heart-shaped foramen magnum
  Temporoparietal overlap at asterion‡

Source: Kimbel *et al.* (1988).

*Variable in *A. africanus*.

†Also in *Homo*.

‡Y. Rak, personal communication.

robust hominins is much too fragmentary, and that the most parsimonious solution is a monophyletic group that includes all robust australopithecines, East African and South African. The alternative would imply that *A. africanus* is the ancestor of South African robust specimens and *A. afarensis* of East African ones.

The conclusion of a volume, edited by Grine (1988) on the evolutionary history of robust australopithecines, is that the discovery of KNM-WT 17000 actually had the opposite effect to that suggested by Walker *et al.* (1986). According to Grine, some traits shared by *P. robustus* and *P. boisei* should be considered synapomorphies (phylogenetically shared homologous characters) and not homoplasies (convergences by analogy). The existence of two clades is more plausible than some authors admit, but the best alternative is one clade: a monophyletic group that includes all robust australopithecines and classifies them within the genus *Paranthropus*. Grine and Martin (1988) reached the same conclusion after examining one of the most obvious features of robust australopithecines, the thickness of their molars: this trait is synapomorphic of the different robust species and distinguishes them from *A. africanus* and any other *Homo* taxon. The considerable thickness of the molars is achieved through similar developmental patterns in South African and East African robust specimens.

The controversy surrounding whether or not the robust australopithecines constitute a monophyletic group took a new turn after the discovery of a very complete *P. boisei* cranium, which retained a considerable part of the face and mandible, in the Konso (Ethiopia) site. The Konso deposits were discovered by Berhane Asfaw and colleagues in 1991 and had previously yielded *H. ergaster* specimens associated with a great abundance of very old Acheulean tools (Asfaw *et al.*, 1992). The new discoveries were made in the same stratigraphic horizon where those previous exemplars had been found.

In 1997, Asfaw and colleagues reported the discovery of several specimens in the Konso KGA10 locality (Table 6.4). This locality is situated between two tuffs (KRT and TBT) estimated, by

Box 6.13 Classification of the black cranium as *P. aethiopicus*

It is still unclear whether or not the black skull (together with KNM-WT 16005, KNM-WT 17400, and the Omo mandible L7A-125) should be classified as *P. aethiopicus*. As Rak (1988) noted, it is risky to use anatomical traits and use them as taxonomic elements independently of

their functional value. Rak argued that the great prognathism shared by KNM-WT 17000 and *A. afarensis* is not reason enough to place the former closer to the latter, taxonomically speaking, than to the orthognath *P. robustus*.

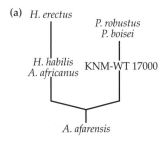

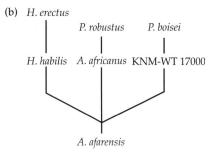

Figure 6.12 Phylogenetic interpretations of KNM-WT 17000, by (a) Eric Delson (1986) and (b) Roger Lewin (1986).

the ^{39}Ar/^{40}Ar method, to be respectively 1.41 ± 0.02 and 1.43 ± 0.02 Ma (Suwa *et al.*, 1997). Thus, the age for the specimens is close to 1.42 Ma. The KGA10 series of specimens includes the KGA10-525 specimen, which provided the first complete association of *P. boisei* cranium and mandible (Figure 6.13). A great part of the face was found, except for the frontal bone and the anterior cranial base. It is an adult specimen, large and presumably male, with a cranial capacity of 545 cm^3, slightly greater than previous South African and East African robust specimens.

Suwa and colleagues (1997) have no doubt that the specimen's morphology corresponds to a robust australopith, although some traits (mostly dental) are shared only with *P. boisei*. However,

Table 6.4 Robust specimens from Konso, Ethiopia

Specimen	Element	Discoverer/year
KGA10-506	Left palate with dentition	A. Amzaye, 1993
KGA10-525	Partial skull	A. Amzaye, 1993
KGA10-565	Right upper M1	G. Suwa, 1994
KGA10-570	Juvenile mandible	Y. Zeleke, 1994
KGA10-900	Molar fragments	H. Nakaya, 1994
KGA10-1455	Left parietal	K. Uzawa, 1994
KGA10-1720	Left lower M3	B. Asfaw, 1996
KGA10-2705	Right lower M2	Y. Haile-Selassie, 1996
KGA10-2741	Molar fragments	K. Gelete, 1996

Source: Suwa *et al.* (1997).

Suwa's team formulated an interesting interpretation of its morphology. KGA10-525 includes some traits, such as the beginning of the sagittal crest, which are not present in the *P. boisei* and *P. robustus* specimens known at the time, and only present in KNM-WT 17000, *P. aethiopicus*. Other traits of KGA10-525 are also at the limits or even beyond the variation ranges of previous robust specimens. According to Suwa and colleagues, certain characters of robust specimens had been considered functionally and adaptively meaningful, to the point of being considered fundamental in the attribution of exemplars to one or another species. But such traits could in fact correspond to polymorphisms resulting from a great variation among different populations. Hence, the *P. boisei* type specimen, OH 5 from Olduvai, would represent an "extreme" specimen (Delson, 1997) of the species, while *P. boisei* of Turkana and those discovered at Konso would belong to individuals whose morphology lies between those from Olduvai and the South African *P. robustus*.

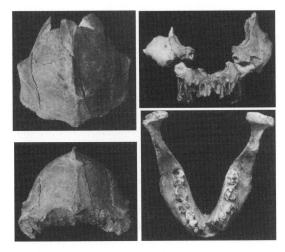

Figure 6.13 KGA10-525, *P. boisei*. Left, cranium; right, associated mandible. Photographs from Suwa *et al.* (1997). Reprinted with permission of Macmillan Publishers Ltd: Nature.

As Delson (1997) noted, Suwa and colleagues could have very well attributed the Konso specimens to a new species. Nonetheless, they chose to suggest that all robust australopithecines (at least *P. boisei* and *P. robustus*) belong to the same species. The conclusions of Wood *et al.* (1994) about KGA10-525, based on the examination of numerous dental traits, suggest that an abrupt change occurred some 2.3–2.2 Ma leading to the transformation of *P.* aff. *boisei* (KNM-WT 17000) into *P. boisei* sensu stricto (KNM-ER 403, KNM-ER 406, etc.). Wood *et al.* (1994) believe *P. boisei* experienced some stasis, so that, for instance, molar size remained unchanged for more than a million years. Cranial traits, judging from KGA10-525, could indicate a great variability and not any evolutionary sequence between several species.

Whether grouped in a single species, or two or three, the distribution of robust australopithecines in East African sites indicates a very broad temporal and spatial presence. The robust lineage persisted from 2.6 Ma (L55-33, Omo 18-18, Omo 18-317, and Omo 84-100) to 1.2 Ma (OH 3 and OH 38 from Olduvai).

The mandible UR 501 extended the geographical range of early gracile hominins towards the south, all the way to Malawi. Similar extension obtains for the robust specimens. In 1999 Kullmer *et al.*

(1999) reported the discovery of the HCRP RC 911 specimen in Malema, Chiwondo Beds (Malawi). It is a maxilla fragment with part of two molars, with features within the size range of *P. boisei*. The age was estimated to be close to 2.5 Ma by associated fauna. Taken together, the Konso and Malema specimens indicate the presence of robust australopithecines in East Africa in a territory extending from Ethiopia to Malawi with a very broad hypodigm.

6.2.3 Omo: the Rift calendar

We have mentioned the formation correlation technique as a method to estimate the age of sediments. Such correlation requires having a well-established sedimentary sequence that can be used as a reference. The geographical area of the River Omo has played a fundamental role. The sedimentary area of Omo is in the southern part of the valley through which the River Omo flows, immediately north of Lake Turkana in Ethiopia (Figure 6.14). The first vertebrate fossils were found in 1902. The expedition led by Camille Arambourg during 1932-1933 retrieved a large number of fossils: up to 4 tons (Coppens, 1980). In 1966 the International Omo Research Expedition was founded, with three sections: the Kenyan, led by Louis and Richard Leakey; the French, led by Camille Arambourg and, after his death, by Yves Coppens; and the American, led by F. Clark Howell.

The first expedition was carried out in 1967. This was the beginning of an intensive study of the site. After the section led by the Leakeys was moved to Lake Turkana in 1968, the American and French sections continued the work until they were interrupted by the Ethiopian civil war (Coppens, 1980).

The sedimentary deposits of the plinth of the lower Omo Valley are a tectonic depression that stretches Lake Turkana's plinth to the north. Omo has eight different formations, three of which have provided hominin remains: Shungura, Usno, and Kibish. Shungura protrudes on the right hand shore of River Omo, extending for more than 200 km² , and has an impressive depth of up to

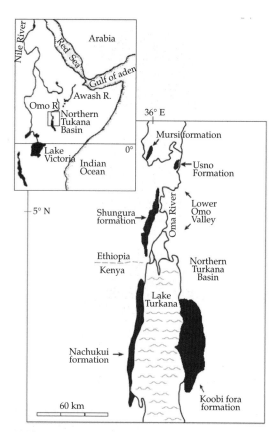

Figure 6.14 Map of the sedimentary area of the Omo Valley and an aerial view of the Shungura Formation (Omo Valley, Ethiopia). Map from Bobe and Behrensmeyer (2004); photograph from http://www.indiana.edu/~origins/images/ShunguraFM.jpg.

850 m. It is ideal for establishing chronostratigraphies because of the perfect continuity of the sedimentation and the presence of many volcanic tuffs (Coppens, 1978a, 1978b). Above the basal

member, from the oldest to the youngest, are the Members A-L (Figure 6.15), an excellent sedimentary sequence from slightly more than 3 Ma to a little less than 1 Ma. The Shungura Formation is useful as reference horizon for the correlation of other nearby sites. If we take into account that the Lothagam Hill site, close to the left-hand shore of Lake Turkana, contains a sequence of strata that spans from the oldest at Omo to nearly 5.5 Ma (Patterson *et al.*, 1970), the Pliocene stratigraphy in the area is completely established for a long period of time.

It is often difficult to correlate different sites, but Shungura's great fossiliferous richness allowed Coppens (1972, 1975) and Coppens and Howell (1976) to define a series of areas, based on the association of large mammals, for later faunal comparison. Three types of faunal association were defined at Omo. Omo 1 extends from the basal member to the top of Member C (3.2-2.4 Ma). Omo 2 ranges from the base of Member C to the top of G (2.6-1.8 Ma) and Omo 3 from the base of Member G to the top of the formation (2.0-1.0 Ma; Coppens, 1978b).

A detailed correlation between the Turkana and Omo plinths has been carried out through successive reinterpretations of the ages of the volcanic tuffs and the use of paleomagnetism. In spite of some doubts regarding the location of the limits between the formations' members (Howell *et al.*, 1987; Feibel *et al.*, 1989), the sequence in all this East African area is reasonably well established and compatible with biostratigraphic and magnetostratigraphic data.

Omo has yielded many early hominins from the A to the L Shungura Members, although subject to very different interpretations and classifications, because the remains are mostly fragmentary. Coppens (1978b) provisionally attributed a great number of isolated teeth, mandibles, maxillas, and very modest (in his own words) cranial and postcranial fragments to *A. africanus*. Coppens (1978b) ascribed the most complete remains to *P. boisei*, including the Omo-323-1976-896 specimen, a partial hominin cranium, dated to *c.* 2.1 Ma, from Member G, Unit G-8 of the Shungura Formation (Alemseged *et al.*, 2002). Omo-323 is made of

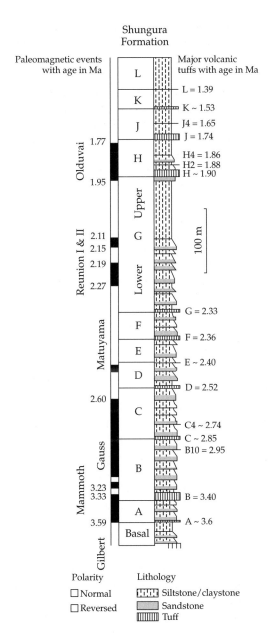

Shungura
Formation

Paleomagnetic events
with age in Ma

Major volcanic
tuffs with age in Ma

L = 1.39
K ~ 1.53
J4 = 1.65
J = 1.74
H4 = 1.86
H2 = 1.88
H ~ 1.90

G = 2.33
F = 2.36
E ~ 2.40
D = 2.52
C4 ~ 2.74
C ~ 2.85
B10 = 2.95

B = 3.40
A ~ 3.6

Polarity

☐ Normal
☐ Reversed

Lithology

▦ Siltstone/claystone
▨ Sandstone
▥ Tuff

Figure 6.15 The stratigraphy of the Shungura Formation.
Picture from Bobe and Behrensmeyer (2004).

Figure 6.16 Right partial maxilla of Omo-323–1976–896,
P. boisei. Photographs from Alemseged *et al.* (2002).

is one of the earliest specimens of that hypodigm.
Alemseged *et al.* (2002) noted that Omo-323 shares
certain traits with KNM-WT 17000, *P. aethiopicus*.

During the late 1970s, Coppens and Howell
provided an overview of the Omo hominins
(Howell and Coppens, 1976; Coppens, 1978a,
1978b, 1980; Howell, 1978), based, as we already
pointed out, on fragmentary and difficult-to-
interpret remains. At the time, comparative evidence regarding *A. afarensis* was not available
(Howell *et al.*, 1987), although it was provided later
by the Hadar and Laetoli sites. The available
hypodigms were *A. africanus*, *P. boisei*, and *H.
habilis*. In accordance with such comparative possibilities, it was argued that there were at least
three types of hominin in the Omo Plio-Pleistocene
sediments: robust specimens, which could belong
to *P. boisei* or akin species; also, a less-robust species related with *A. africanus*, although the authors
were not completely certain about this; finally,
there were gracile specimens linked with *H. habilis*
or, less often, with *H. erectus*.

fragments of the frontal, both temporals, occipital,
parietals, and the right maxilla (Figure 6.16). A
well-developed and completely fused sagittal
crest, heavily worn teeth, and a relatively large
canine suggest the specimen was a male *P. boisei*. It

After a very detailed reanalysis of the hominin remains from the Omo E and F Members, Howell *et al.* (1987) questioned the initial interpretation, *vis-à-vis* the West Turkana discoveries. These findings afforded a more detailed knowledge of robust hominins and the correlation with the Shungura Formation sediments, providing an opportunity to answer three questions related with the hominins present at Omo: (a) do the Omo robust specimens belong to *P. boisei* or another taxon? (b) is *A. africanus* present at Omo? (c) is there evidence suggesting the presence of the genus *Homo* at that place? Let us deal with these issues in order.

1 The study of the microstructure of the Omo fossil sample's enamel carried out by Fernando Ramirez Rozzi (1998) indicated that the specimens belonged to a single species that did not vary more than current apes regarding this trait.

2 Howell and colleagues (1987) argued that most of the identifiable hominins in the sample from Members E and F can be attributed to australopithecines, and that their morphological features correspond to a robust pattern, or at least tend towards it. In consequence, they are different from South African *A. africanus*. But, in what taxon should the specimens be placed? Howell *et al.* (1987) dealt separately with the cranial, postcranial, and mandibular-dental remains. The L-388y-6 cranium, which is fragmentary and corresponds to a juvenile individual, was assigned to *P. aethiopicus* as a "reasonable" option. The L40-19 ulna was classified as *P.* cf. *aethiopicus/ boisei*. The mandibles were assigned to *P. aethiopicus* (the L7A-125 specimen gave the name to this species).

3 Howell *et al.* (1987) saw the dental remains as the most difficult. Most were classified as *P.* aff. *aethiopicus*, but 13 specimens were outside the variability range of robust australopiths and were classified as aff. *Homo* sp. (Howell *et al.*, 1987), which would indicate the presence of the genus *Homo* in the area.

The Omo remains suggest a notable presence of *P. aethiopicus*, extending the existence of the robust hominin clade back to 2.6-2.5 Ma. The gracile indications are more difficult to interpret. Hunt and Vitzthum (1986) saw in the early Omo gracile specimens an opportunity to fill the important gap—of 800,000 years—existing between *A. afarensis* and the *H. habilis* specimens found at Turkana and Olduvai. They argued that the "gracile" Omo specimens from that intermediate period are not similar to *A. afarensis* or *Homo*, but resemble South African *A. africanus*. Consequently, Hunt and Vitzthum (1986) concluded that, contrary to what Johanson's team surmised after the Hadar discoveries, *A. africanus* must be understood as an intermediate step between *A. afarensis* and *Homo*, and as a direct ancestor of the latter genus. More recently, the discovery of *A. anamensis* has provided another alternative interpretation: to relate the Omo gracile specimens with that species.

As we noted previously, Boaz and Howell (1977) linked the L. 894-1 specimen from Member G-28 of the Shungura Formation, estimated to be more than 1.84 Ma, with OH 24 and OH 13 from Olduvai and Sangiran 4 from Java. These exemplars have all been included in *H. erectus* (or *H. ergaster*), which we will study in Chapter 7.

6.2.4 The phylogenetic tree of early hominins

Given the difficulties involved in the assignment of specimens to taxa, we conclude this chapter with a sense of provisionality and great doubts regarding the phylogenesis of Miocene and Pliocene hominins. Figure 6.17 presents

---

Box 6.14 L7A-125

The L7A-125 mandible, initially classified as *Paraustralopithecus aethiopicus*, was later associated with the hyper-robust and very early *Paranthropus aethiopicus* specimen.

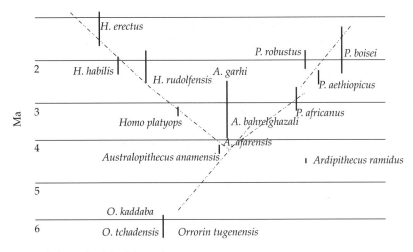

Figure 6.17 A Miocene and Pliocene hominin phylogenetic tree.

the phylogenetic tree that, in our opinion, most accurately reflects what is currently known.

There are many possible alternatives to this proposal, as many as different ways of interpreting the controversial specimens. The tree in Figure 6.17 is based on the following assumptions:

1 A first node, close to 7 Ma, separated the chimpanzee and human evolutionary lineages. The earliest hominin forms (*Orrorin tugenensis*, *Sahelanthropus tchadensis*, and *Ardipithecus kadabba*) could be included in the single genus *Orrorin*. As we saw in Chapter 5, this grouping was made by Haile-Selassie *et al.* (2004) (although not in the *Orrorin* genus).

2 *Orrorin* taxa must be distinguished from australopiths, encompassed by the genus *Australopithecus*, including specimens ranging from the early ones from Hadar (*A. afarensis*) to the later ones from Gona (*A. garhi*).

3 The genus *Ardipithecus* constitutes a separate lineage from australopithecines and other hominins.

4 A Pliocene cladogenesis, about 3.5 Ma, separated the evolutionary lineages of gracile and robust hominin forms, leading to the genera *Paranthropus* and *Homo*. The accentuation of the robust and gracile tendencies coincided with the climatic change that occurred 2.5 Ma.

5 *Paranthropus* includes robust specimens from South Africa and East Africa. Its earliest taxon is the almost 3.5-million-year-old *A. africanus* (now *P. africanus*). This is, therefore, the age of the node separating the robust and gracile lineages.

6 The same cladistic episode requires the appearance of *Homo* close to 3.5 Ma. The first gracile hominin would be *K. platyops*, reclassified as *Homo platyops*. *H. habilis* and *H. rudolfensis* may be a single taxon, but this is a minor problem (Lieberman *et al.*, 1996), given that in the tree presented here they are sister groups. However, this taxonomic solution is not viable if *H. rudolfensis* is included in a separate genus, such as *Kenyanthropus*.

6.2.5 Doubts and alternatives related with the gracile and robust lineages

A taxonomic and phylogenetic alternative to the one in Figure 6.17 is to move *H. rudolfensis* and *H. habilis* to the genus *Australopithecus* (Wood and Collard, 1999a). This option has recently received much attention, but was suggested when Louis Leakey, Tobias, and Napier defined the species in 1964. From a morphological point of view, *H. habilis* retains plesiomorphies shared with

A. africanus, so that it is not unreasonable to consider it an australopith. Taking *H. habilis* out of the genus *Homo* would require excluding traits shared with *H. habilis* as the new genus apomorphies and placing its origin at a more recent time. During the first stage of hominization all the specimens would be classified as *Australopithecus*. However, if *H. habilis* and *H. rudolfensis* are moved to *Australopithecus*, the already considerable paraphyly of this genus is further increased. This will be more so if we also reject the paranthropine genus and we consider the taxa there as *Australopithecus (Paranthropus) aethiopicus, A. robustus,* and *A. boisei*.

When L. Leakey *et al.* (1964) introduced the taxon *H. habilis*, they pointed out certain physical apomorphies, characters that separated it from australopithecines, although they emphasized a functional trait, tool making, as a distinctive feature that justified the awarded name of "skilled being", *habilis*. Stone carving is not just a functional trait, it involves anatomical changes in the brain, as underlined by Phillip Tobias and Dean Falk. *H. habilis* is not just the first hominin associated with stone tools in archaeological sites, but the first one capable of carving them. If this were the case, the very old Oldowan tools found at Bouri would not be linked to *A. garhi*.

Keeping *H. habilis* within the genus *Homo* is currently the most common alternative. From a cladistic point of view, it is more coherent to recognize the existence of the genus *Paranthropus*. We have, therefore, adopted both of these options.

The earliest documented exemplar of the robust lineage is the 2.5-Ma *P. aethiopicus*. If the separation between the gracile and robust branches took place 3.5 Ma, we could expect to find robust hominin forms at that time. Because of its age and morphological differences from early *Homo* (the specimens initially included in *Kenyanthropus*), *A. africanus* could be the first example of the robust lineage. Robustness would have been exaggerated a million years later, with the adaptation to open savannas, which extended throughout Africa when the climatic cooling began 2.5 Ma. This led to a paranthropine lineage specialized in a hard vegetable diet to *Homo*, which began making and using tools for scavenging and possibly hunting.

Within the robust lineage, morphological affinities and geographical variations make interpretation difficult. Because there are no *P. aethiopicus* specimens in South Africa and no clearly identified *A. africanus* in East Africa, it could be argued that the robust lineage is, in actual fact, polyphyletic. If this is the case, the common traits between *P. boisei* and *P. robustus* would be analogous traits or homoplasies. Some interpretations of the Konso specimen favor a monophyletic solution, which seems more reasonable. If the robust lineage is monophyletic, *P. aethiopicus* must be considered as the common ancestor of South African and later East African paranthropines. This is the option that we have adopted in Figure 6.17.

The evolutionary episodes that took place during the Miocene-Pliocene stage of hominin evolution are difficult to define unequivocally. *A. anamensis* is a good candidate for continuing the direct line from *O. tugenensis* to the separation of the robust and gracile lineages. *A. afarensis*, accepted at the time of the Hadar discoveries as the common ancestor of all later hominins, seems to have been displaced to a lateral evolutionary branch by the discovery of *A. anamensis*, whose traits place it closer to *Homo*. *A. bahrelghazali*

Box 6.15 The genus for *A. africanus* and paranthropines

If *A. africanus* and paranthropines belong to the same genus, then, according to the rules of the International Code of Zoological Nomenclature, the genus must be *Australopithecus*, because it was proposed before, when the Taung child was classified.

does not particularly favor any hypothesis except that we must recognize its similarity with *A. afarensis*.

Because the evolutionary changes occurred 2.5 Ma are better documented in East Africa, it is worth searching for *Homo*'s ancestors there. Asfaw *et al*. (1999) and de Heinzelin *et al*. (1999) suggested *A. garhi* as the common ancestor of the gracile and robust lineages based on their geography. However, after the comparison with *A. anamensis*, it seems more appropriate to place *A. garhi* on a divergent evolutionary line.

CHAPTER 7

The radiation of *Homo*

7.1 Asian *Homo erectus*

One of the most important events in hominin evolution, the Pliocene cladogenesis, produced the two lineages just examined: *Paranthropus* and *Homo*. Their fates were quite different. Paranthropines did not evolve much from the time they appeared until their extinction, about 1 Ma. *Homo* forms diversified and dispersed, and they left Africa and occupied other continents. Our own species is part of this lineage of travelers.

The genus *Homo*, including *Homo habilis* and *Homo rudolfensis*, appeared close to 2.5 Ma. We need to add another million years, to nearly 3.5 Ma, if we include in *Homo* the specimens Meave Leakey and colleagues (2001) grouped in *Kenyanthropus*. The hominins belonging to our genus evolved from those initial taxa. Evidence from Olduvai (Tanzania) and Koobi Fora (Kenya) suggests that early specimens were substituted by another kind of hominin with a larger body and greater cranial capacity, which for the moment we'll call *Homo erectus*. *H. erectus* was the first hominin to colonize territories outside the African continent. There are *H. erectus* in Africa and Asia; they have also been found in Europe (Figure 7.1), although the European specimens are not easily classified. Important questions regarding the evolution of *Homo* during the end of the Pliocene and throughout the Pleistocene are: How did the radiation of the genus take place? What species played the main role in the expansion out of Africa? Do African and Asian specimens constitute a single species, or two? If they are two different species, to which do European exemplars belong? Or, do these represent a different taxon altogether, different from the Asian and African ones?

A chronologically ordered description of the evolutionary process of *Homo* would require describing African specimens first, then the departure out of Africa, the subsequent colonization of Asia and, finally, the arrival to Europe. But the first *H. erectus* specimens were found in Asia, and the taxon's profile was established according to their features. Doubts arising from the comparison among Asian, African, and European exemplars would appear later. We will alter the chronological evolutionary sequence, so that we may review, in this chapter, the Asian specimens first and turn later, in section 7.2, to the African ones. Chapter 8 is devoted to the European question, the departure out of Africa, and the cultural transformations that allowed hominins to adapt to climates far from the tropics.

7.1.1 *Homo erectus* sensu stricto and sensu lato

The discovery of a fossil in Java in the late nineteenth century provided the first clue of the presence of very early hominins in southeast Asia. The discovery of the specimen, known as *Pithecanthropus erectus*, was followed, in the first half of the twentieth century, by the discovery of numerous specimens in Java and China. They were all grouped eventually in the taxon *H. erectus*, characteristic of middle Pleistocene hominins. Similar specimens found later in East Africa raised the possibility that another species existed during that epoch: *Homo ergaster*. African specimens showed some slightly different morphological traits and lived, certainly, very far from Javanese and Chinese sites. Those who adopt this double species

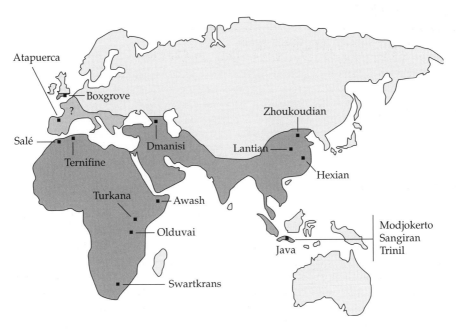

Figure 7.1 The main *H. erectus* sites. Picture from http://anthro.palomar.edu/homo/homo_2.htm.

Box 7.1 The concept of grade

Julian Huxley suggested the concept of "grade" to express the notion that different taxa might have reached a certain evolutionary stage (Huxley, 1958). Bernard Wood and Mark Collard (Collard and Wood, 1999; Wood and Collard, 1999a, 1999b) rescued Huxley's grade to face the difficulties involved in dealing with the genus *Homo*. In their opinion, if the species *H. habilis* and *H. rudolfensis* are included in *Homo*, the latter taxon becomes paraphyletic. This is the same problem that some authors see when grouping *H. erectus* sensu stricto and *H. ergaster* in a single taxon. This problem disappears if the set is considered as a grade rather than a lineage.

The expression *erectus* grade can encompass all middle Pleistocene hominins. The most conspicuous morphological trait of the *erectus* grade is the noticeable increase in cranial capacity, which doubles that of *H. habilis* specimens, around, or just above, 500 cm³, and is much

greater than that of *Kenyanthropus rudolfensis*. The discovery of postcranial remains as complete as those known as the Nariokotome boy, which we will describe later, has led to a reappraisal of the evolutionary significance of this brain increase. To a certain extent it seems to be a proportional (allometric) consequence of an increase in body size. Consideration of allometric increase tones down the hypotheses about the relative increase of brain size during the middle Pleistocene. But it confirms the hominin evolutionary process during that time: there were selective pressures to which hominins responded by increasing their size. There also is illustrative evidence showing that the techniques used to make stone tools became increasingly complex. This suggests that, in addition to cranium size, cognitive capacities also underwent a noticeable improvement.

model refer to Asian specimens as *H. erectus* sensu stricto, as opposed to *H. erectus* sensu lato.

The classification of Plio-Pleistocene and middle Pleistocene *Homo* specimens stands out among the

typical issues that plague the study of human evolution. We go into more detail further on, but for now we will group them as the *erectus* grade, which implies that all those specimens attained a

---

Box 7.2 The single-species hypothesis again

Since the times of the single-species hypothesis—the belief that only one hominin species lived during each epoch—many authors have associated the middle Pleistocene with *H. erectus*. The single-species hypothesis is currently part of the history of the study of hominins, at least with regard to the Pliocene. Nevertheless, as Colin Groves (1989) said, the notion of *H. erectus* as the characteristic prototype of middle-Pleistocene hominins is alive and well.

It is possible to identify that epoch's hominins as cut by a similar pattern, although obviously there are morphological variations. Wolpoff (1999) has argued that the considerable continuity within *Homo* justifies the consideration of a single taxon, *H. sapiens*, whose gradual changes, remarkable as they might seem to us, do not validate the proposal of several species. We will adhere to the alternative point of view.

Table 7.1 Main evolutionary events of the Pleistocene

Era	Duration	Events
Late Pleistocene	130,000–10,000 years ago	*Homo sapiens* arrived in Asia and Europe; the remaining *Homo* species disappear
Middle Pleistocene	730,000–130,000 years ago	Colonization of Europe (towards the end of the early Pleistocene)
Early Pleistocene	1.63 million–730,000 years ago	*Homo* left Africa; extinction of *Paranthropus*

similar stage of morphological evolution. Cultural evolution will also be discussed in the next chapter.

Experts agree that all specimens belonging to the *erectus* grade fit well in the genus *Homo*. (The very small hominins from the island of Flores, which we will describe when we discuss the Neanderthals and anatomically modern *Homo sapiens*, are included in this genus.) We will examine a double history. We will describe the specimens that reached Asia at the beginning of the Pleistocene or even a little earlier. In addition, we will address the issue of the extent to which those specimens are similar to or different from those that remained on the African continent.

The Pleistocene is the geological epoch that ranges from 1.63 Ma to 10,000 years ago. It is usually divided into three stages, corresponding to the early Pleistocene (1.63–0.73 Ma), middle Pleistocene (730,000–130,000 years), and late Pleistocene (130,000–10,000 years). The Pliocene

saw the appearance of different *Australopithecus* forms and the divergence of robust and gracile lineages. There were still paranthropines during the early Pleistocene. However, this diversity disappeared with the subsequent reduction of hominins, which happened in the middle Pleistocene. It reached the point where—leaving aside the Flores specimens—there was only one species during the long lapse of time spanning from the colonization of Asia to the appearance of the immediate ancestors of Neanderthals and anatomically modern humans, or two species at most (Table 7.1).

If similarities are emphasized, a single *H. erectus* species is enough to account for all specimens. It remained in stasis, with hardly any changes, for the taxon's long existence. But if the morphological differences detectable in the *erectus* grade are taken into account, as well as the geographical and temporal differences, then it can be argued that there are two species within *H. erectus* sensu lato.

The alternative between one or two species for *H. erectus* is related to a taxonomic paradox brought to light by G. Philip Rightmire (1986). Until the 1950s, *H. erectus* was defined as a temporal form: the hominins that lived in the intermediate period that, starting with South African australopiths, precedes European "archaic" *sapiens*. Some authors (Delson *et al.*, 1977; Stringer, 1984; Wood, 1984) tried to define the taxon more precisely and not simply by exclusion of others; that is, as a taxon with its own traits, clearly distinguished from *H. habilis* and *H. sapiens*. Thus began the search for the apomorphies of *H. erectus*, its distinctive derived traits, starting with the specimens initially discovered in Java. As apomorphies were described, *H. erectus* gradually became a well-defined and distinctive hominin. But, at the same time, the possibilities of establishing relationships with its ancestors (*H. habilis*) and with ourselves (*H. sapiens*) became gradually weakened. With the description of its characteristic features, *H. erectus* tended to be reduced to an isolated milestone in our tribe's evolution, a mostly Asian lineage.

But if *H. erectus* is a mere lateral branch, like robust australopiths, then we have to admit one of the following possibilities. Either *H. habilis* gave rise to our species directly—which is difficult to sustain—or there was another Pleistocene species that fulfilled the connecting role. The latter alternative implies accepting a model of early and middle Pleistocene in which part of the previously considered *H. erectus* would remain as they were, with very remarkable derived characters. Other specimens initially attributed to that species would now be grouped in another species, a new taxon whose distinctive traits allowed the assumption that it was a true ancestor of current *H. sapiens*. What could this other species be? The best candidate to fill this taxonomic void is *Homo ergaster*.

The taxon *H. ergaster*, defined by Groves and Mazák (1975), meets the desired conditions:

- it was characteristically East African,
- it existed at a later time than *H. habilis*, although there is a certain overlap,
- it exhibited a notable cranial capacity,

- it was (supposedly) the maker of highly sophisticated tools.

If this solution is accepted, then *H. erectus* sensu stricto (with the apomorphies observed in the specimens from Java) would be considered a distinctly Asian species, a lateral lineage in the evolution leading to *H. sapiens*, whereas *H. ergaster* would be the adequate taxon for African specimens. Peter Andrews (1984), Chris Stringer (1984), and Bernard Wood (1984) agreed on this point in their contributions to the 1983 Senckenberg conference.

However, there is yet another possibility, which complicates the evolutionary scene. If *H. erectus* sensu stricto and *H. ergaster* are different species, could the former have also departed to Africa? Colin Groves (1989), one of the authors who introduced the taxon *H. ergaster*, thought that OH 9 from Olduvai (Tanzania)—named *H. erectus olduvaiensis* by Phillip Tobias (1968)—evinces the presence of *H. erectus* sensu stricto in Africa. But if the presence of *H. erectus* sensu stricto in Africa is accepted, then one of the main arguments in favor of adopting two species within the *erectus* grade, their geographical separation, crumbles.

Which is the most reasonable solution, one *H. erectus* species or two, *H. erectus* sensu stricto and *H. ergaster*? There are reasons to support each of the alternatives and there are no conclusive arguments for rejecting either hypothesis. The

---

Box 7.3 The proposal of *H. ergaster*

H. ergaster was proposed by Groves and Mazák (1975) in a study on Villafranchian gracile hominin taxonomy based almost exclusively on dental traits. The examination of the specimens from the upper levels—between 1.4 and 1.8 Ma—of East Rudolf (Lake Turkana, Kenya) led to the definition of the new species, whose holotype is KNM-ER 992: two associated hemimandibles, with a complete dentition except for the first incisors. They were described by R. Leakey and Wood (1973) and classified as *H. habilis* by R. Leakey (1974). The hypodigm constitutes specimens such as KNM-ER 730, KNM-ER 731, KNM-ER 803,

KNM-ER 806–KNM-ER 809, KNM-ER 820, and KNM-ER 1480, all dental or mandibular fragments, the KNM-ER 734 parietal, and the KNM-ER 1805 cranium. According to Groves and Mazák, the premolars and, most of all, the smaller molars, the thick and massive mandible, and the greater cranial capacity represent differences regarding *H. habilis* and justify the proposal of *H. ergaster*.

The taxon has gained credit in recent years, but there are distinguished authors, such as Tobias, Wolpoff, Rightmire, Conroy, and Stringer, who deny the existence of *H. ergaster* as separate from *H. erectus* sensu stricto.

Box 7.4 One species or two?

There are no convincing arguments favoring either only one species taxon—*H. erectus*—or two—*H. erectus* sensu stricto and *H. ergaster*. The principle of parsimony (avoiding the multiplication of taxa) inclines us to favor the use of only one taxon, *H. erectus*. Nevertheless, we will continue to use the concept of *erectus* grade so as to imply the presence within *H. erectus* of substantial heterogeneity in morphology, geography, time, and culture.

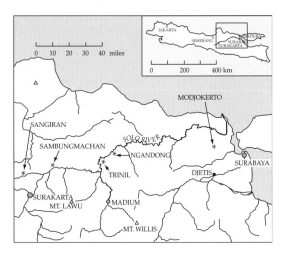

Figure 7.2 *Homo erectus* sites on the island of Java.

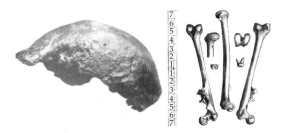

Figure 7.3 Trinil I (left) and Trinil II (right), *Pithecanthropus erectus*. Dubois (1894), drawings by the author.

considerable distance between the Asian Far East and Africa, together with the morphological and cultural differences, support the taxon *H. ergaster*. But, if this is the case, why do specimens very similar to *H. erectus* sensu stricto appear in Africa, such as OH 9 from Olduvai or the Daka cranium? Either the taxon's variability is very high, which is an argument against separating *erectus* into two

species, or the evolutionary episode in Java and China was replicated in East Africa, yielding only the two aforementioned exemplars. The latter scenario assumes that this convergence did not go very far, although the paucity of the African fossil record between 1 and 0.5 Ma hampers reaching a clear-cut decision. Ultimately, the choice depends on how human evolution is understood. One alternative is seeing it as a conservative process, with certain local modifications that do not finally divide populations into species. But it is also possible to understand it as a diversity of lines that gave way to geographically disperse speciations (*H. erectus* in Asia, Neanderthals in Europe, *H. ergaster*, and the first anatomically modern humans in Africa).

7.1.2 The *erectus* grade in Java

Asian *erectus*, considered by some authors to be the only *H. erectus* strictly speaking (*H. erectus* sensu stricto), comes from southeast Asian sites. At the end of the nineteenth century, the Dutch doctor Eugéne Dubois began fieldwork on the island of Java, a Dutch colony at the time, hoping to find the human ancestors described hypothetically by Darwin. The work began at the River Solo, near Kending Hills, in the area named Trinil by Dubois, containing a similar fauna to Siwaliks in Pakistan (Figure 7.2). The work soon paid off. In 1891 a skullcap (technically, this is called a calotte) was found in Trinil, and a femur the year after (Trinil I and Trinil II; Figure 7.3). Dubois attributed both specimens to the same individual, although this identification has often been questioned. *Pithecanthropus erectus* was born.

Only Neanderthals had been identified as possible ancestors of current humans at the time of the

Box 7.5 Naming the Trinil specimens

The first name given to the Trinil specimen was *Anthropopithecus erectus*, because of its similarity with a hominoid found in 1878 in Siwaliks (Pakistan). This name clearly reflected the idea of an intermediate between apes and humans. Dubois changed the genus to *Pithecanthropus* 3 years later, recovering Ernst Haeckel's (1868) proposal of a hypothetical *Pithecanthropus alalus*.

Thus, the Trinil fossil was initially referred to as *Pithecanthropus erectus* (Dubois, 1894). However, the Java specimens and other *H. erectus* known at the time— *Sinanthropus, Meganthropus, Telanthropus*—were placed in *H. erectus* by Ernst Mayr (1944, 1950). The lumping tendency continued with the inclusion of OH 9 and other similar African specimens in *H. erectus* by Le Gros Clark (1964b).

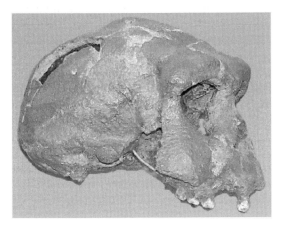

Figure 7.4 Sangiran 17, *H. erectus.* Photograph: *Athena Review,* from cast at AMNH; http://www.athenapub.com/13intro-he.htm.

Trinil discoveries. Not surprisingly, the features of *Pithecanthropus* suggested this was an intermediate being between chimpanzees and us. The Trinil I femur was almost identical to ours, indicating from the first moment that the fossil belonged to a biped. The cranial capacity—slightly over 800 cm^3—was double that of chimpanzees, but much smaller than ours, which led to the specimen being named "ape-man". Subsequent hominin discoveries were made by the Dutch Geological Survey at the Ngandong site: up to 12 skulls classified by W.F.F. Oppennoorth (1932) as *Homo soloensis.* This sample includes massive-looking specimens of large cranial capacity, with an average 1210 cm^3. We will return to this sample when we deal with the dating of the island's sites.

Working with the Dutch Geological Survey, G.H. Ralph von Koegniswald retrieved more specimens at the Sangiran site. We will focus on

some of them. First, a skullcap, Pithecanthropus II (von Koenigswald, 1938), which lacks the cranial base. It is very similar to Pithecanthropus I morphologically. The Pithecanthropus IV specimen (von Koenigswald and Weidenreich, 1939) confirms the posterior part of a cranium including the inferior portion of both maxillae. It was retrieved at the Pucangan Formation, and its features relate it to *erectus* specimens from Java and East Africa, which favored its consideration as the same species. The Pithecanthropus VIII cranium (Sartono, 1971) is the most complete and well preserved of those found at Java, lacking only the left zygomatic region (Figure 7.4). It exhibits typical *erectus* traits, with a distinct supraorbital torus and occipital crest. Four additional remarkable specimens appeared later: the Tjg-1993.05 cranium, the Gwn-1993.09 skullcap, Sbk-1996.02 skullcap, and the Brn-1996.04 occipital bone (Larick *et al.*, 2001).

Thus, towards the 1930s there were enough *H. erectus* specimens to support the notion of an early colonization of Java. When did it take place? As we mentioned earlier, the choice to begin fieldwork at Trinil had to do with the similarity of its fossils with those retrieved at Siwaliks. However, the fossiliferous deposits at Trinil include two different sedimentary beds, and thus the so-called Trinil fauna do not constitute a unit (Day, 1986). The problem of assigning ages to the fossils discovered at Java not only affects Trinil. Four sedimentary formations have traditionally been recognized on the island, each with its corresponding fauna: Kalibeng, Pucangan, Kabuh, and Notopuro (Pope, 1988). At sites such as Sangiran, located at the foot of the Lawu volcano (Figure 7.2), these formations

Box 7.6 Difficulties with the sources of fossils

During the early twentieth century, fossils were often collected by natives and taken to researchers. The source of these specimens is not easily established, so that the exact place where the described fossils had appeared cannot be specified. Only a general description of the area is possible. As stated by Huffman *et al.* (2006), "Past relocation efforts were hindered by inaccuracies in old base maps, intensive post-1930s agricultural terracing, and new tree and brush growth."

(a)   (b)

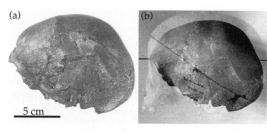

5 cm

Figure 7.5 The Mojokerto child skull. (a) Left-lateral view. (b) Left-lateral view compared to an 11.5-year-old modern child's skull of about the same length. Photographs by von Koenigswald, 1936 and 1938; from Huffman *et al.* (2005). Reprinted from *Journal of Evolution* 48: 4, O. Frank Huffman *et al.* 'Historical Evidence of the 1936 Mojokerto Skull, 321–363, 2005, with permission from Elsevier.

are separated by lava and ash intrusions that allow precise dating. Most of the hominin fossils found at different Javanese sites belong to the Kabuh Formation (middle Pleistocene) but some of them—like Pithecanthropus VII—could belong to Pucangan, which reaches an age of 2 million years. At sites far from the volcano, such as Mojokerto (Perning), dating depends on being able to correlate the sediments with the Kabuh and Pucangan Formations.

It is believed that the oldest Javanese specimen is the Mojokerto child (Mojokerto I or Perning I; Figure 7.5). The specimen was discovered in 1936 in the Perning Valley, west of Trinil, by Tjokrohandojo, a native who worked for Johan Duyfjes. The story of the discovery has been narrated by O. Frank Huffman *et al.* (2005). The Mojokerto child was initially described by von Koenigswald (1938). Carl C. Swisher *et al.* (1994) identified the place of the discovery based on Duyfjes' descriptions and the presence on the specimen's skull of volcanic materials similar to the pumice stone used to estimate the site's age. No other Mojokerto

stratum contains volcanic material. The age of Mojokerto by the ^{40}Ar/^{39}Ar method yields an average of 1.80 ± 0.07 Ma (Swisher *et al.*, 1994). The paleomagnetic study attributed the sediments containing the specimen to a positive event. Swisher and colleagues (1994) chose to assign it to the Olduvai subchron, slightly under 2.0 Ma. If this were the case, the Mojokerto child would be at least as old as the KNM-ER 3733 cranium from Koobi Fora (Kenya). Huffman *et al.* (2006), after investigating the circumstances of the discovery, concluded that it should be younger. The relocated discovery bed would be, they say, about 20 m stratigraphically above the 1.8-Ma horizon.

As de Vos and Sondaar (1994) noted, there is a contradiction between this estimate and that obtained by careful magnetostratigraphical dating studies at Mojokerto by Hyodo *et al.* (1993) that correlate well with the ages obtained using fission-track dating. These techniques place the sediments containing the fossil within the Jaramillo subchron, with an age of 0.97 Ma, very close, hence, to the age of the Trinil cranium (Pithecanthropus I). A review by O. Frank Huffman (2001), however, suggests that Swisher and colleagues' (1994) estimate was correct, and concludes that "recent fieldwork and archival research strongly support the conclusion that the Perning *H. erectus* was found *in situ* in the upper Pucangan Formation, as defined by Duyfjes (1936). Although the excavation spot has not been relocated and detrital materials were used by Swisher *et al.* (1994) to radiometrically date the site, lithologic and paleogeographic evidence from the Pucangan indicates that the *H. erectus* is likely to be 1.81 ± 0.04 Ma." (Huffman, 2001).

If this conclusion is accepted, *H. erectus* would have arrived in southeast Asia before 1.8 Ma. This

Box 7.7 Transcription of Chinese names

We have translated the name of China's capital as Beijing because this is the modern version. For other places we have maintained the names commonly used in the specialized literature. The transcription of Chinese paleontological names can also lead to confusion. Wu Rukang is the same person who signed as Woo Ju-Kang in 1980, and in certain occasions this name is transcribed as Woo Rukang. Alphabetization also varies: sometimes the name is taken as the surname; others it is the middle name that is mistaken. In China the equivalent of the western surname is placed in front.

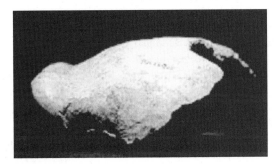

Figure 7.7 Peking Man frontal bone displayed at the Zhoukoudian Peking Man Site Museum in Beijing. Picture from http://www.21stcentury.com.cn/print.php?sid=10548.

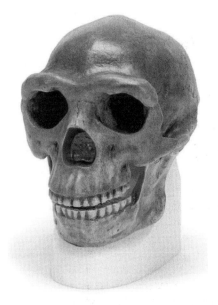

Figure 7.6 *Sinanthropus pekinensis* (Black, 1927); *H. erectus*. The model was developed from a cast of the replica from the collection of the Johann Wolfgang Goethe University of Frankfurt am Main, Institute of Anthropology and Human Genetics for Biologists. Picture from www.3bscientific.com/Datenbank/PDF/VP750_1.pdf-5 Dic 2005.

would mean that the departure out of Africa would have taken place before the emergence of the Acheulean culture (see Chapter 8), given that its earliest manifestations appeared 400,000 years later. Swisher and colleagues (1994) argued that the evolution of *H. erectus* took place out of Africa, but they add that the assignation of the Pucangan Formation specimens to a specific species is not easy.

In regard to *H. soloiensis*—the Ngandong specimens with great cranial capacity—Swisher and colleagues (1997) estimated them to be between 53,000 and 27,000 years old, by means of the uranium-series method and electron spin resonance (ESR) on bovid teeth from the same deposit. Thus, the sample might correspond to a late *H. erectus* population which overlapped with modern humans (Antón, 2002).

The overall picture provided by Java is that of an arrival of *H. erectus* shortly after its appearance in Africa and its persistence on the island for a long period of time. This hypothesis was reluctantly accepted until the Georgia fossils, the role of which we will examine in Chapter 8, provided an important support for that scenario.

7.1.3 The *erectus* grade in China

As we said at the beginning of the chapter, Javanese are not the only Asian *erectus*. Chinese sites have also yielded Pleistocene fossil hominins. The initial identification of the Chinese specimens is an excellent example of the adventures of early twentieth century paleoanthropologists. The story was told in a lively and elegant literary style by von Koenigswald (1981). What follows is taken from that work.

During the early twentieth century, Chinese chemists sold fossils—advertised as "dragon teeth" (*lung tse*) and "dragon bones" (*lung ku*)—that were said to possess efficient aphrodisiac effects. Given that teeth were believed to be more effective than bones, fossil mandibles were destroyed following an understandable commercial objective. The bones were smashed and ground, because they were used in small doses, and their price was very high. K.A. Haberer, a German naturalist who traveled to north China between 1899 and 1901, was able to acquire a considerable collection of such dragon bones and teeth, described in 1903 by Max Schlosser. The specimens gathered by Haberer belonged to many different species. One of them, an upper molar purchased at a Beijing chemist, belonged to a hominin. Schlosser tentatively attributed it to a Pliocene "grad. et sp. indet" being.

Based on the clues found by Haberer, Davidson Black tried to follow the trail of the dragon bones, according to the indications of early Chinese pharmacology, traceable to the Wei period (seventh century BC). In addition, the Swedish Academy sent researchers, such as Johan Gunnar Andersson, to China to collect fossils. Andersson and the Austrian paleontologist Otto Zdansky identified "Dragon Bone Hill" as a possible origin of the fossils (Shapiro, 1976). This hill is currently known as the Zhoukoudian site.

The excavation at Zhoukoudian cave, about 40 km southwest of Beijing, began in 1921 and 1923. It was there that Otto Zdansky found a much worn human molar. This is how the dragon teeth received the desired paleontological context.

In October 1927, Birger Bohlin, one of Davidson Black's colleagues, found a large human molar at Zhoukoudian. It looked like no other specimen known at the time. Black described a new hominin species based on it, *Sinanthropus pekinensis*, also known as Peking man (Black, 1927). In 1940, Weidenreich included these specimens in the same species as the Javanese middle Pleistocene exemplars, in the subspecies *H. erectus pekinensis* (Weidenreich, 1940).

During the 1920s, the cave of Zhoukoudian yielded a considerable number of remains, some of which were very well preserved. Close to 40 different individuals were identified, including 14 crania, 11 mandibles, more than 100 teeth, and a few postcranial remains. Davidson Black described the collection, and included a very detailed study of a juvenile cranium (Black, 1930). After 1936 the description of the findings was continued by Weidenreich (1936), who also made several casts (Figure 7.6). It was fortunate that Black and Weidenreich carried out this work because most of the early Zhoukoudian remains were lost during World War II.

Five additional teeth and two limb-bone fragments were found in 1949 and 1951 (Woo, 1980), and a mandible, together with other dental and postcranial remains, were found later, in 1959 (Woo and Chao, 1959). Finally, in 1966, members of the Institute of Vertebrate Paleontology and Paleoanthropology (Academia Sinica, Beijing) found a frontal and an occipital bone (Figure 7.7). These last specimens, together with two other parts discovered in 1934, were used to reconstruct the so-called 1966 cranium, with the help of casts belonging to the primitive collection. Zhoukoudian Locality I was excavated systematically and regularly from 1979. The report offered by Wu Rukang in 1985 included references to over 5,000 lithic artifacts, but no additional hominin findings (Rukang, 1985).

All Zhoukoudian hominins were discovered at Locality I, a large cavern whose roof has crumbled onto the infill, consisting of a succession of limestone breccias and other materials (sand, clay, and ashes). Wanpo (1960) described six main layers. The first (Layer I) belongs to the early Pleistocene, Layers II and III to the beginning of the middle Pleistocene, and Layers IV–VI to the end of the middle Pleistocene. Liu (1985), however, distinguished 17 layers. The possibility of giving precise datings depends on the correlation of the sediments with those of other sites and, thus, Kurtén (1959) suggested that the second glaciation

(Mindel, nearly 750,000 years ago) is the most appropriate period to place the deposits.

Liu (1985) estimated the age of the 17 layers to between 590,000 and 128,000 years, by means of the correlation with marine fossil coral deposits which contain the isotope ^{18}O. New dating methods rendered absolute estimates summarized by Wu Rukang (1985) and reproduced in Table 7.2. Using the uranium-series method, Tiemei and Sixun (1988) estimated the age of Zhoukoudian Locality I to be 290,000–220,000 years.

There is a significant difference between the ages of the Javanese sites and the Zhoukoudian cave: the latter is much more modern. However, the Zhoukoudian specimens are very similar to the Javanese. The similarity is so striking that many authors, such as Ernst Mayr (1950), Chris Stringer (1984), Bernard Wood (1984), and G. Philippe Rightmire (1990), have felt the need to group all those specimens in *H. erectus*. There are, however, dissenting views (Aguirre and de Lumley, 1977; Aguirre *et al.*, 1980; Rosas and Bermúdez de Castro, 1998; Aguirre, 2000). The greater cranial capacity of Zhoukoudian Locality I specimens, and certain differences in dental traits and other cranial features between the Chinese and Javanese specimens, was interpreted by lumpers as evolution expected by the greater age of the latter. The age of Zhoukoudian is intermediate between other Chinese sites that have rendered *erectus* specimens.

Table 7.2 Zhoukoudian absolute ages

Bed	Age (years)	Dating method
1–3	230,000, 256,000	Uranium series (1, 2)
4	290,000	Thermoluminescence (3)
6–7	350,000	Uranium series (2)
7	400,00–370,000	Paleomagnetism (4)
8–9	420,000, 462,000	Uranium series (1, 2), fission tracking (5)
10	620,000–520,000	Thermoluminescence (3)
12	>500,000	Uranium series (2)
13–17	>730,000	Paleomagnetism (4)

Source: Modified from Rukang (1985); see Rukang (1985) for original sources: (1) Zhao *et al.* (1979); (2) Xia (1982); (3) Pei *et al.* (1979); (4) Qian *et al.* (1980); (5) Guo *et al.* (1980).

Earlier exemplars have been retrieved at Lantian and Longuppo, whereas Hexian and Yunxian have yielded younger ones (see Figure 7.8).

The sites of Chenchiawo or Chen-Chia (or Jia) Wo, as transcribed by Poirier (1987), and Kungwangling or Gongwangling, belonging to the Lantian (Lan-T'ien) district, province of Shaanxi, have also yielded *H. erectus* specimens. The first, a well-preserved mandible (Lantian 1) was discovered in 1963 in Chenchiawo by Wu Rukang (1964). A tooth, a cranium, and part of the facial skeleton (Lantian 2) appeared the following year at Kungwangling hill (Rukang, 1966). Both specimens were attributed to a new species, *Sinanthropus lantianensis*, described by Rukang as morphologically more primitive than the Zhoukoudian specimens, and even than the Trinil (Java) *erectus*. The estimated cranial capacity for the Lantian 2 cranium is around 780 cm^3 (Day, 1986). However, Rukang himself later reclassified these specimens as *H. erectus* (see Rukang, 1980). The dating of the Lantian sites also indicates that these are very early *erectus*. The estimates reviewed by Day (1986) span from the 800,000–750,000 years suggested by paleomagnetism for the sediments belonging to the Matuyama subchron, to 1 million years estimated by the correlation with climatic changes inferred from the isotope content of fossil corals.

It has been suggested that in China there are earlier hominins than the Lantian specimens. They are the exemplars discovered at Longgupo cave, also known as the Wushan site, located 20 km south of the river Yangtze, in the eastern region of the Sichuan province. Two dental fragments and two Oldowan stone artifacts were found in the cave's mid area (excavation levels 7–8) (Wanpo *et al.*, 1995). The levels of the middle area that contained hominin remains were estimated, by means of paleomagnetism, to belong to the Olduvai subchron (1.96–1.78 Ma). Electronic spin resonance applied to dental enamel rendered a much younger age (0.75 Ma). In their commentary about the Longuppo finding, Wood and Turner (1995) argued that the site's fauna seems older than the latter age. Regarding stone tools, given that the Longgupo cave was a scavenger den—possibly

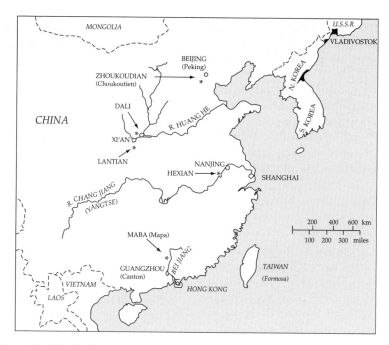

Figure 7.8 Some Chinese sites related with *H. erectus*.

inhabited by the giant hyena *Pachycrocuta*—they are not as significant as if the site had been a hominin habitat.

Wanpo and colleagues' study of the Longuppo dental fragments identified certain affinities with African *H. habilis* and *H. ergaster*. Consequently, these authors thought that the specimens should not be classified as *H. erectus* sensu stricto. According to Wanpo *et al.*, they belong to an almost 1.9-million-year-old pre-*erectus*. This would imply two things: that the genus *Homo* left Africa almost at the time of its diversification, close to 2 Ma, and that the evolution towards *H. erectus* occurred *in situ* within the Asian continent. Schwartz and Tattersall (1996b) doubt that one of the dental fragments belongs to a hominin (it could be close to the orangutan). With regard to the other fragment, Schwartz and Tattersall doubt that it can be compared with any specific hominin species. However, Wood and Turner (1995) accept the hypothesis that the Longgupo hominins preceded *H. erectus* sensu stricto and, thus, that the departure out of Africa was initiated by early members of

the *erectus* grade. This hypothesis had already been suggested by Bernard Wood (1992b).

7.1.4 Late Chinese *Homo erectus*

Other Chinese sites have provided evidence regarding much more recent hominins related to the *erectus* grade, adding to those retrieved at Ngandong, in Java, which document a very prolonged presence of Asian *erectus*. The Hexian site, in eastern China (Figure 7.8), is on the north side of the Wangjiashan hill, within the Taodian commune of Hexian County, north of the River Yangtze (Rukang, 1985). The construction of a canal in 1973 revealed abundant fossil remains of different animals. Scientists from the Institute of Vertebrate Paleontology and Paleoanthropology of Beijing, together with local colleagues and led by Huang Wanpo, unearthed an almost complete cranium, two cranial fragments of a different individual, a mandibular fragment with two molars, and nine isolated teeth during the 1980 and 1981 campaigns. Biostratigraphical evidence

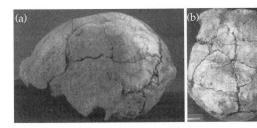

Figure 7.9 The Hexian cranium. (a) Lateral view; (b) superior view. Photographs from http://www.chineseprehistory.org/pics4.htm.

indicates that the clay sediments belonged to the middle Pleistocene.

The Hexian cranium (Figure 7.9) was initially described by Rukang and Dong (1982) as belonging to a juvenile male with many similar morphological traits to Peking man. Consequently, it was classified as *H. erectus*. However, some advanced traits, such as a modest postorbital constriction and a large cranial capacity, close to 1.025 cm³, emphasize its similarity to modern crania, such as number V from Zhoukoudian (Tiemei and Yinyun, 1991). The age of Hexian could also be very modern and similar to the later Zhoukoudian levels (Rukang, 1985). By means of faunal analysis and comparison with evidence from fossil coral deposits, Xu and You (1984), estimated the Hexian cranium at 280,000–240,000 years. Using uranium-series analysis, Tiemei and Yinyun (1991) assigned it an even younger age, 190,000–150,000 years, proposing the possible coexistence in China of *H. erectus* and *H. sapiens*.

The Yunxian site, in the Hubei Province, central China, is formed by deposits of the River Han's terrace. It has provided modern-looking specimens. Two crania, EV 9001 and EV 9002, were discovered *in situ*, embedded in a calcareous matrix. They correspond to middle-Pleistocene adult specimens, with a very large cranial capacity, over 1,000 cm³, and possibly male (Tianyuan and Etler, 1992). In Tianyuan and Etler's opinion, the Yunxian exemplars show *sapiens*-like facial traits which set them apart from typical *erectus*. This corresponds to a period in which African and European specimens exhibited more primitive traits; only female specimens show similarities with the modern facial structure. In spite of this, Yunxian hominins also retain many typical *H. erectus*

traits. In a later study, Yinyun (1998) compared the Yunxian specimens which Zhoukoudian *erectus*. The results showed that the *erectus*-like traits detected by Tianyuan are the result of distortion and damage suffered by the fossils, and that, from a morphological point of view, the Yunxian crania are similar to those of *H. sapiens*. Given their considerable age, 581,000 ± 93,000 years, Yinyun (1998) noted that the presence of *H. sapiens* in Asia could have occurred much earlier than is generally assumed.

Tianyuan and Etler's and Yinyun's opinions are not as different as they might seem. Yunxian hominins suggest, in accordance with Tianyuan and Etler (1992), first, that middle-Pleistocene hominins exhibited a high polymorphism and regional diversity, and, second, that they contributed significantly to the evolutionary lineage that leads to modern humans. Further on, in chapter 9, we will examine this hypothesis in detail, known as multiregional evolution. For the moment, it suffices to note that it contrasts with the view that *H. sapiens* appeared in a single place (East Africa) and later radiated and substituted all other hominin populations in Africa, Asia, and Europe. As Conroy (1997) observed, the "modern" traits of Yunxian hominins have also been identified in other African specimens, so they could be considered primitive. If so, they would lack any significance for determining possible phyletic relationships between Asian *H. erectus* and *H. sapiens*.

7.2 The *erectus* grade in Africa

Koobi Fora is the best place to find support for the argument that somewhere in Africa *H. habilis* and *H. rudolfensis* gave way to hominins belonging to the *erectus* grade. But the first *H. erectus* specimens to be discovered and become part of African paleoanthropology history are not those from the eastern shore of Lake Turkana, but the fragmented mandibles, maxillae, and other bones found by Robert Broom in South Africa, while digging at Swartkrans. Broom and John Robinson (1950) assigned them to a new genus, *Telanthropus*. These hominins are now considered *erectus* grade, together with other specimens such as the SK 847 partial cranium

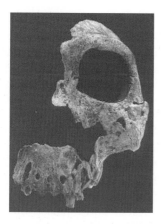

Figure 7.10 SK 847, *H. ergaster* from Swartkrans (South Africa). Photograph from Johanson and Edgar (1996).

(Figure 7.10), reconstructed by putting together different fragments, and assigned by Clarke and colleagues (1970) to *Homo* sp. It seems likely that this specimen belongs to *H. erectus* rather than *H. habilis*, but, in any case, it is anatomically different from australopiths. Something similar can be said about the SK 45 mandible. These are early specimens; doubts regarding their classification in one or another species are understandable.

In a comparative study of Asian and African *erectus* specimens, Tobias and von Koenigswald (1964) defined a series of "grades of hominization". Australopiths represent grade 1, and *H. habilis* grade 2. The Sangrian specimens, together with some exemplars from Olduvai (such as OH 13), correspond to grade 3. Grade 4 is represented by the Olduvai (OH 9), Ternifine, Kabuh, and Zhokoudian specimens. This series of hominization grades leads to paradoxes. As mentioned before, OH 13 had been considered by L. Leakey *et al.* in 1964 as paratype of the new proposed species, *H. habilis*. Thus, it seems contradictory to say that Asian *erectus* are generally considered members of a single species, but when they are compared with African remains it turns out that some of them (the Sangrian specimens) are included even as members of the same grade with *H. habilis* exemplars. This illustrates the difficulties that arise when trying to establish precise frontiers in the transition between habilines and *erectus*.

Findings at Olduvai provided new and more complete data about the morphology of African *erectus*, while at the same time accentuating the controversy. OH 7 (type specimen of the species *H. habilis*), OH 13, OH 6, OH 8, and OH 4 (paratypes of that same species) retrieved from Bed I or the lower part of Olduvai Bed II can be distinguished from other younger specimens, such as OH 9, found in the upper part of Bed II, estimated to be 1.25 million years old (Leakey, 1971). The OH 9 specimen was studied in detail by Rightmire (1979), who later revised his initial conclusions (Rightmire, 1990). OH 9 is a partial specimen, which includes the supraorbital structures and the base of the cranium, but provides no facial information below the nasal bones. Its cranial capacity was estimated to be 1067 cm^3 (Holloway, 1973). OH 12, on the other hand, was found by Margaret Cropper in 1962 on the surface of terrains attributed to Bed IV (Leakey *et al.*, 1971). It is the posterior part of a small cranium, close to 700–800 cm^3 (Holloway, 1973), to which certain facial fragments studied by Susan Antón (2004) were later added. OH 22 is a fairly well preserved right mandible, with premolars and the first two molars. It was found in sediments belonging either to Bed III or Bed IV, and estimated to no less than 0.62 Ma (Rightmire, 1980). There are also other mandibular fragments (OH 23, Bed IV; OH 51, Beds III–IV), and some postcranial remains from Bed IV, such as a left femur and an unidentified bone, which form the OH 28 specimen (Day, 1971). Day classified OH 28 as *H. erectus*, owing to its similarity with the Peking specimens. An almost complete ulna, OH 36, is older, found in the upper part of Bed II, and its morphology is more robust (Day, 1986).

As it is often the case, the differences among those Olduvai remains, which span an interval of 750,000 years, were interpreted either as indicative of sexual dimorphism (Rightmire, 1990), or as evidence of different species (Holloway, 1973). The relation with Asian specimens has also been a constant source of controversy, but before we go into that issue, we must return to a locality that provides valuable information regarding the transition towards the *erectus* grade: Lake Turkana.

7.2.1 The specimens from Turkana

As already mentioned, the sites located on both shores of Lake Turkana provided a broad hominin sample, including specimens belonging to *H. habilis*, *H. rudolfensis*, and the *erectus* grade. The latter include the best-preserved crania, the most complete skeleton, and the earliest remains. If the abundance of remains were the way to resolve the controversies, then Lake Turkana would hold the key for interpreting the evolution of our ancestors that followed *H. habilis*. But that would be a hasty assumption. The Turkana findings provided much information about the *erectus* grade, but they also generated many questions. These concern how many hominin species there were around 2 Ma, the relationship between African and Asian early and middle Pleistocene hominins, and precisely what kind of hominins have really been discovered at Turkana.

We anticipated in section 6.1 that the identification of species at Turkana is a spiny issue. We mentioned Wood's (1985) opinion regarding the presence of three "non-australopith" taxa (in addition to the robust exemplars). Those three taxa can be labeled as type 1470, type 1813, and type 3733/3883.

Those who have tried to provide a coherent and simple view of hominin evolution usually believe that type 1470 and type 1813 are the first representatives of the characteristic genera of the Plio-Pleistocene—that is, *H. habilis*—whereas type 3733/3883 are usually assigned to *H. erectus*. Poirier (1987) and Relethford (1997), for instance, argue in favor of this scheme. This is a reasonable option. However, the classification of the different fossils from Lake Turkana following the usual temporal and morphological criteria produces a rather arbitrary frontier. Let us consider the temporal issue first. The definitive dating of the KBS tuff (McDougall, 1985) suggested that KNM-ER 1470 was 1.8 Ma, whereas KNM-ER 1813 seemed to be slightly younger, between 1.2 and 1.6 Ma (Walker and Leakey, 1978; Walker, 1981). But the oldest specimens from Turkana that Poirier included with KNM-ER 3733 and KNM-ER 3883 in *H. erectus* were the KNM-ER 3228 pelvic fragment and two femoral fragments, KNM-ER 1472 and

KNM-ER 1481. They are all 1.9 Ma; that is, older than KNM-ER 1470 and KNM-ER 1813.

Thus, temporal considerations are not helpful to establish a clear frontier between the *habilis* type and the *erectus* type from Turkana. But the motive that leads lumpers to place those pelvic and femoral fragments within *H. erectus* is their morphology. As Poirier (1987) said, the femora are very reminiscent of Zhokoudian specimens and the latest *H. erectus* from Olduvai. Day (1986), on the other hand, associated KNM-ER 3228 with OH 28 from Olduvai (in addition to other specimens), a relation confirmed by Rose's (1984) detailed study of the specimen. Given that OH 28 had been grouped, as we have just seen, with *H. erectus* from Zhokoudian, Poirier logically concluded that, attending to taxonomic evidence, the earliest of all *H. erectus* are those specimens from Turkana.

But morphological considerations also run into problems when establishing the division among Turkana specimens. A very well-known cranium, KNM-ER 3733, can illustrate this. The KNM-ER 3733 specimen (Figure 7.11) was found *in situ* in the upper Member of Koobi Fora. It is a complete skullcap with the better part of the face—including zygomatic and nasal bones—together with the alveoli of anterior teeth and some molars and premolars (Leakey, 1976; Leakey and Walker, 1976). Richard Leakey and Alan Walker (1976) concluded that the cranial capacity of KNM-ER 3733 was close to 800–900 cm^3 and that all the cranial traits resemble those of Peking *H. erectus*. Not surprisingly, thus, Leakey and Walker (1976) classified KNM-ER 3733 as *H. erectus*. This incidentally, was a very solid argument against the

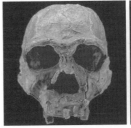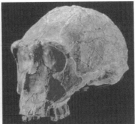

Figure 7.11 KNM-ER 3733, *H. ergaster*. Photographs from http://www.msu.edu/~heslipst/contents/ANP440/ergaster.htm.

single-species hypothesis. Compared with Koobi Fora crania such as KNM-ER 406—which belongs to a paranthropine—KNM-ER 3733 shows that more than one hominin species had lived at the same time and in the same place. In fact, it was the purpose of opposing the single-species perspective that fuelled Leakey and Walker's (1976) article in which they defended KNM-ER 3733 is *H. erectus*.

Many authors have supported the relation between KNM-ER 3733 and the Peking *erectus* noted by Leakey and Walker. We have cited textbooks of human paleontology that agree with that point of view. However, Colin Groves (1989) reached a different conclusion after studying the best cladograms that expressed the relations between the different Turkana specimens attributed to *Homo*. Groves disagreed with the idea that morphometric comparisons between KNM-ER 3733 and the Peking specimens indicated they were similar, and thus that they should be classified in the same species. Groves concluded that KNM-ER 3733 does not show the autapomorphic traits of *H. erectus* and that, therefore, the specimen cannot be classified as such. This had been noted by Wood (1984), who suggested two alternatives: either to change the definition of *H. erectus* to accommodate the Koobi Fora KNM-ER 3733 and KNM-ER 3883 crania, or to exclude these from *H. erectus* sensu stricto.

KNM-ER 3883 (Figure 7.12) is a slightly older and more robust cranium, with certain differences that Wolpoff (1980) attributed to sexual dimorphism. Thus, KNM-ER 3733 would belong to a female and KNM-ER 3883 would correspond to a male. In any case, and as Groves (1989) noted, both their cranial capacities are within the variation

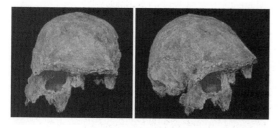

Figure 7.12 KNM-ER 3883 frontal (left) and lateral (right) views, *H. ergaster*. Photographs from http://www.msu.edu/ ∼ heslipst/ contents/ANP440/ergaster.htm.

range of the crania of Chinese *erectus*. The volume of KNM-ER 3883, supposedly a male, is lower than the KNM-ER 3733 specimen, which has a more gracile appearance. But if KNM-ER 3733 and KNM-ER 3883 are not *erectus*, what are they? As we noted earlier, Groves and Mazák (1975) suggested a different taxon, *H. ergaster*, with the KNM-ER 992 mandible as the holotype, for including Turkana specimens such as the KNM-ER 1805 and KNM-ER 1813 crania. But Groves (1989) did not group the KNM-ER 3733 and KNM-ER 3883 crania, nor the South African exemplars from Swartkrans, classified by Broom and Robinson as *Telanthropus*, in the species *H. ergaster*. He simply invoked a *Homo* sp. (unnamed) to include them.

Here we stumble into two opposing requirements. First, if we admit that the African and Asian specimens belonging to the *erectus* grade are different, that their cultures, as we will see in Chapter 8, do not coincide, and that they do not overlap in time, all of this supports their classification in different species. But, there is a need to establish manageable classifications, which plays against the multiplication of the *Homo* species present at Koobi Fora. Adding a third taxon to *Homo erectus sense stricto* and *H. ergaster* seems excessive for the *erectus* grade.

The other shore of Lake Turkana, the western one, has also provided very interesting hominin remains belonging to the *erectus* grade. As Francis Brown *et al.* (1985b) reported, Kamoya Kimeu found a small fragment of a hominin frontal bone on the surface of the Nariokotome III site of the west shore of Lake Turkana, during the 1984 campaign. Many other facial, cranial, mandibular, and postcranial remains appeared after cleaning the terrain and excavating an area of $5 \times 6\,\text{m}^2$. The remains presumably all belonged to the same individual, but they had been dispersed before their fossilization and showed signs of having been transported by water and having been trampled by large mammals, although there are no indications of the action of scavengers. After its reconstruction, the specimen was catalogued as KNM-WT 15000 and was assigned to *H. erectus* (Brown *et al.*, 1985b). It turned out to be one of the most complete early hominin specimens available (Figure 7.13). The KNM-WT 15000 specimen appeared just above a volcanic tuff, part of the Okote tuff from Koobi Fora, which is 1.65 million years old, so the specimen's age was estimated to be 1.6 million years (Brown *et al.*, 1985b).

The discovery of KNM-WT 15000 was published when only a preliminary description of the cranium was available, but some interesting features were already noted. Owing to the degree of dental development, its age at death was estimated to be close to 12 years. Despite this young age, the length of the long skeletal bones is similar to that of current white North American adults (Brown *et al.*, 1985b). This means that the height of the specimen, had it completed its development, would have been above the average for living humans. It is not surprising that Brown and colleagues (1985b) concluded, on the grounds of these and other

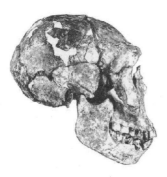

Figure 7.13 KNM-WT 15000 cranium, *H. ergaster*. Photograph from http://www.talkorigins.org/faqs/homs/15000.html.

traits, that KNM-WT 15000 required previous conceptions about the morphology of *erectus* to be modified. A more detailed study (Walker and Leakey, 1993) confirmed the initial impressions. We will go into the morphological details of the postcranial remains below.

7.2.2 Other African specimens belonging to the *erectus* grade

East Africa has yielded other *erectus*-grade exemplars in addition to these from Turkana and Olduvai. Howells (1980) included some middle-Pleistocene partial remains in *H. erectus*, such as the mandible from the Kapthurin beds, in Baringo (Kenya) and the Ndutu cranium (Tanzania). He also included some earlier specimens, such as the Melka Kounturé parietal, from the Gomboré site, Awash (Ehiopia), and a very old humerus fragment, about 1.5 Ma, Gomboré IB. The parietal from Bodo (Ethiopia), VOD-VP-1/1 (Asfaw, 1983), deserves to be mentioned separately. It is modern looking and it is usually considered as an *erectus/sapiens* transition form. Clark and colleagues (1984) estimated the age of this last specimen at 0.6 Ma.

A later finding from the Danakil Formation, belonging to the Afar region close to the village of Buia (Eritrea), revealed a very complete cranium, including parts of the face and the roots of some molars and premolars (UA 31; Figure 7.14), together with two incisors (UA 222 and 369) and two

Figure 7.14 UA 31 calotte, Buia (Eritrea) "erectus-like". Photograph from Macchiarelli *et al.* (2004).

pelvic fragments (UA 173; Abbate *et al.*, 1998). Despite the presence of volcanic material at the site, the attempt to date it by the ^{40}Ar/^{39}Ar method failed due to the contamination of the ashes. Nevertheless, the horizon in which the hominin remains had appeared, close to the upper part of the Jaramillo subchron, was estimated to be close to 1 Ma, by means of faunal comparisons and paleomagnetic methods. The fission-track study of a tephra layer, set between the sediments (Bigazzi *et al.*, 2004), and the comparison of faunas (Martí-nez-Navarro *et al.*, 2004) confirm that estimate.

The UA 31 cranium exhibits features that are characteristic of *H. erectus*, such as elongated shape, thick supraorbital torus, and cranial capacity close to 750–800 cm^3, together with other derived traits, characteristic of *H. sapiens*, such as the thinning at the base and the verticality of the parietal region. Abbate and colleagues (1998) noted that the specimen shows signs of transition towards anatomically modern humans. In a detailed study, Macchiarelli *et al.* (2004) simply assigned the specimen to an *erectus*-like type.

A similar transitional role was attributed to the Bodo parietal VOD-VP-1/1 after Asfaw's (1983) initial proposal. But because the Buia specimen UA 31 is 300,000 years older, the process of the transformation of *H. erectus* to *H. sapiens* was extended further back in time. Wolpoff noted that the Danakil cranium confirms the difficulties involved in establishing a definitive frontier between fossils belonging to the *erectus* grade and *H. sapiens* (see Gibbons, 1998a). We will review the alternative proposals concerning the transition from *erectus* to modern humans in Chapter 9.

Exemplars belonging to the *erectus* grade have also been found north of the Sahara, at Ternifine, or Tighenif (Algeria), and Sidi Abderraman, Rabat, and Salé (Morocco). The Ternifine specimens include three mandibles, one parietal, and some isolated teeth, which, owing to the fauna present at the site, were attributed to the middle Pleistocene (Arambourg, 1955). The mandibles are rather robust, they lack a chin, and their similarities with Peking *erectus* specimens were noticed (Arambourg, 1954), but Arambourg preferred to define a new species, *Atlanthropus mauritanicus*,

which gained little acceptance. As Day (1986) mentioned, the Ternifine specimens are generally considered to belong to *H. erectus*.

The Salé specimen consists of a cranium, lacking the face, and a left maxilla, with some teeth. It was found in 1971 in an open quarry close to El Hamra, north of Salé. In the initial description, Jaeger (1975) noted the specimen's clear affinities with *H. erectus*, although the occipital area is much more prominent and rounded than what could be expected, conferring it a modern appearance. Some have explained this anomaly as a pathology, but Hublin (1985) argued that it should be considered a primitive *H. sapiens* that retained some *erectus* traits. According to Hublin, the North African middle Pleistocene remains show a mosaic of *erectus* and *sapiens* traits, which underlines the need for interpreting the evolutionary relationships between those species. The controversy regarding the evolutionary meaning of the Salé *erectus* specimens, as Hublin (1985) noted sharply, is the same one that could be applied to the Ngandong and Zhokoudian specimens, or even all *H. erectus*.

7.2.3 Daka and Olorgesailie: contradictory evidence

The most recently discovered African *erectus*-like specimens—those of Daka and Olorgesailie—are between 0.5 and 1 Ma, an interval that includes few African specimens. Far from solving the existing controversies, these specimens have provided evidence pointing in two opposite directions. On the one hand, the Daka cranium favors the existence of a single, quite variable, geographically dispersed species during the time when *H. erectus* hominins were present (Asfaw *et al.*, 2002; Gilbert *et al.*, 2003). On the other hand, the Olorgesailie fossil, if it is included in the alleged single taxon, increases its variability excessively.

The Daka specimens, a skull, three isolated hominid femora, and a proximal tibia, come from the Dakanihylo Member—called Daka—part of the Bouri Formation at Middle Awash (Ethiopia). The single-crystal ^{40}Ar/^{39}Ar dating of a pumiceous unit at the base of the member estimated its age at 1.042 ± 0.009 Ma (de Heinzelin *et al.*, 2000).

Paleomagnetism studies of the member assigned it an inverted polarity, so the whole member must be older than 0.8 million years. The skull (BOU-VP-2/66) was discovered *in situ* by W. Henry Gilbert in December, 1997. The postcranial remains were recovered from Daka deposits far from the skull. Abundant fauna embedded in primarily alluvial deposits characteristic of lakeside beaches or shallow-water deposits in distributary channels were also found at Daka. The predominant environment was savanna. Numerous early Acheulean artifacts were also found (Asfaw *et al.*, 2002).

BOU-VP-2/66 is well preserved, although the specimen was found slightly distorted (Figure 7.15). Its vault and supraorbitals exhibit perimortem scraping damage. The frontal and parietals bear multiple sets of subparallel striae, each with internal striations, attributed by its discoverers to animal gnawing. Endocranial capacity is 995 cm^3 (Asfaw *et al.*, 2002). The BOU-VP-2/66 cranial vault is smaller and shorter than Olduvai hominid OH-9 and—according to the discoverers—it is phenetically similar to the partly described Buia cranium, which also comes from the Dakanailio Formation in Eritrea and was mentioned above. Asfaw *et al.* (2002) noted in their description, among other traits:

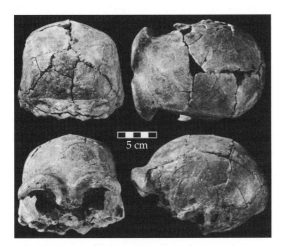

Figure 7.15 BOU-VP-2/66 cranium, *H. erectus*. Posterior, superior, frontal, and lateral views. Photographs from Asfaw *et al.* (2002). Reprinted by permission from Macmillan Publishers Ltd. *Nature* 416: 6878, 317–320, 2002.

- the thick supraorbital tori are strongly arched, with markedly depressed glabellar and supraglabellar regions;
- frontal squama bossed at midline, with weak sagittal keeling there and on the parietals;
- deep and anteroposteriorly short mandibular fossa;
- weak suprameatal and supramastoid crests and angular tori;
- no true occipital torus demarcated superiorly by a supratoral sulcus.

Asfaw *et al.*'s (2002) interpretation showed that the BOU-VP-2/66 cranial metrics overlap with the ranges of both Asian and African samples and fail to distinguish the fossils consistently from either: "the cladistic method, regardless of serious questions concerning its applicability here, fails to support the division of *H. erectus* into Asian and African clades. Whether viewed metrically or morphologically, the Daka cranium confirms previous suggestions that geographic subdivision of early *H. erectus* into separate species lineages is biologically misleading, artificially inflating early Pleistocene species diversity."

The morphological interpretation of the Daka cranium is far from clear. Asfaw *et al.* (2002) saw a relation between the Daka and Buia (Ethiopia) specimens but, as Antón (2003) pointed out, those two specimens differ in some important morphological ways. In the description of the Buia specimen, Abbatte *et al.* (1998) argued that it exhibited a mosaic of *H. erectus* and *H. sapiens* traits. With respect to the Daka exemplar, Antón (2003) noted the absence of some notable *H. erectus* traits. The most conspicuous one is the absence of occipital torus in BOU-VP-2/66. As noted by Asfaw *et al.* (2002): "Rather, the occipital squama rises vertically and curves anteriorly. Viewed posteriorly, the undistorted parietal walls would have been vertical."

The specimen from Olorgesailie (Kenya) raises further questions. The specimen includes the frontal and left temporal bones and nine cranial vault fragments. The whole set is known as KNM-OL 45500 (Figure 7.16). The frontal bone was found *in situ* by the research team led by Richard

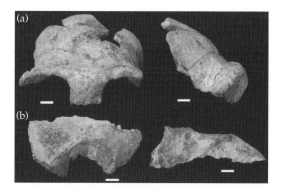

Figure 7.16 KNM-OL 45500, frontal (a) and left temporal bone (b), from Olorgesailie (Kenya), *H. ergaster*. Photographs from Potts *et al.* (2004). Scale bars equal 1 cm. Reprinted with permission from AAAS.

Potts in June 2003. It was embedded in a sediment block that was excavated and removed from the site in August 1999. Other parts of the cranium were discovered close-by between July and August 2003. The same stratigraphic level had previously yielded a large accumulation of Acheulean stone artifacts and mammalian fossils (Potts *et al.*, 2004).

The age of KNM-OL 45500 corresponds to the boundary between Members 6/7 and Member 5 of the Olorgesailie Formation. Pumice clasts from the lower part of Member 5 were dated by the ^{40}Ar/^{39}Ar method, at 974,000 ± 7,000 years (Deino and Potts, 1990). Ashes at the top of Member 8 yielded 747 ± 6 years. A study of the paleomagnetism recorded the Matuyama–Brunhes transition near the base of Member 8 (Tauxe *et al.*, 1992). In consequence, Potts *et al.* (2004) estimated the age of the Member 5/7 boundary and the KNM-OL 45500 specimen to be between 970,000 and 900,000 years.

KNM-OL 45500 was a small individual but developed enough to be considered an adult or near-adult. Different features: the post-toral sulcus relative to the rest of the frontal bone; the convex inter-orbital region; the mid-squama thickness of the frontal vault; the mastoid process of the temporal bone; the root of the zygomatic arch; and so forth, correspond to what can be expected in an adult *H. erectus* (Potts *et al.*, 2004). But, KNM-OL 45500 exhibits the small frontal breadth, supraorbital torus thickness and breadth, and overall temporal bone size of any early- or middle-Pleistocene adult

cranium. Furthermore, some cranial traits are different from those observed in previously known *H. erectus*. The supraorbital torus is thinner vertically than in adult *H. erectus*. The mandibular fossa is the smallest of mid-Pleistocene adult hominins, despite its position beneath the relatively large zygomatic root. The mastoid process is slender and is not associated with a supramastoid crest, which typifies larger adults of mid-Pleistocene *Homo* (Potts *et al.*, 2004).

Despite its very small size, the Olorgesailie cranium possesses characters observed in larger individuals of *H. erectus*: midline keeling of the frontal bone, shelf-like morphology of the posttoral sulcus, lack of torsion in the toral anterior surface, and a short temporal squama with flat superior border. KNM-OL 45500 exhibits a double-arched supraorbital torus, which is more strongly developed in African mid-Pleistocene specimens, such as KNM-ER 3733 from Koobi Fora, Bodo, Daka, or Kabwe, than in Asian *H. erectus*. Moreover, KNM-OL 45500 lacks the deep mandibular fossae typical of Asian *H. erectus* and displays an overall morphology that would broaden the known range of this species. Potts *et al.* (2004) carried out a detailed comparison between KNM-OL 45500 and other *erectus* grade crania (KNM-ER 3733, KNM-ER 3883, and KNM-WT 15000 from Turkana, Kenya; OH 9 and OH 12 from Olduvai in Tanzania, Daka in Ethiopia, Buia in Eritrea, Bodo in Ethiopia, and Ndutu in Tanzania; D2280 and D2282 from Dmanisi in Georgia, Ceprano in Italy, Atapuerca in Spain, Zhokoudian in China, and Sangiran and Ngandong in Java; this is in addition to Kabwe, Zambia and Saldanha, South Africa, which we will review in Chapter 9). Potts *et al.* (2004) argued that KNM-OL 45500 exhibits sufficient similarities with different specimens included in the *erectus* grade to support the scenario of a single polytypical species. But in the commentary that accompanied the specimen's initial description, Schwartz (2004) expressed a different view: "recognizing that 'Homo erectus' may be more a historical accident than a biological reality might lead to a better understanding of the relationships not only of the Olorgesailie specimens, but also of those fossils whose morphology clearly exceeds the bounds of individual variation so well documented in the Trinil/Sangiran sample."

7.2.4 The morphology of *Homo erectus*

We have reviewed a series of Asian and African specimens included in the *erectus* grade. Questions concerning their morphological, taxonomic, and phylogenetic considerations are many. We will now synthesize the morphological traits that characterize *erectus*-grade specimens. We will also discuss the extent to which they indicate the existence of several species.

When Rightmire (1990) defined the hypodigm of *H. erectus*, he stressed that the formal diagnostic of a species must be based on its apomorphic or derived traits, and not on plesiomorphic or primitive characters, inherited from ancestral species. Thus, among all the traits exhibited by *erectus* specimens, the worst choice to characterize one or several species within the *erectus* grade is precisely the one that is emphasized by the name given to the initially discovered exemplar. The condition of "erect" that Dubois saw in the pithecanthropine is a primitive trait, attained by earlier beings in our tribe. If we accept that the locomotion of *H. erectus* is functionally different from the bipedalism of australopiths, or even *H. habilis*, Dubois could be right. But it would be wrong to think that bipedalism itself is a derived trait. Most of the apomorphies observed in the *erectus* grade are located in the cranium. To characterize the species, Stringer (1984), for instance, provided a list of 27 typical *erectus* traits (autapomorphies), among which only one was postcranial. Even though Stringer's purpose was to distinguish plesiomorphies and autapomorphies, to establish which of them constitute derived traits of *H. erectus* sensu stricto, he provided a morphological description which clearly shows what is commonly understood as an "*erectus*". Wood (1984), following different authors, noted 31 traits, most of which are also related to the face and cranium (Table 7.3).

Once the autapomorphies noted by Wood (1984), Stringer (1984), Rightmire (1990), and other authors are reduced to those which are easier to detect, we are left with what could be considered as the typical image of middle-Pleistocene hominins. A different issue is whether there is a single or several species within the *erectus* grade. On the

one hand, the members of the *erectus* grade have greater cranial capacities than *H. habilis*: between 700 and 1250 cm^3. The shape of the cranium is also very striking, with its flattened cranial vaults (platycephalia) and its thick cranial walls, consequence of an incipient pneumatization; that is, the existence of miniscule hollow spaces in the structure of the bone, allowing a greater thickness without increasing its weight. The face of *erectus* exhibits a thick, almost continuous, brow ridge. The Asian specimens also have an angular torus on the occipital bone. The *erectus* grade also retains some primitive traits (plesiomorphies), such as an outwards facial projection (facial prognathism), relatively large molars (although less than australopiths, to the point that Stringer (1987) considered dental reduction as a derived trait), somewhat projected canines, and the absence of chin.

Asian and African specimens have made possible a detailed study of the cranial morphology of *erectus*, but this grade's postcranial features are difficult to establish because of the paucity of that kind of remains in Asian sites. Day (1984) mentioned the Trinil femur that Dubois had associated with the *Pithecanthropus erectus* skull, the Zhoukoudian femoral fragments, the OH 28 femur from Olduvai, the KNM-ER 737 femur from Koobi Fora (Kenya), the KNM-ER 3228 pelvic fragments (Koobi Fora), and a European specimen: Arago XLIV (France). The list is very limited; doubts about the presence of *H. erectus* in Europe would reduce it even further. An ulna, three vertebrae, a rib and part of the innominate bone, and the hand and foot of an *H. erectus* were found in Yingkou (China) in 1984 (Wei, 1984), increasing the sample of postcranial remains available before the discovery of the Nariokotome skeleton.

The appearance of the Trinil remains, attributed by Dubois to *Pithecanthropus erectus*, favored the consideration of *erectus* specimens as hominins with very primitive crania, which below the head would exhibit a similar anatomy to our own. This idea was contradicted by Weidenreich (1941) who studied the Zhoukoudian femur and pointed out a combination of anatomical traits with a very particular and distinct morphology, different from that of *H. sapiens*,

Table 7.3 Autapomorphic traits of *H. erectus* sensu stricto included in Wood (1984)

Cranial shape and size	Specific morphological features
Overall i. Long and low (viz. length/height index ~ 60 and breadth/height index < 75) .... 1,2,4 ii. Maximum breadth across the angular torus or supramastoid crest.............................. 1 iii. Thick valut bones .... 1,2,4 iv. Pronounced postorbital constriction... 2,3,4 **Individual bones** **Frontal** v. Frontal angel $< 60°$ .... 1,4 vi. Frontal keel or metopic ridge................ 1,5 vii. Straight junction of torus and frontal squamae .... 5 viii. Coronal ridge .... 5 **Parietal** ix. Flattented and rectangular parietal........ 1 x. Sagittal angulation or ridge................. 4,5 **Temporal** xi. Low temporal squamae.................... 1,3,4 xii. Small mastoid process............................ 2 **Occipital** xiii. Opisthocranian coincident with inion.. 1,2 xiv. Sharply angulated................................... 1 xv. Upper scale shorter than the lower .... 1,4 xvi. Discrepancy between ectinion and endion... 4 **Base** — — **Face** xvii. Borad nasal bones............................. 1,2	**Frontal** xvii. Large, continuous supraorbital ridges with a supratoral sulcus... 1,2,3,4 xix. Lateral wing to supraobital torus...................... 3,4 **Parietal** xx. Prominent angular torus at mastoid angle .... 1,5 **Temporal** xxi. Marked supramastoid.. 1,5 and mastoid crests.................................... 5 xxii. Occipitiomastoid ridge .... 3 xxiii. No juxtamastoid ridge .... 3 xxiv. Marked suprameatal tegmen........................ 1,3,4 xxv. No vaginal crest... 1,4 **Occipital** xxvi. Occipital torus (with supratoral sulcus above and continuous with angular torus and supramastoid crest) .... 1,3,4 and angular torus and mastoid crest.................. 5 xxvii. Supernumary bones at lambda........................... 1 **Base** xxviii. Thick tympanic plate...................................... 1,2 xxix. Processus supratubalis....................................... 1 xxx. Petrous bone making more acute angle with tympanic.. 3,4 **Face** — — **Endocranial** xxxi. Large posterior branch of middle meningeal artery and vein 4

Source: Wood (1984). Numbering refers to: 1, Weidenreich (1943); 2, Le Gros Clark (1955); 3, Macintosh and Larnach (1972); 4, Jacob, (1976); 5, Santa Luca (1980).

although he assigned the Trinil femur to our own species. Little more could be said about the postcranial features of *erectus* until the discovery of the KNM-WT 15000 specimen, which provided a very complete picture of its morphology.

KNM-WT 15000 was found at the Nariokotome III site, west of Lake Turkana. The surprise generated by its discovery encouraged researchers to examine the available middle-Pleistocene postcranial remains from a different perspective. After studying six African specimens for which the corresponding measurements could be obtained, Walker (1993) concluded that the average height of the species would be around 1.70 m (between 1.58 and 1.85 m) and the average weight would be around 58 kg (between 51 and 68 kg). Thus, *erectus* would be among the top 17% of modern humans, with respect to height and weight. Vandermeersch and Garralda (1994) argued that *H. erectus* with respect to individual and gender variation in height and weight remained unchanged until modern humans, which vary between 1.50 and 1.80 m in height. In relative terms, *H. habilis* and *A. afarensis* would exhibit a broader variability.

Behrensmeyer and Laporte (1981) had noted that the fossil footprints at Koobi Fora, estimated to 1.5–1.6 Ma, are 26 cm long and 10 cm wide, close to the average current North American *H. sapiens*. These estimates suggest a height of 1.685 m if North American men, black and white, are taken as the reference group, and a height of 1.80 m if the comparison of the foot is done with San bushmen. The length of the stride is, however, smaller than that of modern humans, but this was interpreted by Behrensmeyer and Laporte (1981) as due to hesitant steps on a slippery terrain.

If the aforementioned estimates are correct, the height and weight of hominins must have decreased in the last stage of the evolution to modern humans (Ruff, Trinkaus, and Holliday, 1997). Ruff and colleagues studied the size of 163 skeletons ranging from 10,000 years ago to 1.95 Ma. Pleistocene specimens turned out to be up to 7.4 kg heavier on average than current humans (65.6 compared with 58.2 kg), a 12.7% difference. Hominin bodies must have been small during the early Pleistocene, reaching a maximum during the late Pleistocene, and decreasing thereafter. According to Ruff et al. (1997), the brain also decreased in size since the beginning of the Upper Paleolithic, continuing throughout the Neolithic, at least in Europe. This trend reverted in recent centuries, but only at high latitudes is the large body mass maintained, inherited from Paleolithic ancestors. Cohen and Armelagos (1984) attributed this reduction in body mass to diet, a decrease of meat associated with the appearance of agriculture. Formicola and Giannecchini (1999) pointed out that nutritional changes may have impacted height, but without discarding the possibility that genetic drift might have also played a role.

Kappelman (1997) has suggested that the natural selection of smaller bodies, favored perhaps by more efficient cooperation and communication systems, may have resulted in an increase of the brain/body-size ratio simply because the body mass decreased.

CHAPTER 8

Evolutionary characteristics of the *erectus* grade

8.1 Cultural patterns in the migration out of Africa

Grade-*erectus* hominins, after leaving Africa, colonized regions with highly diverse climates, such as South Asian tropical islands and subglacial China. This required solving adaptive problems, such as resistance to cold, and logistical problems, such as crossing deep bodies of water. In this chapter we will discuss the following questions: What kind of hominins left Africa? What kind of culture did they have? How are these hominins related phylogenetically to earlier and later taxa?

8.1.1 Dmanisi

Available evidence suggests that hominins used the Levantine corridor in the Near East as a route out of Africa at different times, with important differences between the cultural traditions of diverse emigrants. The first migration to be identified is evinced in traces of very early human activity, close to 1.4 Ma at Ubeidiya (Israel; Tchernov, 1989). These are primitive-looking lithic tools, which contrast with later Acheulean bifaces, dated to 780,000 years, found at Gesher Benot Ya'aqov (Israel; Goren-Inbar *et al.*, 2000). Another Israeli site that has yielded stone tools, Erk-El-Ahmar, has been dated by magnetostratigraphy, which is somewhat imprecise, between 2 and 1.7 Ma (Ron and Levi, 2001).

The best and earliest evidence of the presence of *Homo* at the doors of Asia are the Dmanisi specimens, to the north of the Levantine corridor. The Dmanisi site, in Georgia, is located under the ruins of a medieval city with the same name, which was

an important urban center between the eighth and twelfth centuries (Figure 8.1). Dmanisi, 85 km southwest of Tbilisi, was built on alluvial deposits with a thickness of nearly 4 m above a basalt base. The sediments contain late Villafranchian fossil mammals (see section 5.3). The basalt base was dated to 1.8 ± 0.1 Ma (^{40}K/^{40}Ar method; Gabunia and Vekua, 1995). The volcanic sediments that contain the fossils belong to the Olduvai subchron (Figure 8.2; Gabunia and Vekua, 1995). The detailed study of the associated fauna, archaeological analysis and paleomagnetic and geochronological studies provide a precise estimate for the Dmanisi specimens of 1.7 Ma (Gabunia *et al.*, 2000).

In addition to many Oldowan tools found earlier, the first exemplar from Dmanisi was discovered towards the end of 1991 at the foundations of a medieval deposit. It was a mandible, D211, with an almost complete body but broken rami (Figure 8.3). Gabunia and Vekua (1995) established that the Dmanisi mandible is a very early *Homo* specimen, with small teeth and a large mandibular body, whose sex was impossible to determine. Its most distinctive feature is the decrease in molar areas: from M1 to M2, and from this to M3. According to Gabunia and Vekua (1995) D211 shows greater similarities with the African *erectus* grade hominins—*H. ergaster*—than with the Asian ones—*H. erectus* sensu stricto. A comparative study by Rosas and Bermúdez de Castro (1998) also emphasizes the similarity between D211 and *H. ergaster*.

The Dmanisi mandible's affinities were soon broadened. Bräuer and Schultz (1996) suggested that D211 shared derived traits with advanced

Figure 8.1 Left, location of Dmanisi. Picture from Vekua *et al.* (2002). Right, aerial view of the medieval ruins. Photograph from http://donsmaps.com/dmanisi.html.

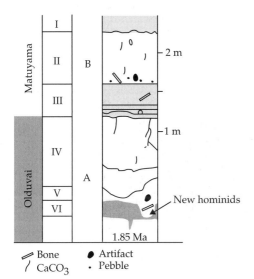

Figure 8.2 Dmanisi paleomagnetic column with the location of the specimens and artifacts. Picture from Vekua *et al.* (2002).

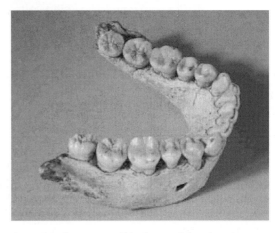

Figure 8.3 The D211 mandible. Photograph from http:/ /www. dmanisi.org.ge/mand.D211.htm.

H. erectus, less than 1 Ma. Gabunia and Vekua (1995) had reached the same conclusion, noting that D211 could be attributed to a *H. erectus* individual that possibly preceded archaic *H. sapiens*, noting also that it represents the earliest indication of hominins out of Africa. Dean and Delson (1995) agreed with the decision of including the specimen in *H. erectus* and added that the Dmanisi mandible suggested a simple phylogenetic model: an ancestor–descendant relationship between *erectus* specimens and archaic *sapiens*, with the latter replacing *H. erectus* from west to east. The Dmanisi specimen would be an example of this replacement.

In the summer of 1999, an almost complete cranium (D2282) and a skullcap (D2280) were discovered in the same site and stratigraphic horizon as D211 (Figure 8.4). Gabunia *et al.* (2000) noted that the morphology of D2282 and D2280 shared essential

similarities with the modern gracile specimens from Koobi Fora. Contrary to the initial attribution of the Dmanisi mandible to *H. erectus*, Leo Gabunia and colleagues believed that the crania were closer to *H. ergaster*. This would represent a great expansion of the geographic range of *H. ergaster*, which could no longer be considered just a local variety within the broad scope of Asian *erectus*. Additionally, doubts were cast on Dean and Delson's model of a single *erectus* taxon, which persisted until replaced by archaic *sapiens* from west to east. Some apomorphies observed in the set of modern humans + Dmanisi specimens + *ergaster* are not observed in exemplars of *H. erectus* sensu stricto.

How is it possible that ancestors of Asian *H. erectus* would share apomorphies with modern humans? Emiliano Aguirre (2000) has suggested that the migration out of Africa may have occurred in several waves. An initial wave led to the Javanese *erectus*, while the Dmanisi specimens correspond to a later wave.

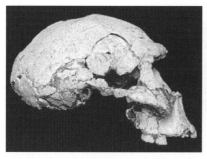
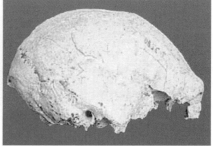

Figure 8.4 Specimens from Dmanisi in Georgia: D2282 (left) and D2280 (right). Photographs from http://www.athenapub.com/13intro-he.htm.

Box 8.1 Speciation in *Homo erectus*

Schwartz *et al.* (2000) have noted that the Dmanisi crania support the notion that, by 1.8 Ma, hominin variability was beyond the dichotomy represented by *H. erectus* sensu stricto and *H. ergaster*. This notion is consistent with the mechanism Schwartz believes was responsible for their speciation, namely, a change in one or very few developmental *homeobox* genes (Schwartz, 1999). Lorenzo Rook (2000) holds a different position regarding the significance of the Dmanisi crania in the determination of the diversity within *H. erectus* sensu lato. Rook believes that the paucity of the fossil record prevents establishing the degree of intraspecific variability. In consequence, distinctive characters are interpreted as intertaxon differences, leading to the definition of new species. This is certainly not a problem inherent to the Dmanisi specimens: it is common in all paleontology, although it is marked in human paleontology. Rook called for new taxonomic criteria, a need many authors agree with, including Schwartz.

8.1.2 A new species for the Dmanisi hominins

The cranial volumes of D2280 and D2282 are small, 775 and 650 cm^3, smaller than what could be expected for *H. ergaster*, and close to the cranial capacity of *H. habilis* specimens from Turkana and Olduvai. Three years later, in 2002, an additional cranium was discovered at Dmanisi, D2700 (Vekua *et al.*, 2002; Figure 8.5), which had a lower volume. The new findings require distinguishing two kinds of hominin at Dmanisi: one that groups D2700, D2280, and D2282; and another including D2735 and D211 (Vekua *et al.*, 2002). Vekua *et al.* reject the possibility that these two kinds represent two different taxa, but it is not easy to determine which species they represent. D2700 shows some similarities with *H. habilis* specimens, such as KNM-ER 1813 from Koobi Fora. However, despite the specimen's small size, Vekua *et al.* (2002) observed

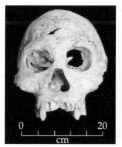
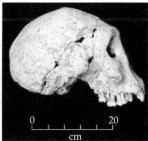

Figure 8.5 Dmanisi specimen D2700: frontal (left) and lateral (right) views. Photographs from Vekua *et al.* (2002). Reprinted with permission from AAAS.

traits which are characteristic in African *H. erectus–H. ergaster*:

- keeling along the sagittal midline,
- depressed appearance of the parietal surfaces,
- the shape of the temporal squama,
- a transverse expansion of the base relative to the low vault.

Box 8.2 Variation within the Dmanisi sample

Lee (2005) has made a statistical analysis of the Dmanisi sample by means of simulated distributions based on comparative samples of modern humans, chimpanzees, and gorillas. The results "suggest no compelling reason to invoke multiple taxa to explain variation in the cranial capacity of the Dmanisi hominids." Nevertheless, the variation within the sample is large. Skinner *et al.* (2006) have compared the variation in mandible size and shape at Dmanisi to that of extant hominoids and extinct hominins. The living taxa in the comparison are *Gorilla gorilla gorilla, G. g. graueri, G. g. berengei, Pongo pygmaeus pygmaeus, P. p. abelii, Pan troglodytes troglodytes, P. t. schweinfurthi, Pan paniscus,* and *Homo sapiens.* For comparison with fossils, Skinner *et al.* (2006) used two taxonomic hypotheses: (1) conservative, accepting the taxa *H. habilis* sensu lato and *H. erectus* sensu lato; (2) more liberal, separating the fossils attributed to *H. habilis* sensu lato into two taxa, *H. rudolfensis* and *H. habilis* sensu stricto. Also, it divides *H. erectus* sensu lato into regionally based taxa including *H. erectus erectus* (Indonesia), *H. erectus pekinensis* (China), *H. erectus mauritanicus* (N. Africa), and African *H. erectus/ergaster* (referring to early-Pleistocene *Homo* fossils from Africa).

According to Skinner *et al.* (2006), the "results indicate that the pattern of variation for the Dmanisi hominins does not resemble that of any living species: they exhibit significantly more size variation when compared to modern humans, and they have significantly more corpus shape variation and size variation in corpus heights and overall mandible size than any extant ape species. When compared to fossil hominins they are also more dimorphic

in size (although this result is influenced by the taxonomic hypothesis applied to the hominin fossil record). These results highlight the need to re-examine expectations of levels of sexual dimorphism in members of the genus *Homo* and to account for marked size and shape variation between D2600 and D211 under the prevailing view of a single hominin species at Dmanisi."

The taxonomic interpretations of Skinner *et al.* (2006) lead to two possible hypotheses for explaining the variation in the Dmanisi sample, as follows.

1 The degree of sexual dimorphism in the Dmanisi *Homo* taxon exceeds expectations, "thus highlighting the need to reconsider conclusions about its inclusion in, and/or the definition of, the genus *Homo.* Dmanisi may represent a *Homo* taxon predating the increase in female body size that characterizes the low levels of [body-size sexual dimorphism] present in other Homo taxa"

2 More than one taxon is present at Dmanisi—exemplar D2600 "possesses morphological features not seen in the other Dmanisi mandibles."

Nevertheless, the majority opinion places the Dmanini sample within a single taxon. The detailed investigation of the Dmanisi exemplars by Rightmire *et al.* (2006) favors a single species, but noting the difficulties associated with D2600: "Although there is variation probably related to growth status and sex dimorphism, it is appropriate to group the Dmanisi hominins together. With the possible exception of the large D2600 mandible, the individuals are sampled from one paleodeme."

Box 8.3 D2700 and D2735

D2700 was found together with a mandible, D2735, and the set was attributed to a young specimen, possibly a female, although the large crowns and massive roots of the upper canines prevent a clear sex attribution (Vekua *et al.*, 2002).

Vekua *et al.*'s (2002) systematic diagnosis recognized the classification difficulties and the authors venture that "The Dmanisi hominids are among the most primitive individuals so far attributed to *H. erectus* or to any species that is indisputably *Homo,* and it can be argued that this population is closely related to *Homo habilis* (sensu stricto) as known from Olduvai Gorge in Tanzania, Koobi Fora in northern Kenya, and possibly Hadar in Ethiopia" (Vekua *et al.*, 2002).

Box 8.4 Gabunia

In the 2002 paper, Leo Gabunia's surname is spelled according to the French style, Leo Gabounia. To avoid confusion, we spell the name as in his other publications: Leo Gabunia.

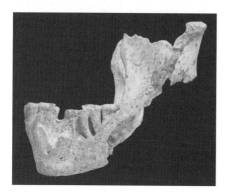

Figure 8.6 Dmanisi specimen D2600: lateral view. Photograph from Vekua *et al.* (2002).

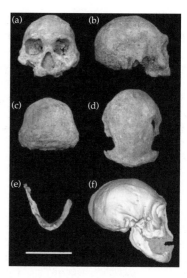

Figure 8.7 Dmanini specimens. Top and middle rows: exemplar D3444. Bottom left: mandible D3900. Bottom right: superposition of a computer tomography scan of D3444, a juvenile, and an adult Dmanisi skull with dentition. Lordkipanidze *et al.* (2005) used this superposition to show that D3444 lost the teeth before dying: see Box 8.5 for more details. Photographs from Lordkipanidze *et al.* (2005). Reprinted by permission from Macmillan Publishers Ltd. *Nature* 434: 7034, 717–718, 2005.

In only 7 years the Dmanisi specimens had gone from being considered *Homo erectus* (Gabunia and Vekua, 1995) to being included in *H. ergaster* (Gabunia *et al.*, 2000) and thereafter to being attributed, with some uncertainty, to *H. habilis* sensu stricto (that is, excluding the *H. rudolfensis* exemplars). The D2600 mandible found in September 2000 (Figure 8.6) led the Dmanisi researchers to take a further taxonomic step. After comparison with other similar-aged *Homo* taxa, Gabunia and colleagues (2002) introduced the new species *Homo georgicus*, with D2600 as a type specimen. The remainder of the Dmanisi sample would be the paratype of *H. georgicus*.

There are two more specimens from Dmanisi: the D3444 cranium and the D3900 associated mandible, discovered during the 2002–2004 campaigns (Lordkipanidze *et al.*, 2005). They are similar to previously described Dmanisi specimens, except that D3900 is edentulous (Figure 8.7). The absence of teeth in D3900 is unusual. It suggests an individual kept alive with the help of its conspecifics. This hypothesis implies advanced

social structures and adaptive strategies in early *Homo*. Lordkipanidze *et al.* (2005) argue that "the consumption of soft tissues such as bone marrow or brain may have increased the chances of survival of individuals with masticatory impairment." This would be possible "by virtue of help from other individuals, which must have exceeded that capable of being offered by non-human primates."

The adaptive strategies of the *Homo* individuals from Dmanisi included the use of tools. D3444 and D3900 were found in sediments that contained stone tools and animal bones with cut and percussion marks. The question of the stone tools available to the first emigrants out of Africa is related to the expansion of the genus *Homo*. As we saw in the previous chapter, the different lithic traditions of *erectus*-grade hominins in Asia and Africa is one main reason favoring the distinction between *H. erectus* sensu stricto and *H. ergaster*.

The earliest and most primitive lithic culture, Oldowan or Mode 1, appears in East African sediments around 2.4 Ma. Around 1.6 Ma, African *erectus*-grade hominins developed a more

Box 8.5 Edentulous mandible D3900

Edentulous means without teeth. As Lordkipanidze *et al.* (2005) note, D3900 maxillary teeth "were lost before death, as evinced by the complete resorption of the tooth sockets and extensive remodelling of the alveolar process. In the mandible all sockets, except those for the canine teeth, have been resorbed and only the left canine persisted at the time of death. The mandibular body has been resorbed down to the level of the mental foramen, and the projection of the symphyseal region is likely to be the result of remodelling following loss of the incisors. Applying clinical comparative standards, the advanced alveolar bone atrophy indicates substantial tooth loss several years before death as a result of ageing and/or pathology" (see Figure 8.7).

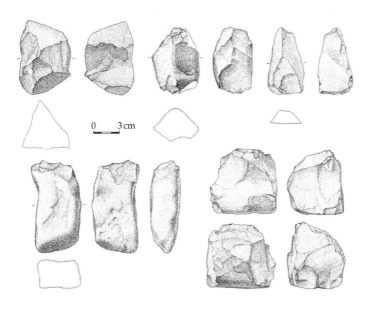

0   3 cm

Figure 8.8 Dmanisi tools. Picture from Gabunia *et al.* (2000). Fig. 5 from L. Gabunia *et al.*, SCIENCE 288: 1019–1025 (May 12, 2000).

Box 8.6 Evolutionary significance of the Dmanisi sample

It is difficult to interpret the taxonomy and phylogenetic position of the Dmanisi sample because the variation is very large, as pointed out in Box 8.2. Moreover, the exemplars proceed from a single locality within a narrow time span. The greatest difficulties arise from the D2600 mandible, which is large and distinct from all other Dmanisi exemplars. The possible alternatives are either to accept that Dmanisi consists of a single population from an undetermined species, or to assume that two species are present, one of which would be *Homo georgicus* that includes D2600.

Rightmire *et al.*'s (2006) investigation highlights the taxonomic difficulties. They assert that Dmanini represents a single but very variable population that resembles *H. habilis* in brain volume and some aspects of craniofacial morphology. However, these resemblances could represent shared primitive traits. According to Rightmire *et al.* (2006), "Other discrete characters and measurements suggest that the Dmanisi skulls are best placed with *H. erectus.*" However, Vekua, one of the authors of that article, prefers to place the D2600 mandible in *H. georgicus*.

We agree with Rightmire *et al.* (2006) that "the evidence from anatomical analysis and measurements supports the hypothesis that Dmanisi is close to the stem from which *H. erectus* evolved."

Box 8.7 Assumptions about tool-making

Caution must be taken when attributing the construction of very early tools to hominins that consist of sympatric species. When tools are found it can often be taken for granted that the makers are the hominins found with them, because similar hominins have been shown to have tool-making capabilities. This is how *Zinjanthropus boisei* was initially identified as the maker of Oldowan culture, which we will soon review. The specimen was found in the same stratigraphic level as the stone tools. Because at the time it was the only known hominin there, it was taken for granted that it was the maker of the lithic tools.

Box 8.8 Tool use by australopiths

The fractured base of baboon crania at Taung and other sites suggest they were opened to eat what was inside. Raymond Dart argued that the bones themselves had been used by australopiths as instruments to injure, crush, and cut, as part of a traditional use of tools with a "natural" origin, osteodontokeratic culture, before stone tool cultures appeared (Dart, 1957).

---

advanced tradition, Acheulean or Mode 2. It seems that the Asian *erectus*-grade counterparts did not have the essential elements of Mode 2. With some exceptions that we will examine later on, no bifaces have been found in Asia. Hence, either hominins arrived there before the African Acheulean tradition had been developed or, for some unknown reason, they lost the ability to make those tools. Before the evidence from Dmanisi was available, Toth and Schick (1993) suggested that during the long migration, and moving through extensive regions lacking adequate materials to construct bifaces, hominins would have lost the ability to make them. If this were the case, their communicative and cognitive capacities must have been too limited to preserve the Acheulean tradition (Toth and Schick, 1993).

Discoveries in Georgia favor another interpretation. More than 1000 artifacts—some choppers and scrapers, chopping tools, and many flakes—have been found in both stratigraphic units of Dmanisi, A and B (Gabunia *et al.*, 2000). They are very primitive instruments, similar to Oldowan industry (Figure 8.8). This seems to indicate that the departure out of Africa occurred before the evolution towards the more advanced techniques of Mode 2.

8.1.3 Culture and adaptation

What is the relationship between different *Homo* taxa and the diverse lithic cultures? Before we proceed, a methodological warning must be made. The assumption that a certain kind of hominin is the author of a specific set of tools is grounded on two complementary arguments: (1) the hominin specimens and lithic instruments were found at the same level of the same site; and (2) morphological interpretations attribute to those particular hominins the ability to manufacture the stone tools.

The first kind of evidence is, obviously, circumstantial. Sites yield not only hominin remains, but also those of a diverse fauna. The belief that our ancestors rather than other primates are responsible for the stone tools comes from the second type of argument. The identification of the authors of lithic instruments is, as we will certainly see, difficult, particularly when several taxa appear at sites containing tools, or when no specimen regarded as capable of creating the artifacts appears to be associated with them.

Moreover, interpretations regarding tool use often involve circular arguments. Members of the genus *Homo* are said to be the authors of tools found in a certain place, because earlier hominins are not

---

Box 8.9 South African lithic instruments

Many lithic instruments have been found at the Sterkfontein Extension Site—handaxes, cores, flakes, and even a spheroid—which are unequivocal signs of the manipulation of raw materials to obtain tools designed to cut and crush (Robinson and Manson, 1957). However, there are doubts regarding the association between stone tools and their authors. The sites that have provided *A. africanus*, Sterkfontein, Makapansgat, and Taung, are not the only ones that have provided samples of an early lithic culture. There is also a stone industry at Swartkrans

(Brain, 1970), although it was found a long time after Dart elaborated his idea of hominization. The interpretation of the possible stone artifacts found at Kromdraai is not easy (Brain, 1958). But even in Sterkfontein, the Extension Site belongs to Member 5, whereas Member 4, older than 5, has provided a great number of *A. africanus* specimens, although it has yielded no lithic tool whatsoever. Thus, the authorship of the South African Plio-Pleistocene lithic instruments needs to be ascertained.

Box 8.10 Hominin adaptation to tropical forests

Paleoclimatological conclusions regarding early hominin taxa suggest they were adapted to tropical forests. This is the case of *Australopithecus anamensis* (Leakey *et al.*, 1995), *Ardipithecus ramidus*, *Australopithecus afarensis*,

and *Australopithecus*—or *Paranthropus*—*africanus* (Rayner *et al.*, 1993; Kingston *et al.*, 1994; WoldeGabriel *et al.*, 1994).

---

believed to have been stone tool-makers. However, doubts about the cultural abilities of australopiths from Taung and Sterkfontein have been raised for some time. For instance, Robinson (1962) argued that, although australopiths did not produce the complex stone instruments found at Sterkfontein, this does not mean that australopiths lacked a culture. They could have used stones, sticks, bones, and any other useful instrument to get food.

However, according to Washburn (1957) the explanation for bone marks and the accumulation of remains in the breccias of South African caves is different. There is a predominance of mandibular and cranial remains because they are the most difficult bones to break, so that they tend to accumulate in the lairs of predators and scavengers. Ancestral hyenas are likely responsible for the accumulation of remains that we now find fossilized, australopiths included. It has been suggested that the Taung child itself was the victim of a predator, probably an eagle (Berger and Clarke, 1996), although this hypothesis has been criticized (Hedenström, 1995).

If the accumulation of bones at Sterkfontein Member 4 was due to scavengers, and if australopiths were the hunted and not the hunters, the question concerning the first tool-makers remains unanswered. The answer will depend on preconceptions regarding cognitive capacities and hominin adaptive strategies. New kinds of evidence have bearing on this issue: the paleoclimate to which different genera and species were adapted; the morphology of certain key elements required for the intentional manipulation of objects, such as hands and the brain; the diet and the taphonomic study of the relationships at the sites of bones and tools.

8.1.4 The colonizers of open savannas

As we have seen previously, the earliest members of our family probably lived on the floors of tropical forests. This argues against Raymond Dart's original hypothesis that related bipedalism, the expansion of open savannas, and the appearance of the first hominins. The link between human adaptation to extensive open savannas and the construction of tools might still be probable even if we accept this primitive forest environment. But, it is doubtful that australopiths were the colonists of open spaces. It is probable that the hominins that invented cultures with manipulated stone tools belonged to a different genus, *Homo*, and specifically to the species *H. habilis*.

Why is it reasonable to think that *H. habilis* was the author of the earliest tools, and not any kind of australopith? Again, we find ourselves with the need to associate fossil remains and lithic instruments found at archaeological sites. With regard to South Africa, the question is uncertain. Sterkfontein Member 5 has yielded the StW 53 cranium, which, as we saw, is considered as either *H. aff. habilis* or an *Australopithecus* of an unspecified species. Swartkrans has also provided some exemplars attributed to *H. habilis* and, regarding Taung, the most widespread opinion argues that the stone tools are much more recent and that they were made by more evolved hominins. But there are solid reasons to associate the species *H. habilis* with stone tool-making at Olduvai. It was this site that provided the *H. habilis* type specimen. It also gave the name to the earliest lithic culture, the Oldowan industry (Figure 8.9).

"When the skull of *Australopithecus* (*Zinjanthropus*) *boisei* was found ... no remains of any other type of hominid were known from the early part of the Olduvai sequence. It seemed reasonable, therefore, to assume that this skull represented the makers of the Oldowan culture. The subsequent discovery of *Homo habilis* in association with the Oldowan culture at three other sites has considerably altered the position. While it is possible that *Zinjanthropus* and *Homo habilis* both made stone tools, it is probable that the latter was the more advanced tool-maker and that the *Zinjanthropus* skull represents an intruder (or a victim) on a *Homo habilis* living site" (Leakey *et al.*, 1964). Here we have a clear example of the argumentative sequence: first, a *P. boisei* cranium and associated lithic instruments were discovered at the F.L.K. I site, Olduvai. Later, hominins with a notably greater cranial capacity, included in the new species *H. habilis*, were discovered at the same place. Finally, stone tools were

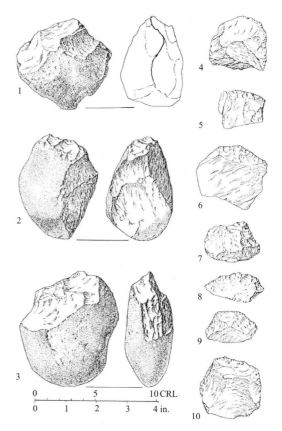

Figure 8.9 Oldowan tools. 1–3, lava choppers; 4–10, quarcite flakes. Drawing by M.D. Leakey (1971).

attributed to *H. habilis*, morphologically more advanced in its planning capacities.

Leakey *et al.*'s (1964) paper included a cautionary note. Even though it is less probable, it is conceivable that *Zinjanthropus* also made lithic tools. John Napier had published an article on the evolution of the hand 2 years before, relating stone tools to the discovery of 15 hominin hand bones by Louis and Mary Leakey at the site where *Zinjanthropus* had been found (Napier, 1962). According to Napier, "Prior to the discovery of *Zinjanthropus*, the South-African man-apes (Australopithecines) had been associated at least indirectly with fabricated tools. Observers were reluctant to credit man-apes with being tool-makers, however, on the ground that they lacked an adequate cranial capacity. Now that hands as well as skulls have been found at the same site with undoubted tools, one can begin to correlate the evolution of the hand with the stage of culture and the size of the brain" (Napier, 1962, pp. 40–41).

Napier's (1962) interpretation of the Olduvai findings exemplifies the risks involved in the correlation of specimens and tools. In this instance, Napier linked the tools, the *Zinjanthropus* cranium (OH 5), and the OH 8 collection of hand and feet bones (with a clavicle), all of them found by the Leakey team in the same stratigraphic horizon. But 2 years later Leakey, Tobias, and Napier himself included the fossils labeled as OH 8 in the paratype of *H. habilis*. Sites yielding tools and fossil samples of australopiths and *H. habilis* require decisions about which of those taxa made the tools. Leakey *et al.* (1964) favor *H. habilis* from Olduvai over *Zinjanthropus*.

8.1.5 The function of lithic tools

Although Olduvai Gorge was not the first place in which early stone tools were found, it gave name to the earliest known lithic industry: Oldowan culture. The excellent conditions of the Olduvai sites provided paleontologists and archaeologists with the chance to carry out taphonomic interpretations for reconstructing hominin habitats. Any lithic culture can be described as a set of diverse stones manipulated by hominins to obtain tools to cut, scrape, or hit. They are diverse tools obtained by hitting pebbles of different hard materials. Silex, quartz, flint, granite, and basalt are some of the materials used for tool-making. In the Oldowan culture, the size of the round-shaped cores is variable, but they usually fit comfortably in the hand; they are tennis-ball-sized stones. Many tools belonging to different traditions fit within these generic characteristics. What specifically identifies Oldowan culture is that its tools are obtained with very few knocks, sometimes only one. The resultant tools are misleadingly crude. It is not easy to hit the stones with enough precision to obtain cutting edges and efficient flakes (Figure 8.10).

Oldowan tools are usually classified according to their shapes, assuming that the differences in appearance involve different uses. Large instruments include:

- pebbles without cutting edges, but with evident signs of having been used to hit other stones, called hammer stones;
- pebbles that have been broken, by percussion, to produce a cutting edge, and serve to break hard surfaces like large bones (to get to the bone marrow, for instance), known as choppers;
- flakes, resulting from impacts to the nucleus, which have a very sharp edge and were presumably used to cut skin, meat, and ligaments of butchered animals. They can be modified or unmodified;
- flake scrapers, modified flakes that have an edge that resembles dented knives, which might have been used to scrape skins to tan them;
- polyhedrons, spheroids, and discoids, which are manipulated cores shaped in different ways, as if flakes had been produced from the whole exterior perimeter. Their function is unknown; they may even be nothing more than functionless residues.

It is not easy to arrive at definitive conclusions regarding the use of Oldowan tools. The idea we have of their function depends on the way we interpret the adaptation of hominins that used them, based on arguments that are often circular. Tool-makers can be seen, as Binford (1981) did, as a last stage in scavenging, when only large bones are available. If this were the case, the most important tools would be the handaxes that allow hitting a cranium or femur hard enough to break

Figure 8.10 Flake production by the Oldowan technique. Picture from Plummer (2004).

Box 8.11 The function of axes and flakes

A functional explanation can be established to distinguish between handaxes, manipulated with power grips, and flakes, which require handling with the fingertips using a precision grip. It is not easy to go beyond this, but some authors, like Toth (1985a, 1985b), have carried out much more precise functional studies. Toth argued that flakes were enormously important for butchery tasks, even when they were unmodified, while he doubted the functional value of some polyhedrons and spheroids.

it. If, on the contrary, we understand that early hominins butchered almost whole animals, then flakes would be the essential tools. The examination of the marks that tools leave on fossil bones provides direct evidence of their function. Taphonomic interpretations of cutmarks suggest that hominins defleshed and broke the bones to obtain food. This butchery function related to meat intake portrays early hominins as scavengers capable of taking advantage of the carcasses of the preys of savanna predators (Blumenschine, 1987).

Several kinds of evidence have been used to resolve the question of how early hominins obtained animal proteins. One is the detailed analysis of the tools and their possible functionality. Microscopic examination of the edge of a lithic instrument allows us to infer what it was used for (Figure 8.11); whether it served as a scraper to tan skin, or as a knife to cut meat, or as a handaxe to cut wood. This affords an explanation of behavior that goes beyond the possibilities of deducing a tool's function from its shape.

By means of the comparative study of the behavior of African apes, ethology has provided some interesting interpretations about how chimpanzees use, and sometimes modify, stones and sticks to get food. Since the first evidence of such behaviors collected by Jane Goodall and Jordi Sabater Pi (Goodall, 1964; Sabater Pi, 1984), many cases of chimpanzee tool use that can be considered cultural have been brought to light. Very diverse cultural traditions have been documented, including up to 39 different behavioral patterns related with tool use by chimpanzees (Boesch and Tomasello, 1998; Vogel, 1999; Whiten et al., 1999). Although their production of flakes is unintentional, there are similarities with those

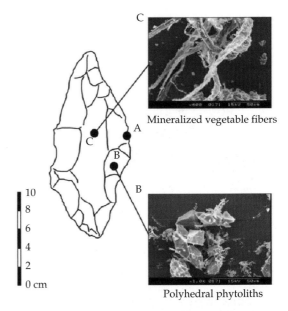

Mineralized vegetable fibers

Polyhedral phytoliths

Figure 8.11 Sections of the handaxes where phytolith samples were found on Peninj bifaces: (A) external edge; (B) internal edge; (C) ventral/dorsal inner surface. Calcium oxalate phytoliths with polyhedral forms (*Acacia* sp.; B, lower photo) and mineralized fibers (C, upper photo) were found on two out of the three handaxes examined. Pictures from Dominguez-Rodrigo et al. (2001).

characteristic of Mode 1. They appear when chimpanzees accidentally hit the anvil with their hammer when cracking nuts to get to the fruit (Mercader et al., 2002).

Taphonomic studies, aimed at reconstructing the process of accumulation of available fossil evidence at a site, have increased our understanding of the behavior of early hominins. Different East African sites (Olduvai, Koobi Fora, Olorgesailie, Peninj) provide samples of hominin living sites with a direct association of hominin fossil remains

Box 8.12 Tools for woodworking

In certain instances lithic tools might have been used as woodworking tools. Indications of the use of wood instruments are not rare in the late Pleistocene. In the middle Pleistocene, the finding of plant microremains (phytoliths, fibers) on the edges of Peninj (Tanzania) Acheulean bifaces (see Figure 8.11) is the earliest proof of processing of wood with artifacts (Dominguez-Rodrigo *et*

al., 2001). Nicholas Toth does not share the view that tool-making hominins depended completely on manufactured tools. After studying the Koobi Fora sites and the distribution of Oldowan tools at those sites, Toth argued that it was very possible that those hominins frequently used unelaborated materials, such as broken shells, horns, wood sticks, and even, citing Brain (1982), bones.

Figure 8.12 Cutmarks on the hominid zygomaticomaxillary specimen StW 53c. The cutmarks are found on the inferolateral surface. Photographs from Pickering *et al.* (2000).

Table 8.1 The earliest cultures

Name	Locality	Age (Ma)
Lokalalei	West Turkana	2.34
Shungura	Omo	2.2–2.0
Hadar	Hadar	2.33
Gona	Middle Awash	2.6–2.5

and manipulated stones. As a result of such studies, Dart's idea of hominins as hunters in the open savanna was followed by the hypothesis that the first lithic tool-makers were scavengers who cooperated to a greater or lesser extent to obtain food. Several authors have attempted to focus the adaptive role of the first tool-making hominins outside the picture of aggressive and bloodthirsty hunters (Isaac, 1978a; Binford, 1981; Bunn, 1981; Blumenschine, 1987). But the role of cooperation and the kind of activity aimed at getting meat are still controversial. It is likely that the first *Homo*

were opportunistic carnivores that took advantage of scavenging and hunting resources.

But in some instances the evidence suggests other hypotheses. Pickering *et al.* (2000) analyzed the cutmarks inflicted by a stone tool on a right maxilla from locality StW 53 at Sterkfontein Member 5. The species to which the specimen belongs is unclear, but it is certainly a hominin (Figure 8.12). They noted that "The location of the marks on the lateral aspect of the zygomatic process of the maxilla is consistent with that expected from slicing through the masseter muscle, presumably to remove the mandible from the cranium." In other words, a hominin from Sterkfontein Member 5 dismembered the remains of another. To what end? Are these marks indicative of cannibalistic practices, or are they signs of something like a ritual? The available evidence does not provide an answer to this question. It is not even possible to determine whether the hominin that disarticulated the StW 53 mandible and its owner belonged to the same species. But cannibalistic behaviors have been inferred from middle-Pleistocene cutmarks. This is how the cutmarks on the Atapuerca (Spain) ATD6-96 mandible have been interpreted (Carbonell *et al.*, 2005). It has also been suggested regarding the Zhoukoudian sample (Rolland, 2004). Cannibalism seems to have been common among Neanderthals and the first anatomically modern humans.

8.2 The earliest cultures

The Oldowan culture was not restricted to Olduvai (see section 5.3). Stone tools have also been found at older Kenyan and Ethiopian sites, although in some places their style is slightly different. These findings have pushed back the estimated time for the appearance of lithic industries (Table 8.1).

Close to 3,000 artifacts were found in 1997 at the Lokalalei 2C site (West Turkana, Kenya), with an estimated age of 2.34 Ma. They were concentrated in a small area, about $10\,m^2$, and included a large number of small elements (measuring less than a centimeter; Roche *et al.*, 1999). The tools were found in association with some faunal remains, but these show no signs of having been manipulated. The importance of the Lokalalei tools lays primarily in the presence of abundant debris, which allows us to establish the sequence of tool-making *in situ* (Figure 8.13). Roche and colleagues (1999) have argued that the technique used by the makers of these tools required very careful preparation and use of the materials, previously unimaginable for such early hominins. This suggested that the

cognitive capacities of those tool-makers were more developed than what is usually believed. One of the cores was hit up to 20 times to extract flakes, and the careful choice of materials (mostly volcanic lavas like basalt) indicates that those who manipulated them knew their mechanical properties well.

The French ethnologist and prehistorian André Leroi-Gourhan (e.g. 1964) tried to integrate interpretations about mental capacities with the usual technical explanations about tool-use: a "paleontology of gesture". Following Leroi-Gourhan's tradition, Schlanger (1994) suggested that the sequence of operations inferred from tools reflects complex intentionality and developed mental skills. Thus, a distinction can be drawn between two kinds of knowledge, as follows.

• Practical knowledge required for any tool-making operation; this is what psychologists would refer to as procedural knowledge, such as what is required to ride a bicycle without falling off.
• Abstract knowledge, or the ability to envisage problems and resolve them; it is closer to declarative knowledge, such as in choosing the safest route to ride a bicycle in a city from one place to another.

One way to distinguish between the two kinds of knowledge is to consider that declarative

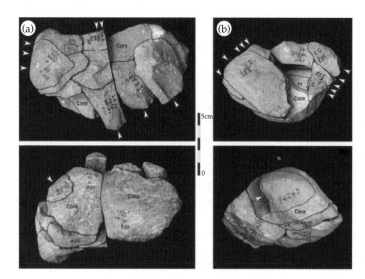

Figure 8.13 Lokalalei sets of complementary matching stone artifacts. (a) R35 refit (two cores and 11 flakes). (b) R9 refit (one core and 14 flakes). The main reduction sequence consists of unidirectional or multidirectional removals flaked on a single debitage surface, from natural or prepared platforms. Arrows indicate the direction of the removals. Pictures from Roche *et al.*, (1999). Reprinted by permission from Macmillan Publishers Ltd. *Nature* 399: 6731, 57–60, 1999.

knowledge can be transmitted by a spoken or written description, whereas procedural knowledge cannot.

James Steele (1999) raised the issue of the cognitive capacities and knowledge of the authors of the Lokalalei 2C tools. Steele admitted that the available evidence does not allow going beyond hypotheses similar to the one which attributed the Olduvai tools to *H. habilis* because of its larger cranial capacity compared with *P. boisei*. The Lokalalei findings indicate that almost 2.5 Ma the motor control of the hands, and thus, the development of the brain, must have been considerable. The identity of the species responsible for manipulating those artifacts is a different issue that is difficult to answer. In his commentary about Roche and colleagues' discovery, Steele (1999) refused to give a definitive answer. He simply argued that we still have similar doubts to those of the authors who, in 1964, associated the tools found at Olduvai with the species *H. habilis*.

The Middle Awash region includes many sites that have yielded Oldowan and Acheulean tools, described for the first time by Taieb (1974). Kalb and colleagues (1982) and J. Desmond Clark and colleagues (1984) provided a general overview of the area's cultural remains. Other authors have confirmed the sequence established at Olduvai: an early Mode 1 industry, lasting for a long period of time. This was replaced by Acheulean tools, although not without a considerable overlap. Acheulean tools usually include a massive presence of handaxes.

As we mentioned in Chapter 6, a *Homo* maxilla (A.L. 666-1) was found in association with Oldowan tools at Hadar (Ethiopia), to the north of Middle Awash. The sediments from the upper part of the Kada Hadar Member were estimated to 2.33 Ma (Kimbel *et al.*, 1996). This was the earliest association between lithic industry and hominin remains (Kimbel *et al.*, 1996). The 34 instruments found in the 1974 campaign (indicative of a low density of lithic remains) are typical of Oldowan culture: choppers and flakes. Additionally, three primitive bifaces, known as end-choppers, appeared on the surface, but it is difficult to associate these tools with the excavated ones.

The earliest known instruments have been found at the Gona site (Ethiopia), within the Middle Awash area, in sediments dated to 2.6–2.5 Ma by correlation of the archaeological localities with sediments dated with the Ar/Ar method and paleomagnetism (Figure 8.14; Semaw *et al.*, 1997). Thus, they are about 200,000 years older than the Lokalalei tools.

Gona has provided numerous tools, up to 2970, including cores, flakes, and debris. Many of the tools were constructed *in situ*. No modified flakes have been found, but the industry appears very similar to the early samples from Olduvai. Semaw and colleagues (1997) attributed the differences, such as the greater size of the Gona cores, to the distance between the site and the places where the raw materials (trachyte) were obtained: these are closer in Gona than in other instances. As hominins have not been found at the site, it is difficult to attribute the tools to any particular taxon. Semaw *et al.* (1997) believed it was unnecessary to suggest a "pre-Oldowan" industry. Rather, the Oldowan industry would have remained in stasis (presence without notable changes) for at least 1 million years. The precision of the Gona instruments (Figure 8.15) led Semaw's team to assume that their authors were not novices, so even earlier lithic industries might be discovered in the future.

Wood (1997) wondered about the authors of the tools found at the site. The great stasis of the Oldowan culture suggested by the tools raises a problem for the usual assignation of the Oldowan tradition to *H. habilis*. Given that the latest Oldowan tools are about 1.5 million years old, this tradition spans close to a million years. This is why Wood (1997) noted that if Oldowan tools had to be attributed to a particular hominin, then the only species that was present during the whole interval was *P. boisei*. This is circumstantial evidence in favor of the notion that robust australopiths manufactured tools. But as we have mentioned several times in this book, there is no need to make a close identification between hominin species and lithic traditions, because cultural sharing must have been quite common. In any case, de Heinzelin and colleagues (1999) attributed the Gona utensils

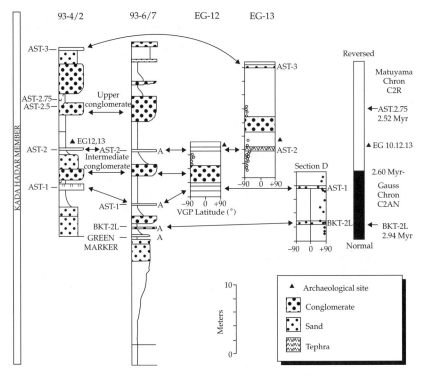

Figure 8.14 Gona (Ethiopia) stratigraphic column with the correlations relative to the magnetic polarity sample. Picture from Semaw *et al.* (1997).

to the species *A. garhi*, whose specimens were found at Bouri, 96 km south of where the tools come from.

The comparison between instruments from different sites has its limitations. As Glynn Isaac (1969) noted, it is not uncommon to find that the differences between the Oldowan techniques found at different locations of the same age are as large as those used to differentiate successive Oldowan stages, or even larger. This problem illustrates that the complexity of a lithic instrument is a function of its age, but also of the needs of the tool-maker.

8.2.1 The Oldowan–Acheulean transition

Mary Leakey (1975) described the transition observable at Olduvai from perfected Oldowan tools to a different and more advanced industry. These tools, made with great care, were identified for the first time at the St. Acheul site (France), and are known as Acheulean industrial complex or

Acheulean culture. Acheulean culture appeared in East Africa slightly over 1.5 Ma, and extended to the rest of the Old World to a greater or lesser degree until around 0.3 Ma. Its most characteristic element is the biface, "teardrop shaped in outline, biconvex in cross-section, and commonly manufactured on large (more than 10 cm) unifacially or bifacially flaked cobbles, flakes, and slabs" (Noll and Petraglia, 2003). But the term biface corresponds to a form of manufacture rather than to a tool. Bifaces led to different utensils, such as those shown in Figure 8.16.

A gradual transition from Oldowan culture to Acheulean culture was justified by the sequence established by Mary Leakey for the Olduvai beds (Table 8.2; Figure 8.17). Louis Leakey (1951) had previously considered the coexistence of cultures and the evolution of Oldowan instruments as evidence of gradual change. However, subsequent studies painted another scenario. Isaac (1969) argued that the improvement of the necessary

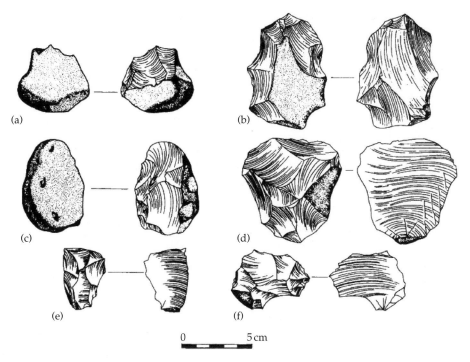

Figure 8.15 Tool assemblages from the Gona (Ethiopia) EG12 and EG10 localities. Flaked pieces: (a) unifacial side chopper, EG12; (b) discoid EG10; (c) unifacial side chopper, EG10. Detached pieces: (d–f) whole flakes, EG10. From Semaw (1997). Reprinted by permission of Macmillan Publishers: Nature.

techniques to go from the Oldowan to the Acheulean traditions could not have taken place gradually. A completely new type of manipulation would have appeared with Acheulean culture, a true change in the way of carrying out the operations involved in tool-making. If so, it would be important to determine exactly when that jump forward occurred and to establish the temporal distribution of the different cultural traditions. Such detailed knowledge is not easy to achieve.

The Olduvai site does not reveal precisely when the cultural change took place. The earliest instruments, from Bed I, are found in a level dated to 1.76-1.7 Ma by the K/Ar method (Evernden and Curtiss, 1965). The later Acheulean utensils appeared at the Kalambo Falls locality at Olduvai, in association with wood and coal materials. The age of these materials was estimated by the ^{14}C method at 60,000 years (Vogel and Waterbolk, 1967). There are other volcanic tuffs between both points, but the 1.6-Ma interval between the most recent level and

Kalambo Falls limits the precision of the chronometry. This period corresponds precisely to the time of the transition between both cultures (Isaac, 1969). If we take into account the evolution within Mode 1, with developed Oldowan tools that overlap in time with Acheulean ones, the difficulties involved in the description of the cultural change increase (Table 8.2; Figure 8.18).

The technical evolution from Mode 1 to Mode 2 can also be studied at other places, such as the Humbu Formation from the Peninj site, to the west of Lake Natron (Tanzania). After the discovery made by the Leakeys and Isaac in 1967, authors such as Mturi (1987) or Schick and Toth (1993) carried out research in the Natron area. Several Natron sites show a transition from Oldowan to Acheulean cultures close to 1.5 Ma (Schick and Toth, 1993). The correlation of the Peninj and Olduvai sediments allows the identification of the Oldowan–Acheulean transition with the upper strata of Bed II from Olduvai. But neither Olduvai

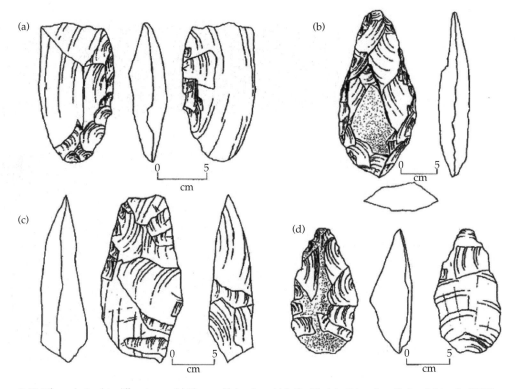

Figure 8.16 Bifaces destined to different uses. (a) Cleaver; (b) handaxe; (c) knife; (d) pick. Picture from Noll and Petraglia (2003).

nor the western area of Lake Natron allow a more precise estimate of the time of the change.

Another site excavated after the works at Olduvai and Peninj, Olorgesailie (Kenya), provided precise dating (by means of the $^{40}K/^{40}Ar$ method) for the Acheulean tools from Members 5–8 of that formation (Figure 8.19), but they are recent sediments, estimated to between 0.75 and 0.7 Ma (Bye *et al.*, 1987). The precise time of the substitution of Oldowan by Acheulean tools cannot be specified. Any group of hominins capable of using Acheulean techniques could have very well employed, on occasions, simple tools to carry out tasks which did not require complex instruments.

An illustrative example is the large number of Acheulean artifacts found at Locality 8 of the Gadeb site (Ethiopia) during the 1975 and 1977 campaigns. Some 1849 elements, including 251 handaxes and knives, were found at the 8A area, a very small excavation; whereas 20,267 artifacts

appeared at 8E (Clark and Kurashina, 1979). The age estimates for the different Gadeb localities with lithic remains are imprecise: they range from 1.5 to 0.7 Ma. These localities contain, in addition to Acheulean tools, developed Oldowan utensils, which led Clark and Kurashina (1979) to conclude that two groups of hominins would have alternated at Gadeb, each with its own cultural tradition. But it is curious that the examination of the bones from Gadeb showed that the butchery activities had been carried out mostly with the more primitive handaxes, those belonging to developed Oldowan. This fact raises an alternative interpretation, namely that tools obtained by advanced techniques are not necessary for defleshing tasks.

Konso-Gardula (Ethiopia), south of the River Awash and east of River Omo, has allowed the most precise dating of the beginning of the Acheulean culture. Since its discovery in 1991, Konso-Gardula has provided a great number of

Table 8.2 Cultural sequence at Olduvai established by Mary Leakey (1975, modified)

Bed	Age (Ma)	Number of pieces	Industry
Masek	0.2	187	Acheulean
IV	0.7–0.2	686	Acheulean
		979	Developed Oldowan C
Middle part of III	1.5–0.7	99	Acheulean
		—	Developed Oldowan C
Middle part of II	1.7–1.5	683	Developed Oldowan A
I and lower part of II	1.9–1.7	537	Oldowan

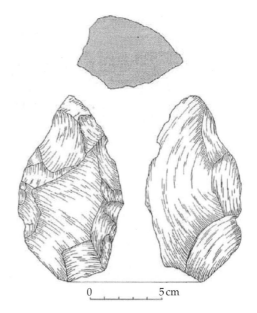

Figure 8.18 Early biface from Peninj (Natron Lake, Tanzania), *c.* 1.5–1.3 Ma. Picture from Wynn (1989).

(a)   (b)

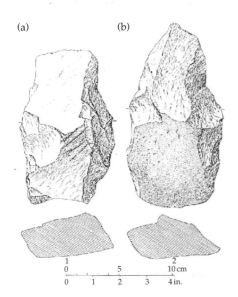

Figure 8.17 Developed Oldowan tradition from the upper part of Middle Bed II, Olduvai Gorge. (a) Cleaver; (b) handaxe. Picture by M.D. Leakey (1971). M.D. Leakey, Foreword by J. D. Clark, *Olduvai Gorge*, (1971), Figure 8.1.9, Figure 8.2.5. Cambridge University Press.

tools, which include rudimentary bifaces, trihedral-shaped burins, cores, and flakes, together with two hominin specimens, a molar, and an almost complete left mandible (Asfaw *et al.*, 1992). The sediments were dated by the ^{39}Ar/^{40}Ar method to 1.38–1.34 Ma. (Asfaw, *et al.*, 1992). Berhane Asfaw and colleagues (1992) associate the

Konso-Gardula hominin specimens with the *H. ergaster* specimens from Koobi Fora, especially with KNM-ER 992.

The cultural sequence identified by Mary Leakey (1975; Table 8.2) involves a three-stage transition. First, the evolution of progressively more sophisticated techniques within the Oldowan culture itself. Second, the coexistence of Oldowan and Acheulean tools. Third, the disappearance of the former, and further development of Acheulean techniques. Isaac (1969) limited that sequence, noting only four cultural–stratigraphical associations during the East African Lower Paleolithic, known, from the oldest to the youngest, as Oldowan, developed Oldowan (both within Mode 1), lower Acheulean, and upper Acheulean (both within Mode 2).

8.2.2 The Acheulean technique

To what extent can the Acheulean tradition be considered a continuation or a rupture regarding Oldowan? Was developed Oldowan a transition phase towards subsequent cultures? Mary Leakey (1966) believed that developed Oldowan was associated with the presence of primitive han-daxes, protobifaces that anticipated Acheulean bifaces. However, protobifaces cannot be strictly considered as a transitional form between Old-owan and Acheulean techniques. Otte (2003) argued that natural constraints (e.g. mechanical laws of the raw materials) force the manufacture of similar forms, which thus may be considered successive stages of a single or very close elab-oration sequence, although this may not always be the case.

The successive manipulation of a core, passing through several steps until the desired tool is

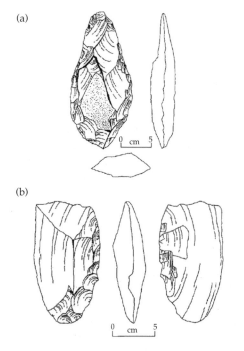

(a)

(b)

Figure 8.19 Tools from Olorgesailie (Kenya). (a) Handaxe; (b) cleaver. Picture from Ambrose (2001).

obtained, is a task that Leroi-Gourhan (1964) named *châine opératoire* ("working sequence"). Whereas a chopper and a protobiface respond to the same *châine opératoire*, the manufacture of Acheulean bifaces is the result of a completely different way of designing and producing stone tools. The most conspicuous novelty is the diver-sity of Acheulean instruments. Sometimes it is difficult to assign a function to a stone tool. We have already seen that Oldowan flakes have been interpreted as both simple debris and valuable tools. However, Acheulean tools include knives, hammers, axes, and scrapers, whose function seems clear. The materials used to manufacture lithic instruments are also more varied within the new tradition. But the most striking difference associated with the Acheulean culture is the tool we mentioned before: the handaxe.

The required technique to execute the Acheu-lean bifaces is different from Oldowan in several features. The first difference is the succession of strikes required to produce a handaxe, which contrasts with the few and unorganized strikes required to manufacture a protobiface. The pro-duction of long oval flakes (more than 10 cm), characteristic of Acheulean techniques, is its key difference from the advanced Oldowan traditions. The shape of those long flakes is not very different from the bifaces themselves. This is why Isaac (1969) suggested that they could be transformed into handaxes without too much effort. The appearance of the technique for producing flakes suddenly changed the possibilities for tool manufacture.

Given that the production of those flakes involves starting from large cores, the availability of quarries with such raw materials can determine important differences in the cultural content of

Box 8.13 The Acheulean–*H. erectus* link

Despite the difficulties involved in assigning tools to specific taxa, and the indications of the presence of cultural exchange among different hominins, the identification of Acheulean culture with *H. erectus*-grade hominins is very common. The strength of the Acheulean–*H. erectus* link led Louis Leakey to conclude that the appearance of Acheulean tools at Olduvai is the result of an invasion of *H. erectus* from other localities (Isaac, 1969).

Box 8.14 Design in Acheulean bifaces

The main objective of Oldowan technique was to produce an edge, with little concern for its shape. However, Acheulean bifaces had a very precise outline, which evinces the presence of design from the very beginning.

The existence of design has favored speculation about the intentions of the tool-makers. It has even been suggested that bifaces have an associated symbolism. We will return to this issue in Chapter 10.

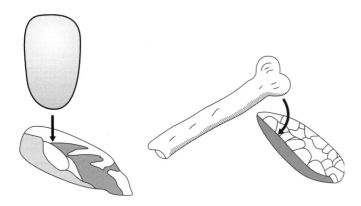

Figure 8.20 Hard-hammer (left) and soft-hammer (right) manufacturing techniques.

different sites. Schick and Toth (1993) noted that bifaces can also be obtained, in the absence of sufficiently large raw material, from smaller cores similar to those that served as a starting point for the manufacture of Oldowan choppers. But the manipulation of large blocks of material (mostly lava and quartzite) to produce long flakes seems to have been the turning point for the development of the Acheulean culture. It would also have involved risk for those who had to manipulate stones of large size (Schick and Toth, 1993).

The most advanced Acheulean stage includes handaxes with such symmetrical and carefully elaborated edges that they must have required the so-called soft-hammer technique (Figure 8.20), which uses softer hammers than the cores themselves, such as wood or bone. Knapping with such a tool allowed more precise control and certainly required more time. Schick and Toth (1993) have provided a detailed description of the process.

8.2.3 First Asian cultures

Three events are usually considered to have taken place together during the evolution of early- and middle-Pleistocene hominins: the appearance of *H. erectus*, Acheulean culture, and the first migration of hominins out of the African continent. The usual interpretation suggests that these three events are related. Leaving Africa confronted hominins with climates colder than the Rift's. Adaptation to those extreme conditions was made possible by cultural novelties, such as the control of fire. The adaptation was achieved by *erectus*-grade hominins, associated with Acheulean culture.

The alternations known as glaciations correspond to cycles, during one stage of which the planet's climate cools down considerably, with the accumulation of great masses of ice on the continents, mainly in the Northern hemisphere. The classical interpretation considered that the Pleistocene lasted 600,000 years, and that there were four great ice ages, known as Gunz, Mindel, Riss, and Würm, according to the sequence observed in the Alps (Table 8.3; Penck and Brückner, 1909).

However, the Pleistocene is currently believed to have lasted much longer, for 1.8 Ma, and included a greater number of climatic alternations (Figure 8.21). The early Pleistocene could have included up to 20 climatic cycles, each corresponding to one glacial or interglacial period. The middle Pleistocene included four cycles, and the late Pliocene two: the Würm glaciation, which began close to 130,000 years ago, and the current interglacial period, which began close to 10,000 years ago (see Gibbar, 2003).

The late Miocene and Pliocene climatic changes had a large influence on the early evolution of hominins. Even though Africa was out of the range of the glaciers (except for its highest peaks), temperature decreases were a global phenomenon. As we saw in Chapter 5, the great glacial period that extended the savannas in the Rift had much to do with the appearance of the genus *Homo*. But hominins were directly confronted with glacial climates during the Pleistocene, not in Java, of course, but in continental China and, later, in Europe.

The climatic changes occasionally turned inland China and central Europe into subglacial zones. The survival of hominins in conditions that were so different from those in tropical environments is often explained in cultural terms, as mentioned before. But a simple correlation between climatic changes, hominin radiation, and tools shows numerous cracks when examined closely. One is the aforementioned difficulty of establishing reliable associations between cultural traditions and hominin species. The coincidence of Oldowan and Acheulean tools for a long period of time suggests that industries were not fixed patterns that were employed consistently. There are examples of reversals; the appearance of simple tools in places and at times that exhibited a more advanced industry.

As Roebroeks (1994) has argued, stone tools lack chronological value. Suffice it to recall that in Europe the Clactonian primitive flake and axe tradition includes elements which are quite similar to Oldowan instruments, but they were manufactured at a much later time. The identification between *H. erectus* and the Acheulean industry is

Table 8.3 A simplified chronology of glacial ages

Epoch	Glaciation	Time (Ma)
Holocene	Postglacial	0.01
	Würm	0.125
	Riss/Würm interglacial	0.275
	Riss	0.375
Pleistocene	Mindel/Riss interglacial	0.675
	Mindel	0.75
	Günz/Mindel interglacial	0.9
	Günz	1
	Donau/Günz interglacial	1.8–1.6
	Donau	?
Pliocene	Biber/Donau interglacial	?
	Biber	3.4
Miocene		

Box 8.15 The consequences of global cooling

As a consequence of the planet's cooling, and increasing glaciation, the ocean water was deposited, frozen, on continental zones in Europe, Asia, and America. Up to 5.5% of the planet's water accumulated in the form of glacial ice, in contrast to the current 1.7% (Williams *et al.*, 1993). Up to a third of the Earth's land was covered by glaciers at their greatest expansion, which coincided with the second glacial stage (Mindel). The ecological consequences of the advancement of the ices are an intertwined chain of events:

• the shallowest seas dry out (the level of the oceans descends 150 m);
• the areas of contact between the continents increase;

• the migration of terrestrial animals, including hominins, is facilitated;
• profound changes in the fauna and flora: great extensions of coniferous forests, open tundras, and the associated fauna of large herbivorous mammals (deers, mammoths, wooly rhinoceros) and their predators.

The annual average temperature in northern Europe was − 2°C during the glaciations. There were only about 30–40 days each year without frost in the zones that were not covered by glaciers. It is easy to imagine the adaptive problems faced by *erectus*-grade hominins, primates that had evolved in tropical forests and adapted to their climate.

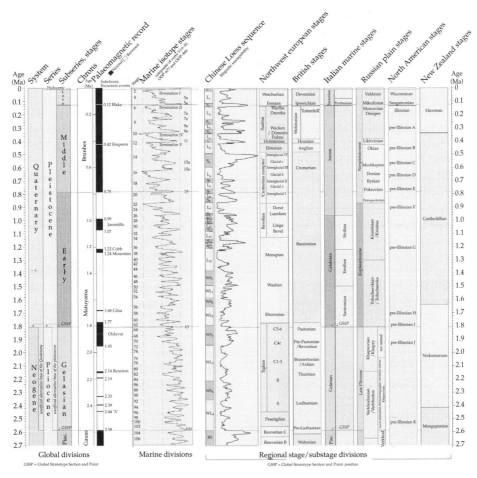

Figure 8.21 The extended chronology of glacial periods. ©2004. Compiled by P.L. Gibbard, S. Boreham, K.M. Cohen, and A Moscariello; http://www-qpg.geog.cam.ac.uk/.

only acceptable if it is done in very general terms. As we saw in Chapter 7, there is the additional problem of specifying which taxa should be included in *H. erectus* sensu lato.

In any case, the truth is that the industry of *erectus*-grade hominins is different in Africa and Asia. In fact, it is so different that the Harvard University anthropologist Hallam Movius (1948) defined a hypothetical frontier, known as the Movius Line, separating the more primitive Asian technology from the more advanced industries found in Africa, the Middle East, and Europe. However, this claim needs to be qualified. Lithic artifacts—handaxes—similar to the primitive pieces

of Acheulean culture were discovered at the Bose site (China; Yamei *et al.*, 2000). This finding put into question Movius' hypothesis. Asian *erectus* produced bifaces after all. They were more primitive than the African and European ones, but they were bifaces nonetheless.

The Bose site is in the Guangxi Zhuang region, southern China. It includes seven fluvial terraces (T1–T7), of which T4 contains the Paleolithic artifacts and dispersed tektites. The age of the stone artifacts was obtained by the $^{40}Ar/^{39}Ar$ method on three suspected Australasian tektites collected *in situ* (Yamei *et al.*, 2000). Yamei and colleagues (2000) considered the overall weighted mean of the isochron ages from the three samples to be the most representative age for the tools: 803,000 ± 3,000 years. The instruments are extensively chipped cobbles of quartz, quartzite, sandstone, and chert and associated flakes (Figure 8.22).

Regarding the Bose evidence, it has been noted that typical Acheulean very elaborated instruments have not been discovered in Asia. This casts doubt on the possibility of a true cultural connection between the African and Asian *H. erectus* sites (Gibbons, 2000). It could be that the Bose bifaces were another example of Marcel Otte's (2003) "similar forms," but maybe the causes of the absence of a true Acheulean culture in Asia, if such an absence is accepted, should not be explained on the grounds of technological delay, but on other

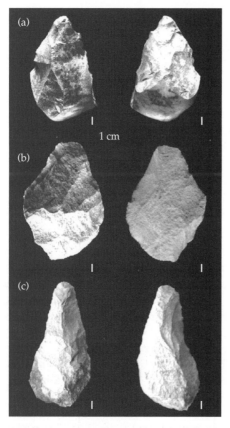

Figure 8.22 Bifaces from Bose (Guangxi Zhuang, China). (a) Bogu 91001, no. 1; (b) Hengshandao 94, no. 3; (c) Yangwu 91003, no. 1. The right and left sides of the figure show the opposite faces of each tool. Picture from Yamei *et al.* (2000).

grounds. In fact, the Java sites yielding *erectus*-grade specimens have provided very few lithic instruments. Stone tools have only turned up at Kabuh (Trinil) and, in very small quantities, at Ngandong. A possible explanation for this absence is that the Javanese *erectus* manufactured tools using other materials, such as wood, bamboo, or bone. The discovery of some canes at Ngandong (Java) and bone and horn tools at Hexian (China; Day, 1986) favors this hypothesis.

Other Chinese sites, such as Zhoukoudian, Dali and, mostly, Lantian have yielded quite a few tools, including choppers, cores, and flakes (Binford and Ho, 1985; Binford and Stone, 1986). These tools were considered different from the Acheulean tradition and, thus, the absence of this tradition in Asia was widely accepted. After the Bose findings, the presence of different technologies has been noted at Chinese sites. Shen and Qi (2004) examined two lithic assemblages—Cenjiawan and Maliang—from middle-Pleistocene sites in the Nihewan Basin (northern China). The results revealed intentional selection of high-quality raw materials, continuously rotating core reduction, and evidence for butchering/meat-processing tool use. Chen and Wei (2004) believe that the Cenjiawan and Maliang lithic assemblages "might represent regional and/or temporal variations of Lower Paleolithic industries in northern China." However, the authors did not propose that such manipulations should be considered Mode 2 or Acheulean. Chen and Wei (2004) simply argued that "core reduction practice—continuously rotating the core and creating new platforms for suitable flake removals—utilized a core nodule to near exhaustion. As a result, flakes at both sites were carefully planned and skillfully removed. Nevertheless, prepared-platform core reduction technique is not evident."

We believe that the reviewed evidence suggests several conclusions. First, the Movius Line still indicates a frontier. The clearly developed Acheulean industries found in the Middle East and Africa lie to the west of this frontier. It is not until close to 670,000 years ago that early Acheulean instruments appear at sites such as Bori and Morgaon, in India (Mishra *et al.*, 1995).

However, to the east of the Movius Line, the evidence from Bose, Cenjiawan, and other sites suggests that the variety of technological alternatives was greater than was assumed at the time of Movius. The most interesting observation may be the finding of tools in localities in the Nihewan Basin (northern China), such as Xiaochangliang and Majuangou (Zhu *et al.*, 2001, 2004; Figure 8.23). The latitude (40°N) and age (up to 1.66 Ma) in Majuangou suggest, according to Rixiang Zhu *et al.* (2004), that "a long yet rapid migration

from Africa, possibly initiated during a phase of warm climate, enabled early human populations to inhabit northern latitudes of east Asia over a prolonged period".

8.2.4 The use of fire

The rapid colonization suggested by Zhu and colleagues (2004) would have been followed by the presence of *erectus*-grade hominins in continental Asia for a long time. Sooner or later they had to

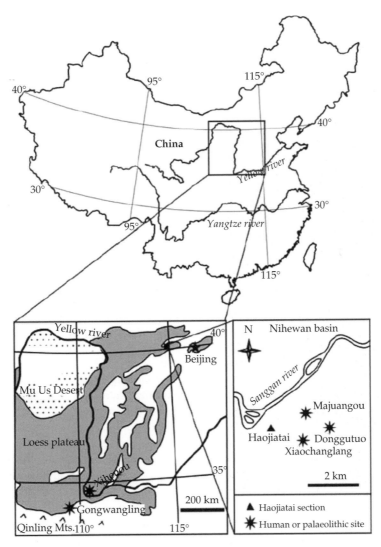

Figure 8.23 Early occupation—1.66 Ma—of the Nihewan Basin (northern China). Picture from Zhu *et al.* (2004). Reprinted by permission from Macmillan Publishers Ltd. *Nature* 431: 7008, 559–562, 2004.

confront the glaciations. During the glacial periods the climate of inland China at 40°N must have been extreme. The *H. erectus* that colonized China survived there, and the most common explanation for their adaptive success is the domestic use of fire. A similar explanation would account for the presence of hominins at European high latitudes. During the mid-twentieth century it was taken for granted that the control of fire by *erectus*-grade hominins had been an important addition to the cultural progress associated with stone tools, which was thought to have occurred around 500,000 years ago (Oakley, 1956). This point of view persisted for several years (see, for instance, Isaac, 1980). It fit the idea of a late colonization of Europe by hominins.

But the extremely cold climate was not restricted to Asia and Europe. The colonization of the high regions of East Africa during the lower Pleistocene, such as the Gadeb Plateau (Ethiopia), must have posed similar adaptive problems for hominins. Thus, it has been suggested that the domestic use of fire must have been the adaptive solution at the Gadeb highland (Ethiopia) from 1.5 to 0.7 Ma (Clark and Kurashina, 1979), although there is no empirical evidence to support this hypothesis. Swartkrans Member 3 (South Africa) has yielded abundant fossil remains of bones which appear to have been burnt. After examining them, Brain and Sillen (1988) concluded that the temperature they had been subjected to was similar to that of campfires. Swartkrans Member 3 might be up to 1.5 Ma, which would imply, if this were the case, a very early use of fire. Even though the site's Members 1 and 2, which are just about as old, contain many robust australopith and some *Homo* remains, Member 3 has only yielded nine specimens, attributed to *P. robustus*. Brain and Sillen (1988) argued that the introduction of fire must have taken place during the brief temporal interval separating Members 2 and 3 and, in any case, while *P. robustus* still existed. However, the authors consider that the available evidence is insufficient to assign the control of fire to any particular species.

The earliest evidence of controlled fire out of Africa comes from the Acheulean site of Gesher Benot Ya'aqov (Israel), where burned seeds, wood, and flint have been found (Goren-Inbar *et al.*,

2004). The locality is above the Brunhes-Matuyama chron boundary and the authors estimate its age at 700,000 years. In continental Asia, Zhoukoudian Locality 1 has provided the best evidence of the early presence of fire. Based on chemical analyses of materials from that site, Black (1931) stated that "It is thus clear beyond reasonable doubt that *Sinanthropus* knew the use of fire". The same view was subsequently endorsed by several authors (for instance, Zhang, 1985; James, 1989), backed by evidence of fire use mainly from levels 4 and 10 from Locality 1. However, after re-examining the evidence from those two levels, Weiner *et al.* (1998) concluded that "on the basis of the absence of ash or ash remnants (siliceous aggregates) and of in situ hearth features...there is no direct evidence for in situ burning."

Weiner and colleagues' (1998) conclusion has been contested by Xinzhi Wu (1999) based on several historical references, starting with Black (1931), who interpreted the signs left by fire at Zhoukoudian as characteristic of anthropic activity. The critical issue is the association of hominin specimens, lithic tools, and evidence of fire use at Locality 1. If the indications of fire are due to human intervention, they should appear at the same places as the tools. But such association is not easy to establish due to the complexity of the taphonomic interpretations in regard to Zhokoudian. The fact that scavengers dispersed the hominin remains makes the task difficult.

By means of a three-dimensional reconstruction of Locality 1, Noel Boaz *et al.* (2004) have provided the best clues to understanding the anthropic activity reflected in the sedimentary sequences of Zhokoudian (Figure 8.24). Locus G of level 7, identified in 1931, was the first to offer an association of hominin specimens and tools (Black *et al.*, 1935). Indications of fire, suggesting human activity, were also detected at the same locus G, or close to it (Pei and Zhang, 1985). However, Boaz and colleagues (2004) argued that "carbon on all the *Homo erectus* fossils from Locus G, a circumscribed area of 1-meter diameter, earlier taken to indicate burning, cooking, and cannibalism, is here interpreted as detrital carbon deposited under water, perhaps the result of hyaenid caching behavior." According to Boaz *et al.* (2004), the

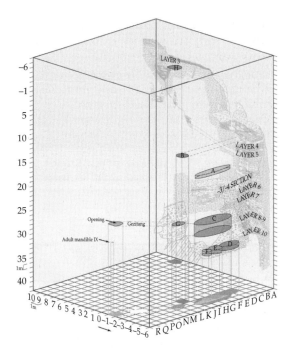

Figure 8.24 A three-dimensional map of Locality 1 at Zhoukoudian, showing Loci A–G. Picture from Boaz *et al.* (2004). Reprinted from *Journal of Human Evolution* Vol. 46: 5. N. T. Boaz *et al* 'Mapping and taphonomic analysis of Homo erectus', 519–549, 2004 with permission from Elsevier.

relative abundance of charred bones is a different issue. These authors believe that "Stone tool distributions indicate transient hominid presence in the cave, as does evidence of burned fresh bone." This possibility had not been excluded by Weiner *et al.* (1998) when examining levels 4 and 10.

8.3 *Homo* in Europe

Europe holds its own paradox related to the culture of the first hominins that arrived in the continent. The central element of Acheulean culture, the handaxe, is absent in many early European sites with signs of human presence (Italy, France, Germany, Czech Republic, and Spain). It was not until a second colonizing wave, which took place about half a million years ago, that Mode 2 handaxes were introduced. Sites corresponding to this time interval include Torralba and Ambrona (Spain), St. Acheul and Abbeville (France), Swanscombe, Boxgrove, and Hoxne (England),

and Torre di Pietra and Venosa-Notarchirico (Italy).

What could have caused such a cultural sequence? Were there failed attempts to colonize the continent at early times, as several authors have suggested (Roebroeks, 1994; Roebroeks and Van Kolfschoten, 1994; Dennell and Roebroeks, 1996)? Were there sporadic attempts that ended up leaving Europe uninhabited until a definitive colonization 500,000 years ago? Or, on the contrary, was there a continuous colonization, but with certain "cultural gaps", as Toth and Schick (1993) suggest?

The earliest European sites showing the presence of hominins (excluding Dmanisi, in Georgia, which is considered by geographers to be part of Europe, at the boundary with Asia) are located at the two farthest ends of the continent. To the east they appear at Ubeidiya (Israel), with an estimated age of 1.4 Ma (Tchernov, 1989), and to the west, at Atapuerca, estimated to about 800,000 years, at the limit of the lower Pleistocene (Carbonell *et al.*, 1995). (Israel is in Asia, but it will be discussed with European hominins, because of its position at the eastern end of the Mediterranean.) The estimated age of the Orce sites (southern Spain)—with industry at Fuentenueva 3—is even older than the Israel sites (Gibert *et al.*, 1994, 1998), although this dating has been contested (Turq *et al.*, 1996; Martinez-Navarro *et al.*, 1997; Palmqvist *et al.*, 1999). Such an early occupation of southern Spain would support the almost simultaneous arrival of hominins at Europe through the Levantine corridor and the strait of Gibraltar. Dennell and Roebroecks (1996) have rejected this interpretation, questioning the Atapuerca and Orce datings, but leaving the door open for other interpretations, such as different dispersal waves across Spain, with different hominins as protagonists.

In a review of *H. erectus*, Howells (1980) emphasized that middle-Pleistocene European remains are, in comparison with the African ones, scarce and younger. Clark Howell (1981) and Stringer (1980), among other authors, argued that the *H. erectus* taxon could not be identified in Europe or that, at the most, its presence would have been sporadic and limited to Mediterranean

Box 8.16 *Erectus*-grade hominins in Europe

The existence of *erectus*-grade hominins on the European continent has traditionally been one of the many issues for discussion in human paleontology. Like most debates, it has been resolved by changing the original formulation rather than by settling the initial dispute.

The issue of whether the taxon *H. erectus* was present in Europe hides two questions. The first one is the time at which hominins arrived at the Old Continent not in a sporadic and provisional way, but as colonizers that would lead, in time, to Neanderthals. The second question is whether those initial human colonists were *H. erectus* or belonged to another taxon, that of the immediate ancestors of *Homo neanderthalensis*. The hypothesis of a late occupation of Europe is compatible with the acceptance of previous migrations: the human occupation of Europe prior to 1 Ma would have been sporadic and intermittent (Turner, 1992; Roebroeks, 1994; Roebroeks and Van Kolfschoten, 1994; Dennell and Roebroeks, 1996).

---

regions (Dennell, 1983). The earliest European crania known at the time, Petralona (Greece), Arago (France), and Vértesszöllös (Hungary)—which we will review in the next chapter—show reminiscent traits of some *erectus*-grade specimens. However, their morphology was more advanced than that of *H. erectus* sensu stricto, a fact that led Howell and Stringer to argue that when hominins arrived in Europe they were at an advanced evolutionary stage compared to the typical Asian *H. erectus*. Stringer's (1984) point of view is illustrative. He argued that, taking the apomorphies (derived traits) into account, European specimens should be classified as *H. sapiens* sensu lato and not as *H. erectus*, whether it be sensu stricto or sensu lato.

Doubts about the status of the European forms arise from both the advanced morphology of those specimens and their young ages. Several European sites with numerous middle Pleistocene lithic instruments, dated to around 500,000 years, but without hominin remains, were identified by the last third of the twentieth century. However, remains of early European hominins finally appeared, and they did so in places that were very far apart. Towards the end of the twentieth century the condition of "first European colonizers" was applied to the Boxgrove specimen (England; Roberts *et al.*, 1994; Figure 8.25), the Dmanisi mandible (Georgia; Gabunia and Vekua, 1995), the Ceprano skull (Italy; Ascenzi *et al.*, 1996), and the early Atapuerca specimens (Spain; Bermúdez de Castro *et al.*, 1997).

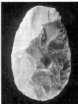

Figure 8.25 Left, Boxgrove tibia. Right, handaxe associated with the finding (the photographs are not at the same scale). Photographs from http://www.archaeology.co.uk/ca/timeline/prehistory/boxgrove/boxgrove.htm.

The Boxgrove specimen appeared on the cover of *Nature* accompanied by the headline "The first European?". It consists of a partial tibia found in 1993 in Amey's Eartham pit, close to Boxgrove (in West Sussex, southern England; Roberts *et al.*, 1994). Lithic instruments were found in association with the specimen. The exemplar's age, dated by faunal comparison, was estimated at 524,000–478,000 years (Roberts *et al.*, 1994). There is evidence of the rodent *Arvicola terrestris* at the Boxgrove site, indicating that it is over half a million years old (Gamble, 1994). This is the frontier that separates this rodent from its descendant, *Mymomis savini*. The age of the Boxgrove tibia was challenged from the beginning (Bowen and Sykes, 1994), but it is clear that the first European colonists would not be found so far from the entry routes to the continent. The Boxgrove tibia comes from a considerable latitude and its age places it at the lower time limit of the generalized presence of hominins in Europe. The finding opened an interesting perspective which went against the general idea that the northernmost regions were colonized a very long time after the arrival of *erectus*-grade hominins in Europe.

The discovery of 32 Mode 1 flints from the Cromer Forest-bed Formation at Pakefield (eastern England; Parfitt *et al.*, 2005) extended the human presence in England even further back in time. The artifacts were dated by means of paleomagnetism, amino acid geochronology, and biostratigraphy to the early part of the Brunhes chron (about 700,000 years ago; Parfitt *et al.*, 2005). This means that *erectus*-grade hominins colonized high latitudes very rapidly after entering Europe, similarly to what happened in Asia. Roberts and colleagues (1994) classified the Boxgrove specimen as *Homo* aff. *heidelbergensis*, based on the exemplar's age and its geographical proximity to the Mauer mandible (see Chapter 9). After a detailed examination of the Boxgrove tibia, Stringer and colleagues (1998) argued that its taxonomic classification cannot be specified beyond *Homo* sp.

The notoriety of the Boxgrove tibia as an indication of the early presence of *Homo* in Europe did not last for long. Much older specimens from the south and east of the continent soon became the center of attention. The new remains from Europe's periphery, which we will soon review, belong to the lower Pleistocene, or even to the Pliocene, so the controversy surrounding them is not about whether they are *H. sapiens*, which they could not be, but about other issues: evolutionary relations and, most of all, reliability of the estimated ages. We will review some of these doubts as we examine the specimens.

How did the first hominins arrive in Europe? There are four possible routes: the Strait of Gibraltar, the bridge between Italy and North Africa, the Bosporus, and the north of the Black Sea. Gibraltar is a very narrow strait (although with violent currents) that at times of low sea levels could be crossed. This might be within reach of middle-Pleistocene hominins. According to Aguirre (1997), three kinds of condition should be met to accept Gibraltar as a migration route: (1) a significant number of terrestrial species on both sides of the strait at the same time, or at overlapping or consecutive times; (2) reduction of the strait's depth, due to tectonic activity or a decrease of the average sea level; and (3) if this route is thought to be the main expansion route, then it should be shown that other immigration routes are impossible or improbable.

This last condition may be ignored for now, because there is no need to assume that the Strait of Gibraltar was the only bridge into Europe. Regarding the issue of the depth of the sea, Aguirre and colleagues have suggested that a tectonic event of short duration made the Mediterranean crossing possible towards the early Pleistocene (Aguirre and de Lumley, 1977; Aguirre *et al.* 1980). That it was possible does not mean that it actually happened. The firmest evidence in regard to the crossing of the Strait of Gibraltar seems to be the fauna on either side of the strait. However, the presence of several

---

Box 8.17 Geochronology by amino acid polarity

Geochronology by means of amino acid analysis is based on racemization effects. The amino acids of the proteins of living organisms are left-handed: if a beam of polarized light is passed through a protein solution, the vibration plane of the light turns to the left. When the organism dies, racemization begins. Amino acids become right-handed, reducing the polarized light's turning effect. If there is enough fossilized protein it is possible to calculate the time elapsed since the organism's death. In Pakefield, a new technique of amino acid analysis was used. This technique "combines a new reverse-phase–high-pressure liquid chromatography (RP–HPLC) method of analysis with the isolation of an intracrystalline fraction of amino acids by treatment with bleach. This combination of techniques results in the analysis of D/L values of multiple amino acids from the chemically protected protein within the biomineral, enabling both smaller sample sizes and increased reliability of the analysis" (Parfitt *et al.*, 2005).

species—hippopotamus, equids, felines (*Meganterion*), and even primates (*Theropithecus*)—in early-Pleistocene sediments from southern Spain is best explained, according to Aguirre (1997), by immigration from Asia. On the contrary, Martínez-Navarro and colleagues (1997) argue that faunas such as those at Venta Micena (Orce, Spain)—with scavengers with an African origin such as the hyena *Pachycrocuta brevirostris*—support a passage across the Strait of Gibraltar.

A better support for the Gibraltar route of migration out of Africa would be provided by hominin remains. A hominin presence is convincingly evinced at Atapuerca by around 800,000 years ago, and suggested by earlier tools found in the south of the Iberian Peninsula, although they are absent in the North African early Pleistocene. Dennell and Roebroeks (1996) suggest that the presence of hominins at North Africa 800,000 years ago is tenuous. According to these authors, substantial occupation of the southern shore of the Mediterranean, both the western and eastern ones, would not have taken place until the middle Pleistocene. This absence also impacts on the arguments in favor of a possible crossing from Africa to Italy. The Ceprano cranium, however, requires that we examine the possible arrival of hominins in Europe through the central Mediterranean.

The Ceprano specimen, named after the place in which it was discovered, 80 km south of Rome, consists of several fragments of a cranium that was fortuitously discovered in 1994, unearthed and damaged by an excavator (Ascenzi *et al.*, 1996). The specimen's first reconstruction suggested high cranial capacity, 1185 cm^3, with modern-looking features that seemingly set it apart from the *erectus* grade. However Ascenzi and colleagues (1996) attributed some of the differences to the specimen's pathological deformations.

After a careful reconstruction by Ron Clarke (2000; Figure 8.26), the supposedly pathological distortions as well as some mistakes made during the initial reconstruction were eliminated. Clarke detected typical apomorphies of *H. erectus* sensu stricto on the Ceprano cranium: prominent occipital crest; continuous and very marked

Figure 8.26 The Ceprano I cranium, after Clarke's (2000) reconstruction: *H. erectus* (Ascenzi *et al.*, 1996; Clarke, 2000); *Homo cepranensis* (Mallegni *et al.*, 2003). Picture from Mallegni *et al.* (2003).

supraorbital torus; and great thickness of the cranial base walls. These traits indicate a great morphological antiquity. Clarke did not calculate its cranial capacity, but it seems to be close to that which Rightmire (1990) considers typical of strict *H. erectus*. Clarke associated the cranium with the most representative specimens of that taxon, such as OH 9 from Olduvai.

Subsequent modifications of the cranium's reconstruction by Marie-Antoinette de Lumley and colleagues prompted Ascenzi *et al.* (2000) to suggest the need for broadening the hypodigm of *H. erectus* sensu stricto if the Ceprano cranium is to be included in that species. The cranium of *H. erectus* sensu stricto is more elongated than the Ceprano cranium, while the latter shows a much wider frontal region above the supraorbital torus than OH 9. The mediolateral expansion of the Ceprano specimen also falls out of the average limits of *H. erectus* sensu stricto. However, Ascenzi and colleagues (2000) accepted the pertinence of assigning the Ceprano specimen to *H. erectus* sensu stricto, more so if its great antiquity is taken into account.

A subsequent morphometric comparison and a cladistic analysis led Mallegni and colleagues (2003) to suggest a new species, *Homo cepranensis*, for the Ceprano cranium. They propose that this taxon had an African origin and migrated to Europe about 1 Ma, but the taxon did not contribute to the evolution of European humans during the middle and late Pleistocene.

Box 8.18 Dating the Ceprano I cranium

The Ceprano sedimentary basin has been associated with the Priverno one, a distance of 25 km, which has volcanic intrusions dated to 1,100,000 ± 110,000 years ago (Sevink *et al.*, 1984). The sedimentary level in which the cranium appeared belongs to sediments located slightly above in the correlation, which led Ascenzi and colleagues (2000) to estimate it to between 900,000 and 800,000 years. If the taxonomic classification suggested by those authors for Ceprano I is accepted, it follows that there were *H. erectus* sensu stricto in Europe at such an early date. Whether or not they arrived in Italy through the chain of islands that separates the two Mediterranean shores remains to be determined.

8.3.1 Atapuerca: *Homo antecessor*

The discoveries at the Atapuerca site have provided abundant evidence regarding the early presence of hominins in Europe and their relation with later taxa found in the continent. The sedimentary area of Atapuerca—about 14 km east of Burgos (Spain)—is very large and lies between the Atapuerca and Ibeas de Juarros regions, and includes 25 catalogued localities (Aguirre, 1995; Figure 8.27). Research at Atapuerca was led by Emiliano Aguirre until 1991 and, thereafter, by Juan Luis Arsuaga, José María Bermúdez de Castro, and Eudald Carbonell.

Towards the end of the nineteenth century, the Sierra Company Limited, which exploited coal and iron minerals in the area, excavated a trench in the mountains to construct a railway. Although the project was eventually abandoned, the trench for the railway revealed some of the Atapuerca sites. Four infill deposits have been excavated to date: Gran Dolina, Galería, and Covacha de los Zarpazos, at Atapuerca, and Sima de los Huesos, at Ibeas. The first fossils were found in 1976 at Sima de los Huesos and belonged to the middle Pleistocene. We will review them in section 9.1. But very early hominins associated with primitive instruments were discovered at another of the Atapuerca localities, Gran Dolina, almost 20 years later.

Gran Dolina is a karstic deposit with a sedimentary depth of 18 m, which includes 11 lithic levels, numbered from bottom to top. Levels TD3–4, 5, 6, 7, 10, and 11 contain abundant lithic tools (Carbonell *et al.*, 1995). More than 30 fossil

Figure 8.27 An aerial view of the railway trench (Trinchera del Ferrocarril) in the Atapuerca hills (Burgos, Spain). Picture from http://cvc.cervantes.es/actcult/atapuerca/geologia.htm.

specimens, including cranial, mandibular, and dental remains belonging to at least four individuals were found in 1994 in level TD6, in a stratum named Aurora (Figure 8.28). The most complete of these specimens is a large frontal piece, ATD6–15, with parts of the glabella and the right supraorbital torus, possibly belonging to an adolescent (Carbonell *et al.*, 1995). By 1996, 80 specimens corresponding to six individuals had been discovered.

Parés and Pérez-González (1995) have estimated the age of the Gran Dolina levels using paleomagnetism, placing the TD6 level within the Matuyama subchron, a meter below the Matuyama–Brunhes frontier, and thus prior to 780,000 years. Sesé and Gil (1987) suggested that the lower TD levels (3, 4, 5, and 6) are older than

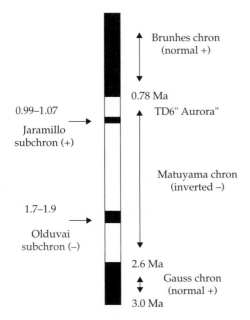

Brunhes chron
(normal +)

0.78 Ma
TD6" Aurora"

0.99–1.07
Jaramillo subchron (+)

Matuyama chron
(inverted –)

1.7–1.9
Olduvai subchron (–)

2.6 Ma
Gauss chron
(normal +)
3.0 Ma

Figure 8.28 Paleomagnetic column of Gran Dolina (Atapuerca) indicating the TD6 stratum, Aurora.

730,000 years after the study of the local microfauna. The discoveries at TD6 required rethinking the age of the initial presence of hominins in Europe. However, the species to which the remains belong had to be determined. A morphological study by Carbonell and colleagues (1995) suggested that the ATD6-15 fragment was larger than the equivalent in *erectus*-grade hominins from Turkana (KNM-ER 3733 and KNM-ER 3883), Sangiran 2, and Trinil, all of them crania with a capacity over 1000 cm^3. It was estimated that the cranial volume of ATD6-15 was similar to that of Sangiran 17 or the Sambungmacan specimens or the smaller ones from Ngandong. The supraorbital torus, which is double arched, was clearly distinguishable from that of typical Asian and African (OH 9) *H. erectus*. Dentition also set the TD6 and *H. habilis* sensu stricto apart, suggesting a continuity with the later European remains, belonging to the middle Pleistocene. Additionally, the lithic industry was pre-Acheulean, without handaxes. Carbonell and colleagues (1995) did not go beyond corroborating the presence of *Homo* in Europe during the early Pleistocene.

The new specimens discovered in 1995 allowed specifying the facial morphology of the hominins from level TD6 (Bermúdez de Castro *et al.*, 1997). A partial juvenile specimen, including the face, part of the cranium and some teeth, ATD6-69, was described as completely modern in its midfacial topography (Figure 8.29). According to Bermúdez de Castro *et al.* (1997), it is necessary to advance to the late Pleistocene to find a similar face, such as Djebel Irhoud 1, Skuhl, or Qafzeh (see section 9.2). Another

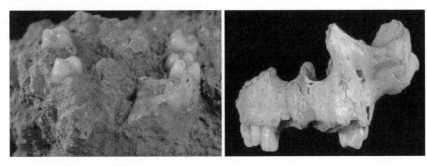

Figure 8.29 ATD6–69. Left, maxilla *in situ*. Right, holotype of *Homo antecessor* (Bermúdez de Castro *et al.*, 1997). Photographs from http://www.ucm.es/info/paleo/ata/albumes/ydolina.htm.

specimen, ATD6-58, attributed to an adult, retained some juvenile traits that are absent in *H. ergaster* (KNM-WT 15000), and later Neanderthals. The latter lack these facial traits even in juvenile specimens (Bermúdez de Castro *et al.*, 1997). Thus, TD6 specimens seemed to show certain derived traits that distinguish them from *H. erectus*, *H. ergaster*, and Neanderthals (Table 8.4 lists such apomorphies). Consequently, Bermúdez de Castro and colleagues (1997) introduced a new species, *Homo antecessor*, with the ATD6-5 specimen (a fragment of the right mandibular body with three molars found in July 1994) and a set of associated teeth belonging to the same individual as the holotype. The paratype of the new species consists of different specimens, up to 38, all from level TD6.

Bermúdez de Castro *et al.* (1997) argued that *H. antecessor* was the common ancestor of *Homo heidelbergensis*—which would later lead to Neanderthals—and anatomically modern humans. The latter would descend from a hypothetical African *H. antecessor* population. The same phylogenetic tree was presented in the final conclusions of an issue of the *Journal of Human Evolution* devoted to the Aurora stratum of Gran Dolina (Bermúdez de Castro *et al.*, 1999). But this is not the only taxonomic option for *H. antecessor*. Aguirre (2000) criticized the study by Arsuaga *et al.* (1999) of the facial skeleton for not giving enough importance to several apomorphic traits, common to the ATD6 sample's fossils (ATD6-38, ATD6-56, ATD6-69) and middle-Pleistocene Chinese fossils. According to Aguirre, such traits are also observed in modern humans but not in Neanderthals or pre-Neanderthals.

A new specimen found in 2003 in the Aurora stratum of Gran Dolina increased the hypodigm of *H. antecessor*, though it also cast new doubts on its evolutionary relationships. It is ATD6-96, the left half of a mandible attributed provisionally to a female (Carbonell *et al.*, 2005; Figure 8.30). The extreme gracility of ATD6-96, in addition to the reinterpretation of the midfacial traits of ATD6-69, led Carbonell and colleagues (2005) to associate the taxon present at Aurora with Chinese *H. erectus* sensu stricto specimens, such as Nanjing I, from the Hulu Cave (Tangshan Hill, eastern central

China). This proposal complicates the phylogenetic reconstruction of middle- and late-Pleistocene hominins. As Carbonell *et al.* (2005) put it, "the present evidence does not support the hypothesis of a phylogenetic relationship between the TD6 hominins and the European lineage leading to the Neanderthals." But the later Atapuerca specimens we will examine in Chapter 10 do show firm indications of a relation with Neanderthals. Are the Gran Dolina and Sima de los Huesos populations unrelated to each other? This seems an unlikely hypothesis, but it also seems unlikely that *H. antecessor* was a direct ancestor of *H. sapiens*, while also closely related to Asian *H. erectus*.

8.3.2 A phylogenetic tree for the members of the *erectus* grade

As we have seen in Chapter 7, neither the accumulation of new findings, cladistic techniques, or metrical studies, nor exhaustive comparisons, have resulted in a widely accepted proposal of the species to be included in the *erectus* grade. It seems sensible to explicitly accept that the problem exists and that the solution is unknown. However, the major discrepancies refer only to whether the taxon *H. erectus* is sufficient to accommodate all the specimens within a single polytypical species or, on the contrary, two species must be defined: *H. erectus* sensu stricto and *H. ergaster*. The authors of this book defend the latter option (Cela-Conde and Ayala, 2003). Beyond this discrepancy, there is virtually unanimous agreement regarding some evolutionary episodes related to the *erectus* grade. Thus, there is an ample consensus that, after the disappearance of *H. habilis*, and within an interval extending up to the late Pleistocene, the following episodes occurred:

- the acquisition of current functional bipedalism,
- an increase in encephalization,
- the reduction of the masticatory apparatus,
- the appearance of longer patterns of ontogenetic development,
- the development of sophisticated cultural solutions, such as Acheulean handaxes and the control of fire.

Table 8.4 Apomorphies of *Homo antecessor*

Cranial traits

1. Midfacial topography shows a fully modern pattern: infraorbital surface is coronally oriented and sloping downwards and backwards (true canine fossa), with a horizontal and high rooted inferior border
2. Supraorbital torus is doubled arched in frontal view
3. Superior border of the temporal squama is convex (arched)
4. Presence of styloid process
5. Cranial capacity above 1000 cm^3

Mandibular traits

6. The mylohyoid groove extends anteriorly nearly horizontal and courses into the mandibular body as far as the level of the M2/M3
7. Thickness of the mandibular body is clearly less than that of *H. ergaster* and *H. habilis* sensu stricto, and specimens from Baringo, Java and OH 22
8. Absence of alveolar prominence at the M1 level
9. Extramolar sulcus is narrow
10. Lateral prominentia is smooth and restricted to the level of M2
11. Design of the inner aspect of the corpus defined by a shallow but well-developed subalveolar fossa and a distinct internal oblique line, similar to that of European middle-Pleistocene fossils

Dental traits

12. Mandibular incisors are buccolingually expanded with respect to *H. habilis* sensu stricto, Zhoukoudian, and specimens such as KNM ER 992 and Dmanisi, although to a lesser degree than *H. heidelbergensis* and *H. neanderthalensis*
13. Postcanine teeth are smaller than those of *H. habilis* sensu stricto, and within the range of *H. ergaster*, *H. erectus*, and *H. heidelbergensis*
14. Maxillary incisors are shovel-shaped
15. Mandibular canine is mesiodistally short
16. Buccal faces of the lower premolars show mesial and distal marginal ridges and grooves, which connect with the shelf-like cingulum
17. Crown shape of the mandibular P3 is strongly asymmetrical
18. Mandibular P3 exhibits a remarkable talonid
19. P3, P4 size sequence for the crown area of the upper and lower premolars
20. Upper and lower premolars are broad buccolingually
21. Mandibular M1 is buccolingually expanded with respect to *H. ergaster*
22. M1, M2 size sequence for the crown area of the upper and lower molar series
23. Mandibular M3 is noticeably reduced with respect to M1
24. Mandibular M1 and M2 show a Y-pattern of the buccal and lingual grooves separating the five principal cusps
25. Maxillary premolars show two, buccal and lingual, well-separated roots
26. Mandibular P3 and P4 exhibit a complex root system, formed by an MB platelike root with two pulp canals and a DL root with a single canal
27. Roots of the mandibular and maxillary molars are well separated and divergent; these teeth present a moderate taurodontism
28. Root system of all teeth is short relative to the crown dimensions
29. Enamel of the occlusal surface of the postcanine teeth is moderate to remarkably crenulated

Source: Bermúdez de Castro *et al.* (1997).

How many species are necessary to account for the different forms related with those evolutionary episodes in Africa, Asia, and Europe? Any taxonomic proposal, whether broad or restrictive, must explain coherently how the transition from the *erectus* grade to modern humans took place.

Box 8.19 The multiregional theory

Emiliano Aguirre is an advocate of the multiregional theory of evolution towards *H. sapiens*, which we will discuss in Chapter 9. Aguirre (2000) argued that, at least as a hypothesis, the facial traits that *H. antecessor* shares with *H. erectus* and modern humans suggest a highly varied genetic input into the origin of modern humans. We will analyze the pros and cons of this proposal when we review the antagonistic multiregional and replacement theories for the origin of anatomically modern humans.

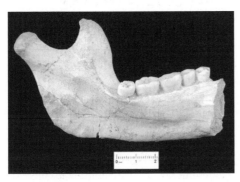 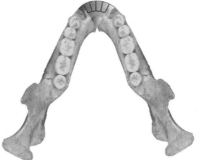

Figure 8.30 ATD6-96. Left, medial view. Right, drawing on a photographic restoration (Bermúdez de Castro *et al.*, 1995). The illustrations are not represented at the same scale. From http://www.pnas.org/cgi/content/figsonly/102/16/5674. Proceedings of the National Academy of Sciences, 102: 5674–5678, 2005. National Academy of Sciences, USA.

Box 8.20 A single *H. erectus* taxon or multiple *erectus*-grade species?

Textbooks, which need to give clear and concise explanations, tend to opt for models of a single *H. erectus* taxon including all middle-Pleistocene hominins, in addition to some belonging to the early and late Pleistocene. Specialized articles, striving for greater detail, tend to multiply the *erectus*-grade species, although not always. Bräuer (1994) studied the differences between African and Asian *erectus* by comparing 40 measurements of angles and indications of diverse anatomical regions, and concluded that sexual dimorphism and regional variability are sufficient to explain the diversity of the Asian and African *H. erectus*. Conversely, Groves (1989), aiming to be a general reference, multiplied middle-Pleistocene hominin species.

This is a question to which we will return in the following chapters, but we must schematically advance now our evaluation of the evolutionary significance of the *erectus* grade. In our view, the following is a reasonable summary of the most likely hypotheses.

1 The *erectus* grade began in East Africa, with the appearance of new hominins that underwent anatomical changes, mostly related to the cranial vault, and a considerable increase in body size. This process began at an early time, around 1.9–1.8 Ma, or even before.

2 Hominins belonging to the *erectus* grade initially kept to the Oldowan cultural traditions of *H. habilis*. Later, significant advances in the manufacture of tools led to a new cultural tradition: Acheulean or Mode 2. However, the climax of this new way of manufacturing stone tools, with handaxes as central pieces, was absent in Asia, with some exceptions, and appeared in Europe at a later time.

3 The migration out of Africa occurred soon after the origin of *H. erectus* and the dispersal through Asia was very fast. The later colonization of Europe—maybe by *erectus* from Asia—also occurred in a short period of time. A reason for the migrations could have been scavenging the victims of great predators that had disappeared from the African continent. Current evidence does not allow a decision on whether the Gibraltar strait was a migration route.

4 Different populations (or species) belonging to the *erectus* grade evolved *in situ*. *H. erectus* sensu stricto specimens are a product of such local evolution and, consequently, exist only in Asia. In accordance with the Atapuerca evidence, it can be argued that at the same time there was a different species in Europe, *H. antecessor*. The absence of remains in Africa does not permit determination of whether *H. antecessor* also existed there during the Matuyama subchron.

5 The transformation of the late Pleistocene forms was different in each continent. It seems that *H. erectus* sensu stricto remained in stasis in Asia. In Europe *H. heidelbergensis* led to *H. neanderthalensis*. In Africa, the species corresponding to contemporary European *H. heidelbergensis* led to *H. sapiens*.

6 There is, however, an alternative interpretation. The Atapuerca specimens may be considered close to *H. erectus* sensu stricto and be ancestors of *H. sapiens*. Such a model would require opting for a multiregional model of the evolution of anatomically modern humans, while the previous one is compatible with the out-of-Africa hypothesis. We will discuss both in the following chapters.

It is not an easy or reliable task to sketch a phylogenetic tree with these brushstrokes for the evolution of *erectus*-grade hominins. One of the first questions is whether the taxon *Homo*, which constitutes the usually attributed species, is a monophyletic group or not. Taxonomy must adjust to monophyletic groups, so any *Homo* genus that is paraphyletic, given the specimens included in the hypodigm, is not acceptable. As we anticipated in Chapter 5, Wood and Collard (Collard and Wood, 1999; Wood and Collard, 1999b) addressed the problem of monophyly by evaluating the adaptive significance of the genus *Homo*. This is an essential issue, given that the ultimate significance of any taxonomic proposal has to do, within the evolutionary paradigm, with the way in which a given taxon adapts to its environment. If several species remain close enough to each other to deserve consideration as members of the same genus, it can be assumed that their adaptive solutions will not differ too much.

Wood and Collard noted that there is considerable size and weight variation within the different species grouped in the genus *Homo*. Not all of them are closer to current humans than australopiths in terms of weight and size. But another trait could be taken as the fundamental apomorphy of the genus: bipedalism, for instance. We have pointed out that australopiths exhibit a peculiar bipedalism, retaining climbing capabilities. *Erectus*-grade hominins developed a bipedalism exclusively adapted to walking. What can be said, in this respect, about the Pliocene species attributed to *Homo*? Wood and Collard (1999b) mentioned the structure of the hand of OH 7 and the size of the arms of OH 62, both Olduvai specimens, as arguments in favor of an incomplete bipedalism in *H. habilis*.

Cranial capacity, the notable increase of which identifies the taxon *Homo*, raises similar questions. Any evolutionary consideration about cranial size must make clear whether its variation is related to the size of the brain or, rather, whether it is the whole body that grows, the head included. This requires estimation of brain size relative to the size of the body, which is not easy when there are few postcranial remains. Wood and Collard used the orbital area as an indication of body size, which is available for all cases except *A. afarensis* (species which the authors referred to as *Praeanthropus africanus*, based on the criterion of temporal priority of names given to fossil species). According to Wood and Collard's calculations, both *K. rudolfensis* and *H. habilis* are closer to australopiths than to modern humans, while African *H. erectus*—that is, *H. ergaster*—is midway. The study of the masticatory apparatus revealed

Box 8.21 Oscillations of the taxon *H. habilis*

The suggestion of returning *H. habilis* to an australopithecine genus prompted an ironic comment by Phillip Tobias (spoken out loud at one of the colloquia held prior to the V World Congress of Human Paleontology,

Barcelona, November 1999) regarding the pendular oscillations of the taxon *H. habilis*, taking into account its rejection and support during the 1960s when Louis Leakey *et al.* (1964) introduced the species.

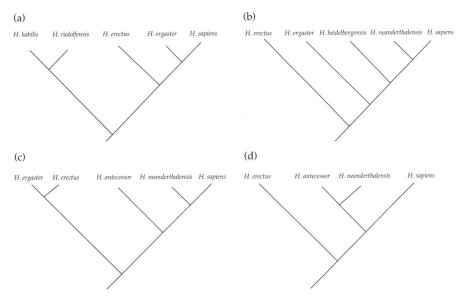

Figure 8.31 Four different cladograms depending on the phylogenetic position of *erectus*-grade hominins and their immediate ancestors. (a) Early migration from Africa, with the evolution of *H. erectus* from a taxon previous to *H. ergaster*. (b) *Homo* taxa following Wood and Collard (1999b). (c) *H. antecessor* as the ancestor of *H. neanderthalensis* and *H. sapiens*. (d) *H. antecessor* as the ancestor of *H. neanderthalensis*, but not of *H. sapiens*, and without *H. ergaster*, which would be included within *H. erectus* (see section 7.1).

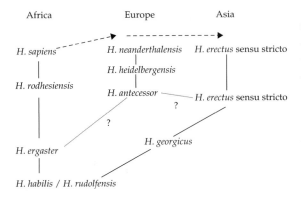

Figure 8.32 Phylogenetic tree showing the evolution of the *erectus* grade in the three continents, admitting chronospecies. It is unknown whether *H. antecessor* evolved from Asian or African specimens.

similar differences within the genus *Homo*. Wood and Collard (1999a, 1999b) concluded that if the aforementioned taxa are included in the genus *Homo*, this group becomes paraphyletic. They suggested limiting the species included in the genus *Homo* to those which are closer to current humans than to australopiths; that is to say, to take *H. habilis* and *H. rudolfensis* out of the genus. However, this solution leads to thorny alternatives. It requires either including those two taxa in the genus *Australopithecus*, making this one extremely paraphyletic, or defining a new genus for *H. habilis*, *H. rudolfensis*, and *H. platyops*, without a clear adaptive significance.

The alternatives derived from considering whether *H. habilis* and *H. rudolfensis* belong to the genus *Homo* or not, taking into account *H. antecessor* or not, and considering *H. ergaster* to be the first emigrant out of Africa or not, are reflected in four different cladograms depicted in Figure 8.31.

One more issue remains. When we explained cladistic methodology, we mentioned that the Hennigian clades do not include anagenetic species—or chronospecies—characteristic of a lineage that evolves without ramifications. The consideration of chronospecies allows accepting species like *H. antecessor* and *H. heidelbergensis* as parts of the Neanderthal lineage (Figure 8.32). A rigorous cladistic perspective would require eliminating them from the phylogenetic tree. The same goes for *Homo georgicus*, considered as an anagenetic ancestor of *H. erectus* sensu stricto.

A possible alternative, if chronospecies are not accepted, is to consider *H. neanderthalensis*, in accordance with the rules of transformed cladistics, as a daughter species of *H. erectus* sensu stricto. If this were the case, the date of separation between *H. neanderthalensis* and *H. sapiens* should be the cladogenesis event—or the date of separation between the Asian and African populations, if only one species is accepted, as we have done—between *H. ergaster* and *H. erectus* sensu stricto, given that the Neanderthal lineage would derive from the latter lineage. The possibilities of this being the correct scenario depends on how the phylogenetic transition that involves all those species is interpreted. This is one of the central themes that we will review in the following chapters.

CHAPTER 9

The late-Pleistocene transition

9.1 Archaic *Homo sapiens*

9.1.1 Hypotheses regarding the evolution towards modern humans

The evolution of *H. erectus*-grade hominins can be seen as leading, in very general terms, to anatomically modern humans. However, as is often the case in human paleontology, specialists disagree on the specific details of this phylogenetic process. They all usually agree on at least one issue, nonetheless: the existence of hominins that cannot be classified comfortably as either *H. erectus* or *H. sapiens* during the second interglacial period—between the Mindel and Riss glaciations—and maybe even earlier. These fossils have been found in Europe (Swanscombe, England; Heidelberg, Germany; Petralona, Greece), Asia (Solo, Indonesia), and Africa (Omo, Ethiopia; Kabwe, Zambia; Saldanha, South Africa). These fossils are usually conceived as so-called transitional forms. Here is where the consensus ends. There is no agreement with regard to the outcome of that transition in each particular instance. It all depends on assumptions about the appearance of our own species.

There are at least two contrary explanations for the genesis of *H. sapiens*. The first is known as the multiregional hypothesis or hybridization hypothesis (Figure 9.1). It suggests that evolutionary changes happened contemporaneously in different regions of the world. It also posits genetic exchanges between regions, so that the unity of the species would be preserved without divergence. The evolution from *erectus*-grade hominins to anatomically modern humans can be traced in the different regions; all those populations contributed to the appearance of our species.

The alternative perspective, known as the out-of-Africa hypothesis or replacement hypothesis, suggests that the transition from *H. erectus* to *H. sapiens* occurred in a fairly localized population in East Africa. *Erectus*-grade Asian hominins would not have contributed genetically to the appearance of *H. sapiens*. Asian *H. erectus* sensu stricto remained relatively unchanged on that continent, until they were displaced by modern humans, or simply disappeared, leaving their territories to be occupied by *H. sapiens*. A different species, *Homo*

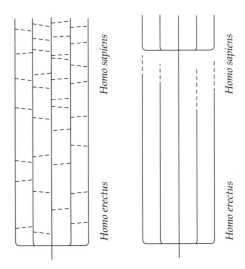

Figure 9.1 Left: representation of the multiregional-evolution hypothesis, with parallel *H. erectus*–*H. sapiens* transitions on different continents. There were continuous hybridizations that preserved the species' unity. Right: representation of the out-of-Africa hypothesis. The transition took place in Africa, beginning with a small population. In the long run, populations in other continents were substituted, without hybridization.

neanderthalensis, inhabited Europe and the Near East. Eventually, it was also displaced, or substituted, by anatomically modern humans.

9.1.2 The evolution of the *erectus* grade

The Mindel–Riss interglacial period began close to 675,000 years ago and ended about 375,000 years ago. The initial steps from the *erectus* grade towards *H. sapiens* most probably began in this time frame. These early dates and the dispersion of the remains across three continents have favored the classification of the fossils belonging to this interglacial period in very different species. For instance, European remains were assigned to *Homo heidelbergensis* (Schoetensack, 1908), *H. erectus petraloniensis* (Murrill, 1975), and *Homo swanscombensis* (Kennard, 1942). During the same period we find *Homo rodhesiensis* (Woodward, 1921) and *Homo helmei* (see Box 9.2) in Africa and *Homo* (*Javanthropus*) *soloensis* (Oppennoorth, 1932) in Asia, among other taxonomical proposals. In addition to being relatively contemporary, all these taxa exhibit certain similar morphological features. In comparison with *H. erectus*, their crania show:

• greater capacity, in most instances,
• a higher cranial vault,

• expansion of the parietal region,
• a reduction in prognathism, the frontal projection of the face.

There are, of course, differences among the diverse aforementioned taxa. Moreover, given that many of the fossils were found a long time ago and in places difficult to date with the techniques available at the time, there is a fair amount of doubt regarding their age. These circumstances account for their assignation to different species at the time of their discovery. But, as it became apparent that they shared certain common characteristics, more parsimonious solutions were suggested. The first of them, historically speaking, was to place them in *H. erectus*, noting that each case was an evolved form of *H. erectus*, but not different enough to justify a new species. This taxonomic solution is part of the single-species (*Homo erectus*) proposal put forward by Milford Wolpoff (1971a).

However, Stringer (1984, 1985) and other authors opposed the existence of *H. erectus* in Europe. The different specimens from the second interglacial period were considered as predecessors of *H. sapiens*, without further taxonomic detail. Thus, given the lack of such specification, they were designated informally as "archaic"

Box 9.1 Multiregional or out-of-Africa?

Emiliano Aguirre, Eric Trinkaus, and Milford Wolpoff, among other authors, have advocated the multiregional hypothesis. Christopher Stringer, Ian Tattersall, and Bernard Wood, for instance, are supporters of the out-of-Africa model. There is not a broad consensus about which of the two hypotheses describes the transition from *erectus*-grade hominins to our species with greater accuracy. In section 9.3 we will examine molecular evidence in favor of each hypothesis.

Box 9.2 *Homo helmei*

As Sally McBrearty and Allison Brooks (2000) noted, the taxon *Homo helmei* was introduced in an irregular way, because the species has not been described formally and its apomorphies have not been noted explicitly. Marta Mirazón Lahr and Robert Foley (1994) grouped archaic African *H. sapiens* in *H. helmei*, although European and Asian specimens were added later (Stringer, 1996; Foley and Lahr, 1997).

Box 9.3 The single-species hypothesis once more

A single species does not only involve eliminating taxa corresponding to transitional specimens. Examination of the differences between hominoids and humans led Darren Curnoe and Alan Thorne (2003) to reduce the species located in the direct line of modern humans to four, leaving a total of five in the human lineage. This meant including *H. erectus*, *H. neanderthalensis*, and all the corresponding chronospecies in *H. sapiens*.

Box 9.4 Early and late archaic *H. sapiens*

Günther Bräuer (1989) distinguished between "early archaic" *H. sapiens*, including the specimens attributed at the time to developed *H. erectus* (Bodo, Hopefield, Broken Hill) and "late archaic" *H. sapiens*, including specimens like Laetoli H 18, Omo II, Florisbad, and KNM-ES 11693 from West Turkana (Kenya). They all were, according to Bräuer (1989), previous to anatomically modern humans.

H. sapiens (Stringer, 1985; Bräuer, 1989). It must be understood that a grade was being introduced with the "archaic" proposal (McBrearty and Brooks, 2000), by means of an operation similar to the one we described when talking about *erectus*-grade hominins.

Are there other alternatives besides introducing a grade for archaic *H. sapiens*? One possibility is to include the transitional specimens in the first species that was named, *H. heidelbergensis*. *H. heidelbergensis* as a transitional taxon is meaningful only if it includes the last common ancestors of Neanderthals and modern humans (Rightmire, 1997; Ward and Stringer, 1997). Regarding the latter, there is little doubt that their direct ancestors were African. But, what about Neanderthals? Did their ancestors also come from Africa? If this were the case, then the common ancestor could belong to the time interval attributed to the different transitional specimens. But if Neanderthal ancestors were Asian—as the Gran Dolina (Atapuerca, Spain) specimens seem to suggest—the common ancestor existed before the last hominin migration to Asia.

The latter scenario leads to a taxonomic difficulty if the distinction between the Neanderthal and modern human lineages is accepted. Given that, by definition, a species can never be paraphyletic, the African specimens leading to the human lineage cannot be included in *H. heidelbergensis*, which belongs to the Neanderthal lineage under this scenario. These African specimens would constitute a different taxon for which different names have been proposed, such as *H. helmei*. However, *H. rodhesiensis* (Wood and Richmond, 2000) holds the taxonomic priority for the African taxon corresponding to archaic *H. sapiens*. The discoveries at Atapuerca cast, in any case, doubt on the validity of the European *H. heidelbergensis* taxon. We will deal with this problem in section 9.1.4, where we examine the Sima de los Huesos specimens.

9.1.3 European archaic *H. sapiens*

Let us turn to the description of the first European remains belonging to the Mindel–Riss interglacial period. The Mauer mandible was the first specimen from this period to be discovered. It is also the earliest specimen of those traditionally considered as archaic *H. sapiens*. It was discovered in 1907 by a worker at the Rösch gravel pit, close to Mauer (Germany). It is a very complete specimen, though the left premolars and the first molar are broken (Figure 9.2).

The contradiction between some robust mandibular traits, characteristic of *erectus*-grade hominins (broad ramus, absence of a chin), and relatively modern teeth (whose size and crowns are small) soon became apparent. Otto

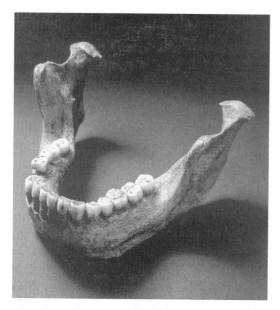

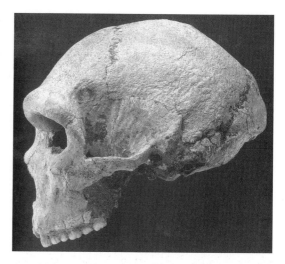

Figure 9.3 The Petralona cranium is one of the most representative specimens of the archaic *H. sapiens* grade. Photograph from Johanson and Edgar (1996).

Figure 9.2 Mauer mandible, *H. heidelbergensis.* Photograph from Johanson and Edgar (1996).

Schoetensack, the first author to describe the mandible (although Michael Day doubted that Schoetensack ever saw the fossil *in situ*; Day, 1986), defined a new species: *H. heidelbergensis* (Schoetensack, 1908). The mixture of the Mauer mandible's archaic and modern traits makes it difficult to decide whether to assign it to an evolved relative of *H. erectus* or to an incipient Neanderthal. Ernst Mayr (1963) suggested the former alternative, whereas Clark Howell (1960) favored the latter. Beause there is no direct dating, the fossil's age has been estimated from the fauna present at the site, which suggests an adaptation to mild temperatures. It is usually placed at the end of the first interglacial period, or at the beginning of the second (Day, 1986).

A cranium was discovered in exceptional circumstances in the Petralona Cave, close to Thessalonika (Greece) in 1959. It was hanging from a stalactite, with the rest of the skeleton—which was later lost—on the ground together with some primitive Mousterian stone tools (Poulianos, 1971). The Petralona cranium is among the most noteworthy fossils from the Mindel–Riss interglacial period (Figure 9.3). It is very well conserved, and it is

large, with a capacity of about 1,200 cm³. It exhibits traits reminiscent of Neanderthals (Kokkoros and Kanellis, 1960), the earliest *H. sapiens* (Stringer *et al.*, 1979), and even advanced *H. erectus* (Hemmer, 1972). It is not surprising that it is among the most "distinguished" members of the archaic *H. sapiens* grade.

The arguments about the specimen's age have been summarized by Day (1986). Given that there are no sedimentary references, its estimation had to be indirect. Based on the fauna present at the cave, it was initially assigned to the third interglacial period (Riss–Würm). Different ages have subsequently been proposed. Poulianos (1971) suggested 70,000 years. However, electronic spin resonance (ESR) applied to the stalagmites rendered an age of over 700,000 years (Poulianos, 1978), which would place the specimen in the lower Pleistocene. According to Day there is some consensus around 400,000–350,000 years, which is rejected by Poulianos.

In 1935 a cranium with most of the face and the upper molars and a premolar, corresponding to a young specimen, distorted by fossilization, was found in fluvial deposits of the Steinheim gravel pit (Germany), close to the River Murr (Figure 9.4). The sediments are attributed quite unanimously to

---

Box 9.5 Electronic spin resonance

Electronic spin resonance (ESR) is based on the measurement of the radiation from the site's radioactive isotopes. This measurement is usually carried out on the hydroxyapatite of fossil teeth. The difference between the total radiation and the one that might come from cosmic rays is calculated to estimate the internal radiation. Thermoluminescence is a similar technique, although it is usually applied to stalactites or burnt materials.

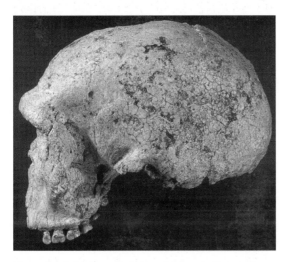

Figure 9.4 Steinheim cranium, *H. steinheimensis.* Photograph from Johanson and Edgar (1996).

the second interglacial period (Mindel–Riss). Fritz Berckhemer (1936) created a new species for the Steinheim specimen, *Homo steinheimensis.* In his reorganization of hominin taxonomy, Bernard Campbell (1964) lowered the specimen's category to subspecies, *H. sapiens steinheimensis.* Some of the Steinheim specimen's traits are shared with *H. erectus,* such as the low cranial capacity, around 1,100 cm^3, and Neanderthals, such as the very marked supraorbital torus and the wide nose. In Day's (1986) opinion, the small teeth, the elevated location of the maximum cranial width, and the shape and thickness of the cranial vault highlight the relation between *H. steinheimensis* and the British findings at Swanscombe.

An occipital bone, as well as the left and right parietals of the same cranium, were discovered in Barnfield Pit (Swanscombe, Kent, England) between 1935 and 1955, by Alvan Marston, an amateur archaeologist. These fossils were associated with a large number of Acheulean tools. The fauna and tools at the site suggested that the soils correspond to the second interglacial period: they were dated by thermoluminescence to 225,000 years (Bridgland *et al.,* 1985). The bones have a modern appearance, with thick walls and a high maximum cranial width, with an occipital protuberance.

The Swanscombe specimens were placed in a new species, *H. swanscombensis* (Kennard, 1942). A committee in charge of studying the first two bones, led by Le Gros Clark, suggested its classification as *Homo* cf. *sapiens* (Le Gros Clark *et al.,* 1938). However, Wolpoff (1971b) argued that the Swanscombe specimen's modern human-like traits are explained by the fact that it is a female exemplar. According to Wolpoff, the differences among the Swanscombe, Vértesszöllös (Hungary), Petralona (Greece), Steinheim, and Bilzingsleben (Germany) specimens are sexual dimorphisms, so that all of these specimens belong to *H. erectus.* Other authors have related the Swanscombe cranium with Neanderthals (Howell, 1960). After examining these interpretations, Day (1986) concluded that the Swanscombe specimens belong to a female transitional between *H. erectus* and *H. sapiens,* which could be placed at the base of the lateral branch leading to European Neanderthals.

The so-called "Tautavel man" refers to the 1964 discovery made by Henry de Lumley's team in the Arago Cave (close to the village of Tautavel, some 20 km from Perpignan, France). It is a partially deformed cranium (Arago XXI; Figure 9.5) including the face, zygomatic, and maxillary bones (Arago XLVII; de Lumley and de Lumley, 1971). The mandible included five teeth. Two additional partial mandibles (Arago II and XIII) were

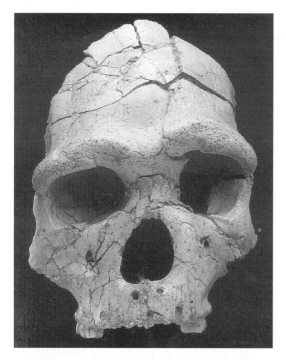

Figure 9.5 Arago XXI, Tautavel man. Photograph from Johanson and Edgar (1996).

retrieved from Tautavel (de Lumley and de Lumley, 1971). The hominin remains were associated with a great amount of tools, belonging to an early tradition, similar to Clactonian (see Chapter 10), although smaller in size. The most recent levels contain Acheulean tools, but few handaxes.

The estimated age of the Arago Cave is uncertain. By the associated fauna (micromammals), Henry and Marie-Antoinette de Lumley (1973) assigned it to the beginning of the Riss glaciation. de Lumley (1979) later suggested it might be older, possibly from the Mindel glaciation, some 400,000 years ago. The morphology of the Tautavel specimens is intermediate, exhibiting Neanderthal-like and *H. erectus*-like traits (de Lumley and de Lumley, 1973). Its similarity to the Steinheim cranium is striking, although its brain capacity is larger (1,100–1,200 cm^3). However, the vault appears to be lower and longer, like Asian *H. erectus*. The femoral fragments (Arago XLVIII, LI, and LIII) and the remains of the left hip (Arago XLIV) are reminiscent of the Olduvai *H. erectus* OH

28 (Day, 1982). Day (1986) has argued that, again, this was a European transitional form between *H. erectus* and *H. sapiens*.

9.1.4 The Sima de los Huesos site at Atapuerca

The Sima de los Huesos site at Atapuerca has yielded numerous hominin remains that are about 300,000 years old. The pit, with a 13-m vertical shaft within the Cueva Mayor, ending in a sloping, 15-m tunnel, is very difficult to access. Many hominin fossils have been found at Sima, together with fossils of many other mammals. Retrieving them requires extenuating work and the use of speleological techniques. Even so, more than a ton and a half of reddish clay has been removed, without defined stratification levels, which included hominin remains, often jumbled by amateur geologists who had been there earlier. After sieving and selecting, the remains of at least 27 different individuals appeared, including every part of the human skeleton (Arsuaga, 1994; Aguirre, 1995). Gamma-ray measurements from uranium isotopes suggest that the Sima de los Huesos mandibles are 300,000 years old (Aguirre, 1995). Faunal comparison with large mammals estimates the site's age to between 525,000 and 340,000 years (Aguirre, 1995).

Sima de los Huesos has yielded an enormous number of specimens. Postcranial remains found at this site amount to 70% of all known middle-Pleistocene specimens (Arsuaga, 1994). Furthermore, they correspond to geographically and temporally bound individuals, who possibly could have been members of the same family, which has encouraged paleodemographic studies that reveal the distribution of sizes, sexual dimorphisms, and polymorphisms existing in a single population (Bermúdez de Castro, 1995). An example is the comparative analysis of the two segments of the dental arcade, the anterior (incisors and canines) and the posterior (premolars and molars), which led Bermúdez de Castro and Nicolas (1996) to argue that the Asian middle-Pleistocene *H. erectus* sample is easily distinguishable from the European one, which exhibits more similarities with the African sample.

Box 9.6 Reasons for the abundance of specimens at Sima de los Huesos

All the remains from Sima de los Huesos belong to young adults, the populational segment of least expected mortality. The explanation for the presence of so many hominin remains in such an inaccessible place is still tentative. The first taphonomic interpretation ventured was that they might have been dragged there by floods. However, some members of the Atapuerca research team have later noted that it might be due to some kind of intentional burial (see http://www.pagina12.com.ar/2001/suple/Futuro/01–04/01–04–28/nota_a.htm).

Atapuerca has provided new indications regarding the way in which *erectus*-grade hominins evolved into subsequent species. It was clear from the beginning that the Sima de los Huesos specimens could not be classified within *H. erectus* (Aguirre and de Lumley, 1977; Aguirre *et al.*, 1980). These specimens were more evolved. The available taxonomic alternatives for the European specimens of that age were the archaic *H. sapiens* grade or the species *H. heidelbergensis*.

Three crania were found at Sima de los Huesos in July 1992: a skulltop (cranium 4), a virtually complete cranium (cranium 5; Figure 9.6), and a more fragmented infantile one (cranium 6; Arsuaga *et al.*, 1993). These specimens have high cranial capacities, although they are not all the same: the volume of cranium 4 is 1,390 cm^3 and that of cranium 5 is 1,125 cm^3. Arsuaga and colleagues (1993) noted that the Sima's sample lacks the apomorphies of *H. erectus* sensu stricto. The supraorbital arch and other cranial traits suggest the Sima de los Huesos specimens and those from Bilzingsleben, Steinheim, and Petralona are close to one another. Consequently Arsuaga and colleagues (1993) placed the Sima de los Huesos specimens within archaic *H. sapiens*.

But such a classification only indicates they belong to a grade. What about the species? If *H. erectus* is discarded, could the Sima specimens be considered as *H. heidelbergensis*? That would have been a reasonable decision. However, because so many specimens have been retrieved from the Sima de los Huesos, the variability of their mandibular and cranial traits is very high. Such variability covers part of the distance between many *H. erectus* sensu lato and Neanderthal traits. This is what we call the abundance paradox. Based on the facial projection and the incipient suprainiac fossa of

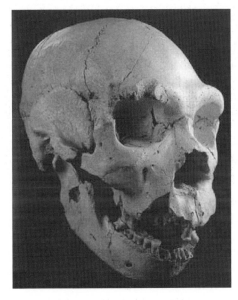

Figure 9.6 Cranium 5 from Sima de los Huesos (Atapuerca, Spain), archaic *H. sapiens*. Photograph from Arsuaga *et al.* (1993). Reprinted by permission from Macmillan Publishers Ltd. *Nature* 362: 6420, 534–537, 1993.

crania 4, 5, and 6, Stringer (1993) suggested the fossils belong to the Neanderthal taxon. But other traits, such as the cranial capacity, quite low in comparison to that of *H. neanderthalensis*, do not allow their straightforward classification as such. In any case, Stringer (1993) argued that the great variation of the Sima de los Huesos sample raises doubts about the validity of the taxon *H. heidelbergensis*. The Atapuerca sample brings the overlap between this species and Neanderthals to light.

The earliest Atapuerca specimens, those from Gran Dolina, *Homo antecessor*, show a modern midfacial morphology, reminiscent of the most advanced *H. erectus* from Dali (China) and

Box 9.7 The abundance paradox

The fossil remains from Atapuerca raise what we will call the abundance paradox, one of the difficulties current human systematics has to face. Different taxa are defined easily if there are few specimens from different periods. A significant difference in mandibular size or robusticity, or a higher cranial volume, for instance, would justify the proposal of a new species. But as specimens continue to appear, the distinction becomes problematic. For instance, if we began with two taxa, A and B, and new specimens whose morphology is intermediate are subsequently discovered, to which of the two taxa should they be ascribed? Should a new species be introduced? The abundance of specimens broadens any trait's degree of variability, transforming the initial distance between A and B into a set of small variations along a continuous scale.

The same can be said about the temporal distribution. If A and B are considered chronospecies separated by a certain time interval and new specimens of intermediate age appear, establishing a frontier between both becomes difficult. Ultimately, if a large number of specimens were available, reflecting the population's characteristics at each time, it would be unreasonable to seek to determine in which generation the species transition occurred. Speciation processes are not instantaneous, not even in cladogenesis. Regarding anagenetic evolution, it is notable that chronospecies are taxonomic artifices to name populations separated in time. As we noted in section 1.3, the cladistic criteria for distinguishing biological species cannot be applied to chronospecies.

Florisbad (South Africa) and early modern humans from Skuhl and Qafzeh (Bermúdez de Castro *et al.*, 1997). But, the crania and mandibles from the Sima de los Huesos are akin to European specimens (Arago, Swanscombe) and African ones (Bodo, OH 22 and OH 24 from Olduvai Bed IV; Arsuaga *et al.*, 1993). Overall, these similarities would support the hypothesis of a multiregional evolution of *H. erectus* to *H. sapiens*.

Atapuerca's abundance paradox does not end there. Carbonell and colleagues (2005) associated the ATD6-96 specimens from Gran Dolina, classified as *H. antecessor*, with *H. erectus* sensu stricto. An explanation for that morphological proximity would be the arrival at Atapuerca of a human population derived from Asia. Archaeological evidence also supports this hypothesis, because the tools found in the Aurora stratum of Gran Dolina belong to Mode 1, whereas by that time African *erectus*-grade hominins had been manufacturing Mode 2 tools for more than half a million years. However, the Sima de los Huesos specimens show clear affinities with archaic *H. sapiens*, and even with Neanderthals. Thus, it seems that Atapuerca permits establishing a phylogenetic sequence *Homo erectus* sensu stricto → *H. antecessor* → *H. heidelbergensis* → *H. neanderthalensis*, though, according to Stringer (1993), the taxon

H. heidelbergensis should be eliminated from the sequence. Carbonell *et al.* (2005) rejected this sequence in the presentation of ATD6-96: "the present evidence does not support the hypothesis of a phylogenetic relationship between the TD6 hominins and the European lineage leading to the Neanderthals."

Thus, there are very serious difficulties to overcome to understand the evolutionary history told by Atapuerca. If *H. antecessor* is not part of the lineage leading to Neanderthals, then there cannot be an evolutionary continuity between the Gran Dolina and the Sima de los Huesos specimens, which weakens the support of Atapuerca for the multiregional hypothesis. The hominins from Gran Dolina and Sima de los Huesos would correspond to two different European colonizations, with no direct relation between them.

Once *H. antecessor* has been removed from the Neanderthal lineage, its affinities with modern humans, suggested by ATD6-96, can be explored. These similarities have been used to suggest that this is an ancestral species of *H. sapiens* (Carbonell *et al.*, 2005). However, the cradle of the modern human lineage is in Africa (see Chapters 10 and 11) but no *H. antecessor* specimens have been found in Africa. Moreover, the possible Asian origin of the Gran Dolina specimens casts serious doubts on their inclusion in the lineage leading to modern humans. It is not surprising, thus, that Carbonell *et al.* (2005) wrote "that more information on the Gran Dolina and other contemporaneous hominins will be necessary before revising the phylogenetic position of *H. antecessor.*"

The significance of *H. heidelbergensis* is not easy to interpret, either, in the light of the Sima de los Huesos sample. If this taxon were eliminated, then *H. neanderthalensis* would span a very long span of time throughout which notable morphological changes would have taken place. According to David Dean *et al.* (1998), there are up to four stages within the anagenetic evolution of Neanderthals (Table 9.1), although the purpose of these authors was not taxonomical. The question of how many chronospecies should be included in the Neanderthal lineage to account for these stages remains unclear. In a later chapter we will comment on some indications that allow distinguishing *H. heidelbergensis* from *H. neanderthalensis.*

Table 9.1 Stages within the anagenetic evolution of Neanderthals

Stage	Specimens
Neanderthal 1 (early pre-Neanderthals)	Arago, Mauer, Petralona
Neanderthal 2 (pre-Neanderthals)	Bilzingsleben (similar in part to those from stage 1), Vértesszöllös (similar in part to those from stage 1), Atapuerca (Sima de los Huesos), Swanscombe, Steinheim, Reilingen
Neanderthal 3 (early Neanderthals)	Ehringsdorf, Biache (1), La Chaise Suard, Lazaret, La Chaise Bourgeois-Delaunay, Saccopastore, Krapina (most of them; not completely clear), Shanidar (some of them; not completely clear)
Neanderthal 4 (classic Neanderthals)	Neanderthal, Spy, Monte Circeo, Gibraltar Forbes Quarry (not completely clear), La Chapelle-aux-Saints, La Quina, La Ferrassie, Moustier, Shanidar, Amud

Source: Modified from Dean *et al.* (1998).

9.1.5 African specimens from the Mindel–Riss interglacial period

The European specimens we have reviewed, which are only part of those available, highlight the rather ambiguous status of the hominins from the Mindel–Riss interglacial period. They share traits with *erectus*-grade hominins, Neanderthals, and even anatomically modern humans. But there are African remains of similar age and appearance that introduce additional complexities.

Whereas *H. heidelbergensis* is considered the European species that corresponds to archaic *H. sapiens*, its African equivalent is *H. rodhesiensis*, which is the taxon grouping African *H. erectus–H. sapiens* transitional specimens. This scheme is not free of problems. Day (1973) suggested separating the middle- and late-Pleistocene African specimens in three different grades: early, intermediate, and modern. Following this proposal, McBrearty and Brooks (2000), carried out an extensive classification of African specimens in each of these groups (Figure 9.7). McBrearty and Brooks place *H. rhodesiensis* in group 1, together with specimens such as OH 9 from Olduvai, prior to the true transition to *H. sapiens*.

African middle- and late-Pleistocene hominin systematics is highly complex and does not allow such a straightforward identification of the transitional forms as in Europe. The paucity of specimens dated close to 300,000 years ago is an important problem, given that these would be the best exemplars to illustrate the process of transition. However, focusing on the Mindel–Riss interglacial period and the African transition to *H. sapiens*, there are some especially significant specimens. These can be seen as representing the notion of modern human ancestors.

In 1957, the Kenyan team of the International Paleontological Research Expedition found a partial skeleton (Omo I), a cranium (Omo II), and some fragments of another cranium in Member I of the Kibish Formation, at Omo (Ethiopia). The age of the Kibish Formation is rather uncertain: it could range from 3,100 to 130,000 years (Butzer, 1971). The Omo I specimen, the most complete of the set, has a rounded cranial vault and an overall

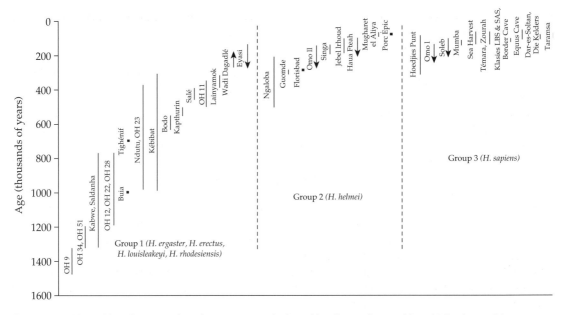

Figure 9.7 Middle- and late-Pleistocene African hominin groups 1 (early grade), 2 (intermediate grade), and 3 (modern grade). Vertical lines indicate time span; the arrows refer to possible displacements due to uncertainty. From McBrearty and Brooks (2000).

appearance similar to modern humans, although with rather robust teeth. Day (1969, 1972) classified it as *H. sapiens*, although maybe of a slightly archaic kind. All the specimens from the Kibish Formation, assumed to be the same age, were equally classified. But the Omo II cranium has traits that are reminiscent of *H. erectus*. Day and Stringer (1982) reinterpreted the Omo sample, keeping the classification of *H. sapiens* for Omo I but assigning Omo II to *H. erectus*. Following a similar criterion, McBrearty and Brooks (2000) placed Omo I in group 3 (modern) and Omo II in group 2 (intermediate, although McBrearty and Brooks placed the specimens most similar to *H. erectus* in group 1). Such a classification suggests that not all the Omo specimens are of the same age, but it provides an interesting documentation of the process of evolution *in situ* from *H. erectus* to *H. sapiens*.

During the 1976 campaign, the Rift Valley Research Mission led by Jon Kalb found a cranium at Bodo (Middle Awash, Ethiopia), popularly known as Bodo man (Conroy *et al.*, 1978; Figure 9.8). A second specimen, a left parietal

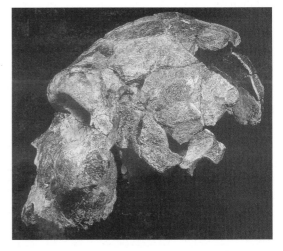

Figure 9.8 Bodo man. Photograph from Johanson and Edgar (1996).

fragment, BOD-VP-1/1, was found in 1981, 350 m from where the first one had been discovered (Asfaw, 1983). Another specimen, the BOD-VP-1/2 humeral fragment, was discovered in 1990 (Clark *et al.*, 1994). The age of Bodo man was initially estimated at 350,000 years. However, with the $^{40}Ar/^{39}Ar$ method, Clark and colleagues (1994) estimated a volcanic tuff located in unit u to 0.64 ± 0.04 Ma. The specimens were found in a different unit, t, but because of the correlation between the sediments, Clark *et al.* (1994) considered Bodo man to be that same age, close to the lower limit of the Mindel–Riss interglacial period.

Moving to southern Africa, the Kabwe fossils include remains of three or four individuals, with a well-conserved cranium, a parietal, a maxilla, and several postcranial remains. They were found in the Broken Hill mine, Kabwe (Zambia), and were attributed to the new species *H. rhodesiensis* (Woodward, 1921; at the time Zambia was part of Rhodesia). In addition to abundant fauna, some lithic and bone tools, belonging to the African tradition known as Still Bay points, were also found in the mine. It is a characteristic culture of the middle Paleolithic, developed from Acheulean techniques. The age of the Kabwe specimens was estimated, by the study of the fauna and culture, to about 40,000 years, the late Pleistocene. The age of the specimens has since been increased. Day (1986) suggested that they belong to the end of the middle Pleistocene. McBrearty and Brooks (2000) believed that the Kabwe specimens are between 0.78 and 1.33 million years old, following the comparison of the site's fauna with that of Olduvai Bed IV.

The dental and postcranial morphology of the Kabwe specimens is modern, to the extent that several authors have argued that it is a subspecies of *H. sapiens* (Campbell, 1964; Rightmire, 1976; Kennedy, 1984). Both the Kabwe and Saldanha (southern Africa) specimens, which are very similar, share morphological similarities with European hominins from the second interglacial period, although McBrearty and Brooks (2000)

placed the Kabwe and Saldanha specimens in group 1 (early grade).

9.1.6 Asian specimens from the Mindel–Riss interglacial period

There are specimens in Java and China that show similar features to the European and African specimens from the second interglacial period. In the early 1930s the Ngandong site, in the River Solo valley, yielded twelve cranial remains (Ngandong 1–12; Figure 9.9), the most complete of which is Ngandong 7, and two tibias. The exemplars were attributed by Oppennoorth (1932) to the species *H. (Javanthropus) soloensis*, and are popularly known as Solo man. Between 1960 and 1980 Teuku Jacob carried out excavations at Ngandong and other Javanese sites. As we saw in section 7.1, the uranium-series and ESR techniques yield estimates

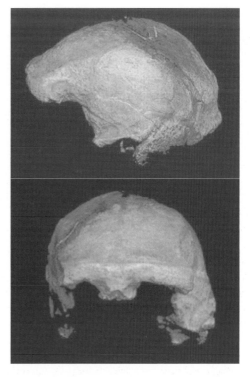

Figure 9.9 Ngandong 6 specimen, *H. soloensis*. Top: lateral view; bottom: frontal view. Photographs from http://www.mnh.si.edu/anthro/humanorigins/ha/erec.html.

Box 9.8 Transitional forms from Olduvai

Olduvai has also provided specimens prior to the second interglacial period that could be considered transitional forms between *H. erectus* and *H. sapiens*, such as OH 22 (Bed IV) and OH 23 (Masek).

for the Ngandong sediments of between 53,000 and 27,000 years (Swisher *et al.*, 1996).

The large cranial capacity of some Ngandong specimens has led some to argue that they are *H. erectus–H. sapiens* transitional specimens. This hypothesis is supported by the late ages of the sedimentary deposits. However, similar to other transitional cases, there are alternative points of view. The initial idea of a new species was followed by the attribution of the specimens to a Neanderthal variant (Vallois, 1935; von Koenigswald, 1949). Weidenreich's (1933) detailed examination rejected such an interpretation, and suggested that the Solo specimens were close to the *erectus* grade, although he did not propose any specific classification. After a thorough study, A.P. Santa Luca (1980) argued that the Ngandong specimens are similar to those from Trinil and Sangiran and should be classified as *H. erectus*. The differences in size and brain capacity would be due to sexual dimorphism (Ngandong 6, considerably larger, would belong to a male, and the smaller Ngandong 7 to a female).

The re-examination of the ages attributed to the Ngandong and Sambungmachan sites led Swisher and colleagues (1996) to consider the possibility of the coexistence of *H. erectus* and *H. sapiens* in Java. Using ESR and uranium-series mass spectrometry dating techniques applied on fossil bovids from levels that had yielded hominins, Swisher and colleagues (1996) estimated the ages of those sites to be between 27,000 and 53,000 years. According to Swisher *et al.*, this means that *H. erectus* survived in Java at least 250,000 years after its extinction in China. Thus, they must have overlapped with modern humans in southeast Asia in a similar way to Neanderthals and modern

humans in Europe. Swisher *et al.* (1996) suggested that *H. erectus* and *H. sapiens* might have exchanged genes.

With regard to China, we mentioned in section 7.1 that some of the remains from Hexian and Yunxian exhibited advanced traits that moved them close to modern hominins. The specimen found in Jinniu Shan, province of Liaoning, northeast China, is similar. It consists of a modern-looking cranium and numerous postcranial remains. Although the specimen was discovered in 1984, its age was estimated later, by ESR methods, to 200,000 years (Chen *et al.*, 1993). This age opens the possibility that *H. sapiens* (represented by the Jinnui Shan specimen) and *H. erectus* (Zhoukoudian cranium V and the Hexian cranium) coexisted in China (Conroy, 1997). This situation would be similar to the one Swisher and colleagues (1996) suggested for Java. However, the classification of the Jinniu Shan specimen as *H. sapiens* is not generally accepted. Poirier (1987), for instance, included it in *H. erectus*.

The Dali cranium (Figure 9.10), which lacks the lower part of the face, was found in a gravel pit near the city of Jiefang (Shaanxi Province, China; Wang *et al.*, 1979). Its initial description suggested that it was another transitional form, but it was classified as *Homo erectus*. However, Xinzhi Wu (1981) emphasized its advanced traits and placed it as a subspecies of *H. sapiens*, *H. sapiens daliensis*. The specimen's marked supraorbital arches do not form a continuous line, as they do in *H. erectus* sensu stricto.

9.1.7 The Flores enigma

In October 2004 *Nature* published a finding made by a team of the Indonesian Centre for

---

Box 9.9 The ages of Javanese sites

In section 7.1 we mentioned the difficulties involved in estimating the ages of the sites on Java. The possible damage to the specimens hampers direct dating. This is why Swisher's team used bovid bones instead of hominin ones, which were not available for electronic resonance studies. But a French research team has managed to use non-damaging experimental techniques to date the Javanese transitional specimens at close to 300,000 years, although questions have been raised concerning this estimate (Gibbons, 1996).

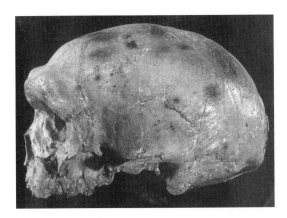

Figure 9.10 A specimen from Dali (China), *H. erectus* (Wang *et al.*, 1979); *H. sapiens daliensis* (Wu, 1981). Photograph from Johanson and Edgar (1996).

Archaeology in Jakarta, led by Mike Morwood and R.P. Soejono. A miniscule hominin specimen, LB 1 (Brown *et al.*, 2004) was found in the Ling Bua site of the island of Flores (Indonesia), close to Java (Figure 9.11). Its stature and endocranial volume were close to 1 m and 380 cm^3, respectively, similar in size to the smallest australopithecines.

Even if it corresponds to a female, LB 1 contrasts sharply with common understanding of human evolution. Despite the many controversies reviewed in previous chapters, specialists agree that after *H. habilis*, the process of human evolution led to larger beings with larger brains. The increasing complexity of lithic instruments was attributed to increasing brain size. Suddenly the LB 1 specimen appears, similar in body mass and cranial volume to the smallest Pliocene hominins, but in late-Pleistocene terrains, almost belonging to our time.

The Ling Bua site is a dolomitic cave located in the Wae Racang valley (Figure 9.12). The first excavations at the cave began in 1965, led by the priest Theodor Verhoeven, while R.P. Soejono excavated 10 sectors between 1978 and 1989 (Morwood *et al.*, 2004). In 2001 Mike Morwood's team began the excavation of sectors I, III, IV, and VII.

The age of LB 1 was estimated to 18,000 years by the ^{14}C method applied to two samples (18,700/ 17,900 and 18,200/17,400 years). Sector VII, where the specimen appeared, was dated by thermoluminescence to between $35,000 \pm 4,000$ and $14,000 \pm 2,000$ years. These estimates imply that 18,000 years ago there were hominins with body sizes and cranial volumes comparable to 4 million-year-old australopiths. The time is so recent that LB 1 can only be considered a fossil in a broad sense. The skeleton has not yet undergone any fossilization.

Other specimens found in Ling Bua sectors IV, VII, and IX in 2004 complete the currently available hypodigm of *Homo floresiensis* (Morwood *et al.*, 2005). The new specimens, from at least nine individuals (Table 9.2), seem to exclude the possibility that LB 1 might represent a pathological individual (Lieberman, 2005). Moreover, according to Morwood and colleagues (2005), "*H. floresiensis* is not just an allometrically scaled-down version of *H. erectus*. Other *H. floresiensis* morphological traits, for example in the humeral torsion and ulna, are not shared with any other known hominin species" (Morwood *et al.*, 2005).

What kind of hominin is the Flores hobbit, as the popular press christened it? Figure 9.13 shows the cranium, femur, and tibia of LB 1. Brown *et al.* (2004) note their morphological similarity to *H. erectus*, if size is not taken into account. The indices of cranial shape follow closely the pattern in *H. erectus*. Viewed from behind, the parietal contour is also similar to *H. erectus* but with reduced cranial height. The cranial base angle is relatively flexed in comparison with both *H. sapiens* and Sambungmacan *H. erectus*. The overall facial morphology of LB1 is similar to that of members of the genus *Homo*. The masticatory apparatus lacks most of the characteristic adaptations of *Australopithecus*. The overall femoral anatomy is most consistent with the broad range of variation in *H. sapiens*, although it exhibits some departures that "may be the result of the allometric effects of very small body size" (Brown *et al.*, 2004). Brown *et al.* (2004) named the new species *H. floresiensis*, with LB 1 as the holotype and a single referred material: the LB 2 isolated left mandibular P3. Other specimens have been found subsequently (see Table 9.2).

To sum up, except for size, LB 1 is, according to Brown *et al.* (2004), very closely related to *H. erectus* and completely different from *Australopithecus*, the taxon with a comparable body size. What is the

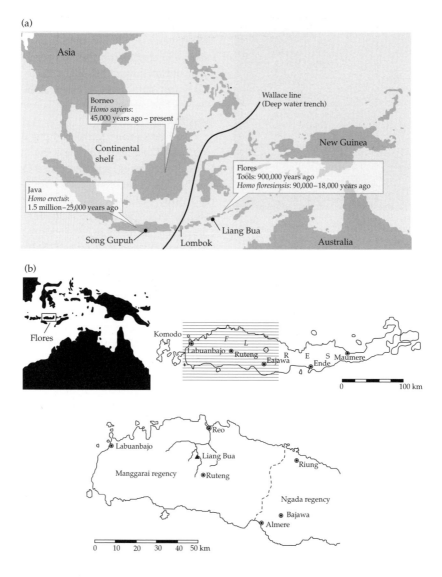

Figure 9.11 Location of the island of Flores (Indonesia) and the Ling Bua site. Top: picture from Dalton (2005); Reprinted by permission from Macmillan Publishing Ltd. *Nature* 434: 7032, 432–434, 2005. Bottom: pictures from Morwood *et al.* (2004). Reprinted by permission from Macmillan Publishing Ltd. *Nature* 431: 7012, 1087–1091, 2004.

explanation for this strange condition? Is it the result of some kind of achondroplasia? A pathological condition—microcephaly—was not at all discarded by Weber *et al.* (2005) after examining the reconstruction of the brain of LB 1 by Falk *et al.* (2005; see below). However, the view of Brown *et al.* (2004) was categorical: the specimen's morphology is not due to any pathology. "Dwarfing in

LB1 may have been the end product of selection for small body size in a low calorific environment, either after isolation on Flores, or another insular environment in southeastern Asia" (Brown *et al.*, 2004).

However, both Maciej Henneberg and Teuku Jacob suggested another possibility: that the Flores specimen corresponded to a population of microcephalic, pygmylike modern humans rather than

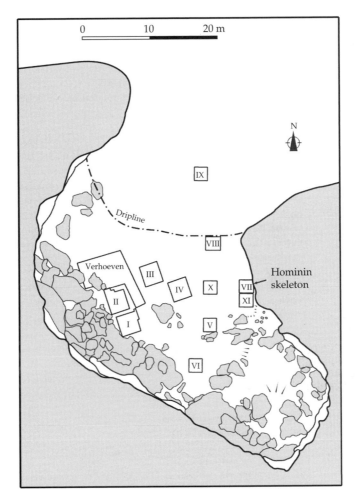

Figure 9.12 Sediments from the Ling Bua Lake. Picture from Morwood *et al*. (2004). Reprinted by permission from Macmillan Publishers Ltd. *Nature* 431: 7012, 1087–1091, 2004.

to a new species (cited by Balter, 2004a, 2004b). In fact, Henneberg related LB 1 with a 4,000-year-old microcephalic modern human skull found on the island of Crete, although he did so in a letter to the Adelaide *Sunday Mail* (Balter, 2004b). However, in their commentaries to the finding, Marta Mirazón Lahr and Robert Foley (2004) and Jared Diamond (2004) remarked on the greater proximity of LB 1 to *H. erectus* than to any *H. sapiens*.

A recent investigation of the Flores sample by Jacob *et al*. (2006) explores again the question whether the taxon *Homo floresiensis* should be rejected. They assert that "LB1 is drawn from an

earlier pygmy *H. sapiens* population but individually shows signs of a developmental abnormality, including microcephaly." Jacob *et al*. (2006) endorse placing the Ling Bua exemplars in *H. sapiens* because, "Anomalies aside, 140 cranial features place LB1 within modern human ranges of variation, resembling Australomelanesian populations. Mandibular and dental features of LB1 and LB6/1 either show no substantial deviation from modern *Homo sapiens* or share features (receding chins and rotated premolars) with Rampasasa pygmies now living near Liang Bua Cave." However, Argue *et al*. (2006) have explored the affinities of LB1 using

Table 9.2 *Homo floresiensis* specimens found at Ling Bua in 2004

Bone	Description
Radius: child, left (LB4/1)	Proximal epiphyses unfused, articular surfaces not recovered, and distal quarter of shaft incomplete. Maximum length of fragment 101 mm.
Tibia: child, right (LB4/2)	Distal and proximal epiphyses unfused and articular surfaces not recovered. Maximum length 117 mm. Distal end recovered from Spit 44.
Cervical vertebra C1 (LB5/1)	Incomplete, represented by two fragments.
Metacarpal: adult (LB5/2)	Proximal end broken, length 58 mm.
Mandible: adult (LB6/1)	With incomplete left ramus and right coronoid process and condyle. Originally with a fracture through the corpus between right P_3 and P_4 and left M_1 and M_2. Subsequently broken at the sysphysis when removed from the Centre for Archaeology in Jakarta. This has altered the original arch dimensions occlusion and the morphology of the symphysis. At the same time, cut marks, fill and gule altered the morphology of the lateral corpus and ramus.
Radius: adult, right (LB6/2)	Complete. It has an angulated, healed fracture in the distal third with compensatory remodeling and extensive callus development. The forearm would have been bowed and distorted and movement of the hand restricted. Maximum length 157 mm.
Ulna: adult, left (LB6/3)	Proximal shaft, Maximum length 137 mm.
Scapula: adult, right (LB6/4)	With incomplete superior border, medial spine, inferior angle and coracoid process. Maximum breadth of glenoid cavity 25.1 mm, length of the auxiliary border 82 mm.
Metatarsal (LB6/5)	Articular surfaces not preserved.
Phalanx: 1st of foot (LB6/6)	Articular surfaces not preserved.
Phalanx: 3rd of hand (LB6/7)	Complete. Maximum length 10 mm.
Phalanx: 1st of the hand (LB6/8)	Complete. Maximum length 30.5 mm.
Phalanx: 2nd of the hand (LB6/9)	Complete. Maximum length 16 mm.
Phalanx: 2nd of the hand (LB6/10)	Distal end not preserved.
Phalanx: 1st of the hand (LB6/11)	Complete. Maximum length 10.5 mm.
Phalanx: 3rd of the hand (LB6/12)	Complete. Maximum length 12.5 mm.
Phalanx: 1st of the foot (LB6/13)	Complete. Maximum length 16 mm.
Incisor: mandibular I_1 (LB6/14)	
Phalanx: 1st of hand (LB7)	Proximal end not preserved.
Humerus: adult, right (LB1)	Lateral epicondyle and capitulum are incomplete, and the greater and lesser tubercles are not preserved. Maximum length 243 mm.
Ulna: adult, right (LB1)	Distal end of shaft and head preserved. Estimated maximum length 205 mm.
Fibula: adult, left (LB1)	Complete but with fracture through distal end of shaft. Maximum length 226 mm. This pairs with the right fibula of LB1 recovered from Sector VII in 2003 and is the same length.
Tibia: adult, right (LB8)	Medial condyle incomplete and medial malleolus not preserved. Estimated maximum length 216 mm. This tibia duplicates that found with the skeleton and is smaller. It is therefore from another individual. Reassembled when removed from the Centre for Archaeology in Jakarta. Now has altered morphology and adhering glue.
Ulna: adult, left (LB1)	Both epiphyses missing. 167×16.5 mm. (Note: from baulk collapse).
Femur (LB9)	Fragment. Shaft spilt longitudinally. 91×17 mm.

Source: Morwood *et al.* (2005).

Box 9.10 *Homo floresiensis*: microcephalic?

Dean Falk *et al.* (2007) have made a virtual comparison—by three-dimensional computer tomographic reconstruction of the internal brain case—between 9 microcephalic humans and 10 normal humans. They have identified two traits that distinguished between them with 100 percent accuracy. When these are applied to the virtual endocast from LB1, this has the features of normal humans rather than those of microcephalics.

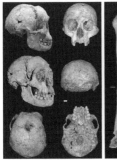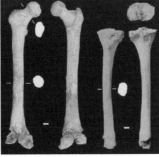

Figure 9.13 LB 1 cranium (left) and femur and tibia (right), *H. floresiensis* (not on the same scales). Photographs from Brown *et al.* (2004). Bar = 1 cm. Reprinted by permission from Macmillan Publishers Ltd. *Nature* 431: 7012, 1055–1061, 2004.

cranial and postcranial metric and non-metric analyses, by comparing it to early *Homo*, two microcephalic humans, a "pygmoid" excavated from another cave on Flores, *H. sapiens* (including African pygmies and Andaman Islanders), *Australopithecus*, and *Paranthropus*. They affirm, contrary to Jacob *et al.* (2006) that "it is unlikely that LB1 is a microcephalic human, and it cannot be attributed to any known species. Its attribution to a new species, *Homo floresiensis*, is supported."

9.1.8 The culture of Flores

The size of *H. floresiensis* is not exceptional at Flores. Other mammals on the island experienced an identical adaptive process of size reduction. The dwarf elephant, *Stegodon florensis*, is particularly striking, with the added interest that *H. floresiensis* might have hunted it. The kind of stone tools used by *H. florensiensis* is another controversial issue. Two different kinds of tool and two different kinds of hominin are found at Ling Bua. Simple flaked stone artifacts have been found in older deposits— 95,000–74,000 years old—at Ling Bua (Morwood *et al.*, 2005). Other Flores early- and middle-Pleistocene sites have also provided primitive lithic instruments. But Ling Bua contains, additionally, "points, perforators, blades and microblades that were probably hafted as barbs" (Morwood *et al.*, 2004; Figure 9.14).

The scarcity of tools in sector VII—32 were found in the same level as the hominin skeleton—contrasts with the abundance in sector IV. Up to 5,500 artifacts/m^3 were found associated with *H. floresiensis* specimens. The ones found close to *Stegodon* remains are the most advanced among those included in Figure 9.14.

With regard to hominins, the *H. floresiensis* specimens were recovered at sectors IV, VII, and IX. But modern humans, *H. sapiens*, have appeared in Ling Bua sector XI. It is always difficult, as we have already mentioned, to relate tools and specimens. However, Morwood *et al.* (2004) believe that "The chronologies for Sectors IV and VII show that *H. floresiensis* was at the site from before 38 kyr [thousand years] until at least 18 kyr—long after the 55 to 35 kyr time of arrival of *H. sapiens* in the region. None of the hominin remains found in the Pleistocene deposits, however, could be attributed to *H. sapiens*. In the absence of such evidence, we conclude that *H. floresiensis* made the associated stone artifacts."

Several researchers (Maringer and Verhoeven, 1970; Sondaar *et al.*, 1994; Morwood *et al.*, 1998; O'Sullivan *et al.*, 2001; Brumm *et al.*, 2006) have a different interpretation of the Flores tools. Several deposits in the center of the island, such as Mata Menge, Boa Lesa, and Kobatuwa, have stone tools associated with a diversity of fossils, which include the Komodo dragon, rat, and the dwarf elephant *S. florensis*. The age of the sediments where the stone tools are found is 840,000– 700,000 years ago (Brumm *at al*, 2006). The excavations by Brumm *et al.* (2006) in Mata Menge in 2004–2005 yielded a total of 487 stone artifacts *in situ*. The stone-artifact technology is simple, based on the removal of small- to medium-sized flakes from cobbles and flake blanks (see Figure 9.15).

The most primitive-looking tools associated with *H. floresiensis* at Liang Bua and those found in Mata Menge are, according to Brumm *et al.* (2006), remarkably similar, even though separated by nearly 700,000 years, which makes difficult the interpretation of the hominin presence in Flores. According to Brumm *et al.* (2006), "We still do not know the species identity of the Mata Menge knappers, as no associated hominin remains have been recovered so far, but the age of the site clearly

Box 9.11 What species is LB 1?

Jacob *et al.*'s (2006) interpretation is opposite to that of Lahr and Foley (2004), Diamond (2004), and Argue *et al.* (2006) with respect to the microcephaly of LB 1 and also its relationship to *H. erectus* and *H. sapiens*. Although the fossil is fairly young, it seems likely that the analysis of its mitochondrial DNA (see section 9.3) would be difficult owing to the decay of the fossils in the hot and humid climate of the island of Flores. If it were accomplished, it would help decide whether the new taxon *H. floresiensis* is appropriate or, rather whether the Ling Bua population is one of pygmoid Australomelanesian modern humans. For now, we will retain the name *H. floresiensis* for these exemplars.

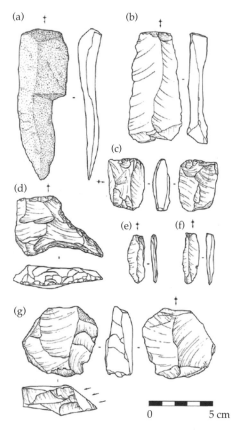

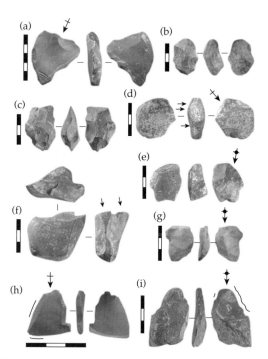

Figure 9.14 Tools from sector IV from Ling Bua (Flores, Indonesia) associated with *H. floresiensis* and *Stegodon*. (a, b) Macroblades; (c) bipolar core; (d) perforator; (e, f) microblades; (g) burin core for producing microblades. Arrows indicate striking positions, where knappers detached the flakes from cores by direct percussion using hammerstones. Picture from Morwood *et al.* (2004). Reprinted by permission from Macmillan Publishers Ltd. *Nature* 431: 7012, 1087–1091, 2004.

Figure 9.15 Stone artifacts from Mata Menge. (a, b) Chert bifacial radial cores; a is made on a flake; (c, d) volcanic/metavolcanic bifacial radial cores; d is made on a flake and features three truncation scars (indicated by small arrows); (e) volcanic/metavolcanic retouched flake; (f) volcanic/metavolcanic cobble radial core with two 'burination' scars (indicated by arrows); (g) chert flake; (h) chert flake with microwear in the form of edge rounding (a) and edge scarring (b); (i) volcanic/metavolcanic flake with microwear including abrasive smoothing and a grainy polish (a) and small scars and striations (b). Scale bars, 10-mm increments. Picture from Brumm *et al.* (2006). Reprinted by permission from Macmillan Publishers Ltd. *Nature* 441: 7093, 624–628, 2006.

precludes modern humans. At Liang Bua, however, the skeletal remains of at least nine individuals are represented in finds from the Pleistocene levels, and all diagnostic elements are of *H. floresiensis*. The most parsimonious explanation for this is that the stone artifacts from Mata Menge and Liang Bua represent a continuous technology made by the same hominin lineage." This is a parsimonious interpretation that, if correct, would reject the inclusion proposed by Jacob *et al.* (2006) of the Ling Bua specimens in *H. sapiens*. The age of the Mata Menge tools corresponds to *H. erectus*, but no specimens of this taxon have been found so far on the island.

The hypothesis that a hominin with such a small brain as *H. floresiensis* could manufacture the Ling Bua sector-IV complex tools contradicts the usual hominization models. However, the study of the LB 1 specimen by Falk *et al.* (2005) suggests the presence of derived frontal and temporal lobes and a lunate sulcus in a derived position. Falk *et al.* (2005) believe these traits are consistent with capabilities for higher cognitive processing. The same study included the comparison of a virtual endocast of LB 1 with endocasts from great apes, *H. erectus*, *H. sapiens*, a human pygmy, a human microcephalic, specimen number Sts 5 (*A. africanus*), and specimen number KNM-WT 17000 (*P. aethiopicus*). As Falk *et al.* (2005) note: "Morphometric, allometric, and shape data indicate that LB1 is not a microcephalic or pygmy. LB1's brain/body size ratio scales like that of an australopithecine, but its endocast shape resembles that of *Homo erectus*."

Three interpretations of the Lingua Bua fossils are possible, as follows.

1 Flores was colonized by a population of *H. erectus* that evolved into *H. floresiensis*. This population produced the diverse stone tools found in the island.

2 Flores was colonized by a population of *H. erectus* that evolved into *H. floresiensis*. This population produced the older stone tools. A second colonization by *H. sapiens* produced the more recent tools found in Ling Bua.

3 If it were confirmed that the Ling Bua fossils belong to a population of pygmoid Australomelanesian *H. sapiens*, they would be the makers of the tools found there, but it would remain unsettled who were the creators of the older stone tools found at Mata Menge.

9.2 Neanderthals

There is broad agreement among scientists that in Europe hominins belonging to the archaic *sapiens* grade led to Neanderthals. However, there is no general consensus regarding the best taxon in which to include the Neanderthals. In this book we consider Neanderthals to be a separate species, *H. neanderthalensis*. This would be an inadequate taxon if Neanderthals were a subspecies within *H. sapiens*, as supporters of the multiregional hypothesis generally propose. We'll present the reasons for our choice in section 9.3. As pointed out earlier, there are serious difficulties in offering a consistent phylogenetic interpretation of the

---

Box 9.12 Why two cultures in Flores?

Mark Moore and Adam Brumm (2007) have proposed the following explanation for the two cultures found in Flores: "Our research indicates that Pleistocene Knappers on Flores processed large cobbles into large flake blanks, abandoned the large cobble cores, and transported the blanks across the landscape. This produced two spatially segregated assemblage variants: (1) those containing large cores, and (2) those in which the blanks struck from large cores were reduced. Large-sized artifacts (typologically 'core tools') and small-sized artifacts were both produced from one reduction sequence." According to Moore and Brumm (2007) the same pattern is found in other islands of Southeast Asia: "large-sized 'core tool' assemblages are in fact a missing element of the small-sized flake-based reduction sequences found in many Pleistocene caves and rock-shelters."

 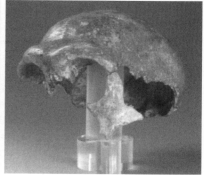

Figure 9.16 Left: a contemporary engraving of the Feldhofer Cave (Bongard, 1835; picture from http://www.ateliereigenart.de/ historie.htm). Right: Neander calotte and zygomatic, *H. neanderthalensis*. Photograph from Schmitz (2003).

Box 9.13 Early impressions of the Neander specimens

The discovery of the Neander specimens took place 3 years before the publication of Darwin's *Origin of Species*. Lacking the evolutionary perspective we have today, it is not surprising that contemporary explanations of the morphology of the specimens seem outlandish to us. On February 4, 1857, Hermann Schaafhausen, professor of anatomy at the nearby University of Bonn, presented the set of fossils before the Lower Rhine Medical and Natural History Society in Bonn. Later that year, on June 2, Fulhrott and Schaafhausen delivered a detailed description of the Neander specimens before the Natural History Society of Prussia, Rhineland, and Westphalia (Trinkaus and Shipman, 1993). Schaafhausen's anatomical interpretation was rigorous when compared with subsequent ones. He attributed the shape of the cranium, deformed in comparison with that of modern humans, to a natural condition, a shape that was unknown "even in the most barbarous of races", to put it in his own words. Thus, two years before the publication of Darwin's *Origin of Species*, Schaafhausen suggested the possible existence of humans prior to Germans and Celts, in a "period at which the latest animals of the diluvium still existed". He believed that the bones belonged to some savage tribe from northeastern Europe subsequently displaced by Germans. He denied that the deformations, both traumatic (the discovered radius exhibited an improperly healed fracture) and anatomical, could be due to pathologies such as rickets.

Despite Schaafhausen's cautions, the idea that the Neander bones belonged to a pathologically deformed human was widespread during the years following the discovery. For instance, after examining a replica of the cranium, C. Carter Blake (1862) concluded that it corresponded to "some poor idiot or hermit" with pathological malformations that indicated an anomalous development and that had died in the cave he had used as a shelter. Trinkaus and Shipman (1993) emphasize Blake's strange statement that the Neander remains could not be attributed to a different species other than *H. sapiens*, given that no one had yet suggested anything like it.

---

different *Homo* taxa (*H. ergaster*, *H. erectus*, *H. antecessor*, *H. heidelbergensis*, *H. rhodesiensis*, *H. neanderthalensis*, *H. sapiens*). None of the alternatives is fully compatible with the available morphological, taphonomic, and genetic evidence. In section 9.3 we will advance the interpretation that, in our view, better fits current data. The present section is devoted to the general study of the taxon *H. neanderthalensis*.

Few hominins have received as much scientific and popular attention as the Neanderthals, which is one reason why few other taxa have also been subjected to such diverse interpretation. The Neanderthal story begins, as Trinkaus and Shipman (1993) described, in the Neander Valley, close to Düsseldorf (Germany). Mining at the Feldhofer calcareous caves, about 20 m above the Düssel River, led to a paleontological discovery of great

Box 9.14 King's interpretations

Trinkaus and Shipman (1993) attributed the way in which William King interpreted the Neander cranium's morphology, tinted with that time's emphatic language, to influences from phrenology, in vogue at the time. This style was blatant regarding the "moral obscurity" of the organism whose remains he was interpreting. It should not be forgotten that Darwin himself manifested a belief in the moral inferiority of savage tribes in his *Descent of Man*, published in 1871.

importance. In August 1856 the entrance to the caves was blasted to extract materials for construction. While cleaning the debris produced by the explosion some fossil bones appeared, including a skullcap, hipbones, ribs, and part of an arm, which seemed to belong to an animal more robust than living humans (Figure 9.16). The foreman believed these were the remains of a cave bear, and had the bones set aside for the teacher at the Elberfeld school. This teacher, Johann Fulhrott, immediately identified them as belonging to a very primitive and robust human.

At a meeting of the British Association for the Advancement of Science, William King, professor at Queen's College, Galway (Ireland), suggested (in a footnote) the classification of the Neanderthal remains as *H. neanderthalensis* (King, 1864). King argued, however, that the differences between this organism and humans were so considerable that it not only merited a separate species, but another genus altogether. In fact, King thought the Neander specimens exhibited a greater similarity with chimpanzees.

Despite his reservations, history remembers William King as the author of the proposal of *H. neanderthalensis*. Up to 34 different species and six genera have been proposed to accommodate Neanderthals since that first taxonomic solution (Heim, 1997). Today the only widely accepted alternatives are to regard Neanderthals as *H. neanderthalensis* or as a subspecies of our own species, *H. sapiens neanderthalensis* (Campbell, 1964).

9.2.1 The first Neanderthals

The Neander Valley fossils were the first to receive a taxonomic classification. However, some

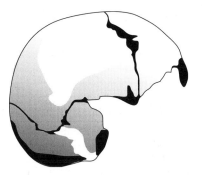

Figure 9.17 Engis infantile cranium (Belgium), *H. neanderthalensis*. Picture from http://aleph0.clarku.edu/huxley/CE7/FosRem.html.

specimens currently included within this taxon had appeared three decades earlier, although they were not associated with Neanderthals at the time. In 1829 Phillip Charles Schmerling, a Belgian doctor and anatomist, found three crania in Engis Cave (Belgium). The first was considerably deteriorated, and was from a modern human, but the third belonged to a 2 or 3-year-old Neanderthal child (Conroy, 1997). It is remarkably well preserved, as its state is the finest among currently available infantile Neanderthal crania (Figure 9.17). Another very complete Neanderthal cranium, which was not identified as such at the time, was discovered in 1848 at Forbes Quarry during the Gibraltar fortification works (*Homo calpicus*; Keith, 1911). After the Feldhofer discovery, other European Neanderthals were found in Moravia, Croatia, France, Germany, Italy, and Spain; the Neanderthal occupation of the continent excluded the northernmost countries. As expected, the specimens were attributed to several different species (*Homo primigenius*, Schaaffhausen, 1880; *Homo transprimigenius*, Forrer, 1908; *Homo breladensis*,

Marett, 1911). The Near-East Neanderthals were first discovered in 1929, during a campaign led by Dorothy Garrod (see below). The Shanidar (Iraq) Neanderthal site has been excavated by Ralph Solecki and colleagues since 1951.

In chronological terms, the oldest known specimens identified as Neanderthals—if we do not consider the Sima de los Huesos exemplars as such—were retrieved at Ehringsdorf, not far from Weimar (Germany). The specimens found at Ehringsdorf between 1908 and 1927 include a cranium, a mandible, and some postcranial fragments belonging to different individuals, of which at least one is an infant (Weidenreich, 1927). The cranium restored by Weidenreich shares many common features with the Würm glaciation Neanderthal specimens, although it has a high and vertical forehead, which, in Kennedy's (1980) opinion, could be the result of incorrect reconstruction. ESR and uranium-series dating suggest that the Ehringsdorf fossils are about 230,000 years old (Cook *et al.*, 1982; Grün and Stringer, 1991). This is the age given by Stringer and Gamble (1993) for the separation of the Neanderthal and modern human lineages.

Colin Groves (1989) classified the Ehringsdorf specimens as Neanderthals, but not everyone agrees. Conroy (1997), to cite a recent example, considers the Ehringsdorf fossils as "archaic" *H. sapiens*; that is to say, as predecessors of *H. neanderthalensis*. Poirier (1987) noted the proximity between the Ehringsdorf sample and the Swanscombe and Steinheim specimens, although he noticed that the Ehringsdorf specimens exhibit greater similarity with later Neanderthals. As mentioned earlier, in light of the Sima de los Huesos variability, Stringer (1993) suggested that the species *H. heidelbergensis* specimens should be considered Neanderthals. Thus, there are different ways of viewing the origin of the Neanderthal clade:

1 an early appearance, if "archaic" European fossils belonging to the initial stages of the Mindel–Riss interglacial period, or even earlier specimens, are considered to be Neanderthals;
2 an intermediate appearance, if the Ehringsdorf specimens, and maybe others from similar aged sites, such as Pontnewydd in Wales, UK (Green *et al.*, 1981) or Saccopastore in Italy (Condemi, 1988), are considered as the first Neanderthals;

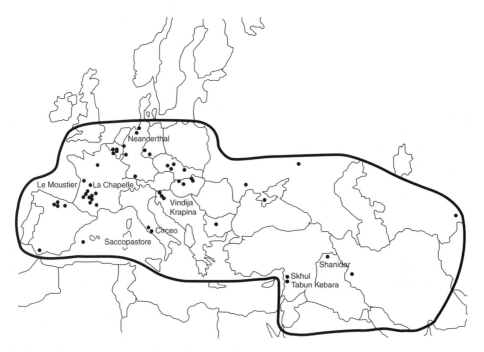

Figure 9.18 Geographical range of the sites with Neanderthal specimens.

3 a late appearance, if "classic" Neanderthals are conceived as hominins living in southern and eastern Europe during the beginning of the Würm glaciation; this point of view, supported by Clark Howell (1952, 1957) is considered the traditional one.

Before we attempt to throw light on the Neanderthal phylogenetic scenario—which we will pursue in the following sections and chapters—we will analyze some of the main specimens. This will allow us to illustrate Neanderthal general morphology and the variability within their clade. It is not possible to include a complete list of Neanderthal specimens and sites because they are very numerous. Remains from the Riss–Würm interglacial period and the Würm glaciation are very abundant in southern, central, and eastern Europe, North Africa, and the Near East (Figure 9.18). We have selected some of the most complete remains, which provide a reliable picture of the Neanderthals.

9.2.2 Morphology of "classic" Neanderthals

Earnest Hooton's distinction in 1946 between "classic" Neanderthals—that is to say, those from

Figure 9.19 "Classic" Neanderthal; Forbes Quarry specimen, Gibraltar. Photograph from http://www.msu.edu/~heslipst/contents/ANP440/neanderthalensis.htm.

western and southern Europe during the Würm glaciation—and "progressive" Neanderthals, including certain specimens from Near-East sites which show some similarities with anatomically modern humans, was for a time widely accepted. This viewpoint reflects the idea that, although the Neanderthal lineage is prior to ours, it contributed genetic material—by means of hybridizations—to *H. sapiens*. As we will soon see, the Neanderthal story is not simple. But it is true that European specimens correspond adequately to what is considered as their typical and distinctive morphology, attributed very early on to Neanderthals in general. Howell (1952) observed that those typical features are characteristic of the latest Neanderthals, survivors of a broader general population. Thus, there is not much sense in applying the term classic to specimens that constitute the last stage of a species. Therefore, Clark Howell used quotation marks to refer to "classic" Neanderthals (Figure 9.19), a practice we will continue here. We will do so to distinguish them from "progressive" Neanderthals, another doubtful term, as we will also see.

According to Clark Howell's (1952) description, "classic" Neanderthals exhibit the following distinctive traits, among others:

- long, low, and wide cranial vault, which represents the preservation, to a lesser degree, of the typical platycephaly of *erectus*-grade specimens;
- broad facial skeleton, with a prominent zygomatic bone and large nose;
- thick semicircular and separate supraorbital tori, which do not extend laterally and are considerably pneumatized by the frontal air sinuses, which lighten these structures that appear so massive;
- absence of chin;
- strong mandible, with a retromolar diastema between M3 and the mandibular ramus;
- large cranial capacity, on average greater than that of *H. sapiens*. Poirier (1987) attributed this high brain volume to a heavier body; Trinkaus (1984) suggested it could be the result of a longer gestation period;
- short and massive vertebral column;
- robust and short limbs in relation to total height.

Can these traits be considered as apomorphies, derived traits of the Neanderthal taxon? To answer this question, Trinkaus (1988) analyzed the traits usually associated with Neanderthals. For instance, Trinkaus (1988) considered the features related to the mandible and dentition, such as the remarkable traits of the retromolar diastema and the absence of chin, as the result of a combination of two factors. First, the facial prognathism retained from *erectus*-grade hominins. Second, the posteriorly placed masticatory musculature. The first factor is a plesiomorphy—the persistence of an ancestral character—while the second is a synapomorphy shared with anatomically modern humans. Neither of these traits, thus, is an exclusively Neanderthal apomorphy.

Trinkaus (1988) believed that the short limbs constitute a derived trait, which differentiated Neanderthals from middle-Pleistocene hominins. However, he considered this trait to fall within the variability range of current humans and, consequently, suggested it should not be considered a Neanderthal autapomorphy. According to Trinkaus, the same could be said about such traits as the high encephalization and the shape of the cranium's occipital region. Hence, Trinkaus (1988) only accepted as true Neanderthal apomorphies a few traits relative to the cranium's temporal and occipito-mastoid regions. This point of view is not unrelated, in our opinion, to Trinkaus' view of Neanderthals as a variety of our own species, *H. sapiens*. If this were the case, we would expect Neanderthals and anatomically modern humans to share many synapomorphies.

Lieberman and McCarthy (1999) described and interpreted the lack of facial projection and the prominent chin of anatomically modern humans in a very different way, as the result of the early ontogenetic reduction of the sphenoid bone. In their view, this is a favorable argument for the classification of Neanderthals and modern humans as two separate species. Schwartz and Tattersall (1996a) also classified Neanderthals in a different species from *Homo sapiens* on the grounds of certain autapomorphies of the internal nasal region. Additionally, the study of the upper part of the mandibular ramus (the coronoid process, the condylar process, and the notch between them) led Yoel Rak and colleagues (2002) to conclude that: "Neanderthals (both European and Middle Eastern) differ more from *Homo sapiens* (early specimen such as Tabun II, Skhul, and Qafzeh, as well as contemporary populations from as far apart as Alaska and Australia) than the latter differs from *Homo erectus*."

Leaving aside whether they are plesiomorphies, synapomorphies, or autapomorphies, the traits mentioned by Clark Howell have often been used to characterize the morphology of Neanderthals. It must be noted, however, that they are not present in all specimens. The Neanderthal sample exhibits a broad variability range, which Wolpoff (1980) attributed to sexual dimorphism.

The morphology of "classic" Neanderthals has generally been considered to reflect adaptation to a cold climate (Brose and Wolpoff, 1971). Thus, the structure of the nasal cavity (very wide) was considered by Schwartz and Tattersall (1996a) as a

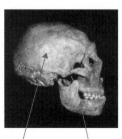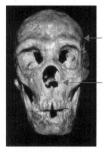

High cranial volume (*c.*1,600 cm³)   Retromolar space

Separated supraciliar torus

Wide nasal passages

Figure 9.20 Some apomorphies of *H. neanderthalensis*. Specimen photographs from http://www.msu.edu/~heslipst/contents/ANP440/neanderthalensis.htm.

reliable way to distinguish Neanderthals and modern humans (Figure 9.20). Such a trait is believed to be an anatomical accommodation to very cold and dry environmental conditions. According to Schwartz and Tattersall (1996a), increased surface for mucus/ciliated membranes and a very large sinusal cavity would have warmed and humidified the cold and dry air. Heim (1997) disagreed with this argument, noting that there were Neanderthals in warm and humid climates, such as the Near East. With regard to body shape, Trinkaus (1981) acknowledged that the shape of Neanderthals' lower limb bones reflect an adaptation to cold climate. However, Trinkaus also considered a complementary hypothesis: that such a trait could be related with biomechanical aspects resulting from an increase in running power, although at the price of loosing speed. This interpretation is supported by Churchill's model (see Holden 1999). Churchill made a cast of the human nasal cavity to study the aerodynamic flux produced by a fluid passing through them, and suggested that the very wide nose of Neanderthals could be an adaptive improvement to achieve considerable air flow while minimizing the turbulence. Churchill believed the need for a great amount of oxygen was related to Neanderthal metabolism, which

would have required close to double our own energetic needs.

The hypothesis that at least "classic" Neanderthals had an anatomy adapted to rigorous climatic conditions had already been put forward by F. Clark Howell (1952), relying on abundant paleoclimatic support. Howell argued that the conditions of periglacial areas had quite an effect on the morphology and distribution of Neanderthals. In fact, the Neanderthal remains belonging to the initial stages of the Würm glaciation appear for the most part in sheltered places, in caves close to water currents in the valleys of southern France and the Italian and Iberian peninsulas (Howell, 1952). This does not mean that Neanderthals did not venture far from those protected places. Howell believed that they would do so, mainly during the summer months, in hunting expeditions towards northern territories which lasted for limited periods of time. One of the migrations assumed by Howell would have taken advantage of the descent of the sea level, starting out from Italy, moving along the north of the Adriatic, towards the central European corridor. The fauna isolated in central Europe by the subglacial Alps would have constituted a hunting reserve that Neanderthals exploited (Figure 9.21). However, the subsequent rise of the sea level

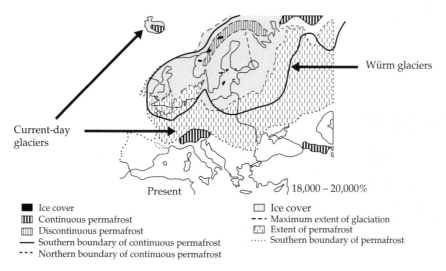

Figure 9.21 Maximum range of the ices during the Würm glaciation.

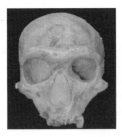
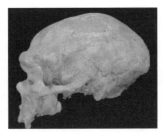
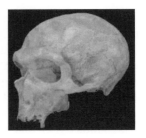

Figure 9.22 The old man from La Chapelle-aux-Saints (France). Pictures from www.mnh.si.edu/anthro/humanorigins/ha/lachap.htm.

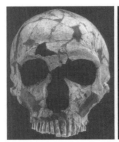
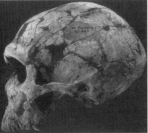

Figure 9.23 Specimen La Ferrassie I (France), *H. neanderthalensis.* Left: front; right: side. Pictures from http://www.msu.edu/~heslipst/contents/ANP440/neanderthalensis.htm.

prevents us from accessing the sediments containing the proof of these population movements.

Neanderthal specimens exhibiting classic morphology have appeared in many places: France, Italy (Guattari), Croatia (Krapina and Vindija), Slovakia (Ochoz, Sipka, Sala), Hungary (Subalyuk), in addition to the aforementioned Engis (Belgium), Forbes (Gibraltar), and Neander (Germany) specimens. Those found in the southwest of France are the most complete and revealing of them all.

An almost complete skeleton of an old adult male, lacking most of its teeth, was found buried in the floor of a cave at La Chapelle-aux-Saints (Corrèze, France) in 1908 (Boule, 1908; Figure 9.22). Since Marcellin Boule's studies (1911–1913) the specimen, known informally as the old man from La Chapelle-aux-Saints, is often used to illustrate the morphology of "classic" Neanderthals. From the first, it was associated with the Neander cranium, and was classified consequently as *H. neanderthalensis* (Boule, 1911–1913). The remains

---

Box 9.15 Neanderthal dentition wear

In relation to the use of dentition among Neanderthals, Vandermeersch and Garralda (1994) suggested that not all specimens show incisor wear. They noted that the Qafzeh and Skhul specimens, in the Near East, show no such trait despite living in the same environmental conditions and using the same objects as Neanderthals. Some authors attribute the Qafzeh and skull exemplars to modern humans (see Section 9.2.5).

---

from La Ferrassie (Dordogne, France), found between 1909 and 1973 in a rocky shelter, bear similarities with the skeleton from La Chapelle. They are two almost complete adult skeletons (a male, La Ferrassie I, and a female, La Ferrassie II; Figure 9.23) and several fragmentary exemplars of at least a dozen infantile or unborn individuals. Boule (1911–1913) associated them with the Neanderthal from La Chapelle, although Day (1986) argued that the La Ferrassie I mandible, which has a slight chin, could be more recent. The conspicuous dental wear observed on many Neanderthal specimens has often been seen as a consequence of the use of dentition as tools, helping the hands during work. But the possible use of teeth as work aids is a very contested hypothesis, and the question has not been answered definitively (see Wallace, 1975; Brace *et al.*, 1981).

Work to widen a road in St. Césaire (Charente-Maritime, France) in 1979 turned up new specimens. They included the right part of a cranium and mandible, together with some postcranial

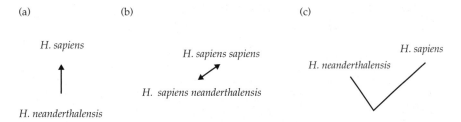

Figure 9.24 Models of the relationship between Neanderthals and modern humans. (a) Neanderthals evolve by anagenesis into modern humans. (b) Neanderthals and modern humans are two subspecies that exchanged genes. (c) Neanderthals and humans are sister groups, species that appeared by cladogenesis (though possibly with earlier intermediate chronospecies in both cases).

> ### Box 9.16 Subspecies or local populations?
>
> Trinkaus and Howells (1980) argued that the term subspecies is unnecessary and suggested that it is better to speak of the evolution of local populations, some of which are those considered as Neanderthals.

remains (Lévêque and Vandermeersch, 1980). Their traits are clearly Neanderthal (Day, 1986), but their relevance lies mainly in their age. The chatelperronian tools found at the same site are very modern. Radiocarbon dating estimates the localities bearing chatelperronian tools to be between 34,000 and 31,000 years. If the skeleton is the same age, this represents one of the last Neanderthals (ApSimon, 1980).We will discuss other late specimens, such as the so-called Lapedo child, in the next section.

9.2.3 The Neanderthal question

Treatises on Neanderthals usually include discussion of their phyletic relations with anatomically modern humans. Three main perspectives can be distinguished. The first argues that Neanderthals are a prior and ancestral species to *H. sapiens*, such that anatomically modern humans evolved from them. This idea developed during the early twentieth century (Day, 1986), but it reappeared with Loring Brace (1964). This view proposes that typical Neanderthal morphological traits disappeared as they evolved towards *H. sapiens*. Wolpoff (1980) adhered to this

hypothesis, while considering Neanderthals and *sapiens* as a single species (Figure 9.24a).

The second perspective considers Neanderthals as a subspecies of *H. sapiens*, *H. sapiens neanderthalensis*, which may have contributed genetically to the appearance of *H. s. sapiens*. This hypothesis is accepted by specialists, such as Trinkaus (Trinkaus and Smith, 1985). Supporters of this model not only believe that genetic exchange was possible between the two subspecies, but that fossil evidence shows that it actually took place (Figure 9.24b).

The third hypothesis suggests that Neanderthals and anatomically modern humans were two different species, sister groups in a cladogram (Figure 9.24c). Neanderthals are conceived as a lateral branch holding no ancestry relation to *H. sapiens*. In fact, the latter might have had an earlier origin than *H. neanderthalensis*. This is Stringer and Gamble's (1993) view, which adds that anatomically modern humans displaced Neanderthal populations wherever they overlapped.

The late presence of classic Neanderthals, such as those found in southern France and the Iberian Peninsula, is an argument against the first hypothesis, that there was a "Neanderthal phase" prior to modern humans and ancestral to our species. It could be said in response that this evolutionary transition took place locally, maybe in the Near East, while the Neanderthal forms remained unchanged at other places.

The Neanderthal question is currently resolved by accepting one of the two following scenarios. The first one suggests a rapid substitution of prior Neanderthal populations by anatomically modern humans. This is the replacement hypothesis, supported by advocates of the out-of-Africa model of the transition of *erectus*-grade hominins to modern humans. The second scenario assumes a gradual transition from archaic to modern populations, in many different places. The Neanderthals would have contributed with their genetic pool to the genes of modern humans (the hybridization hypothesis, followed by those who favor the multiregional model). In the first case Neanderthals are considered as a

different species (as in Figure 9.24c), whereas the second requires their consideration as *H. sapiens* specimens exhibiting populational differences or, at the most, as another subspecies (as in Figure 9.24b).

Morphological considerations are not enough to resolve the controversy. A test for the hybridization and replacement hypotheses is the detailed study of those localities that have yielded both Neanderthal and modern-human exemplars. These are sites located in the Near and Middle East.

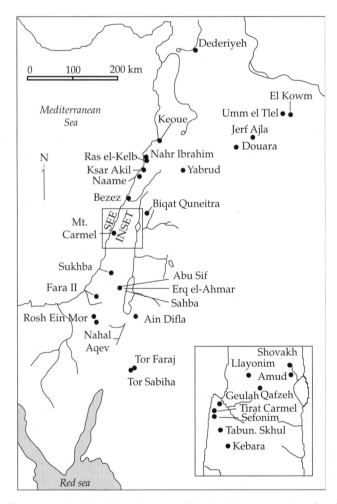

Figure 9.25 Near- and Middle-East sites with "classic" and "progressive" Neanderthal specimens. Picture from http://www.athenapub.com/8shea1.htm.

Box 9.17 Neanderthals 28,000 years ago.

Clive Finlayson *et al.* (2006) have described the presence of Neanderthals in Gorham's Cave, Gibraltar, with an estimated date of 28,000 years, the most recent date known: "Our results show that the Neanderthals survived in isolated refuges well after the arrival of modern humans in Europe."

9.2.4 The Near East: contact between Neanderthals and anatomically modern humans

The easternmost territories occupied by Neanderthals extend from Israel to Iraq. Sites in this region have not only yielded "classic" Neanderthals, similar to the Europeans. Two different kinds of Neanderthal specimens have been found in nearby sites, or even in the same one. The specimens found in Tabun (Mount Carmel, Israel), Shanidar (Iraq), Amud (Amud Valley, Israel), and Kebara (Mount Carmel, Israel) are similar to "classic" Neanderthals. A different kind of specimen, exhibiting greater morphological resemblance to *H. sapiens*, has appeared at Skhul (Mount Carmel, Israel), Qafzeh (close to Nazareth), and Ksar Akil (Lebanon; Figure 9.25). The question is whether these specimens are highly variable members of a single species ("classic" and "progressive" Neanderthals) or whether they belong to

two different species (*H. neanderthalensis* and *H. sapiens*).

To assess the hypotheses regarding the evolution of the Near-East specimens, it is necessary to settle two issues. First, the morphological relationships among different exemplars must be determined by comparative studies. Second, the correlation of the sedimentary sequences must be established by dating the sites in which indications of both species (or populations) appear.

The joint expedition, led by Dorothy Garrod, of the British School of Archaeology of Jerusalem and the American School of Prehistoric Research, carried out a 5-year excavation plan in the Skhul, Tabun, and el-Wad Caves, on the hills of Mount Carmel, close to Wadi el-Mugharah (Valley of the Caves). Two of the many caves in these slanted calcareous terrains very near each other, the large Tabun Cave (*Mugharet el-Tabun*, "Cave of the Oven") and the smaller Skhul Cave (*Mugharet es-Skhul*, "Cave of the Kids"), have provided hominin specimens, as well as Acheulean and Mousterian tools. A nearly complete skeleton of an adult female (Tabun I), a mandible (Tabun II), and some other postcranial remains were found at the Tabun Cave (Garrod and Bate, 1937). The Skhul Cave, deteriorated and reduced to a rocky shelter, contained a very broad collection of human remains which appeared to have been buried deliberately. This collection included six adult specimens (Skhul II–VII; one of them partial; Figure 9.26), one skeleton belonging to a child (Skhul I), the remains of lower limbs (Skhul III and VIII), and a mandible together with a femoral fragment, both of them infantile (Skhul X; McCown and Keith, 1939).

The Qafzeh Cave is close to Nazareth. The initial excavations carried out between 1933 and 1935 by René Neuville, from the Institute of Human Paleontology in Paris, and Moshe Stekelis, from the Hebrew University, led to the discovery of several fossil hominins (Qafzeh 1–7) close to the cave's entrance, although their description was not published at the time. Since 1966 the team led by Bernard Vandermeersch has discovered the remains of about 14 individuals, at least three of which seem to have been buried intentionally. In some cases, such as the Qafzeh 11 infantile

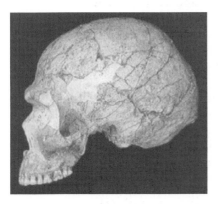

Figure 9.26 Skhul V cranium. Photograph from http://www.msu.edu/~heslipst/contents/ANP440/neanderthalensis.htm.

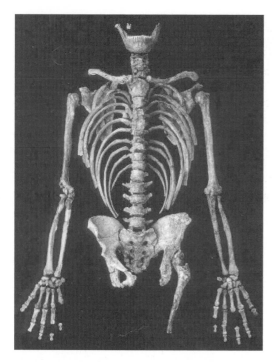

Figure 9.27 Kebara skeleton, Mount Carmel (Israel). Photograph from http://www.msu.edu/~heslipst/contents/ANP440/neanderthalensis.htm.

specimen, the remains appeared accompanied in the same tomb by objects that could have a ritual meaning, such as a deer cranium with Qafzeh 11 (Vandermeersch, 1981).

At Wadi Amud, north of Tiberias and about 50 km northeast of Haifa, the Amud Cave has been excavated since 1959 by the Tokyo University Scientific Expedition to Western Asia, under the direction of H. Suzuki. Amud contains two stratigraphic levels, A and B. The specimen known as Amud I, an incomplete cranium, a mandible in excellent condition, and some postcranial remains (femur and tibia) were found in level B in 1969 (Suzuki and Takai, 1970).

The Kebara Cave, on the western hillside of Mount Carmel, has been excavated since 1982 by Ofer Bar-Yosef and Bernad Vandermeersch (Arensburg et al., 1985; Bar-Yosef and Vandermeersch, 1991, 1993) with the main objective of identifying the behavioral patterns of hominins discovered in Israel. The research was a great

success. In 1983 Kebara yielded one of the most complete Neanderthal specimens to date: a male adult lacking the cranium but not the mandible (Arensburg et al., 1985; Figure 9.27). Some authors (Bar-Yosef and Vandermeersch, 1993; Conroy, 1997) have tentatively attributed the absence of the cranium to funerary practices. The excellent conservation of this specimen—it was buried—has allowed a revision of the theories concerning the Neanderthal capability for speech, based on the specimen's hyoid bone. The Kebara hyoid, together with the one recently discovered in Dinika, Ethiopia (section 4.2), plus another one from Atapuerca's Sima de los Huesos, are the only hominin hyoids known, other than those from modern humans.

Leaving aside the Teshik-Tash site (Uzbekistan), Shanidar Valley, in the Zagros mountains of northern Iraq, has revealed indications of the easternmost Neanderthal occupation. The Shanidar Cave has yielded the greatest number of Neanderthal remains in the entire East. Between 1951 and 1960 the team led by Ralph Solecki retrieved up to 28 burials with remains of nine individuals. The site and its contents were described by Solecki (1953), Stewart (1958) and, more extensively, by Trinkaus (1983). According to Trinkaus, the Shanidar I–VI fossils correspond to adults (Figure 9.28), VII to a youngster, whereas Shanidar VIII and the one known as the Shanidar child are infantile specimens that had not reached the age of 1. Shanidar I, a skeleton, includes the best-preserved cranium of the entire sample. Table 9.3 gives a summary of late-Pleistocene sites in the Near and Middle East.

9.2.5 One or two populations in the Near East?

The morphological study of the late Pleistocene at the Near East revealed the existence of conspicuous differences among the specimens from nearby sites. McCown and Keith (1939) suggested that the Tabun I female was similar to European Neanderthals, with separate and large supraorbital tori, large face and nose, and no mental protuberance. However, the specimen was small, and its cranial capacity was very low for a

Figure 9.28 Burial sites of Shanidar (Iraq). Left, Shanidar III; right, Shanidar IV and VI.

Table 9.3 Specimens from the late Pleistocene in the Near and Middle East.

Site	Location	Specimen
Skuhl	Mount Carmel, Israel	Modern human
Qafzeh	Nazareth, Palestine	Modern human
Amud	Amud Valley, Israel	Neanderthal
Tabun	Mount Carmel, Israel	Neanderthal
Kebara	Mount Carmel, Israel	Neanderthal
Shanidar	Iraq	Neanderthal

Neanderthal (1,270 cm^3). In contrast, the Tabun II mandible showed indications of a chin.

The Skhul specimens were somewhat different to those from Tabun. Skhul V, the best preserved cranium, has a high and short cranial vault with vertical lateral walls (Figure 9.26). The specimen exhibits marked supraorbital tori, in contrast with Skhul IV, which lacks them, although they are not of the typical Neanderthal shape (Stringer and Gamble, 1993). To a greater or lesser extent, the Skhul specimens also show a mental protuberance—which is incipient in Skhul V—and lack midfacial projection. Overall, Skhul specimens appear more modern than those from Tabun. Even so, McCown and Keith (1939) considered that they were all highly variable members of the same population, and attributed the anatomical differences to a gradual evolutionary trend.

Box 9.18 Shanidar burial?

Some aspects of the Shanidar specimens, like the possible deliberate burial and even the presence of associated floral offerings, suggest the existence of funerary practices at the site. These indications will be reviewed in Chapter 10.

The idea of a single population at Tabun and Skhul was supported by the widespread notion at the time that the caves bearing these names were contemporary. If they were quite close and contemporary, it would seem unlikely that two species or even two subspecies were present. When Tabun was estimated to be older (Higgs, 1961), the differences between the specimens collected at the caves were interpreted differently. The Skhul specimens were considered younger and more modern-looking than those from Tabun. These samples seemed to be an excellent testimony of the gradual evolution from Neanderthal forms to modern humans.

The remaining sites in the Near and Middle East can be classified consonantly with the two populations from Skhul and Tabun, as "classic" and "progressive" Neanderthals (Figure 9.29). The morphology of the Qafzeh specimens is similar to that of Skhul exemplars. Some of the crania within the Qafzeh sample exhibit small supraorbital tori (Qafzeh 9), whereas others exhibit protruding

"Classic" Neanderthals, similar to European specimens

Close-to-modern-humans specimens; "progresive" Neanderthals

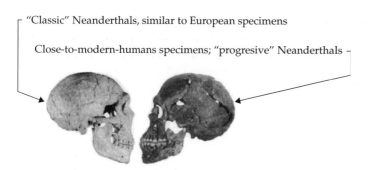

Figure 9.29 The two forms of "Neanderthal" from the late Pleistocene in the Near and Middle East.

Box 9.19 One population or two?

The general view with regard to the eastern Mousterian specimens is to assume that there were two populations. But some authors, such as Wolpoff (1991), continued to support McCown and Keith's (1939) initial idea that all Near-East human remains constitute a highly variable single population. Nevertheless, the division between the Tabun, Kebara, Amud, and Shanidar samples on one side, and those from Skhul and Qafzeh on the other, is evident. Whether this difference has phylogenetic significance is a different matter.

supraorbital tori but a shape different from that of classic Neanderthals (Qafzeh 6). Qafzeh 9 exhibits modern traits, such as the elevated vault and the supraorbital torus, together with some primitive ones: prognathism and large teeth. In spite of this, the Qafzeh exemplars have been classified together with those from Skhul as part of the same population, considered to be different from the one in Tabun.

In contrast to the more modern Skhul and Qazfeh specimens, the Amud exemplar is similar to those from Tabun and Shanidar. In fact, materials from the latter and other sites were used to complete the missing parts during the reconstruction of the cranium. This shows some modern traits, such as an incipient mental protuberance or a modest supraorbital torus, but Suzuki and Takai (1970) as well as Howells (1974) included it in the Near-East Neanderthal population found at Tabun and Shanidar. This idea has been accepted widely.

The affinities shown by the Kebara specimen cannot be doubted either. Despite the absence of the cranium, its remaining traits are reminiscent of the Tabun Neanderthals (Bar-Yosef and

Vandermeersch, 1991). Arensburg (1989) preferred not to use the term Neanderthal and referred to the Kebara specimen as "Mousterian", without a taxonomical label. However, Arensburg (1989) associated some of the specimen's traits, such as the pelvis, with those of European and Shanidar specimens.

Regarding Shanidar, the tendency is to include its specimens in the category of Near-East classic-like Neanderthals. The Shanidar I cranium, which exhibits such traits as a low and long vault, marked supraorbital torus, facial prognathism, and occipital torus, constitutes a very typical Neanderthal example. The mandible, which is complete, shows the typical Neanderthal traits: robusticity and lack of mental protuberance.

We have mentioned that the hybridization hypothesis suggests that Neanderthals were a subspecies, *H. s. neanderthalensis*, and could have contributed with their genetic material to the origin of *H. s. sapiens*. If this scenario is accepted, then the transition of "classic" Neanderthals (Tabun, and so on) to "progressive" forms (similar to Skhul specimens) could be excellent evidence in favor of the gradual evolution that transformed Neanderthals into anatomically modern humans. The hypothesis requires that the ages of the caves coincide with the evolutionary sequence; that is to say, that the remains of the Neanderthal-looking specimens are also older than the modern-looking exemplars. Thus, we need to turn to the chronological sequence of the Near-East sites.

9.2.6 The ages of the Near-East caves

We will refer to the estimates obtained during the early studies of the caves, prior to the application of methods such as ESR or thermoluminescence, as initial ages.

Tabun includes a series of sedimentary levels, A–G. Tools have been found at levels Tabun B and C, whereas human remains appeared in sediments 1 and 2 of level C. The initial age of Garrod and Bate (1937) assigned levels C and D to the Riss–Würm interglacial period (between 275,000 and 125,000 years ago). Later, Garrod (1962) considered that level C was younger, towards the second half of the Würm glaciation (less than 100,000 years ago). Radiocarbon dating estimated the minimum age of Tabun at 51,000 years (Farrand, 1979). However, these ages are too old for the use of carbon isotopes to yield reliable results.

With regard to Skhul, T.D. McCown identified three stratigraphic levels: A, B, and C. The specimens were buried in level B, whereas A, above it, contained stone tools. The fauna and the tools at Skhul correspond well to those at Tabun level C; so, Garrod and Bate (1937) assigned it the same initial age, the Riss–Würm interglacial period. Subsequent macrofaunal studies suggested that Skhul was at least 10,000 years younger than Tabun C (Higgs, 1961), which allowed the

establishment of an evolutionary sequence for the specimens found at both caves. If Tabun was prior to Skhul, then the evolutionary sequence of "classic" to "progressive" specimens from both caves fitted nicely. The cave with other modern-looking remains, Qafzeh, also seemed to be younger than Tabun. Qafzeh includes up to 24 sedimentary levels, with human remains (which in some instances are difficult to associate stratigraphically) found in level XVII. Amino-acid-racemization studies estimate that that level is 39,000–32,000 years old (Farrand, 1979). This age was subsequently raised, by the same technique, to 78,000–39,000 years old (Masters, 1982).

Hence, the chronological scenario suggested that the Tabun, Amud, and Kebara set were older than the Skhul and Qazfeh Caves. Sites that yielded similar remains to classic Neanderthals were older than those containing modern-looking specimens. McCown and Keith's (1939) gradual evolution hypothesis was supported by such a chronology. However, that hypothesis became difficult to sustain after Bar-Yosef and Vandermeersch (1981) reassessed the situation. These authors carried out a comparative study of the stratigraphic relations, paleoclimatic data, and microfaunal and cultural sequences. They concluded that Qazfeh was about as old as Tabun D, close to 100,000 years old. This questioned the evolution of "classic" towards "progressive" Neanderthals.

The application of more modern and precise dating techniques—ESR and thermoluminescence—towards the end of the 1980s has confirmed Bar-Yosef and Vandermeersch's results. The thermoluminescence study of 20 burnt flint stones from Qafzeh yielded an average of $92,000 \pm 5,000$ years, with a range of 105,000–85,000 years (Valladas et al., 1988). With regard

---

Box 9.20 A shared culture?

The so-called Tabun C2 industry could be older than the estimated dates of the Tabun specimens. By means of thermoluminescence, similar tools have been dated in Tabun C and Hayonim E at 150,000 or 170,000 years (Bar-Yosef, 2000). The question is who manufactured the instruments, given the lack of fossil specimens. In section 10.1 we will explore the issue of the possibly shared culture in the Near and Middle East between Neanderthals and modern humans.

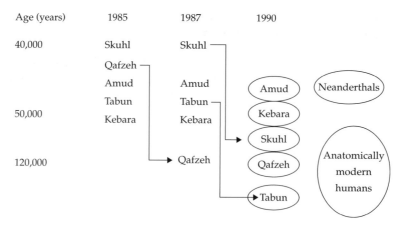

Figure 9.30 Chronological sequence of late-Pleistocene sites in the Near East, according to Stringer and Gamble (1993).

Box 9.21 Difficulties with proving hybridization

Vandermeersch (1996) considered that cultural practices shared by the two populations present in the Near East since the beginning of the Riss–Würm interglacial period exclude the hypothesis that there was an alternated occupation of the area. If both populations were present at the same time, then cohabitation can be understood as either (a) corresponding to two separate species, with no possibility of hybridization, or (b) the contact between varieties of a single species, which would allow genetic exchange. He noted that the available data do not justify denying that such exchange occurred, although the possibility must be tentative because of the great difficulties involved in its demonstration. A way of doing so would be to identify specimens that were morphologically intermediate between the two populations living in the area. However, this kind of evidence is controversial because of the broad variation among Neanderthals, as well as among our own species.

---

to Skhul, the study of mammal teeth with ESR estimated its age to be between 80,000 and 100,000 years, much older than previously thought (Stringer *et al.*, 1989). Thermoluminescence assigned to Skhul level A, believed to be the most modern, an even greater age, about 120,000 years (Mercier *et al.*, 1993). Given that one of the sites containing "classic" Neanderthals, Kebara, was estimated by means of thermoluminescence to be about 60,000 years old (Valladas *et al.*, 1987), it seemed clear that the chronological sequence assuming that "classic" Neanderthal sites were old and that "progressive" Neanderthal sites were young could no longer be upheld. The age of the broader sedimentary sequence at Tabun was estimated by ESR to be 200,000 years for the oldest

level to 90,000 years for the youngest (Grün *et al.*, 1991). Level C, the sister of Tabun I, was estimated by Grün and colleagues (1991) to be between $102,000 \pm 17,000$ and $119,000 \pm 11,000$ years old, a similar age to the Skhul specimens.

These dates, which completely upset the assumed archaeological sequences of Mount Carmel, were criticized by many authors. However, the very precise technique of uranium-series mass spectrometry applied to samples of bovid teeth from Tabun, Skhul, and Qafzeh confirmed the new scenario (McDermott *et al.*, 1993). Tabun level C was estimated to be about 100,000 years old (three samples yielded 97.8 ± 0.4, 101.7 ± 1.4, and 105.4 ± 2.6 thousand years). The Skhul specimens are difficult to date using this technique, owing to

Box 9.22 Reinterpreting the Tabun C1 skeleton

The view of an alternation in the occupation of the Levantine region could change substantially if the discovered remains turn out to belong to levels different from those currently believed. This is what Henry Schwarcz *et al.* (1998) argued after reinterpreting the Tabun C1 specimen, the almost complete female skeleton found by Garrod and Bate (1937). Schwarcz and colleagues described the location of the specimen so close to the surface of level C that it might actually correspond to level B. Because the specimen had been buried, it would

artificially have been included in the lower stratigraphic level, as suspected by Garrod (Bar-Yosef, 2000). By means of gamma-ray spectrometry Schwarcz *et al.* (1998) estimated the Tabun C1 skeleton at 34,000 ± 5,000 years, which would suggest that the ages of the last Near-East Neanderthals are similar to the European ones, contrary to what is usually assumed. Bar-Yosef (2000) has discarded the implication of the results obtained by gamma-ray spectrometry because of its unreliable results.

the presence of two different faunas in level B. McDermott and colleagues (1993) suggested that there are human specimens in both, but they also expressed the need for further detailed studies to confirm this point. In any case, the coincidence of different dating techniques (ESR, thermoluminescence, uranium series) in the attribution of a greater (but similar) age to Tabun and Skhul seems conclusive, taking into account the fact that the results rendered by these techniques are very similar. Figure 9.30 summarizes the dates of the various Neanderthal sites.

The most satisfactory explanation for the alternative presence of "classic" Neanderthal-like hominins preceding and following those resembling anatomically modern humans is the one put forward by F. Clark Howell in 1959. He suggested that Near-East "progressive" specimens belong to our species, and named them proto-Cro-Magnons. As Vandermeersch (1989) noted, this idea involves the presence of two different populations in the area during the first part of the Würm glaciation. Stringer and Gamble (1993) are the leading advocates of the current perspective favoring the presence of the two species in the Near East: *H. neanderthalensis*, which includes the specimens resembling exemplars from La Ferrassie, and *H. sapiens* for modern-looking examples.

9.3 Molecular data

The morphological, taphonomic, geological, and cultural evidence have not definitively settled the

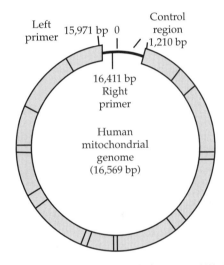

Figure 9.31 A schematic of mitochondrial DNA, or mtDNA. At the top is the noncoding control region (hypervariable region, or HVR). It contains 7.5% of all the mitochondrion's genetic information.

disagreements over the origin of modern humans between supporters of the hybridization (multiregional) and replacement (out-of-Africa) hypotheses. Furthermore, there is no broad consensus as to whether Neanderthals and modern humans belong to separate species. In this section we will continue to examine this controversy using a new type of evidence, provided by sequencing of fossil DNA.

The cloning of ancient DNA has been achieved on several occasions using nucleic acids present in mitochondria, mtDNA. Mitochondria are cellular

Table 9.4 Criteria for obtaining and sequencing ancient DNA

Criterion	Comments
1. Cloning of amplification products and sequencing of multiple clones	This serves to detect heterogeneity in the amplification products, due to contamination, DNA damage, or jumping PCR.
2. Extraction controls and PCR controls	At least one extraction blank that does not contain any tissue but is otherwise treated identically should be performed. During each PCR blank, PCR controls should be performed to differentiate between contamination that occurs during the extraction and during the preparation of the PCR.
3. Repeated amplifications from the same or several extracts	This serves two purposes. First, it allows detection of sporadic contaminants. Second, it allows detection of consistent changes due to miscoding DNA lesions in extracts with extremely low numbers of template molecules.
4. Quantification of the number of amplifiable DNA molecules	This shows whether consistent changes occur or not. If consistent changes can be excluded (roughly for extracts containing $> 1,000$ template molecules), a single amplification is sufficient. Quantitation has to be performed for each primer pair used, as the number of amplifiable molecules varies dramatically with the length of the amplified fragment, the sensitivity of the specific primer pair used, and the base composition of the amplified fragment.
5. Inverse correlation between amplification efficiency and length of amplification	As ancient DNA is fragmented, the amplification efficiency should be correlated inversely with the length of amplification.
6. Biochemical assays of macromolecular preservation	Poor biochemical preservation indicates that a sample is highly unlikely to contain DNA. Good biochemical preservation can support the authenticity of an ancient DNA sequence.
7. Exclusion of nuclear insertions of mtDNA	It is highly unlikely that several different primer pairs all select for a particular nuclear insertion. Therefore, substitutions in the overlapping part of different amplification products are a warning that nuclear insertions of mtDNA may have been amplified. A lack of diversity in population studies can also be taken as an indication that nuclear insertions may have confounded the results.
8. Reproduction in a second laboratory	This serves a similar purpose to 2 and 3, i.e. to detect contamination of chemicals or samples during handling in the laboratory. This is not warranted in each and every study, but rather when novel or unexpected results are obtained. Note that contaminants that are already on a sample before arrival in the laboratory will be reproduced faithfully in a second laboratory.

Source: after Pääbo *et al.* (2004).

organelles that possess their own genetic material, separate from the DNA in the cell nucleus. Their mtDNA is inherited through the maternal lineage, transmitted through the ovule. There are hundreds of mitochondria in each cell, which increases the chances of obtaining their DNA from fossils. Nuclear DNA is present only once in each cell.

Human mtDNA is comprised of around 16,000 nucleotides, a very small number compared to the 3 billion nucleotides in each nuclear genome,

where most genes reside (Figure 9.31). The cloned sequences usually belong to a noncoding control region, usually referred to by the acronym HVR, for hypervariable region, and which contains 1,210 base pairs (bp) and exhibits great variability.

In 1984, Allan Wilson and collaborators managed to retrieve ancient DNA for the first time, from the quagga, an extinct relative of horses (Higuchi *et al.*, 1984). Shortly thereafter Svante Pääbo (1984, 1985) retrieved human mtDNA from a 2,500-year-old

---

Box 9.23 Procedures for retrieving fossil genetic material

A very precise experimental protocol specifies the procedures that must be followed to retrieve fossil genetic material (see Table 9.4). The protocol includes, among other precautions, analyzing the samples in two independent laboratories so that contaminating mtDNA can be detected. For a detailed description of the technique and its limitations, see O'Rourke *et al.* (2000), Hofreiter *et al.* (2001), and Pääbo *et al.* (2004); for a review of the different techniques involved see Cipollari *et al.* (2005).

Box 9.24 Cloning ancient nuclear DNA

In his commentary to the cloning of the Mezmaiskaya specimen, Höss (2000) noted that the fact that none of the examined Neanderthals contributed mtDNA to the human lineage does not mean that they did not contribute nuclear DNA. To find out whether they did, it will be necessary to clone nuclear DNA, something that remains quite difficult in the case of fossils. Fossil nuclear DNA is usually degraded and mixed with microbial contaminants. DNA is degraded rapidly by enzymes such as lysosomal nucleases. In addition, bacteria, fungi, and insects feed on and degrade DNA macromolecules (Eglinton and Logan, 1991).

Nuclear DNA cloning was already achieved in 2003 by sequencing a single gene locus through the direct use of PCR on the material to be sequenced (Huynen *et al.*, 2003). Using a different technique, known as the metagenomic approach, Noonan and colleagues (2005) were able to clone a 26,000-bp-long DNA sequence from two 40,000-year-old extinct cave bears. The procedure involved obtaining the ancient DNA first, composed of a mixture of genome fragments from the ancient organisms and sequences derived from other organisms in the environment. Thereafter, the metagenomic approach was used, in which all genome sequences are anonymously cloned into a single library. To avoid likely contamination, the operation must be carried out in a laboratory into which modern carnivore DNA has never been introduced. The next step was to sequence the library, which gave way to chimeric inserts due to the presence of DNA from different organisms. Finally, the inserts were compared with GenBank nucleotide, protein, and environmental sequences, which includes those of bears. These precedents suggested that nuclear DNA might be retrieved from well preserved Neanderthal fossils.

Egyptian mummy. By the year 2000 mtDNA had been cloned from up to 18 fossil or extinct organisms. The DNA of these organisms was cloned using the technique known as the polymerase chain reaction (PCR), which allows millions of identical copies to be obtained from a single DNA molecule (Pääbo *et al.*, 2004).

Obtaining useful genetic material from a fossil depends on the conservation of the specimens, which is a function of their age and the conditions of the site. Recent specimens and a dry and cold environment offer the greatest possibilities for successful cloning. One difficulty is the possible contamination of the material to be analyzed, which requires that the samples be handled with great care. Even so, contamination happens frequently. Table 9.4 lists some criteria for obtaining and sequencing ancient DNA.

The first Neanderthal mtDNA to be cloned came from the Feldhofer cranium (Neander, German; Krings *et al.*, 1997). The same 379-bp-long sequence was obtained by the two laboratories that carried out the analysis, and the cloned mtDNA was compared with the same sequence corresponding to a current reference population (Anderson *et al.*, 1981), yielding 27 differences. Comparison of the mtDNA segment between individuals from different living human populations revealed an average of between five and six differences. The results obtained by Krings and colleagues (1997) indicated that the difference between Neanderthal and modern human mtDNA was several times

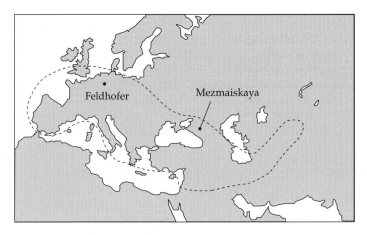

Figure 9.32 The Feldhofer and Mezmaiskaya sites, sources of the two first Neanderthal specimens whose mtDNA was cloned. The dashed line encompasses the region where Neanderthal specimens have been found. Picture from http://www.nature.com/nature/journal/v404/n6777/fig_tab/404490a0_F1.html#figure-title.

Figure 9.33 LM3 from Lake Mungo (Australia), *H. sapiens* (Bowler and Thorne, 1976). Left, lateral view; middle, superior view; right, frontal view. Photographs from http://www-personal.une.edu.au/~pbrown3/LM3.html.

larger than that among living humans. The comparison of humans and chimpanzees renders twice the number differences between Neanderthals and modern humans. Krings *et al.* (1997) concluded that these results support the two-species hypothesis.

The second cloning of Neanderthal mtDNA was carried out with material from a rib belonging to a 29,000-year-old juvenile specimen found in the Mezmaiskaya Cave (northern Caucasus; Ovchinnikov *et al.*, 2000). A 345-bp-long sequence was obtained, corresponding to the same mtDNA segment as that of the Feldhofer specimen. The comparison of the Mezmaiskaya sequence with the Anderson *et al.* (1981) reference and the Feldhofer sequence revealed 22 and 12 differences respectively. The two Neanderthals share 19 substitutions relative to the reference sequence.

Igor Ovchinnikov and colleagues (2000) arrived at several conclusions based on the comparison of the two Neanderthal mtDNA sequences with each other and with those of modern humans. The genetic differentiation between the two Neanderthal sequences is greater than among 300 individuals in a sample from current Caucasoid and Mongoloid populations, but comparable to a similar sample from Africans. The Feldhofer and Mezmaiskaya sites (Figure 9.32) are far apart and thus may be taken as representative of Neanderthals as a whole. (Later on we will see to what extent this is so.) A second conclusion was that "these data provide further support for the hypothesis of a very low gene flow between the Neanderthals and modern humans. In particular, these data reduce the likelihood that Neanderthals contained enough mtDNA sequence diversity to encompass modern human diversity." Third, based in their results, Ovchinnikov *et al.* (2000) estimated the age of the common ancestor of western (Feldhofer) and eastern (Mezmaiskaya) Neanderthals to be between 151,000 and 325,000 years. The divergence between Neanderthals and

Box 9.25 Dating the LM3 specimen

The age of the LM3 specimen from Lake Mungo was initially established on the basis of geomorphological criteria and stratigraphic association with Mungo 1, between 28,000 and 32,000 years (Bowler and Thorne, 1976). However, by a combination of different dating systems (see below), Thorne and colleagues (1999) determined the age for LM3 as 62,000 years. This date was soon criticized. It turned out to be 20,000 years older than the lower level of the Mungo Unit from which LM3 was recovered, reliably dated at 43,000 years; see Bowler and Magee (2000) and Gillespie and Roberts (2000), who argued that, given that LM3 had been buried—and thus deposited in an inferior sedimentary level—it had to be younger than 43,000 years.

The dating techniques used by Alan Thorne *et al.* (1999) were: ESR on dental enamel; uranium series on calcite crust covering the skeleton; and optically stimulated luminescence on the sediment surrounding the skeleton. The latter technique examines individual grains of minerals, which absorb radiation from the sediment. The mineral releases this energy when erosion exposes it to sunlight. The energy still retained by the mineral indicates when it was last exposed to sunlight. But Gillespie (2002) noted, "whether LM3 remains turn out to be 40, 50, or 60 ka [thousand years old] does not seem to matter for the human origins debate, with *modern* people living in Africa > 100 ka there is plenty of time to reach Europe or Australia by any of those dates." An earlier estimate for LM3 could run against the credibility of the mtDNA study carried out by Gregory Adcock *et al.* (2001), because "DNA is not expected to survive for this length of time outside of cold environments" (Cooper *et al.*, 2001).

Box 9.26 Cloning Neanderthal nuclear DNA

The First cloning of Neanderthal nuclear DNA has been carried out by two research teams using samples about 38,000 years old from Vindija (Croatia), where the conditions for preservation are excellent.

Richard E. Green *et al.* (2006) used the recently developed technique of large-scale parallel 454 sequencing: "In this technology, singlestranded libraries, flanked by common adapters, are created from the DNA sample and individual library molecules are amplified through bead-based emulsion PCR, resulting in beads carrying millions of clonal copies of the of the DNA fragments from the samples. These are subsequently sequenced by pyrosequencing on the GS20 454 sequencing system" (Green *et al.*, 2006).

James P. Noonan and colleagues used an amplification-independent direct cloning method to construct a Neanderthal metagenomic library. Neanderthal genome sequence was covered from this library "through a combination od Sanger sequencing and massively parallel pyrosequencing" (Noonan *et al.*, 2006).

Nevertheless, the issue whether Neanderthals and modern humans exchanged genes remains unsettled: "If Neanderthal admixture did indeed occur, then this could manifest [. . .] as an abundance of low-frequency derived alleles in Europeans where the derived allele matches Neanderthal. No site in the data set appears to be of this type. In order to formally evaluate this hypothesis, we extended our composite likelihood simulations to include a single admixture event 40,000 years ago in which a fraction of the European gene pool was derived from Neanderthals. We fixed the human-Neanderthal split at 440,000 years ago (the split time estimate for Europeans). With these assumptions, the maximum likelihood estimate for the Neanderthal contribution to modern genetic diversity is zero. However, the 95% CI for this estimate ranges from 0 to 20%, so a definitive answer to the admixture question will require additional Neanderthal sequence data" (Noonan *et al.*, 2006).

Green *et al.* (2006) also accept the possibility of gene exchange between Neanderthals and modern humans. The large number of derived alleles in Neanderthals is incompatible with a simple population-split model when the split times are inferred from the fossil record. Nevertheless, there may have been some "gene flow between modern humans and Neanderthals." Given that the Neanderthal X chromosome shows a higher level of divergence than the autosomes [. . .], gene flow may have occurred predominantly from modern human males into Neanderthals. More extensive sequencing of the Neanderthal genome is necessary to address this possibility" (Green *et al.*, 2006).

Gene exchange between species may occur during the early stages of specification, when reproductive isolation may not yet be complete (See Box 9.27.).

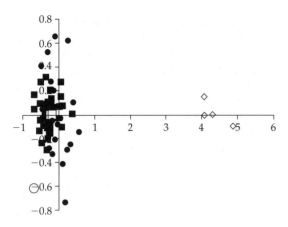

Figure 9.34 Multidimensional scaling of HVRI sequences of 60 modern Europeans (filled squares), 20 modern non-Europeans (filled circles), four Neanderthals (open diamonds), the Australian LM3 (open circle), and two Cro-Magnons (open squares). The axes have different scales. Picture from Caramelli *et al.* (2003).

modern humans would have taken place between 365,000 and 853,000 years ago (Ovchinnikov *et al.*, 2000). Ward and Stringer (1997) had estimated that date as no less than 500,000 years, based on the Feldhofer results of Krings *et al.* (1997), which falls within that interval.

Adcock and colleagues (2001) cloned mtDNA from a very early modern human, Lake Mungo 3 (LM3; Australia), described by Bowler and Thorne (1976; Figure 9.33). The comparison of the mtDNA from LM3 with the modern human reference sequence yielded 13 differences. This is a large difference since, on the grounds of its morphology and the tools found at the site, there is no doubt that LM3 is a modern human. The authors concluded that the mtDNA of LM3 eventually disappeared from the human lineage. If this were so, it could also be that the same might have happened in other cases, such as the Neanderthals. If the absence of the genetic peculiarities of LM3 in current populations does not justify excluding the Lake Mungo specimen from the modern human lineage, then Neanderthals need not be excluded either, on the grounds of the results from the cloning of their mtDNA. This argument has been used by the supporters of the multiregional hypothesis to invalidate the evidence of the separation deduced from Neanderthal mtDNA cloning.

The controversy regarding the significance of the cloned material from Lake Mungo illustrates some possibilities and limits afforded by molecular tests for the understanding of human evolution. The specialists in mtDNA cloning were quick to criticize the work by Adcock *et al.* (2001) for three reasons. First, the findings were not replicated by

---

Box 9.27 Genetic exchange between sister species

Genetic exchange between two nearby populations, such as sister species, is not always impossible. Reproductive isolation is, in some circumstances, a permeable barrier.

With time, the separation becomes unbreachable. But during the process of species differentiation, limited genetic exchange can take place (Cela-Conde, 1988).

Box 9.28 The similarity of Cro-Magnon and modern-human mtDNA

David Caramelli and colleagues (2003) cloned mtDNA from two Italian Cro-Magnon specimens. Their results suggested that the specimens were well within the variation range of current humans. However, their work was criticized, on the grounds of a methodological issue. If the mtDNA that is being cloned is very similar to that of current humans, it is not possible to determine which part

of the obtained results correspond to endogenous material and which is due to contaminations (Nordborg, 1998; Trinkaus and Zilhao, 2002; Pääbo *et al.*, 2004; Serre *et al.*, 2004). In Pääbo's words, "Cro-Magnon DNA is so similar to modern human DNA that there is no way to say whether what has been seen is real" (cited by Abbott, 2003).

Table 9.5 MtDNA genetic diversity of African apes, modern humans, and Neanderthals.

	Gorillas	Chimpanzees	Modern humans	Neanderthals
MtDNA genetic diversity (substitutions/ 100 nucleotides)	18.57 ± 5.26%	14.82 ± 5.70%	3.43 ± 1.22%	3.73%

The numbers indicate substitutions for every 100 nucleotide sites. The sequences of current humans and apes correspond to 50,000 randomly chosen among 5,530 humans, 359 chimpanzees, and 28 gorillas. The Neanderthal sequences are from the three Feldhofer, Mezmaiskaya, and Vidija 75 specimens. The probability that the Neanderthal sample's diversity corresponds to the highest diversity in the complete population is 50%. Data taken from Krings *et al.* (2000).

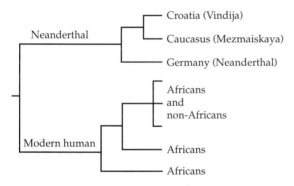

Figure 9.35 Phylogenetic tree with current humans and Neanderthal specimens from Feldhofer (Germany), Mezmaiskaya (Caucasus), and Vidija 75 (Croatia). Figure from Hofreiter *et al.* (2001).

an independent laboratory (Cooper *et al.*, 2001; Gillespie, 2002). The second difficulty is the specimen's doubtful age (Cooper *et al.*, 2001). The third line of criticism embraces the same arguments leveled to criticize the cloning procedures used with Cro-Magnon specimens (see Box 9.22). Endogenous mtDNA and contaminating current mtDNA cannot be distinguished in early modern human remains, a circumstance which affects LM3 (Pääbo *et al.*, 2004; Serre *et al.*, 2004). However, it is not completely fair to apply this caution to the Lake Mungo specimen because, in this case, it highlights precisely the differences regarding the

reference sequence. They could not be produced by contaminating current mtDNA, which lacks the peculiarities of LM3. Pääbo *et al.* (2004) have argued that it is virtually impossible to exclude all modern human DNA sequences as possible sources of contamination, "including excavators, museum personnel, or laboratory researchers."

The fairest criticism to the interpretation of LM3 as supporting the multiregional hypothesis has to do with the distance of the specimen with regard to modern humans and Neanderthals. Caramelli and colleagues (2003) carried out a comparative analysis by means of multidimensional scaling, which graphically shows the genetic distances between sequences. The analyzed sample included current humans (60 Europeans, 20 non-Europeans), Neanderthals (four), Cro-Magnons (two), and the LM3 specimen. The results, which appear in Figure 9.34, provide a clear idea of the situation. Even though some mitochondrial peculiarities of the Lake Mungo specimen are absent in modern populations, it falls within the variation range of *H. sapiens*, whereas Neanderthals do not. A similar argument was advanced by Cooper *et al.* (2001).

Following the cloning of the Mezmaiskaya Neanderthal mtDNA, ten additional specimens have been sequenced: Feldhofer 2 (Germany; Schmitz *et al.*, 2002), Vindija 75 (Croatia; Krings *et al.*, 2000), Vindija 77 and 80 (Serre *et al.*, 2004), Engis 2 (Belgium; Serre *et al.*, 2004), a specimen from La Chapelle-aux-Saints (France; Serre *et al.*, 2004), RdV 1 from Les Rochers-de-Villeneuve (France) (Beauval *et al.*, 2005), El Sidrón (Spain; Lalueza-Fox *et al.*, 2005) Scladina Cave (Belgium; Orlando *et al.*, 2006) and Monte Lessini; (Italy; Caramelli *et al.*, 2006). The comparisons between the cloned and reference sequences have produced very similar results to the ones we have already discussed. But some of the studies merit further attention.

Krings *et al.* (2000) obtained a 357-bp HVRI sequence and a 288-bp HVRII sequence from the Vindija 75 specimen. Comparison with the Anderson reference sequence produced very similar results to the two previous cases: it is "highly unlikely that a Neanderthal mtDNA lineage will be found that is sufficiently divergent to represent an ancestral lineage of modern European

Box 9.29 The *H. heidelbergensis* and *H. neanderthalensis* chronospecies

It is generally quite difficult to pinpoint the time at which one chronospecies was replaced by another. The genetic data and the estimation of the different dental growth patterns in *H. heidelbergensis* and *H. neanderthalensis* (Ramirez Rozzi and Bermúdez de Castro, 2004) would allow, exceptionally, establishment of a precise frontier

between these two chronospecies. However, the abundance paradox could blur a scheme that for the moment seems clear due to the lack of intermediate specimens. The sample of *H. heidelbergensis* in Rozzi and Bermúdez de Castro's (2004) study included only specimens from Sima de los Huesos (Atapuerca, Spain).

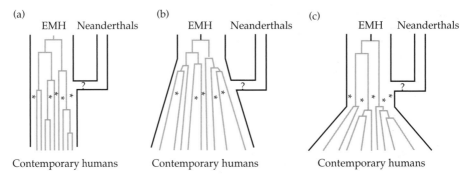

Figure 9.36 (a) Under the assumption of a constant effective population size of 10,000 for modern humans, contemporary mtDNAs trace back to approximately five mtDNA lineages 25,000 years ago. The modern-human fossils represent five additional samples from around the time of putative admixture (stars). The contemporary and early modern human (EMH) samples reject a Neanderthal contribution of 25% or more to modern humans about 30,000 years ago ($P > 0.05$). (b) Under the more realistic scenario of an expansion of the human population during and after the colonization of Europe, a smaller Neanderthal contribution can be excluded because the number of ancestors of the current human gene pool was larger 30,000 years ago. However, the contribution that can be excluded would depend on when and how the expansion occurred. (c) Under the scenario that population size was constant before a putative merging with the Neanderthal population and expanded only thereafter, the Neanderthal contribution could have been larger, but this similarly depends on how the expansion occurred. Figure and explanation taken from Serre *et al.* (2004).

mtDNAs." The authors analyzed Neanderthal genetic diversity, based on the Feldhofer, Mezmaiskaya, and Vindija 75 specimens. When compared to African apes and current humans, Neanderthals are more similar to modern humans than to apes in having a low species-wide mtDNA diversity (Krings *et al.*, 2000; Table 9.5). According to Krings *et al.*: "If the Neanderthals, similar to humans, had a diversity lower than that of the great apes, in spite of inhabiting a region much larger than the apes, this may indicate that they also had expanded from a small population." A phylogenetic tree corresponding to these three Neanderthal specimens is illustrated in Figure 9.35.

The study carried out by David Serre *et al.* (2004) is especially interesting, because it takes into

account the two main problems noted by the supporters of the multiregional hypothesis: (1) possible Neanderthal contribution to the gene pool of modern humans might have been erased by genetic drift (Nordborg, 1998) and (2) if some Neanderthals carried mtDNA sequences similar to contemporaneous humans, such sequences may be regarded erroneously as modern contaminants (Trinkaus, 2001).

Serre *et al.* (2004) retrieved mtDNA from four Neanderthals and five early modern humans. To minimize the problem of contamination of current genetic material on early modern humans, the authors concentrated on a region that contains two particular substitutions found in previously studied Neanderthals. Their results showed that "All

Figure 9.37 Specimen Lagar Velho 1; Lapedo child's burial from the Abrigo do Lagar Velho (Portugal). Picture from http://artsci.wustl.edu/~anthro/blurb/LagarVelho1.jpg.

Box 9.30 Was the the Lapedo child a Gravettian child?

In their study on the Lapedo specimen, Lagar Velho 1, Tattersall and Schwartz (1999) stated: "The archaeological context of Lagar Velho is that of a typical Gravettian burial, with no sign of Mousterian cultural influence, and the specimen itself lacks not only derived Neanderthal characters but any suggestion of Neanderthal morphology. The probability must thus remain that this is simply a chunky Gravettian child, a descendant of the modern invaders who had evicted the Neanderthals from Iberia several millennia earlier." As we will see in section 10.1, Mousterian culture is common in Neanderthal sites, although it is also present in some of the Near-East modern-human sites. Gravettian is characteristic of modern humans that arrived at Europe.

---

four Neanderthals yielded 'Neanderthal-like' mtDNA sequences, whereas none of the five early modern humans contained such mtDNA sequences, even though they were as well-preserved as the Neanderthals" (Serre *et al.*, 2004). Serre and colleagues used these results to elaborate a statistical model based on coalescence theory, which "excludes any genetic contribution by Neanderthals to early modern humans larger than 25%. However, any direct evidence of such a contribution has yet to be found, so it is quite possible that no such contribution took place" (Serre *et al.*, 2004).

In addition to serving as synthesis of ancient mtDNA cloning studies, the work performed by Lalueza-Fox and colleagues (2005) offers new information on the episode that saw the replacement of *H. heidelbergensis* by *H. neanderthalensis*. The two explanatory hypothesis for the substitution of the former chronospecies by the latter assume (1) a gradual emergence of distinctive Neanderthal features through chronospecies continuity and (2) emergence of Neanderthals as the result of a clearly defined speciation event, occurring around 300,000–250,000 years ago. Lalueza-Fox *et al.* (2005) argue that "The present genetic data support the latter hypothesis that *H. neanderthalensis* emerged as a distinct biological entity after a speciation event, c. 250,000 years. This event not only coincides with the TMRCA [time to the most recent common ancestor] estimates of the Neanderthal mtDNA variation but also with the appearance in Europe of the cultural Mode 3 industry and a decrease in the morphological variation observed in *H. heidelbergensis*".

9.3.1 Genes and populations

Importantly, molecular methods may contribute to the understanding of the evolutionary relations of late-Pleistocene hominins. The retrieval of ancient nuclear DNA affords the opportunity of determining whether Neanderthals contributed to the genetic pool of modern humans and, if the answer is positive, to what extent. The mtDNA only offers some indications, although these consistently show that "there is no valid evidence of any mtDNA gene flow between Neanderthal and early modern humans" (Serre *et al.*, 2004). That is, the Neanderthal mtDNA sequences cloned up to now are not present in anatomically modern humans. This could be for two reasons. First, Neanderthal mtDNA might have evolved in a separate lineage to that of anatomically modern humans, a scenario that favors the out-of-Africa hypothesis and considering the Neanderthals and modern humans as two different species. In theory at least, it could be that Neanderthal mtDNA disappeared in modern humans due to genetic drift. As several authors have noted (Kahn and Gibbons, 1997; Höss, 2000; Cooper *et al.*, 2001; Paunovic *et al.*, 2001; Beauval *et al.*, 2005), some genetic exchange between Neanderthals and modern humans cannot be discarded on the grounds of current evidence. However, assuming that human populations have consisted of 10,000 reproductive individuals per generation, it is possible to exclude the possibility that the amount of DNA coming from Neanderthal ancestors would have been greater than 25% of the total (Serre *et al.*, 2004; Figure 9.36).

The results of mtDNA cloning studies contrast with the anatomical study of some late Neanderthal specimens, which have been interpreted as evidence of hybridization. This is the case of the almost complete skeleton of a Neanderthal child discovered in 1998 in the late site (24,500 years) of Abrigo do Lagar Velho (Portugal; Duarte *et al.*, 1999). Morphological examination of

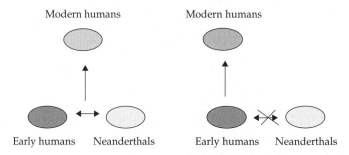

Figure 9.38 Alternative views of Neanderthal genetic contribution to modern humans. Left: early humans and Neanderthals interbred and both populations contributed their DNA to modern humans. Right: early humans and Neanderthals did not interbreed; rather, only early humans contributed their DNA to modern humans. The fact that there was no hybridization does not necessarily imply that early humans and Neanderthals belonged to different species.

the specimen (Lagar Velho 1, named the Lapedo child; Figure 9.37) reveals, in the opinion of the discoverers, mosaic evolution. Some traits, such as the incipient chin, the width of the molar crowns, the relative size of the thumb's proximal and distal phalanxes, are close to traits considered derived in anatomically modern humans. Other traits, such as the size of the incisors or the proportions of the femur, are characteristically Neanderthal. Trinkaus *et al.* (1999) concluded that Neanderthal and anatomically modern human populations interbred, at least on the Iberian Peninsula.

Solid evidence of such interbreeding would speak against the hypothesis of a replacement of one population by another without genetic exchange. But the authors who interpreted the Lagar Velho finding have advanced an additional argument. Trinkaus *et al.* (1999) believe that such genetic exchange was made possible by the similarities in Neanderthal and modern-human behavioral traits. In their opinion, the most important implications of Lagar Velho 1 are not taxonomic or phylogenetic, but behavioral. These authors argue that the morphological mosaic demonstrates that when modern humans dispersed throughout the Iberian Peninsula they encountered Neanderthal populations. Both groups would have recognized each other as humans with similar behavioral strategies, social systems, communication structures, and adaptive strategies. As Trinkaus and colleagues (1999) said about the two populations that they believe coincided at Lapedo: "There may well have been significant cultural contrasts (one was Middle Paleolithic and the other was Upper Paleolithic), but the fundamental differences must have been relatively subtle. In the perspective of mid last glacial humans in Iberia, they were all people."

The perspective taken by Trinkaus and colleagues (1999) is not, however, the only way to interpret the Lapedo specimen. The interpretation of anatomical traits is often controversial, more so when the specimen is infantile, as is the case with Lagar Velho 1. Tattersall and Schwartz (1999) argued that it was an anatomically modern human (see Box 9.25).

Anatomical analyses cannot settle whether the infantile specimen from Lagar Velho was a Neanderthal or a modern human. But we must not forget William Goodwin's warning: it is possible that Neanderthals and humans interbred but produced sterile offspring (Chang, 2000). Thus, proving that the Lapedo child was a hybrid of a Neanderthal and a modern human would not necessarily mean that these populations belonged to the same species. This would be shown if Lagar Velho 1 specimen were a fertile hybrid.

9.3.2 Populations and species

For reasons related to the logic of the scientific method, positive particular hypotheses, such as "Neanderthal mtDNA is present in current humans" can be demonstrated simply by finding a case that proves them. To date, this has not been achieved. But, it is not possible to demonstrate universal negative statements, such as "Neanderthal genetic material is not present in current humans". Even if in all tests the negative hypothesis has been supported, its validity is only provisional. It is always conceivable that a future experiment might prove it wrong. However, the accumulation of experiments that systematically show differences between Neanderthal and modern human mtDNA represents strong support for the hypothesis that they were two separate groups, with no shared genetic flow between them.

Were they two populations or two species? The most likely evolutionary scenario suggests that, coinciding with the age of the Dmanisi fossils, a hominin population left Africa to rapidly colonize Asia. The two hypotheses, out-of-Africa and multiregional, differ in their account of what happened after this episode. The former argues that the new populations of Asian hominins remained separated from the early African ones. The latter assumes there was a continuous genetic flow.

Let us assume that Neanderthals are descendants from Asian populations after their migration to Europe. (If they were successors of some new immigrants to Europe, but from Africa and not Asia, this scenario would not change much

Figure 9.39 Carolus Linnaeus.

except for the timing of the episodes.) Let us also assume that modern humans are descendants of an African population. We will refer to the Neanderthal population arriving at Europe as Np, and the modern human population that evolved in Africa as Mhp. The two alternative hypotheses are: H_1, that there was no genetic flow between Np and Mhp (replacement); H_2, that such flow did in fact occur (hybridization; Figure 9.38).

How can we test these hypotheses? It can be done, to a certain extent, by genetic comparison between Neanderthals and modern humans. In this respect, mtDNA studies support H_1. But, for reasons that have to do with negative universal hypotheses, this support can only be provisional.

Let us now assume that, in spite of the provisional nature of the argument, we are convinced that there was no such genetic flow. The next question is why this happened. Np and Mhp might not have shared genetic flux because they evolved to become two different species after their geographic separation. But it could also be the case that, while belonging to the same species, there was no genetic flow because the populations remained isolated in different places. Such

flux could not have taken place without physical contact between them. Thus, the possible alternatives are:

1 Np and Mhp constitute two different species, *H. neanderthalensis* and *H. sapiens*,
2 the subspecies hypothesis, *H. s. neanderthalensis* and *H. s. sapiens*,
3 Np and Mhp are just two populations of the same taxon.

Choosing one of these alternatives is a decision that cannot be made strictly in terms of ancient DNA retrieval, for the moment. It depends on other factors, especially in common paleontological practice. These practices impose the use of the concept of biological species to name fossil taxa and, thus, throughout this book we have referred to *Orrorin tugenensis*, *Australopithecus afarensis*, *Paranthropus robustus*, and *Homo erectus*, among many others. But, as we know, the technical means available today do not allow testing the criteria relative to biological species (whether they can produce fertile progeny) for these fossil "species".

The case of Neanderthals is similar to that of any other fossil group. If we compare Neanderthals with modern humans, mtDNA studies support the hypothesis of two different populations, Np and Mhp, which did not exchange genetic material. Serre and colleagues' (2004) estimation of a flow below 25% between both populations does not mean that there was that amount of genetic exchange. It only establishes that, if indeed there was genetic exchange, it could not have been above that limit. Negative evidence suggests, thus, there was no or very little exchange. This fact would not help us decide whether they were two populations or two species.

Why do discussions such as the one concerning the Lagar Velho 1 specimen become heated? As Clark (1997a, 1997b) noted, this might be a paradigmatic problem. The hybridization and replacement hypotheses are grounded on radically different preconceptions about humans' remote past. Hence, the variables researchers measure and the methods they employ to do so, such as mtDNA comparisons, cannot resolve this issue. The debate has to do more with implicit concepts about what a

Box 9.31 The number of species in the genus *Homo*

Not all anthropologists recognize *H. neanderthalensis*, or even *H. erectus*, as valid taxa, different from *H. sapiens*. Most notable among authors who greatly reduce the number of recognized species among the hominids is Milford Wolpoff (see Wolpoff *et al.*, 1994).

A reduction of the number of species within the genus *Homo*—or within any other genus—becomes necessary if chronospecies are not recognized. In that case, a monophyletic lineage may be considered as only one species. If an author holds that no cladogenetic events have occurred in the lineage of the genus *Homo*, then all *Homo* exemplars would need to be included in *H. sapiens*, which was the first taxon identified within the genus *Homo*. In this book we don't hold such extreme reductionism.

Box 9.32 Reproductive isolation

Why are all people from different ethnic groups considered humans, from the Inuit Eskimos to African Hazda and to Japanese, Amerindians, and whites? The criterion of reproductive isolation for recognizing species is applicable in this case. It is obvious that substantial hybridization occurs between living human populations. This criterion, however, is not readily applicable to the fossil record. It, therefore, becomes necessary to identify the apomorphies of modern humans so as to have a valid criterion for identifying ancestral modern humans.

Table 9.6 *Homo sapiens* apomorphies

Feature	Reference	Annotations
Globular braincase	Lieberman (1998)	Features related to a reduction of prognathism (Spoor *et al.*, 1999)
Vertical forehead	Lieberman (1998)	
Diminutive browridge	Lieberman (1998)	
Canine fossa	Lieberman (1998)	
Pronounced chin	Lieberman (1998)	
Universal loss of robustness	Howells (1989)	
Small size of dental crowns	Kraus *et al.* (1969), Hillson (1996)	
Tendency to reduce the number of cusps and roots	Kraus *et al.* (1969), Hillson (1996)	
More elongated distal limb bones	Trinkaus (1981) Holliday (1995)	The sharp definition of postcranial contrasts between modern humans and Neanderthals has more
Limbs that are long relative to the trunk		to do with the uniqueness of
Narrow trunk and pelvis	Ruff *et al.* (1997)	Neanderthal morphology than
Low body mass relative to stature	Ruff *et al.* (1997) Pearson (2000)	with the ability to adequately define the characteristic features of
Pelvic shape including a short, stout pubic ramus, and relatively large pelvic inlet		*H. sapiens* (Pearson, 2000)
Aesthetic values		Functional features
Symbolic creative language		
Moral judgments and moral values		

The annotations in the third column refer to Wood and Richmond's (2000) explanations, but we have added the functional features (for their phylogenetic justification, see Chapter 10). *Source*: after Wood and Richmond (2000).

human being is than with the origins of our own species. For reasons that have to do with the concept of biological species and with currently available molecular evidence, we favor that Neanderthals and modern humans are separate species. Placing Neanderthals in another species does not imply viewing them pejoratively. Neanderthals are no longer reduced to the brutish image they were associated with for a time (see Tattersall, 1998). Molecular methods do not provide information about what sort of humans the Neanderthals were, or about their apomorphies. The archaeological record is the best source of information as to what Neanderthals were.

9.4 The origin of modern humans

Carolus Linnaeus (Figure 9.39) placed humans in the taxon *Homo sapiens*, which would be the starting point for investigating modern humans from a taxonomic perspective. As we have said throughout this book, our species belongs to the order Primata, superfamily Hominidea, family Hominidae, tribe Hominini, and genus *Homo*.

If we apply to *H. sapiens* the criteria that we have used with other hominin species, a starting point would be to identify the holotype and paratypes that constitute the hypodigm of modern humans. This is not necessary or possible. Linnaeus did not define a holotype for *H. sapiens*, nor is this a matter of concern for any anthropologist. We know who are the members of the taxon *H. sapiens*. Yet, on reflection, we realize that it is far from easy to list the apomorphies that identify our species.

Nevertheless, if we include in *H. sapiens* only modern humans and their ancestors that are not ancestors of *H. neanderthalensis* as well, we need to define the autapomorphies of *H. sapiens*. However, the abundance paradox becomes apparent. It seems easy to identify the traits characteristic of modern humans, because we have an immense number of exemplars. But their abundance entails enormous intraspecific variation, which handicaps any effort to establish quantitative measurements. Consider, for example, the paradox faced by Stringer *et al.* (1984) seeking to specify the cranial measurements of

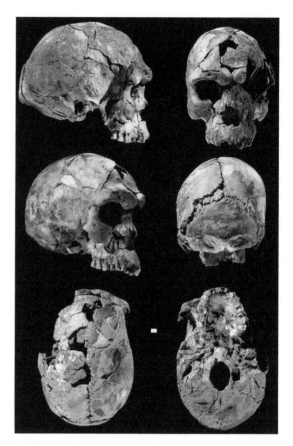

Figure 9.40 The Herto BOU-VP-16/1 adult cranium in (from top left) lateral, frontal, three-quarter, posterior, superior, and inferior views. Scale bar, 1 cm. Photograph from White *et al.* (2003). Reprinted by permission from Macmillan Publishers Ltd. *Nature* 423: 6941, 742–747, 2003.

H. sapiens. If bounds are used that would distinguish modern humans and their direct ancestors from other taxa, those bounds would exclude many living humans from the taxon. The issue is not, of course, how to know who is human and who is not. Fortunately the ethnocentrism of past generations that classified the so-called inferior races as arboreal primates disappeared long ago. The issue, however, is how to identify the taxon of modern humans so that we can decide whether particular fossils belong or not within the taxon, when this is not readily obvious.

The observation of modern humans makes apparent their distinctive apomorphies, anatomical

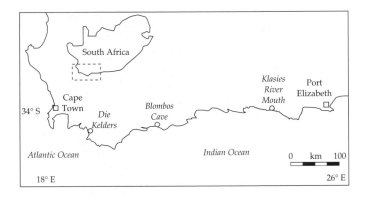

Figure 9.41 The South African sites of Die Kelders, Bolombos, and Klasies. From Grine *et al.* (2000).

as well as functional. But distinctive with respect to which other species? We have a large brain, bipedal gait, use clothing as well as an advanced technology, which includes airplanes and computers, and a well-developed culture, which includes art, literature, legal codes, and political institutions. Certain traits that seem distinctive in comparison with other primates, such as the absence of hair from much of our body—we are the naked ape—occur in other mammals, such as mole rats and whales. However, these mammals are not bipedal, nor do they have an advanced technology or literature and art. But bipedal gait, even though it distinguishes *H. sapiens* from other apes, was already present in *H. erectus*. More generally, direct comparison of modern humans with chimps and other close relatives is likely to yield plesiomorphies, primitive features inherited from ancestral hominins, in addition to autapomorphies that would characterize *H. sapiens*.

To identify valid human autapomorphies, we need to compare the traits of modern humans with those of our sister taxon, *H. neanderthalensis*, assuming, of course, that this taxon is a different species. William Howells (1973, 1989) made that comparison with respect to cranial measurements. More generally, Wood and Richmond (2000) have listed the apomorphies of *H. sapiens* (Table 9.6). These apomorphies, as noted by Wood and Richmond, seem to fit best modern humans from hot, arid climates. This is hardly surprising since modern humans evolved in tropical Africa and their earlier expansion was through tropical or subtropical lands.

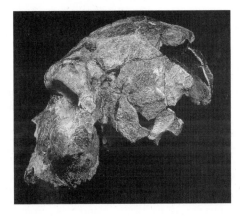

Figure 9.42 Bodo Man; *H. sapiens*? From Clark *et al.* (1994).

9.4.1 The first modern humans: fossil evidence

Modern genetic (molecular) methods have provided the best estimates of the time when modern humans first evolved. But seeking to find the fossil remains of the first modern humans remains a challenge. As we saw in section 9.1, specimens found in Skuhl and Qafzeh (Israel), dated to about 100,000 years ago, are thought to be early representatives of modern humans. But if *H. sapiens* evolved some 150,000 years ago in Africa, where are their fossil remains?

There are several, including two crania (Irhoud 1 and 2) and a juvenile mandible (Irhoud 3) from Djebel Irhoud, Morocco. Although their age is not well known, it seems consistent with generally accepted molecular dates (but see below). These fossils were first classified as Neanderthals

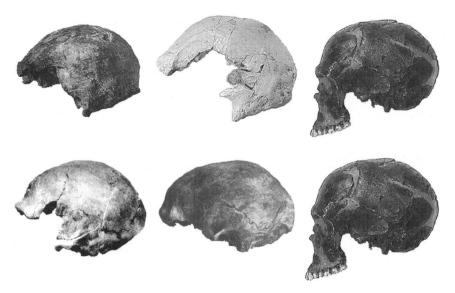

Figure 9.43 Top, left to right: Ngandong 1, *H. erectus*; WLH-50, modern human; Qafzeh 9, modern human. Bottom, left to right: Spy 2, Neanderthal; Mladec 5, modern human; Qafzeh 9, modern human. From Wolpoff *et al.* (2001).

(Ennouchi, 1963), but later classified as *H. s. sapiens* (Hublin and Tillier, 1981). Also from Morocco are a number of fossils dated between 60,000 and 90,000 years (McBrearty and Brooks, 2000): several crania, mandibles, and maxilla from Dar-es-Soltan (Ferembach, 1976), and mandibles found in Temara (Vallois, 1960) and Zouhrah (Close, 1984). A partial juvenile skeleton has been found in Taramusa, Egypt (Vermeersch *et al.*, 1998), as well as several cranial and mandibular fragments from three individuals in Soleb, Sudan (Giorgini, 1971).

Other possible early *H. sapiens* fossils include the following. One partial skeleton (Omo I), about 130,000 years old, and an older cranium (Omo II) found in Omo, Ethiopia, in the members KHS and PHS of the Kibish formation (Day, 1969), although their attribution to *H. sapiens* has been questioned. There are one infantile and two adult crania, dated to 160,000–154,000 years ago from Herto, Ethiopia (Figure 9.40); several molars, dated to 130,000–110,000 years ago from Mumba Rock Shelter, Tanzania (Bräur and Mehlman, 1988); and cranial and mandibular fragments from two individuals of uncertain age from Kabua, Kenya (Whitworth, 1966). There are a number of fragments found in South African sites, such as Border Cave (de Villiers,

1973), Klasies River Mough (Singer and Wymer, 1982), Equus Cave (Grine and Klein, 1985), Witkraus Cave (McCrossin, 1992), Hoedjiespunt (Berger and Parkington, 1995), and Die Kelders Cave (Grine *et al.*, 1991; Figure 9.41). Yet, in spite of this wealth of African remains, the perception exists that the oldest known modern humans are from the Near East.

Grine's (2000) analysis of the exemplars from Kelders Cave 1 provides some insight about this matter. There are at least 27 exemplars—24 teeth, a mandibular fragment (AP 6276) and two phalanges (AP 6267 and 6289)—proceeding from 10 individuals, many of them infantile, dated to 80,000–60,000 years ago. The teeth's morphology seems modern, but the morphological variants shared with modern humans are, for the most part, plesiomorphies (Grine, 2000). They do not allow determination of their phylogenetic position, nor confirm that they are indeed *H. sapiens*. Hilary Deacon (1989), nevertheless, based on the available evidence in Klasies River, placed the presence of modern humans in South Africa at 100,000 years. The evidence came from artifacts and decorated objects from Klasies River, which display, according to Deacon, social and cognitive practices similar to those of our species. The use of symbolism

Box 9.33 Coalescence theory

Coalescence theory examines the genealogical relations between genes (Griffiths, 1980; Hudson, 1990). According to this theory, all genes (alleles) present in extant populations must have descended from a single gene, to which they coalesce. That is, the phylogeny of individual genes is star-like, with the most recent common ancestor at the vertex of the star, as shown in the figure (Slatkin and Hudson, 1991).

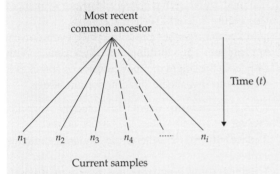

The theory was first formulated for neutral or nearly neutral genes, which are genes that do not modify the welfare of the organism. In a randomly mating population at equilibrium, the mean coalescence time of neutral genes is given by

$$T = 4N_e[1 - (1/i)] \text{ generations}$$

where T is the number of generations to coalesce, N_e is the effective size of the population, i is the number of sampled genes, and the variance is large (Kingman, 1982a, 1982b; Tajima, 1983; Tavaré, 1984; Takahata and Nei, 1985). For any two genes ($i = 2$), the mean coalescence time reduces to $T \approx 2N_e$ generations; for a large number of genes, the mean coalescence time is $T \approx 4N_e$. Thus, in a population with $N_e = 1$ million

individuals, genes are expected to converge to their one ancestor 4 million generations earlier.

The coalescence equation can be used in the opposite direction, so that we can estimate population size if the coalescence time is known. To determine the time to the coalescent (the most recent common ancestor or MRCA), we need to know the rate of neutral mutation. This can sometimes be determined by the number of neutral substitutions between the genes of two species, of which the time of divergence is known. Under the assumptions of coalescence theory and ignoring the possibility of multiple hits at individual sites; the number of neutral polymorphisms that we observe in a sample of multiple genes will have a Poisson distribution with a mean that depends on the neutral mutation rate, the time elapsed, and the number of lineages examined. The expected number of polymorphisms is

$$\lambda = \mu t \, \Sigma \, n_i l_i$$

where μ is the neutral mutation rate, t is the time since the MRCA, n_i is the number of lineages sampled at the ith locus, and l_i is the number of neutral sites at the ith locus. Solving for t and replacing λ with S, the observed number of polymorphisms:

$$t = S/\mu \, \Sigma \, n_i l_i$$

If S is assumed to have a Poisson distribution, the 95% confidence intervals can be estimated.

When examining genes that are not located in the autosomes one needs to take into account that the relative effective population size of autosomes/X chromoromes/nonrecombining Y/mtDNA is 4:3:1:1. Accordingly the estimated value of N based on mtDNA, for example, is N_f, the estimated effective population size of females, so that $N_e = 2N_f$.

---

and artifacts would account for the adaptive success of Klasies River at a time when climate cooling reduced the availability of resources.

In conclusion, there are quite a number of African fossils, dated around 100,000 years, attributed to *H. sapiens*. But there are older fossils that might possibly be *H. sapiens*. Bräuer *et al.* (1997) proposed that the oldest modern humans would be a

cranium (KNM-ER 3884) and a femur (KNM-ER 999) from Koobi Fora (Kenya), dated by uranium series at 270,000 and 300,000 years, respectively. The 640,000-year-old specimens from Bodo (Clark *et al.*, 1994) would make the origin of modern humans much older, if this Bodo man, or *Hombre de Bodo*, is thought to be a modern human (Figure 9.42). If the exemplar from Danakil (Abbate *et al.*,

1998) is considered a modern human, the date would climb to about 1 million years. These dates are considerably older than those derived from most molecular analyses.

Proponents of the multiregional hypothesis of the origin of modern humans disagree with the proposals just reviewed. According to Hawks *et al.* (2000) and Wolpoff *et al.* (2001), a suitable test of the out-of-Africa hypothesis is provided by the specimen WLH-50 from Villandra Lakes, Australia. These two sets of authors compared this 15,000-year-old modern human with several Skuhl and Qafzeh specimens from the African late Pleistocene, as well as to *H. erectus* from Ngandong (Figure 9.43). Their statistical analysis of all available cranial measurements does not show differences that would justify classifying them in different species. These authors concluded that their results reject the out-of-Africa hypothesis, support the multiregional hypothesis, and require that the Ngandong specimens be classified as *H. sapiens*. However, the multivariate analysis of inter- and intra-group variation of African and European specimens from the early, middle, and late Paleolithic and of the Iberian Mesolithic, carried out by Turbon *et al.* (1997), results in a strong conclusion that Neanderthals and anatomically modern humans belong to two separate monophyletic lineages. A similar conclusion had earlier been reached by Bräuer and Rimbach (1990).

9.4.2 Insights from molecular evolution

We have stated repeatedly in this book that molecular genetics has become over the last few decades a powerful method for investigating evolutionary questions. We saw in section 1.2 how Fitch and Margoliash (1967) were able to reconstruct the evolutionary history of 20 very diverse species, from yeast to insects to vertebrates, including humans, by comparing their cytochrome *c*, a small protein consisting of only 104 amino acids, involved in energy production in the cell and present in virtually all organisms. The phylogenetic relationships in the tree of Fitch and Margoliash (1967) were not all accurate, but it seemed astonishing at the time that one small protein,

without any other evolutionary information, would provide so much phylogenetic information about organisms that had started diverging more than 1 billion years ago. It became immediately apparent that additional protein or DNA sequences would provide valuable phylogenetic information, so that the evolutionary history of all living organisms could be reconstructed just by obtaining their protein or DNA sequences, and that greater and greater precision could be obtained by studying more and more sequences (see Figure 1.21 and section 1.2).

Molecular genetics provides other evolutionary information, in addition to phylogeny. With respect to the origin of modern humans, it makes it possible to investigate three important issues:

1 time of origin: when did modern humans evolve?
2 place of origin: where did modern humans evolve?
3 demography: how large was the original population of modern humans?

We explored these questions in section 9.3, when comparing modern humans with Neanderthal fossils. We will now explore these questions further. As we shall see, these questions are more complex than just the simple formulation given above. Thus modern humans may have evolved primarily in tropical Africa, as seen in section 9.3, but some molecular data suggest that other populations may have contributed to the gene pool of modern humans. This, in turn, raises the question of whether the origin of modern humans may have more than one time frame, if contributions to their gene pool from non-African populations (or from African populations other than the main population where the transition to modern humans took place) occurred at different times. There is also the related question of the time when other continents were colonized by modern humans. And did the colonization occur primarily at one time or did it occur in several waves, at different times? The question of demography is particularly complex. At issue is not only the size of the original African population where modern humans originated, but also the size of the popu-

The evolutionary cradle of humanity

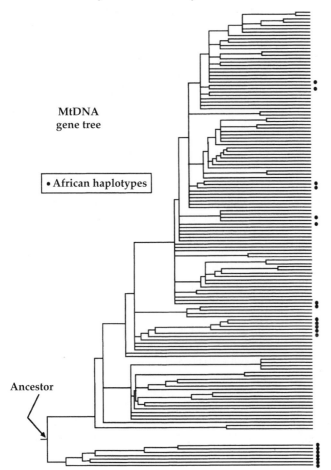

**MtDNA
gene tree**

• **African haplotypes**

Ancestor

Figure 9.44 MtDNA phylogeny of modern humans. The dots mark mtDNAs from native Africans; the others are from native Europeans, Asians, Australians, and New Guineans. From Avise (2006), redrawn after Cann *et al.* (1987).

lations that colonized other continents. Also, were there any bottlenecks in Africa after the origin of modern humans or in other continents after their colonization? Molecular genetics makes it possible to detect not only population bottlenecks but also patterns of demographic expansion.

We shall now turn to these questions, starting with the early mtDNA investigations, which placed the origin of modern humans in Africa.

9.4.3 Out of Africa

As stated in section 9.3, the molecular evidence favors the out-of-Africa hypothesis over the multiregional (or hybridization) model of the ori-

gin of modern humans. We will now review the evidence and explore two questions: (1) whether the migration out of Africa occurred as a single event or whether it involved successive colonizing waves; and (2) whether non-African populations may have contributed to the gene pool of modern humans, through hybridization with the African colonizers (Clark 1997a, 1997b; Garrigan and Hammer 2006).

The multiregional hypothesis suggests that the transition from *H. erectus* to modern *H. sapiens* took place concurrently in several regions of the Old World, involving several intermediate popula-tions, and with frequent genetic interchanges that maintained the species' unity. The out-of-Africa hypothesis states, on the contrary, that modern humans appeared somewhat before 100,000 years ago in Africa, from where they dispersed to the rest of the world. Previous hominin populations of the rest of the world (*H. neanderthalensis* and *H. erectus* and its descendant species) either disappeared on their own without leaving descendants, or were replaced by modern humans migrating out of Africa.

The multiregional hypothesis was formulated on the basis of fossil evidence. Advocates of this model underline what they interpret as fossil continuity in the transition from *H. erectus* to "archaic" *H. sapiens*, and thereafter to modern humans, in Australasia, the Middle East, and other regions. These authors postulate that there were periodic genetic exchanges between populations of different regions, such that, despite geographical dispersion, the species evolved as a single genetic pool. Nevertheless, some geographical differentiation gradually emerged, which is currently reflected in genetic and morphological differences between ethnic groups (Wolpoff *et al.*, 1988; Clark and Lindly 1989; Bräuer and Mbua, 1992; Clark, 1992; Thorne and Wolpoff, 1992; Waddle, 1994; Templeton, 2002).

The out-of-Africa hypothesis was formulated largely on the basis of mtDNA molecular data. It proposes that anatomically modern humans evolved in Africa about 200,000–150,000 years ago. Starting about 100,000 years ago, they dispersed from there to the rest of the world, replacing any pre-existing human populations, whether *H. erectus* or "archaic" *H. sapiens* or *H. neanderthalensis*, although the replacements may have occurred at different times in different regions (Cann *et al.*, 1987; Stringer and Andrews, 1988; Stoneking *et al.*, 1990; Vigilant *et al.*, 1991; Stringer, 1992; Ruvolo, 1993; Cavalli-Sforza *et al.*, 1994; Goldstein *et al.*, 1995; Horai *et al.*, 1995; Rogers and Jorde, 1995).

The reconstruction of the mtDNA genealogical tree places its roots—the origin of ancestral mtDNA—in Africa (Cann *et al.*, 1987; Stoneking *et al.*, 1990; Vigilant *et al.*, 1991; Ruvolo, 1993; Horai *et al.*, 1995; Figure 9.44). Early mtDNA studies focused on the control region, which represents less than 7% of all mitochondrial genetic information and does not have a coding role. A study of the complete mtDNA (16,500 nucleotides in length) from 53 individuals confirmed the same African origin (Ingman *et al.*, 2000). The mtDNA evidence would not be conclusive by itself, given that mtDNA constitutes a tiny fraction of the total human DNA. Each of the two nuclear genomes of a human consists of about 3 billion nucleotides, which is about 250,000 times more than the mtDNA. But in the 1990s chromosome DNA microsatellites (Goldstein *et al.*, 1995) and of a large sample of nuclear genes spread throughout the entire human genome (Cavalli-Sforza *et al.*, 1994) also yielded genealogical trees rooted in Africa. Recent investigations of DNA from the autosomes and X chromosomes yield a more complex picture, as we shall see below.

In the mtDNA tree, ancestral African populations are set apart from all non-African populations, which are located on a single branch emerging from the multi-branched African tree. The most profound divergence of non-African populations in the genealogical trees is calculated at about 150,000 years ago (with a possible error of tens of thousands of years). The time estimates to the most recent common ancestor of modern humans vary from one study to another; for example, Ingman *et al.* (2000) set the time at 175,000 years. In any case the first divergence between African and non-African populations would mark the earliest possible point in time at which modern humans would have dispersed from Africa to the rest of the world. Ethnic differentiation among modern populations would be a relatively recent event, a result of divergent evolution among populations separated only for the last 50,000 or 100,000 years. This conclusion, emerging from the genealogical trees, is consistent with extensive studies of genetic polymorphism, showing that living human populations from different parts of the world are not greatly differentiated genetically (see below).

Advocates of the multiregional hypothesis have, however, presented supporting mtDNA data that are inconsistent with the out-of-Africa hypothesis. As we saw in section 9.3, the mtDNA sequenced from 10 fossil specimens of anatomically modern humans, retrieved from two different regions of Australia, most dated to 2,000–15,000 years old, but one, LM3, around 60,000 years old, have shown a mtDNA sequence in specimen LM3 which is absent from the other ancient specimens, as well as from present-day modern humans (Zimmer, 1999b). The inference is that the genetic diversity of this sample is much higher than expected under the scenario of a recent modern-humans origin, thereby supporting the multi-regional hypothesis.

However, alternative interpretations have been suggested (section 9.3). In particular, Pääbo (see Holden, 2001; Zagorski, 2006) has argued that the Australian investigators failed to maintain the necessary precautions for avoiding contamination, and that, in any case, what the results would show is that the mtDNA polymorphism of anatomically modern humans is higher than previously thought (see Zimmer, 1999b).

9.4.4 The ancestral ZFY gene

The Y chromosome is the genetic counterpart of mtDNA in that it is inherited only from fathers to sons. There are regions on chromosome Y that are not homologous to chromosome X and thus are transmitted only through the paternal line. A DNA fragment of 729 nucleotides of the *ZFY* gene (probably involved in testicle or sperm maturation) found on chromosome Y was sequenced in 38 men representative of major ethnic groups by Dorit and colleagues (1995). These authors concluded that the origin of modern-human Y chromosomes dates back to a Y chromosome close to 270,000 years ago (but with a confidence margin extending from zero to 800,000 years).

The ancestral Adam from whom all living men have inherited the Y chromosome was not, however, our only male ancestor in his own or any other generation. Similarly, the woman from whom all modern humans have inherited their mtDNA was not the single woman of her generation ancestral to modern humans. The rest of our genes other than ZFY and mtDNA come from many other different male and female ancestors

(Ayala, 1995). This is an important point to which we now turn.

9.4.5 The myths of the mitochondrial Eve and the ZFY Adam

Most of the human genetic information is found in the chromosomes and is inherited from both parents (except for the non-recombining segment of the Y chromosome). In contrast, the amount of DNA in the mitochondria is relatively small and follows a matrilineal inheritance pattern. The nonrecombining segment of the Y chromosome is also small and paternally inherited. The ancestral mtDNA sequence was unfortunately named the mitochondrial Eve (Cann *et al.*, 1987; see also Stoneking *et al.*, 1990; Vigilant *et al.*, 1991). This Eve, however, is not the only woman from which all present day humans descend, but an mtDNA molecule from which all current mtDNA molecules descend. Similarly, the ancestral *ZFY* is not the only man from which all modern humans descend, but a gene from which all current *ZFY* genes descend.

Coalescent theory shows that all sequences of a given gene coalesce back in time to a single sequence, from which all living sequences derive. However, the media and even some scientists (e.g. Brown, 1980; Lowenstein, 1986) made the erroneous inference that the mtDNA data showed that all women descend from a single woman. This erroneous inference stems from the confusion between a gene genealogy and a genealogy of individuals. This may be illustrated with an analogy. A present-day surname can be shared by many people, even on different continents, but it may have a singular origin centuries ago. If we accept that the surname is transmitted only from father to sons, all those carrying the surname will be descendants, by paternal line, from the "founder", the family's "Adam", but those people will also descend from many other men and women who lived in the same generation as the founder, as well as before and after the founder. Similarly, many contemporary women to the mitochondrial Eve have left descendants in modern humanity, contributing with nuclear genes.

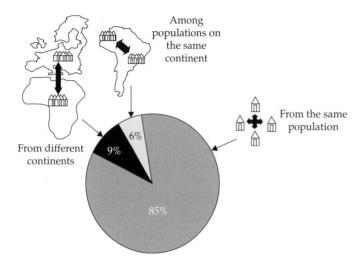

Among populations on the same continent

From the same population

From different continents

6%

9%

85%

Figure 9.45 Genetic diversity in human living populations. Most (85%) of human genetic variation can be found within a single village. Populations from other villages of the same continent contribute an additional 6%, and those from different continents an additional 9% of humankind's total genetic variation.

The legitimate conclusion of the mtDNA analysis is that the mitochondrial Eve is the matrilineal ancestor of modern humans. Everyone has a single matrilineal ancestor in any given generation. Everyone inherits the mtDNA from the mother and, in turn, from the maternal grandmother and from the maternal lineage great-grandmother, and so on. But every person also inherits other genes from the other three great-grandmothers and from their four great-grandfathers. The mtDNA we inherit from the mitochondrial Eve represents a small fraction of our total DNA. The rest of DNA has been received from other individuals contemporary or not of the mitochondrial Eve.

The coalescence of the mtDNA of modern humans into a single ancestor is a feature that necessarily occurs for any one gene or genetic trait. As one proceeds back in time, at any gene locus (or DNA segment) all 2N genes of a species with N individuals derive from fewer and fewer ancestral genes, eventually converging into a single gene ancestor to all 2N descendants. But the ancestral genes for different gene loci occur in different generations and, of course, different individuals. The genome of each living human individual derives from many ancestors. The converse of this is the nonintuitive inference that any human who lived a few thousand generations ago and who has living descendants is an ancestor of all living

individuals (Rohde *et al.*, 2004; Hein, 2004), although he/she would have contributed different genes to different living individuals.

Cavalli-Sforza and colleagues (1994) pointed out discrepancies between the calculated bifurcation time between African and non-African populations based on nuclear genes (about 100,000 years ago) and mtDNA (close to 200,000 years ago). This is hardly surprising. Divergence time estimates in such studies show great variation, largely due to the limited data-set they are based on. It is unsurprising, therefore, that mtDNA polymorphism coalescence has been estimated at 143,000 years by Horai and colleagues (1995) and at 298,000 years by Ruvolo (1993), with confidence intervals ranging from 129,000 to 536,000 years. The differences between estimates based on mtDNA, Y chromosomes, and other nuclear genes are also due to gender and social differences in migration patterns (Cavalli-Sforza, 2003). For example, patrilocal marriage has historically been more common than matrilocal, which can explain differences between mtDNA and Y chromosome patterns in different populations. Demographic differences between the sexes, such as greater male than female mortality, the greater variance in reproductive success of males than females, and possibly the greater frequency of polygyny than polyandry, may explain the discrepancy between estimated dates obtained from the nonrecombining part of the Y chromosome and from mtDNA.

In conclusion, at least until recently (see below), molecular evolution data favor an African origin for modern humans, but there is no reason to assume that a severe population bottleneck occurred at the time of origin of modern humans.

Studies of genetic diversity in living human populations have revealed information consistent with a recent origin of all living human populations, as proposed by the out-of-Africa hypothesis. When the genetic diversity of human populations is mapped out geographically, it is found that 85% of it is present in any local population; this is to say, in any village or city of any continent (although the genes contributing to this 85% vary from one population to another). Some 5–6% additional genetic variation is found when local populations on the same continent are compared, and an additional 10% when populations from different continents are compared (Barbujani *et al.*, 1997; Jorde *et al.*, 1997; Kaessmann *et al.*, 1999). This seems at first surprising when we consider how easy it is to distinguish a Congolese, a Swede, and a Japanese. The explanation is that ethnic differences, such as the colour of the skin and other observable morphological features, are associated with a small number of genes, which became differentiated because of their high adaptive value in response to different latitudes and climates. The distribution of genetic variation among populations does not give the time of dispersion of modern humans throughout the world, but suggests that the dispersion could not be very ancient, given the relatively little genetic differentiation existing among continents (Figure 9.45).

9.4.6 Demographic features

The theory of gene coalescence makes it possible to estimate the number of ancestors that were contemporaries of the mitochondrial Eve. The mtDNA is inherited as a single copy, from only one parent, and we may assume that the mtDNA polymorphism is largely neutral. The theory says that the coalescence into a single ancestral molecule is expected to be $T = 2N_f$ generations ago, where N_f is the number of mothers per generation (N_e is the

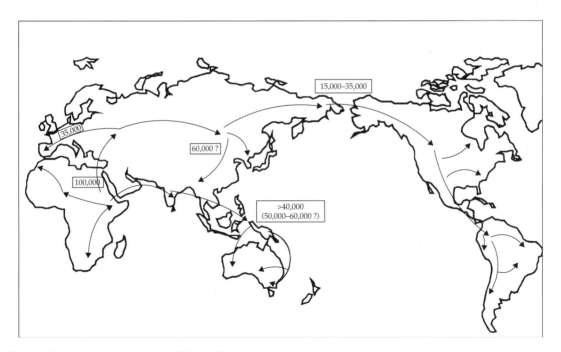

Figure 9.46 Modern human colonization of the world's continents.

effective population size, which includes males and females). This inference assumes constant population size and other conditions unlikely to be the case in reality, so that the conclusions reached are only rough approximations. If we assume 20 years per generation and 200,000 years for the mtDNA coalescence, $T = 10,000$ generations, and therefore, $N_f = 5,000$ mothers, or $N_e = 10,000$ individuals, which is almost certainly an underestimate (Wills, 1995).

Other estimates of the coalescence of the mtDNA are 143,000 and 298,000 years ago (Ruvolo, 1993; Horai *et al.*, 1995). An analysis that takes into account variable substitution rates for different sites of the mtDNA genome, and is supported by computer simulations, yields a coalescent estimate of 889,000–622,000 years ago (Wills, 1995) and corresponds to $N_e = 31,100$–44,450 individuals. As noted above, the estimates of mean coalescence time as a function of N_e, and vice versa, have large variances. When the sample of genes is large, the standard deviation of the mean for mitochondrial DNA is larger than $N/2$ (see Nei, 1987, p. 395, eqn 13.74). The 95% confidence interval coalescence will correspondingly extend at the upper end to more than 88,900 generations and an equal number of individuals. Thus, in spite of considerable uncertainty, the mtDNA results yield a mean population size that ranges between 10,000 and more than 50,000 individuals throughout the Pleistocene. This finding is consistent with estimates based on the histocompatibility locus antigen (HLA) polymorphisms that the ancestral population of modern humans would not have been smaller than about 10,000 individuals (Ayala, 1995).

The expected coalescence of a DNA polymorphism that is transmitted in single copy and paternally inherited, such as *ZFY*, is $T = 2N_m$, where N_m is the number of males. If the human generation is 20 years, the coalescence to the ancestral *ZFY* gene yields an effective population size of 6,750 fathers, or 13,500 humans, with a 95% confidence upper limit of $N_e = 40,000$ individuals. If we account for the standard deviation of the mean coalescence, the 95% upper limit for N_e would increase to 80,000 individuals.

9.4.7 Effective size versus census populations

Two features of coalescence theory should be kept in mind for interpreting N_e as a measure of ancestral human populations: first, N_e is an harmonic mean and, second, it measures the number of reproductive individuals, not the census population.

The harmonic mean of a sample is strongly influenced by the small numbers in the sample. When the number of individuals in a population oscillates from generation to generation, small numbers, such as during a population bottleneck, have a disproportionably large effect on the value of N_e. Consider a sequence of 10 numbers that includes nine 100s and one 10. The arithmetic mean is 91, while the harmonic mean is only 52.6.

Moreover, the parameter N_e that we have used for estimating the size of human populations is a theoretical construct corresponding approximately to the number of reproducing individuals at a given time. The census number is likely to be about four or five times greater. In humans and in other primates, a number of individuals, which may be one-third of the total, do not reproduce at all. Of the females who reproduce, only about one-third are actively reproducing at any one time, the others are juveniles or beyond reproductive age. With the use of these rough approximations, we conclude that N_e is about two-ninths of the census population. A long-term effective population size of $N_e = 10,000$ corresponds, therefore, to a census population of 40,000–50,000 individuals.

Whether or not a population is geographically structured is another factor that impacts N_e. If the human ancestral population was subdivided into several subpopulations, N_e may actually overestimate the size of the population. Population structure also impacts the distribution of rare polymorphisms in an expanding population. We will return to this matter below.

9.4.8 One or several out of Africa migrations?

The prevailing consensus emerging from molecular studies up to a few years ago was that anatomically modern humans evolved about

200,000 years ago and that their speciation involved a bottleneck of some 10,000 reproductive individuals (N_e). An African subpopulation would have migrated out of Africa starting about 100,000 years ago, which colonized Asia and reached Australasia about 60,000–50,000 years ago and Europe only about 35,000–30,000 years ago (Figure 9.46). The reduced variation observed in non-African populations' mtDNA would be attributed to the small size of the African population that colonized the other continents. Expansion of the human population would have occurred fairly gradually in Africa as well as the rest of the world. Gene exchange would have also occurred frequently enough to maintain the overall genetic uniformity of *H. sapiens*.

Recent investigations of nuclear genes, whether from the autosomes or from the X chromosome, generally support an African origin of modern humans. Takahata *et al.* (2001), for example, found that nine out of 10 autosomal gene trees root unambiguously in Africa. One important exception, however, is the study of Garrigan *et al.* (2005) of the ribonucleotide reductase M2 polypeptide pseudogene 4 (*RRM2P4*), which is sex linked. The coalescent analysis of 42 samples, 12 from Asia and 10 each from Africa, Europe, and America, places the root of the tree in Asia with a probability of $P = 0.92$, and $P = 0.05$ for Africa, 0.01 for Europe, and 0.02 for America. Summary statistics of polymorphism in these samples indicate that this region of the genome derives from a large Asian population (Garrigan *et al.*, 2005). But any simple model of *H. sapiens* origins has been challenged recently as a consequence of several analyses of autosomal and X-chromosome-linked polymorphisms that present a different picture from the earlier studies of haploid polymorphisms (mitochondrial and nonrecombining Y chromosome).

If the human population had expanded gradually throughout the Pleistocene after the colonization from Africa, the expectation is that there would be an excess of rare-frequency polymorphisms. This is because as the population becomes large, the opportunity for new mutations to appear increases, and this would have happened more

and more as the population became larger and larger. However, this is not what the chromosomal polymorphisms show; on the contrary, there is an excess of older mutations (Harding, 1997; Hey, 1997; Fay and Wu, 1999; Garrigan and Hammer, 2006). The hypothesis that this excess of older mutations is due to balancing selection (Harpending and Rogers, 2000; Excoffier, 2002) has been rejected because the distribution pattern of the polymorphisms is inconsistent with that explanation (Fay and Wu, 1999; see also Templeton, 2005).

Voight *et al.* (2005) have examined simultaneously the linkage disequilibrium and the distribution spectrum of rare polymorphisms in 50 noncoding autosomal loci in two populations, one Italian and the other Han Chinese. Their conclusion is that there was a period of reduced N_e about 40,000 years ago, such that the bottleneck would have been more severe for the Asian than the European population. Similar studies have led to the conclusion of (at least) two out-of-Africa bottlenecks, ranging 52,000–27,000 years ago (Reich, 2001) and 112,000–58,000 years ago (Marth *et al.*, 2004). Wen-Hsiung Li and his collaborators (Yu *et al.*, 2000, 2002; Zhao *et al.*, 2000, 2006) have investigated five noncoding DNA regions of about 10 kb in length in different chromosomal regions, which yield very different estimates for the age and size of the ancestral population of modern humans. Combining three autosomal regions, they estimate that the long-term effective human population size to be $11,000 \pm 2,800$ or $17,600 \pm 4,700$, depending on the method used. The time of the most recent common ancestor was estimated to be $860,000 \pm 250,000$ years ago (Zhao *et al.*, 2006).

A number of other studies have also challenged the original out-of-Africa model in two respects. First, some polymorphisms are very old and thus inconsistent with the recent origin of modern humans from a single population in Africa. These old polymorphisms have been found on the X chromosome (Harrison and Hey, 1999; Zietkiewicz *et al.*, 2003; Hammer *et al.*, 2004; Garrigan *et al.*, 2005) as well as on the autosomes (Baird *et al.*, 2000; Barreiro *et al.*, 2005; Hardy *et al.*, 2005;

Stefansson *et al.*, 2005; Hayakawa *et al.*, 2006). One notable example is the X-chromosome noncoding locus Xp21.1, which shows two haplotypes that diverged from one another nearly 2 Ma (Garrigan *et al.*, 2005). Two highly diverse haplotypes of the *CMAHp* pseudogene on chromosome 6 are estimated to be about 2.9 million years old (Hayakawa *et al.*, 2006).

One possible explanation for these observations is that other populations may have contributed some of their genes to the main population that gave rise to modern humans. The old age of the haplotypes makes it possible that the contributions may have occurred in Africa or from populations descended from earlier out-of-Africa migrants (*H. erectus*). A disturbing inference is that these old genetic contributions seem to imply that anatomically modern humans were not fully reproductively isolated from other hominin populations; that is, that full speciation had not occurred so that populations of anatomically modern humans were able to incorporate genes from other populations descended from archaic *H. sapiens* or even *H. erectus* (Garrigan and Hammer, 2006). New molecular genetic investigations may make it possible to decide whether or not these challenges to the single out-of-Africa model are confirmed. The predominant role of African populations in shaping the human genome is not challenged, but rather whether genetic contributions from other populations may have occurred (Templeton, 2002, 2005).

Another challenge to the out-of-Africa model is whether there was only one original migration or whether the appropriate model is "out of Africa again and again" (Templeton, 2002). Several recent analyses are consistent with relatively small migrations (gene exchanges) between populations, even from different continents, persisting after the first migrations of *H. erectus* out of Europe until the time of origin of anatomically modern humans, plus a small nonnegligible genetic contribution from pre-*H. sapiens* populations to modern *H. sapiens* after their colonization and population expansion outside Africa (Garrigan and Hammer, 2006). Harding and McVean (2004) have, however, pointed out that sustained population structure (subdivision) may account for the data, without

requiring genetic input from non-*H. sapiens* populations into anatomically modern humans.

Templeton (2002) has drawn two conclusions from an analysis of human haplotype trees for six autosomal regions, two X-linked regions, mtDNA, and Y-chromosome DNA. His analysis shows (1) two major migrations out of Africa, one around 800,000–400,000 years ago and the other around 150,000–80,000 years ago; and (2) persistent genetic interchange between human populations of the same and different continents, implying in effect other migrations out of Africa in addition to the two major ones noted. In a more extensive recent analysis of 25 DNA haplotype regions, Templeton (2005) concludes that an out-of-Africa expansion occurred about 1.9 Ma. This would correspond to the common view of the first out-of-Africa colonization by *H. erectus*. But Templeton further concludes that there was a pattern of migrations between African and Eurasian populations that started 1.5 Ma, without detectable interruptions until the present. The same analysis indicates that there was a second major expansion out of Africa about 700,000 years ago, which involved interbreeding with preexisting Eurasian populations; and that a third out out-of Africa event occurred around 100,000 years ago, which also involved interbreeding.

Other analyses show that a major demographic expansion, out of a small African population, may have occurred first within Africa and then out of Africa, within the range of 80,000–60,000 years ago (Mellars, 2006a, 2006b). This expansion, which happened about 100,000 years after the generally accepted emergence of modern humans, was made possible, according to Mellars (2006a, 2006b), by a major increase in the complexity of the technological, economic, social, and cognitive behavior in the originally small African population where modern humans had evolved. This author argues that the out-of-Africa colonization was a unique event and that the dispersal across Asia into Australasia was rapid, with only a secondary and later dispersal into Europe. Be that as it may, it seems apparent that the availability of large sets of DNA sequence data and powerful mathematical methods of analysis, have not settled matters concerning the origin of modern humans. It seems that neither the out-of-Africa nor the multiregional hypotheses can be maintained in their original formulation. More DNA sequence data and additional analyses are needed before a picture emerges of our origins that would represent a consensus among anthropologists.

One additional point is that positive selection has now been demonstrated in the evolution of modern humans. Wang *et al.* (2006), by means of powerful genomics methodologies, have analyzed 1.6 million single-nucleotide polymorphisms (SNPs) available from recently published genotypic data (Hinds *et al.*, 2005) searching for evidence of positive Darwinian selection. The analysis consists of the search for the expected decay in adjacent SNP linkage disequilibrium, which arises as a consequence of positive natural selection, as the selected site carries along any alleles linked to it. The authors could show evidence of natural selection for about 1,800 genes. The selected genes are involved in major biological adaptations such as reproduction, DNA metabolism, protein metabolism, neuronal function, and host–pathogen interactions. Other investigations of large DNA regions or haplotypes also indicate that positive natural selection may have been pervasive in the evolution of modern humans (Templeton 2005; Garrigan and Hammer, 2006; Zhao *et al.*, 2006).

CHAPTER 10

The uniqueness of being human

10.1 Culture and mental traits during the late Paleolithic

In a previous chapter we discussed the Neanderthals' morphology and their possible evolutionary relationship with anatomically modern humans. What kind of beings were the Neanderthals? What did they think? How did they resolve their adaptive problems? To what extent were their solutions similar to those of our direct late-Paleolithic ancestors? It is becoming increasingly clear that, to understand human evolution, these questions are much more important than the issue of whether or not Neanderthals and modern humans belong to the same species (Clark and Lindly, 1989; Arsuaga, 1999).

The contrast between Neanderthal "brutality" and our "humanity" is sometimes based on a misunderstanding. Cannibalism is a good example. Several Mousterian sites, among which Moula-Guercy (France) stands out, have provided reasonably firm evidence of anthropic action on Neanderthal bones. Moula-Guercy contains several stratigraphic levels (IV–XX), one of which (XV, estimated to 120,000–100,000 years by associated fauna) contains some lithic tools, mammal bones, and some skeletal remains of six Neanderthals. The bones of Neanderthals and the other animals present marks of anthropic action (Figure 10.1).

By means of a taphonomic analysis of the site, Defleur and colleagues (1999) compared the patterns of ungulate and Neanderthal bones, the skeletal parts that were found, and the tool marks on the bone fragments. They concluded that all the fragments with marks indicate prey that had been subjected to butchery. This suggests that cannibalistic practices existed among the Neanderthals. However, the study of the Combe-Grenal Neanderthals carried out by Garralda and Vandermeersch (2000) suggested a different interpretation of the marks. The striae on the bones revealed an obvious defleshing produced by other Neanderthals by means of silex instruments, but the reason for those practices is not so clear. Garralda and Vandermeersch (2000) argue that the use of fire has not been detected on the remains from Combe-Grenal, or those from L'Abri Moula, or any others from different Mousterian sites showing evidence of defleshing. Hence, they argue that the practice was probably unrelated to cannibalism,

---

Box 10.1 The mental divide between Neanderthals and humans

If we leave behind last century's romantic view of Neanderthals as brutes, clumsy and deformed, and instead we dressed them up in any of our neighbor's clothes, would we pick a Neanderthal out among a group of human beings? Maybe not. But would that make him one of us? The intuition that distinctive mental processes, rather than physical traits, characterizes humans among other living primates is very common nowadays. Only human language is characterized by dual patterning, only our species uses ethical codes, and expresses aesthetic values. What would be the case for a Neanderthal?

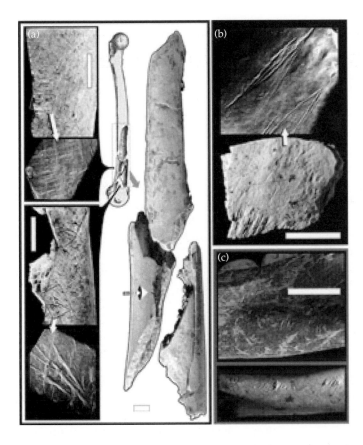

Figure 10.1 Photographic and scanning-electron-microscope images of Neanderthal skeletal remains (Moula-Guercy, France). (a) Distal left femur. White arrows indicate percussion impact scar; anvil striae on the opposite side (upper left), and internal conchoidal scars indicate defleshing before fracture. (b) Ectocranial surface of the left parietal bone with cut marks. (c) The mandibular corpus of a juvenile Neanderthal (top) and a red deer (bottom) to show the similar position and form of cut marks made by stone tools. Scale bars, 1.0 cm. Pictures from Defleur *et al.* (1999).

and suggest it might be part of a funerary ritual. Some of the striae on the bones from Combe-Grenal (III and 567) might have been made for no specific reason.

The notion of Neanderthal cannibalism appeared in the nineteenth century because of a paradox. Trinkaus and Shipman (1993) commented with irony the consequences of Edouard Dupont's report on the discovery made in 1866 of a mandible at the site of La Naulette. Dupont expressed his absolute rejection of the idea that Neanderthals were cannibals. Although no one had made that suggestion until that moment, Dupont's remark sparked the contrary opinion, and Neanderthals were attributed cannibalism (Blake, 1867).

Trinkaus and Shipman (1993) were very skeptical about Neanderthal cannibalism, and thought that considering them as brutes was due to prejudice. But the question of cannibalism can be

understood in another way. When Wolpoff was asked about the meaning of the evidence found at Moula-Guercy, he replied with a question: why should modern humans be the only violent ones? Arsuaga went even further in believing that Neanderthal cannibalistic behavior actually constituted a very human behavior, which revealed a human mind (both cited by Culotta, 1999). Hence, the presumed brutality of Neanderthals would not be evinced by cannibalism, but by its absence, because so-called humanity includes this behavior.

Leaving ethnological documentation aside, archaeological evidence of cannibalistic behavior has been provided by the detection of human myoglobin in coprolites from an Anasazi site (Pueblo Indians) from Colorado (USA; Marlar *et al.*, 2000). To what extent was cannibalism common among early modern humans?

The study carried out by Simon Mead and colleagues (2003) on the polymorphism of the human

---

Box 10.2 Cannibalistic behavior at Atapuerca

Fernández-Jalvo and colleagues (1996) have also studied marks left on bones. They documented indications of cannibalistic behavior in Atapuerca level TD6, which has

yielded remains attributed to *H. antecessor*. Whether this was only a ritual practice, as Emiliano Aguirre suggested, is difficult to ascertain.

---

prion protein gene (*PRNP*) provides a possible answer. The high incidence of the Kuru disease among the Fore from the Papua New Guinea Highlands is caused by a prion transmitted during endocannibalistic feasts (Mead *et al.*, 2003). Kuru was common in this cultural and linguistic group until authorities prohibited cannibalism in the mid-1950s. Heterozygosity for *PNRP* confers relative resistance to prion diseases. Consequently, Kuru imposed strong balancing selection on the Fore, essentially eliminating *PRNP*-nonresistant homozygotes (Mead *et al.*, 2003). But the *PRNP* polymorphism is not exclusive of the Fore group. Mead and colleagues (2003) believe that worldwide *PRNP* haplotype diversity "suggests that strong balancing selection at this locus occurred during the evolution of modern humans." The authors admitted that the prion gene could have been subjected to other unknown forms of selection. However, they argued that "available evidence appears consistent with the explanation that repeated episodes of endocannibalism-related prion disease epidemics in ancient human populations made coding heterozygosity at *PRNP* a significant selective advantage leading to the signature of balancing selection observed today" (Mead *et al.*, 2003). As part of the "strong evidence for widespread cannibalistic practices in many prehistoric populations," Mead and colleagues (2003) cited Derfleur's (1999) work on Moula-Guercy Neanderthals.

It would be unreasonable to reduce the behavioral similarities between Neanderthals and modern humans to cannibalism. The measure of the distance that separates us from Neanderthals, or brings us close to them, has to do importantly with matters such as their capacity for symbolism, religion, and language. These are matters that Trinkaus and colleagues (1999) had in mind when

they pointed out that the most important considerations regarding the child from Lagar Velho were behavioral. Unfortunately, such traits are not easily ascertained. Many arguments in favor of and against Neanderthal mental capacities are based on speculation to the extent that Lindly and Clark (1990) argued against any attempt to identify hominin taxa with symbolic capabilities. They believe that there is no evidence whatsoever of symbolism before the upper Paleolithic, which means that anatomically modern humans would have also been paleocultural for a large portion of their existence. However, it is worth wondering about the origin of this behavioral dimension, even if only so that we can reflect upon the significance of our own mental processes.

10.1.1 Mousterian culture

Mousterian culture is, strictly speaking, the lithic tool tradition that evolved from Acheulean culture during the middle Paleolithic. The splendor of the Mousterian culture occurred in Europe and the Near East during the Würm glaciation, the last one. Mousterian techniques changed in time. Geoffrey Clark's (1997b) study of the middle- and upper-Paleolithic cultural stages convincingly demonstrated how wrong it is to speak about "Mousterian" as a closed tradition, with precise limits; as a unit with precise temporal boundaries. Even so, we will talk about a Mousterian style, as Clark himself did, which becomes apparent when compared with the upper-Paleolithic Aurignacian technical and artistic explosion. Aurignacian tools and decorated objects contrast sharply with the earlier ones. The reasons behind this sudden change and its mental correlates are some of the matters we will deal with in this chapter. But first we must extend the consideration of Mousterian

Box 10.3 Prions and natural selection

Soldevila *et al.* (2006) studied the polymorphism of the human prion protein gene (*PNRP*) and found no evidence of balancing selection over the last half million years of human evolution. They conclude that there is no evidence that any prion disease related with cannibalism led to the current *PNRP* polymorphism.

culture from lithic tools to other products and techniques that appear at Mousterian sites. In a broad sense, Mousterian culture includes controversial features, such as objects created with a decorative intention and indications of funerary practices.

Let us begin with the Mousterian tool-making techniques. They were used to produce tools that were much more specialized than Acheulean ones; the Mousterian tools were given a form before sharpening their edges. The most typical Mousterian tools found in Europe and the Near East are flakes produced by means of the Levallois technique, which were subsequently modified to produce diverse and shaper edges (Figure 10.2). Objects made from bone are less frequent, but up to 60 types of flake and stone foil can be identified, which served different functions (Bordes, 1979).

The Levallois technique appeared during the Acheulean period, and was used thereafter. Its pinnacle was reached during the Mousterian culture. The purpose of this technique is to produce flakes or foils with a very precise shape from stone cores that serve as raw material. The cores must first be carefully prepared by trimming their edges, removing small flakes until the core has the correct shape. Thereafter, with the last blow, the desired flake, a Levallois point for instance, is obtained (Figure 10.3). The final results of the process, which include points, scrapers, among many other instruments, are subsequently modified to sharpen their edges. The amazing care with which the material was worked constitutes, according to Bordes (1953), evidence that these tools were intended to last for a long time in a permanent living location.

Tools obtained by means of the Levallois technique are, as we said earlier, typical of European and Near-East Mousterian sites. Bifaces, on the

Figure 10.2 Pengelly's original photograph of a Mousterian biface flint tool found and documented on 1871. Image copyright Torquay Museum.

contrary—so abundant in Acheulean sites—are scarce. The difference has to do mostly with the manipulation of the tools; scrapers were already produced using Acheulean, and even Oldowan, techniques. The novelty lies in the abundance and the careful tool retouching.

Most European and Near-East sites belonging to the Würm glacial period contain Mousterian tools. The name comes from the Le Moustier site (Dordogne, France), and was given by the prehistorian Gabriel de Mortillet in the nineteenth century, when he divided the Stone Age known at the time in different periods according to the technologies he had identified (Mortillet, 1897). Mortillet introduced the terms Mousterian, Aurignacian, and Magdalenian, in order of increasing

(a)

(b)

(c)

(d)

(e)

(f)

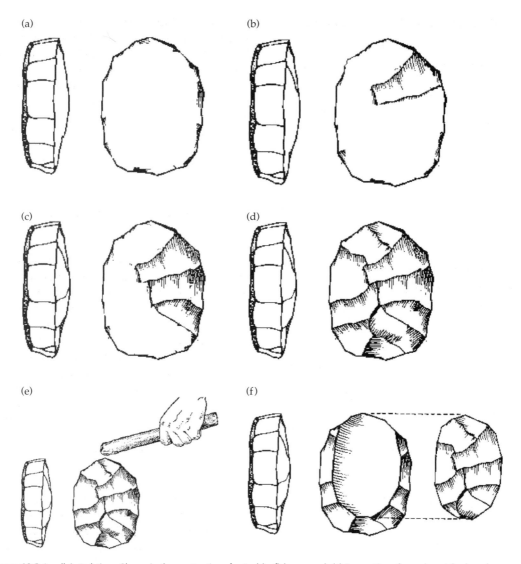

Figure 10.3 Levallois technique. Phases in the construction of a tool by flake removal. (a) Preparation of an adequately shaped core. (b, c) Removal of flakes. (d) The prepared platform is obtained. (e, f) A last blow with a soft hammer separates the tool from the core. Drawings from http://www.hf.uio.no/iakh/forskning/sarc/iakh/lithic/LEV/Lev.htm.

complexity, to designate the tools from the French sites of Le Moustier, Aurignac, and La Magdalene. However, as we have said, almost all the sites belonging to the Würm glacial period mentioned in the previous chapter contain Mousterian tools. In many instances, their lower archaeological levels also show the transition of Acheulean to Mousterian tools, and even from the latter to Aurignacian ones. The archaeological richness and sedimentary breadth of some of these sites, like La Ferrassie, La Quina, and Combe-Grenal, grants them a special interest for studying of the interaction between cultural utensils and adaptive responses. Similar Mousterian utensils have appeared in the Near East, at Tabun, Skuhl, and Qafzeh.

10.1.2 Neanderthals and Mousterian culture

The period known as the middle Paleolithic is not only manifest in Europe, but also in many places in Africa and Asia, where there is evidence of a development of Acheulean tools equivalent to the Mousterian culture. But the type of tools does not completely coincide: there are local variations of the Levallois technique. The North African Aterian tradition provides an example near Europe. Scrapers also appear in Africa and Asia, but they are considerably less abundant than at European and Near-East sites. Hence, both in spatial and temporal terms, the Mousterian culture coincides with Neanderthals.

This identification between the Mousterian culture and *H. neanderthalensis* has been considered so consistent that, repeatedly, European sites yielding no human specimens, or with scarce and fragmented remains, were attributed to Neanderthals on the sole basis of the presence of Mousterian utensils. Despite the difficulties inherent in associating a given species with a cultural tradition, it was beyond a doubt that Mousterian culture was part of the Neanderthal identity. This perception changed with the reinterpretation of the Near-East sites (Bar-Yosef and Vandermeersch, 1993). Scrapers and Levallois points, which were very similar to the typical European ones, turned up there. Neanderthals also existed there, of course (Figure 10.4), but in contrast with European sites, a distinction could not be drawn between localities that had housed Neanderthals and anatomically modern humans solely on the grounds of the cultural traditions. The more or less systematic distinction between Neanderthal–Mousterian and Cro-Magnon–Châtelperronian (or Aurignacian, or Magdalenian) helped to clarify the situation in Europe. But it could not be transferred to the Near East, where sites occupied by Neanderthals and those inhabited by anatomically modern humans, proto-Cro-Magnons, yielded the same utensils of the Mousterian tradition.

This implies several things. First, that cultural sharing was common during the middle Paleolithic, at least in Levant sites. Second, that during the initial stages of their occupation of the eastern shore of the Mediterranean, anatomically modern humans made use of the same utensils as Neanderthals. Hence, it seems that at the time Skuhl and Qafzeh were inhabited, there was no technical superiority of modern humans over Neanderthals. The third and most important implication has to do with the inferences that can be made because Neanderthals and *H. sapiens* shared identical tool-making techniques. As we have already seen, the interpretation of the mental processes involved in the production of tools suggests that complex mental capabilities are required to produce stone tools. We are now presented with solid proof that Neanderthals and modern humans shared techniques. Does this mean that Neanderthal cognitive abilities were as complex as those currently characteristic of our own species? Many authors, headed by Trinkaus, Howells, Bar-Yosef, and Vandermeersch, believe so. But some authors arguing in favor of high cognitive capacities in Neanderthals went beyond showing that lithic culture was shared in the Near East. They presented other kinds of items which, in their opinion, were indications of Neanderthal aesthetic, religious, symbolic, and even maybe linguistic, capacities.

10.1.3 The origin of religious beliefs

The possibility that Neanderthals buried their dead is the best basis to attribute transcendent thought to them. Voluntary burial is indicative of respect and appreciation, as well as a way to hide the body from scavengers. This may also imply concern about death, about what lies beyond

---

Box 10.4 Mousteroid

The term Mousteroid is occasionally used to highlight differences between the middle-Paleolithic tools found Africa and the Far East, where there are no Neanderthals, and the European Mousterian tradition (Bever, 2001).

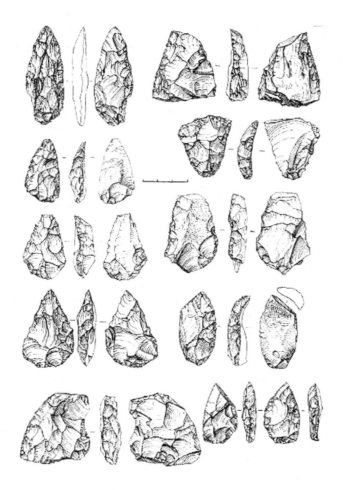

Figure 10.4 Mousterian handaxes from Mezmaiskaya Cave (Caucasus), layers 3-2b. Figure from Doronichev and Golovanova (2003).

Box 10.5 Cultures of Neanderthals and anatomically modern humans

In spite of the evidence of cultural sharing, it is common to identify Mousterian culture with Neanderthals. *H. neanderthalensis* is also usually associated with "transitional" cultures: Châtelperronian, Szeletian, and Uluzzian. Anatomically modern humans are usually associated with the Aurignacian culture, generally thought to be contemporary with the Châtelperronian, and with the most advanced lithic industries of the late Paleolithic: Gravettian, Solutrean, and Magdalenian. We will say more about the relations among these cultures in the next section.

---

death, and the meaning of existence. The argument for religiousness is convincing when burial is accompanied by some sort of ritual.

Neanderthal burials have been located in several areas, mainly in southern France, Italy, northern Balkans, the Near East (Israel and Syria), and central Asia (Iraq, Caucasus, and Uzbekistan). In most cases these burials seem to be deliberate. Hence, the "old man" from La Chapelle-aux-Saints appeared in a rectangular hole dug in the ground of a cave that could not be attributed to natural processes (Bouyssonie *et al.*, 1908). In regard to La

Ferrassie and Shanidar, the possible evidence of the existence of tombs led Michael Day (1986) to remark, in a technical and unspeculative treatise, that these exemplify the first intentional Neanderthal burial that has been reliably determined. Trinkaus' (1983) taphonomic considerations point in the same direction. The abundance and excellent state of Neanderthal remains at those sites, together with the presence of infantile remains, are a proof that the bodies were out of the reach of scavengers. Given that there is no way natural forces could produce those burials, Trinkaus believes the most reasonable option is to accept that the remains were intentionally deposited in tombs. However, Noble and Davidson (1996) argued that, at least in the case of Shanidar (Iraq), it is probable that the cave's ceiling collapsed while its inhabitants were sleeping.

Some of the aforementioned remains were not only buried intentionally, but there seems to be evidence of rituals. This is the case of the Kebara skeleton (Israel), which, despite being excellently preserved—it even includes the hyoid bone—is lacking the cranium (see Figures 9.27 and 10.5). Everything suggests that the absence of the cranium is due to deliberate action carried out many months after the individual died (Bar-Yosef and Vandermeersch, 1993). It is difficult to imagine a different taphonomic explanation. Bar-Yosef and Vandermeersch (1993) wondered about the reasons for such an action, suggesting that the answer might lie in a religious ritual.

A Neanderthal tomb with an infantile specimen was found in the Dederiyeh Cave (Syria), 400 km north of Damascus. Akazawa and colleagues (1995) interpreted the burial as an indication of the existence of a ritual. The reason behind this argument is the posture in which the specimen was deposited in the tomb. The excellently preserved skeleton was found with extended arms and flexed legs. Mousterian lithic industry also turned up in the cave, which Akazawa et al. (1995) associated with that from Kebara and Tabun B, although there were few tools at the burial level. An almost rectangular limestone rock was placed on the skeleton's cranium, and a small triangular piece of flint appeared where the heart had once been (Figure 10.6). Although Akazawa and colleagues (1995) did not elaborate an interpretation of these findings, they suggest that these objects had ritual significance.

The Shanidar IV specimen is one of the most frequent references in relation to ritual behaviors. The discovery of substantial amounts of pollen at the tomb was interpreted as evidence of an intentional floral offering (Leroi-Gourhan, 1975). If this were the case, it would represent the beginning of a custom that has lasted until today. It must not be forgotten either that two of the Shanidar crania, I and V, show a deformation that was attributed to aesthetic or cultural motives. However, Stringer and Trinkaus (1981) indicated that the specimens had been reconstructed incorrectly and that the shape of the first one was due to pathological circumstances. In his study of the Shanidar IV burial, Solecki (1975) argued that there is no evidence of an intentional deposit of flowers at the burial. The pollen must have been deposited there in a natural way by the wind. Supporting the notion of an unintentional presence, Gargett (1989) suggested that the pollen could have been introduced simply by the boots of the workers at the cave's excavation. Paul Mellars (1996) believes that the accidental presence of objects at French burial

---

Box 10.6 The Lapedo child

Another possible instance of Neanderthal burial is the "Lapedo child", from Lagar Velho (Portugal). However, there are concerns about this specimen (see section 9.3).

Box 10.7 Burial and scavengers

The intention of protecting the dead from scavengers might be an immediate explanation for Neanderthal burials. However, Gamble (1989, 1993) has noted that the presence of well-preserved Neanderthal skeletons is greater in areas with low scavenging activity.

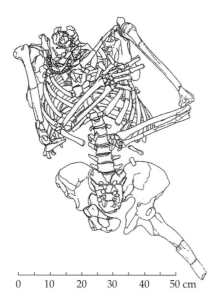

Figure 10.5 Neanderthal burial in the Kebara Cave. Redrawn from http://www.hf.uio.no/iakh/forskning/sarc/iakh/lithic/AmudNet/Kebara.html.

Figure 10.6 Neanderthal child skeleton from Dederiyeh Cave (Syria). Picture from Gargett (1999). Reprinted from *Journal of Human Evolution* Vol 37:1, R.H. Gargett, 'Middle Palaeolithic burial is not a dead issue' 27–90, 1999, with permission from Elsevier.

sites, such as La Ferrassie or Le Moustier, is inevitable: the tombs were opened at places in which faunal remains and Mousterian utensils were abundant.

The Teshik-Tash site (Uzbekistan), located on high and precipitous terrain, contains an infantile burial associated with wild-goat crania. According to Hallam Movius (1953), the horns formed a circle around the tomb (Figure 10.7). This would support a symbolic purpose and a ritual content associated with the burial. Currently, however, even those who favor Neanderthals as individuals with remarkable cognitive capacities are quite skeptical about the presumed intentional arrangement of the crania (Trinkaus and Shipman, 1993; Akazawa *et al.*, 1995; Mellars, 1996).

Neanderthal burials can be interpreted as a functional response to the need of disposing of the bodies, even if only for hygienic reasons. But they could also be understood as the reflection of transcendent thinking, beyond the simple human motivation of preserving the bodies of deceased loved ones. According to Mellars (1996), "we must assume that the act of deliberate burial implies the existence of some kind of strong social or

emotional bonds within Neanderthal societies". However, Mellars believes that there is no evidence of rituals or other symbolic elements in those tombs. The appearance of such evidence would demonstrate that Neanderthals were capable of religious thinking. Similarly, Gargett (1989) argued that the evidence of Neanderthal burials is much more solid than the evidence of offering or rituals.

Neanderthal burials contrast sharply with the burials made by modern humans, living approximately at the same time. The differences are especially illustrative in the Near East. The only intentional, and potentially symbolic, funerary middle-Paleolithic objects are the bovid and pig remains found in burials at Qafzeh and Skuhl (Mellars, 1996; Figure 10.8). Both appeared in modern-human sites. Taking into account that humans and Neanderthals living at

Figure 10.7 Reconstruction of a Neanderthal (*H. neanderthalensis*) burial site based on remains discovered at Teshik-Tash, Uzbekistan, dating back 70,000 years. Image from http://piclib.nhm.ac.uk/piclib/www/image.php?img=48196andsearch=burial.

Box 10.8 Modern-human burials in the upper Paleolithic

Julien Riel-Salvatore and Geoffrey A. Clark (2001) have noted that applying Gargett's criterion to the early upper Paleolithic would also lead to doubting the intentionality of the first modern human burials. They believe that there is a continuity, regarding the tombs, between the middle- and early upper-Paleolithic archaeological records. True differences do not appear until the late phase of the upper Paleolithic (20,000–10,000 years ago).

---

those sites shared the same Mousterian tradition, this is a significant difference. It not only has to do with the manufacture of objects, but with much more subtle aspects, which are associated with mental processes like symbolism, aesthetics, or religious beliefs.

William Noble and Ian Davidson (1996) stressed that Neanderthal burials have not been found outside caves. In contrast, there are examples of very early human tombs in open terrains at places such as Lake Mungo (Australia), Dolni Vestonice (Czech Republic), and Sungir (Russia). In Noble and Davidson's (1996) view, the appearance of a Neanderthal tomb outside the caves would be proof that this is an intentional burial. For now,

known tombs provide no conclusive clues about Neanderthal self-awareness, not to mention their religion.

10.1.4 Symbolic thought

Given the lack of persuasive evidence related with tombs, a possible key to the symbolic thought of *H. neanderthalensis* could come from stone and bone objects belonging to the Mousterian tradition.

What distinguishes a symbolic object from others? One usual way to identify symbolism in human paleontology and archaeology is to divide objects into those that have a practical use (knives,

(a)

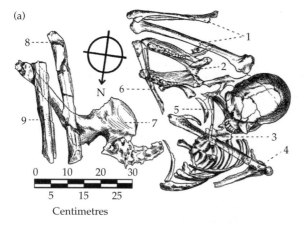

(b)

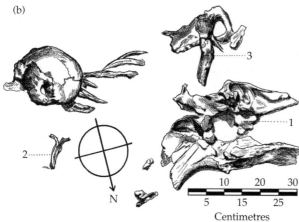

Figure 10.8 Position of a pig mandible (2 in panel a) and a crushed bovid skull and maxilla (1 in panel b) in the Skhul V and IX burials. Plan of the contracted burial of a tall male, Skhūl V.1, right arm; 2, Pig's mandible; 3, dorsal vertebrae; 4, left scapula and humerus; 5. left clavicle; 6, left radius; 7, right ilium; 8, left femur; 9, left tibia and fibula. The incompletely preserved skull and skeleton of an adult male Skhūl IX. 1, crushed bovid skull and maxilla; 2, spine of the left scapula; 3, left femur, with adjacent parts of the left pelvis. Picture from Grün *et al.* (2005). Reprinted from *Journal of Human Evolution* Vol 49:3 R. Grün *et al.* 'U-series and ESR analyses of bones and teeth', 316–334, 2005, with permission from Elsevier.

Box 10.9 What is it like to be a Neanderthal?

If we consider that *H. neanderthalensis* belonged to a different species than our own, how can we know what a symbol, a belief, or a burial ceremony means for someone like that, given the very scarce hints provided by the fossil record? In a subtle article—"What is it like to be a bat?"—Nagel (1974) warned about the impossibility of knowing what it is like to be a bat. Not "to be a bat" for a human, but to be a bat for the bat itself. Although there are few doubts that the mind of a Neanderthal was much closer to ours than that of a bat or any other mammal, the species barrier introduces insoluble problems when dealing with something like religious beliefs. Symbols are a different matter, because they at least leave visible traces, although they may be difficult to interpret.

---

axes, scrapers, and so on) and those that do not. Objects that have no direct use might be considered to be symbolic.

Making tools to use them in one way or another requires a capacity to formulate objectives and anticipate behaviors. When Louis Leakey and colleagues (1964) introduced *H. habilis*, they associated cognitive capacities (inferred from cranial volume) with tool-making. In our species there is great individual variation in regard to brain size,

Box 10.10 Encephalization

The increase of brain size disproportionally to body size is known as encephalization. If body growth is part of the evolution of a species, it is to be expected that the cranium and, therefore, the brain, will also grow. Blue whales have the largest brains among sea mammals, and elephants among land mammals. But this is because they are the largest animals. But increase in brain size may occur beyond what is expected for body size (extra-allometric increase). Indices are used to calibrate its extent, such as the encephalization quotient, which expresses the relative increase of the brain's size compared to the body's. Although, it is not easy to calculate it in fossil species, relevant information is sometimes available. (See also Box 2.6 and Figure 2.15.)

but it does not correlate with intelligence measured by IQ or any other way. If cranial size and cognitive capacities are currently not correlated, why should we accept such a correlation in our ancestors?

The difference is that paleontologists are not dealing with comparisons between individuals belonging to the same species, but with the cranial size of different species. They resort to statistical averages, when possible. If the average encephalization quotient of a fossil species is significantly greater than another's, the cognitive capacities of the former are assumed to be more complex (Tobias, 1975; Popper and Eccles, 1977). As pointed out in Box 2.6 New World monkeys that feed on insects have larger brains than those that feed on leaves, which is accounted for by the greater need for processing environmental information (Jerison, 1977a). A similar assumption underlies the attribution of the manufacture of the earliest tools to *H. habilis*. The encephalization quotient estimated for *H. habilis* is larger, according to many authors, than that of *P. boisei*.

Ethological information derived from the comparative behavior of primates is relevant here. The first hominins had cranial capacities similar to those of current chimpanzees, as we saw earlier (Tobias, 1975; Johanson *et al.*, 1978). Nevertheless, the latter are capable of using simple tools, such as sticks or stones, to obtain termites from a termite nest or to crack nuts open (Goodall, 1964; Sabater Pi *et al.*, 1984). Different colonies exhibit different traditions of learned tool use (de Waal, 1999; Whiten *et al.*, 1999, 2005). Yet, they do not seem able to anticipate objectives, as would seem

necessary for designing and executing the tools of the Oldowan culture.

Acheulean handaxes deserve particular attention if we are trying to identify the earliest indications of a possible symbolic capacity, which was, to some extent at least, independent from the functional use of tools. Advanced bifaces are beautiful objects. Taking into account their age, these handaxes exhibit a surprising symmetry, and it seems that their careful elaboration is a reflection of an aesthetic purpose (Figure 10.9). Thus, Schick and Toth (1993) argued that bifaces belonging to the latest stages of the Acheulean tradition are often undeniably and surprisingly beautiful. They possess a bilateral symmetry in both dimensions, left to right and back to front, without loosing their edges and cutting efficiency. Washburn and Lancaster (1968) observed that symmetrical bifaces constitute the first beautiful manufactured objects.

Granted that Acheulean handaxes are beautiful objects, were these tools created with the *intention* of being beautiful? Enquist and Arak (1994) argued that the preference for symmetry appeared phylogenetically as a subproduct of the need to recognize objects independently of their position or orientation regarding the visual field. This comment was in reference to sexual preferences (see Johnstone, 1994), but it can apply to symmetry in general. Bifaces could be an early manifestation of the evolution of preferences for lateral symmetry.

To attribute aesthetic content to Acheulean bifaces, we would need to show that they do not owe their symmetry to any useful purpose. It could be that *H. erectus* produced them for

Figure 10.9 Current appreciation of the beauty of bifaces. This biface is described as "Elegant cordate handaxe from Hoxne, Suffolk, England. Late Acheulean, about 350,000 BCE." Picture from http://www.personal.psu.edu/users/w/x/wxk116/axe/.

purposes other than creating beautiful objects. It might be the case that the purpose was to obtain very sharp edges and cutting efficiency (Schick and Toth, 1993), enough to justify the care involved in Acheulean tool manufacture. Also, as Washburn and Lancaster (1968) noted in their study on the origins of hunting, the symmetry of Acheulan handaxes and hunting tools served the purpose of achieving improved aerodynamic performance for hitting prey at a distance. Nevertheless, as Washburn and Lancaster added, the capacity to appreciate these tools must have evolved together with the competence to manufacture and use them. If so, the symmetrical tool would be, rather than a symbol, a simple tool.

The hypothesis of Washburn and Lancaster (1968) is not backed directly by empirical facts. It is true that the Acheulan symmetrical tools turned into beautiful objects, artistic representations for us, who live hundreds of thousands of years after the objects were manufactured. But were they also so perceived, at least to some extent, by those who manufactured them? Can a gradual and slow evolution towards more advanced symbolic objects be documented? Or, rather, did symbolic expression and perception come about relatively

suddenly, late in human evolution? We deal with these questions in the following section.

10.2 The origins of symbolism

10.2.1 Gradualism versus explosion

Before discussing the origin of symbolism we need to clarify the meaning of origin in the present context. As we have seen, the symmetry of Acheulean handaxes could be considered as a possible indication of symbolism (Figure 10.10 and previous chapters). The earliest of these instruments are about 1.5 million years old (Leakey, 1975). Realistic representations, such as the Altamira Cave paintings (from Spain) or the bison high-relief examples from Le Tuc d'Audobert (France; Figure 10.11), were made towards the end of the upper Pleistocene, about 14,000 years ago (Bahn, 1992). There is an enormous time gap between these two cultural manifestations (symmetry and realistic painting). Within this gap, there is the Mousterian culture, characteristic of the European middle Paleolithic, ranging from about 100,000 to 40,000 years ago.

We will discuss this sequence by dividing it into three stages:

1 a long period, lasting for more than 1 million years, characterized by Acheulean symmetrical bifaces (in Africa and later in Europe);
2 a period of about 100,000 years, coinciding for the most part with the last glacial period (Würm), in which tools believed by some authors to have symbolism, perhaps even religious symbolism, are present;
3 the period ranging from 40,000 to 10,000 years ago, with abundant objects widely regarded as created with an aesthetic intention.

The almost unanimous agreement that drawings, engravings, and paintings from period 3 are symbolic does not hold for objects from periods 1 and 2. But before we move on to the so-called artistic explosion of the European Aurignacian, Solutrean, and Magdalenian cultures, we need first to consider the very origin of symbolism. Was it present in periods 1 and 2, or did it appear only during period 3?

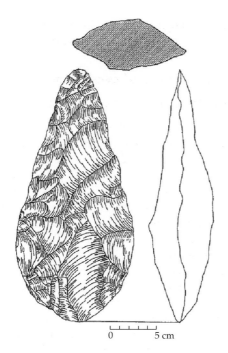

Figure 10.10 Handaxe from Isimila (Tanzania, c. 300,000 years ago). As Tomas Wynn observed, this artifact has congruent symmetry in three dimensions. Illustration from Wynn (2002). Thomas Wynn 'Archaeology and cognitive evolution', *Behavioral and Brain Sciences.* Vol. 25(3): pp 389–402, (2007), Figure 7. Permission by Cambridge University Press.

There are two mutually exclusive hypotheses about the process that led to the massive production of artistic representations unquestionably charged with symbolism: the gradual and explosive models. The former argues that the capacity to appreciate Acheulean "beautiful forms" evolved gradually and continuously, leading to the great abundance of late-Paleolithic artistic objects. This gradual model does not refer to *an* origin of art. This origin is thought to be fuzzy, widespread in space and continuous in time. According to this model, the initial manifestations of that origin were scarce; slowly, over a long period of time, they became progressively generalized. The explosive model of the appearance of the symbolism characteristic of art argues that it appeared fairly suddenly during the late Paleolithic, and that it is exclusively an attribute of modern humans.

The earliest undisputed signs of beauty appreciation are ornamental elements. One instance of

this might be the choice of stones that contained fossils to construct tools. Manufacturing bifaces for handaxes that have fossils in the center is, in Oakley's (1981) view, beyond functional purposes (Figure 10.12). However, Noble and Davidson (1996) noted that the presence of fossils in flint is very common. What is surprising, according to these authors, is that there are not more handaxes containing fossils. Handaxes with fossils in the middle are relatively rare. Noble and Davidson (1996) suggested that fossils that happened to be on the sides of the nuclei were lost during flaking.

Neanderthal perforations and engravings on bone or stone objects are also often interpreted as evidences of an incipient art. This is the way Marshack (1988, 1990, 1995), Bednarik (1992, 1995, 1997), Hayden (1993), and Bahn (1996) interpreted those objects. d'Errico and Villa (1997) analyzed evidence presented in favor of their being artistic manifestations, carrying out, with the aid of an electron microscope, taphonomic analysis of the marks and holes present in the bones. They concluded that these objects, characteristic of the lower Paleolithic or even earlier, do not offer solid evidence of artistic intention. The holes in the bones, which had often been interpreted as decorative pendants, are all very similar to those produced by the gastric juices of scavengers, such as hyenas. According to d'Errico and Villa (1997) natural foramen suffices for digestive acids to widen the hole, round off its edges, and generate a very similar result to the perforated bones that allegedly were designed to serve as pendants.

d'Errico and Villa's (1997) study was one-sided and did not take into account the engravings made on stones. However, Chase and Dibble (1987), Davidson (1990), and d'Errico himself (1991) have attributed such marks to natural erosion processes or the action of scavengers. Even so, d'Errico and Villa (1997) admitted that the strongest and most ordered marks could be the result of intentional engraving. This seems to be the case with the Bilzingsleben designs. Some of the pieces from the Bilzingsleben site (Germany), which has yielded *H. erectus* specimens and has been dated to around 350,000 years ago, show geometrically arranged lines (Bednarick, 1995; Figure 10.13).

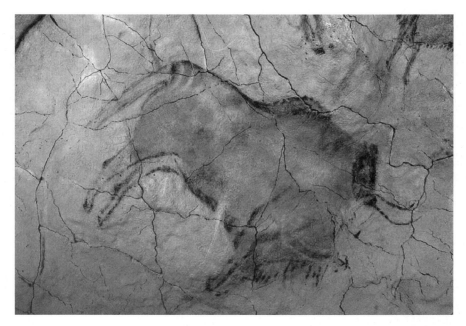

Figure 10.11 Magdalenian bison from Altamira (Spain, *c.*15,000 years ago). Image from http://www.educarchile.cl/ntg/mediateca/1605/articles-94222_imagen_0.jpeg.

K.P. OAKLEY

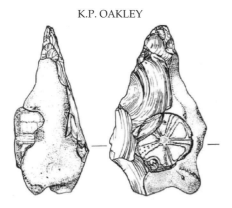

Figure 10.12 Handaxe with an embedded upper-Cretaceous echinoid, *Conulus* sp., from the Swanscombe site (UK). Illustration ©Oakley (1981), figure 3. Figure 3 in Oakley, K. (1981). 'Emergenge of higher thought 3. 0–0.2 Ma B.P. Phil. Trans. *R. Soc.* London, B292, 205–211.

> **Box 10.11 Art or nature?**
>
> According to Francesco d'Errico and colleagues (1998), the examination under electron microscope of the grooves that appear on bones such as those from the Bois Roche and Roc de Marcamps sites (France), suggests that they are not anthrogenic. Rather, they seem to be marks left by the arterial and venous systems on the surface and interior of bones.

Figure 10.13 Geometrical marks performed on an elephant bone from the Bilzingsleben Acheulean site (Germany, *c.* 350,000 years ago). Illustration from Bednarik (1995).

Geometrical motifs appear again in certain isolated objects estimated to be 300,000 years younger in diverse prehistoric sites. An artifact found in a Mousterian site at Quneitra (Golan Heights, between Israel and Syria) is a beautiful example of this. It is an approximately triangular plate of flint cortex, about 7.2 cm high (Goren-Inbar, 1990; Figure 10.14). The site's age, estimated by means of ESR applied on bovid dental enamel, is close to 54,000 years (Ziaei *et al.*, 1990). The Quneitra

Figure 10.14 Quneitra artifact (Syria, c .54,000 years ago). Illustration from Goran-Inbar (1990).

Box 10.12 Cognitive ability and the Quneitra artifact

Marshack (1995) believes that the engraving process indicates a surprising cognitive complexity of the author of the lines on the Quneitra artifact. Far from being a random carving, it shows an intentional trend to center the semicircles that had to be kept in mind while the stone was being turned for making the marks. According to Marshack, the required technique was complex and required precise coordination of hand movements, under the supervision of the visual system and following a preconceived plan. Very sophisticated cognitive capacities are necessary to carry out a designed plan by means of hand coordination. Marshack talked about a true "gestalt" to produce the lines of the engraving in accordance with the shape and size of the flint.

---

fragment shows a set of marks that were described as the earliest sample of a representation in the form of an engraving (Goren-Inbar, 1990). The microscopic examination of the piece revealed a set of four concentric semicircles carefully carved and surrounded by angular lines that roughly follow their form, together with other vertical lines on the right hand side. A sinuous line traces the shape of the broken right side of the flint (Marshack, 1995).

Marshack's (1995) interpretation of the Quneitra fragment was based on how fast and how many incisions were made to produce the different lines on the flint plate. According to Marshack, the vertical lines indicate a clear intention of covering the entire available surface. Some of them were made by means of a long blow followed by a shorter one to reach the edge of the plate. The sinuous line of the plate's right hand border shows that the intention was to mark the trajectory of the flint's side (Figure 10.15).

The importance of the marks left on the Bilzingsleben bones and the Quneitra artifact resides in that they show that geometrical designs were created by hominins throughout a long span of time. As we have noted, some explanations for the origin of the earliest iconic representations

Figure 10.15 Schematic representation of the lines on the Quneitra fragment. Illustration from Marshack (1995).

suggest they could be randomly produced marks. Only accidentally are they reminiscent of a figure, and only later are they recognized as such. Neither the Bilzingsleben nor the Quneitra signs correspond to any stylized image of animals or human beings. As Bednarik (1995) noted, the random distribution of marks left by natural processes can produce geometrical patterns. But in the case of the Bilzingsleben objects, the relation between the spatial distribution of the marks and the available space on the bone argue against the absence of intentionality (Bednarik, 1995). This is even truer of the Quneitra plate, which in no way could be considered a personal decorative object, such as a pendant or a piece of a necklace. The concentric circle motif did not appear again until the European upper Paleolithic, 25,000 years later, and only occasionally. The Quneitra artifact is unique in that respect, unparalleled at the time.

The cognitive complexity necessary for manipulating a fragment such as the Quneitra plate must not be interpreted as artistic capability. The visual–motor coordination required to carry out the hand movements for engraving the lines on the plate is also a necessary prerequisite to create Acheulean bifaces and, to a certain extent, the most primitive Oldowan instruments. What distinguishes the

Bilzingsleben bone engravings and the Quneitra stone engravings from other manufactured objects is, as we said above, the absence of any kind of useful value, the fact that their only purpose seems to be symbolic.

If this is the case, we have identified objects that satisfy the following conditions: they are very old (from 350,000 to 54,000 years), they are anthropogenic, they lack specific utility, and they show a geometrical arrangement that suggests aesthetic intention.

If the interpretations made by Bednarik and Marshack are correct, there is no doubt that these characteristics formally meet the requirements for establishing artistic behavior. However, Bednarik himself raised an interesting question: to what extent can such partial evidence as that from Bilzingsleben be taken as the characteristic norm of middle-Pleistocene hominins? Bednarik (1995) believes that the Bilzingsleben marks do not constitute a sufficiently solid argument to posit the involvement of concepts. Further evidence corresponding to that period is required.

This is the most serious criticism that the gradual model, or the very early presence of intentionally aesthetic objects, has received. The paucity of remains contrasts sharply with the artistic explosion during the upper Paleolithic. Moreover, the Aurignacian, Solutrean, and Magdalenian traditions of the upper Paleolithic include not only tools that are much more precise and sophisticated than those from the earlier Mousterian culture. There also are abundant representations of real objects in the form of engravings, paintings, and sculptures, which represent a much more advanced cognitive level than the prior geometrical drawings.

In any case, it seems that the final word, about the evolution of human cognitive faculties and the significance of artifacts constructed, cannot yet be said. In 1988 Yellen and colleagues (1995) found a harpoon made from bone, reminiscent of the European points common around 14,000 years ago, at the site of Katanda, by the River Semliki (Zaire). In itself, this would not be surprising at all. But Brooks *et al.* (1995) estimated the site's age to at least 75,000 years, maybe even 90,000 years, by means of thermoluminescence and ESR techniques applied to hippopotamus teeth. The existence of such a technique at this early time was unimaginable at the time of the discovery. Gibbons (1995) raised questions such as: why there is no evidence of advanced techniques at that time in Africa? How can we be certain that the hippopotamus teeth and the harpoon are of the same age?

The Vogelherd site (Germany) has provided evidence of aesthetic intention beyond reasonable doubt. The site has been estimated to be 32,000 years old, and was inhabited by anatomically modern humans. It has yielded delicate samples of Aurignacian art, such as figurines carved in mammoth ivory. The Vogelherd horse (Marshack, 1990) is a small piece, about 5 cm long, which leaves no doubt whatsoever about what it is that its author wanted to represent. The Vogelherd figures are comparable in representational significance to the characteristic drawings and engravings of the later Magdalenian tradition, the most developed within the upper Paleolithic, such as those from Les Combarelles and Limeuil (France).

The Vogelherd horse shows evidence of having been polished, which has been attributed to its use, possibly as a pendant in a necklace. The accuracy of its features is more striking yet than its possible decorative role. It is evident that Aurignacian figures and Magdalenian drawings and statuettes share a common intention: representing in a realistic way animals common in the environment. The representational effectiveness from the late upper Paleolithic is more advanced than the one from the early stages of that period, but in any case all those animal reproductions denote a common qualitative jump compared to the middle-Paleolithic geometrical engravings.

Apparently, something that happened around 40,000 years ago led to the sudden and abundant manifestation at many different places of aesthetic expressions, which had been restrained and isolated before that time. What is the cause for this explosion of new ways of expression? The most reasonable hypothesis is a cognitive change, the acquisition of new cognitive capacities.

10.2.2 The explosive model: cognitive correlates of art

The explosive model of the appearance of art (sometimes called, a bit ironically, *big bang*) posits a sudden cognitive emergence. Many prestigious researchers have lent support to the abrupt appearance of aesthetic experience, including Ofer Bar-Yosef (1988), Iain Davidson and William Noble

---

Box 10.13 The spread of Aurignacian culture

Some authors have suggested that Aurignacian culture appeared in the Near East and extended gradually towards the west, slowly substituting Mousterian culture. Vandermeersch and Garralda (1994) contest this account.

They present evidence of Aurignacian objects from the north of Spain which were as old as those from the Near East (close to 40,000 years).

(1989; Noble and Davidson, 1996), Chris Stringer and Clive Gamble (1993), and Paul Mellars (1989, 1996). Ian Tattersall (1995b) has said that, if there ever was a great leap forward in the history of human culture, it is the one that occurred between the middle and upper Paleolithic. Tattersall (1995b) believes that such a large step forward involved different sentiments and capacities. Leaving aside for the moment possible interpretations, the case is that ornaments, engravings, and naturalistic paintings are very abundant for the last 38,000 years, the age of the site that yielded the earliest unequivocal samples of artistic objects, Kostenki 17 (Russia; Appenzeller, 1998).

The astoundingly high level of the pictorial technique that appears in very early Cro-Magnon caves has been noted very often. Hence, the Aurignacian masterful technology required to produce such objects as the Vogelherd horse, was followed by the Gravettian, known for its representations of women with exaggerated sexual traits, though lacking a face, such as those from Laussel (France) and Dolni Vestonice (Moravia), and the Solutrean, limited to a very specific area between southern France and northern Spain, which includes exquisitely worked flint tools. However, it is the Magdalenian tradition that manifests the maximum explosion of symbolic and artistic expression. This culture encompasses such magnificent examples as the color paintings on cave walls and ceilings representing animal scenes at Lascaux, Niaux, and Chauvet (France) and Altamira (Spain; Figure 10.11). The humans who produced those "Sistine Chapels" of Paleolithic art did not just create grandiose artworks, they also incorporated an aesthetic sensibility into their everyday life. Commonplace objects also are decorated profusely with geometrical motifs and representational illustrations. With the Magdalenian tradition the naturalist interpretation of the world reaches levels similar to those of current artists and craftsmen.

The time spans of the characteristic cultures of the artistic explosion are from somewhat more than 30,000 years for the Aurignacian period, to 28,000–18,000 years for the Gravettian, 22,000–18,000 years for the Solutrean, and 18,000–10,000 years for the Magdalenian (Tattersall, 1995b). The Solutrean and Magdalenian caves are restricted to a very limited area. The time spans for the other technologies are adjusted accordingly.

The location of these cultures is impacted by climate. The consequences of the Würm glacial period were still being felt at the time of the Aurignacian culture. The valleys in southern France and northern Spain offered milder living conditions. In any case, there is no tight correlation between artistic tradition and age. The discovery of the Magdalenian Chauvet Cave, dated at 32,000 years, shattered the temporal sequence of the different kinds of technical and artistic manifestation that placed Magdalenian as the last of the aforementioned periods.

The presence of such exquisite and early art as that from Chauvet can be interpreted so as to justify a sudden emergence of artistic capacities or their gradual evolution. The brusque and broad appearance of richly colored representational art in European caves at the same time that decorative objects were proliferating outside the continent (such as those from Enkapune Ya Muto, Kenya) seem to suggest a fairly sudden phenomenon. But elaborate technical abilities and evidence of deep aesthetic feelings are also interpreted, by such authors as Mellars or Bahn, as an indication of gradual evolution over time. Which of the two scenarios, gradual or explosive, portrays better

---

Box 10.14 The purpose of Paleolithic paintings

There has been much discussion regarding the purpose for the polychrome paintings at Altamira and Lascaux from the simply artistic to religious and magical ones. Some understanding might be achieved by thinking of them as similar to ourselves and reflecting on why we wear necklaces, bracelets, and earrings, or why we choose the shape and design of our dinner sets.

such an important event in the history of our species as the acquisition of our current cognitive capacities? We will return to this topic in Section 10.2.4. Before that, we will consider the Neanderthals capacity for artistic expression.

10.2.3 Separate traditions or cultural sharing?

It seems that after their contact with anatomically modern humans, Neanderthals borrowed tool-making techniques from our species, as well as decorative elements, which is of great importance when dealing with the question of the origin of symbolism. In addition to Châtelperronian tools constructed *in situ*, a site inhabited by Neanderthals (Hublin *et al.*, 1996), the Grotte du Renne, Arcy-sur-Cure (France), has yielded a series of up to 36 objects such as carved ivory pieces and perforated bones, the sole purpose of which must have been decorative (Figure 10.16).

From 1949 Leroi-Gourhan carried out studies that revealed important differences between the engraving techniques used to produce the Arcy-sur-Cure Châtelperronian artifacts and the latest Aurignacian utensils that were found in the most modern strata of the same cave (Leroi-Gourhan, 1958, 1961). Hence, the Châtelperronian (Neanderthal) and Aurignacian (modern human) cultures were different. But the decorative objects from the Grotte du Renne raised doubts about these

differences existing between modern humans and Neanderthals. Thus, Hublin and colleagues (1996) interpreted the Arcy-sur-Cure artifacts as the result of cultural exchange. d'Errico and colleagues (1998) arrived at a different conclusion: those objects were the result of an independent and characteristically Neanderthal cultural development, which had managed to cross the threshold of the symbolism inherent in decorative objects. There is no reason to assume that the biological differences between Neanderthals and modern humans necessarily translated into differences between their intellectual capacities. Bahn (1998) also believed the Arcy-sur-Cure objects merited attributing Neanderthals a sophisticated and modern symbolic behavior.

Figure 10.16 Perforated bones and teeth from the Châtelperronian site of Arcy-sur-Cure (France). Photograph from Bahn (1998). Reprinted by permission from Macmillan Publishers Ltd. *Nature* 394: 6695, 719–721, 1998.

Box 10.15 Interpreting the Grotte du Renne objects

Randall White (2001) has offered an alternative interpretation of the decorative objects from the Grotte du Renne: "It seems implausible that... Neanderthals and Cro-Magnons independently and simultaneously invented personal ornaments manufactured from the same raw materials and using precisely the same techniques." Consequently, he argues that the Châtelperronian ornaments from the Grotte du Renne are Aurignacian and were produced by modern humans. The question of whether the authors of the Châtelperronian culture were Neanderthals, modern humans, or both, has sparked numerous discussions. The evidence from Saint-Cesaire

(France), with both middle- and upper-Palaeolithic strata, allowed *in situ* studies of the association of specimens and tools, as well as the cultural transition (Mercier *et al.*, 1991). Mercier *et al.* (1991) used thermoluminescence to estimate the age of the Neanderthal specimens found in levels with Châtelperronian industry. Their results suggest that they were 36,300 ± 2,700 years old. Mercier *et al.* (1991) argued that there was contact between Neanderthals from Western Europe and the first modern humans that arrived there. They also noted something we have said on several occasions: the straightforward identification of cultures with taxa is not possible.

Arcy-sur-Cure suggests that Neanderthals were capable of producing decorative objects; other sites provide evidence of cultural sharing. Karavanic and Smith (1998) documented the presence of two contemporary sites at Hrvatsko Zagorje (Croatia) that are close to each other. The Vindija Cave has yielded Neanderthals, while Velika Pécina has only produced remains of anatomically modern humans. The authors believed that the coincidences exhibited by the tools from both sites are due to imitation or even commercial exchange. These Croatian sites do not include ornaments, but they provide remarkable indications of cultural exchange. This is corroborated beyond a doubt by *H. neanderthalensis* and *H. sapiens* coincidence at Palestinian caves. Although the shared Near-East Mousterian culture could be interpreted as the maximum horizon Neanderthals could reach, the Arcy-sur-Cure objects, assuming they were constructed or used by Neanderthals, suggest this was not the case. They seem to support the notion that Neanderthals appreciated pendants enough to identify them as "beautiful objects". Thus, what is the explanation for the upper-Paleolithic symbolic explosion?

Any chronological table of the cultural sequences reveals the difficulties we are encountering. Direct correspondences are usually drawn between cultural manifestations and species, associating Mousterian with Neanderthals and Aurignacian with modern humans. Hence, it seems clear that attributing or not to Neanderthals sufficient cognitive capacities for aesthetic experience is heavily influenced by a given author's point of view about the Mousterian evidence. Those who argue that Neanderthals and *H. sapiens* belong to different species generally reject the presence of symbolism in the former's contrivances, and vice versa.

There are difficulties inherent in both points of view about Neanderthal aesthetic experience. If the Neanderthals' cognitive capacities were not advanced enough, how could they have manufactured—or exchanged—decorative objects such as those at Arcy-sur-Cure? But if their mental architecture was similar to that of anatomically modern humans, why does the upper-Paleolithic artistic explosion associated with anatomically modern humans, such as Cro-Magnons, coincide with the disappearance of Neanderthals?

We started by asking what sort of a creature was a Neanderthal. The short answer is rather ambiguous. Keeping to what it seems they *could do*, Neanderthals must have been quite similar to ourselves; for example, they could bury their dead, and sometimes they did so. But if we only take into account what Neanderthals usually *did*, things change. The key issue here is that Neanderthals left behind few manifestations that would be informative. For instance, no Neanderthal necropolis has ever been found. The scarcity of evidence of Neanderthal aesthetic or transcendent thought is relevant for understanding a very important event that occurred after they disappeared. Anatomically modern humans transformed sporadic manifestations of symbolic activity into a common component of their behavior. Why did this leap forward take place? Did it have to do with the notion favored here that Neanderthals and anatomically modern humans were two separate species?

Even if we accept that the genetic makeup of *H. sapiens* provided them with superior cognitive capacities, unresolved problems remain. Our species appeared at least 100,000 or 150,000 years ago, or even earlier. So, we are faced with a true compromise: the species endowed with the emerging cognitive processes that characterize aesthetic capacities and symbolism appears much earlier than the upper-Paleolithic artistic explosion. So, how do we explain this sudden event in the manifestations of anatomically modern humans?

10.2.4 The leap forward

The great cognitive transformation evinced in the upper-Paleolithic artistic explosion must have included different capacities for adaptation in our ancestors. Several authors have suggested co-evolutionary sequences of cultural manufactures and communicative abilities. Some of these models have been proposed within a paleontological framework (Washburn, 1960; Isaac, 1978b; Lewin, 1984), whereas others take a sociobiological point of view (Lumsden and Wilson, 1983). Davidson and Noble (1989) put forward an

Box 10.16 Neanderthal cultures

Neanderthals used at least four different cultures: Mousterian, Châtelperronian, Szeletian, and Uluzzian (Vandermeersch and Garralda, 1994). For simplicity, here we will only discuss the Mousterian and Châtelperronian.

The hypothesis that Neanderthal decorative elements found in the Châtelperronian deposits are imitations of Aurignacian objects made by modern humans implies that both cultures were contemporary or that the Aurignacian culture was older. Zilhao *et al.* (2006) have investigated the sequence of sediments and the archeological association of the Grotte des Féees at Châtelperron (France) and reject the Châtelperronian–Aurignacian contemporaneity: they assert that "its stratification is poor and unclear, the bone assemblage is carnivore accumulated, the putative interstratified Aurignacian lens in level B4 is made up for the most part of Châtelperronian material, the upper part of the sequence is entirely disturbed, and the few Aurignacian items in levels B4-5 represent isolated intrusions into otherwise *in situ* Châtelperronian deposits." Their conclusion is that "as elsewhere in southwestern Europe, this evidence confirms that the Aurignacian postdates the Châtelperronian and that the latter's cultural innovations are better explained as the Neandert[h]als' independent development of behavioral modernity." This hypothesis deserves attention, but to be accepted similar studies should be carried out at places other than the Grotte des Féees.

interesting hypothesis that relates the origin of art itself, and not just the general cultural sequence, with the origin of language. They believe that drawing requires prior communication. Drawing later transformed communication into language. According to Davidson and Noble, a painting of a bison is not a bison. The word "bison" is not a painting of a bison or the animal bison. Both, painting and word, *represent* the reality of the animal. Whereas language cannot be similar to what it represents, except for occasional onomatopoeias, drawings can be. This is why Davidson and Noble (1989) argue that pictorial representations are halfway between reality and language.

Davidson and Noble's (1989) hypothesis refers to drawings which represent reality, not to middle-Paleolithic geometric lines. This is precisely the issue, because the upper-Paleolithic artistic explosion is abundant with iconic representations of real objects. Davidson and Noble (1989; Noble and Davidson, 1996) sustain the notion of cognitive emergence as a barrier that separates "art" from "non-art". They believe that the middle-Paleolithic geometrical engravings are not continuous with the subsequent proliferation of representational drawings. Neanderthals cannot belong to the same tradition that later developed throughout the upper Paleolithic. This would require that the older and the more modern object shared meanings, but meaning is absent in early geometric engravings.

Davidson and Noble believe that the definitive step towards humanity was the episode involving the appearance of the earliest art and, bound to it, language. Hominins began using things (drawings) to represent other things (objects, concepts and, later on, words), that is how hominins became humans. But even accepting this scenario, the process must be specified in greater detail. Were the first *H. sapiens* that appeared more than a hundred thousand years ago similar to contemporary humans? Or was there a subsequent "cognitive revolution", coinciding with the Aurignacian culture in Europe?

McBrearty and Brooks (2000) carried out a detailed study of the available evidence of such revolution as well as of the models that interpret such evidence. They believe that the proposal of a cognitive revolution repeats a scenario introduced in the nineteenth century with the "age of the reindeer" (Lartet and Christy, 1865–1875). Around the 1920s the upper Paleolithic was generally characterized by the presence of sculptures, paintings, and bone utensils. But according to McBrearty and Brooks, the evidence used to determine the changes between the lower, middle, and upper Paleolithic was always taken from the western-European archaeological record. During the last glacial period, the human occupation of

...at area was irregular, as Howell (1952) pointed out, with populations periodically reduced, or even extinguished. McBrearty and Brooks (2000) argued that the "revolutionary" nature of the European upper Paleolithic is mainly due to the discontinuity in the archaeological record, rather than to cultural, cognitive, and biological transformations, as suggested by advocates of the so-called human revolution. Instead, they suggest that there was a long process that gradually led to the European Aurignacian richness.

The traditional consideration of the African archaeological record was influenced by the scheme used to create the European sequence of lower, middle, and upper Paleolithic stages. The cultural phases of Africa were consonantly grouped into the early, middle, and late Stone Ages, following the sequence of tools found in South Africa. The African middle Stone Age began with the disappearance of Acheulean handaxes, and lasted until the appearance of microliths. Despite the lack of a straightforward equivalence, the middle Stone Age is usually associated with European middle Paleolithic. The sites that have rendered tools belonging to the middle-Stone-Age phase have also yielded the earliest anatomically modern human specimens or their immediate ancestors, the Klasies River and Border Cave among them. Because they are considered less advanced in cultural terms than those belonging to the European upper Paleolithic, they seem to support further the mystery of the Aurignacian artistic revolution that came out of nowhere.

But why did our species take so long to manifest its cognitive capacity through the manufacture of instruments? McBrearty and Brooks (2000) argue that this is simply not the case. *H. sapiens* exhibited important cognitive differences detectable in their archaeological manifestations from the very start. They list ecology, technology, economy, social organization, and symbolic behavior among the kinds of evidence that are usually employed to detect modern human behavior, which is identified with the European cognitive revolution. McBrearty and Brooks find indications of all of these kinds in the earliest African sites with modern humans (Figure 10.17). With respect to symbolic behavior, the subject of the present

section, McBrearty and Brooks (2000) provide evidence from deliberate burials in Border Cave (South Africa; Beaumont *et al.*, 1978), Taramsa (Egypt; Vermeersch *et al.*, 1998) and Mumbwa (Zambia; Dart and Del Grande, 1931). These are controversial, although no more than those from other European and Near-East localities. Sites estimated at 100,000 years in Border Cave and Zhoura (Morocco; Debénath, 1994) have yielded decorative shells, ostrich eggshells, and perforated bones. Ornamental stones have turned up in Aterian sites estimated to be between 130,000 and 40,000 years old, such as Seggédim (Niger; Tillet, 1978). Numerous beads, and engraved and perforated objects appear between 40,000 and 20,000 years ago in many sites. Hence, Africa precedes Europe with regard to personal decoration. Metallic ochre pigments—iron and manganese oxides—are abundant in sub-Saharan Africa, especially in sites classified, owing to their instruments, in the middle Stone Age. The Apolo 11 (Namibia) site is known for its remarkable animal paintings. Although the estimated age of Apolo 11 is recent (about 28,000–26,000 years), McBrearty and Brooks point out the much earlier date, close to 60,000 years, attributed to the ostrich eggshells retrieved from the same stratigraphic horizon as the paintings (Miller *et al.*, 1999).

J. Desmond Clark and Hiro Kurashina (1979) published a study on the culture found at the Gadeb site (Ethiopia), 1.5 million years old. They noted the presence of worn basalt fragments which, when scraped, leave a red pigment. They speculated that contemporary *erectus*-grade hominins might have used color pigments experimentally. However, Clark and Kurashina admitted that no evidence of the scrapes produced by the basalt stones has appeared at the site. If we leave this extremely hypothetical possibility aside, the earliest remains of red ochre found at archaeological sites are not over 300,000–250,000 years old.

Only about a dozen sites have yielded reddish pigments prior to the upper Paleolithic (Knight *et al.*, 1995). But the use of pigments cannot be due to anything but decorative purposes: they are useless for anything else (although decoration by itself, as we will see immediately, can have an

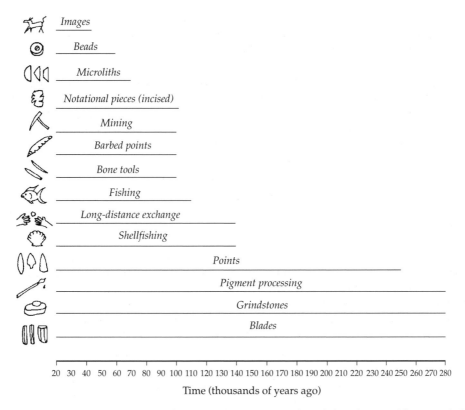

Figure 10.17 Evidence of early symbolic activity in Africa. Picture from McBrearty and Brooks (2000). Reprinted from *Journal of Human Evolution* Vol 39: 5, S. McBrearty and A.S. Brooks 'The revolution that wasn't', 453–563, 2000, with permission from Elsevier.

Box 10.17 Early evidence of intentional transportation

Kenneth Oakley (1981) argued that the presence of a dark red jasperite pebble at Makapansgat Member 4 (South Africa), dated around 4 Ma, is the earliest evidence of intentional transportation. The stone appeared far from its place of origin. Given that its shape is remotely reminiscent of a humanoid face, it could also suggest that the hominin that transported it there was capable of some sort of aesthetic appreciation. Following Ernst Wreschner, Oakley wondered whether the interest in mineral pigments, the beginning of social organization, and the origins of language coincided in time. If this were the case, those transported reddish stones would indicate the beginning of the process.

added function to the purely aesthetic one). However, early sites with remains of red ochre contain neither paintings on rocks nor colored instruments. If the inhabitants of those places bothered to produce color paintings and did not paint their surrounding rocks for more than 70,000 years, what did they use them for?

The first artistic activities related to the production of what we could consider decorative objects can be assumed to coincide with the moment at which ochre pigments appear at the sites. But if this is the case, we have to conclude that these objects have not survived in the fossil record. Why?

The most abundant object that can be decorated but is not permanent is the body itself. If these first modern humans appreciated decorations with purely aesthetic purposes, and they painted their bodies with red ochre, the lack of evidence should not be a surprise. This proof would have disappeared. Knight and colleagues (1995) have suggested that the decoration of male bodies could be a symbolic representation of female menstruation and, thus, have social organization functions coinciding with the beginning of the division of labor. Drawings and colors adorning the bodies of current hunter-gatherers often serve as a hierarchical symbol beyond the mere decorative intention (Figure 10.18).

The evidence provided by McBrearty and Brooks (2000) suggests three things. First, that modern behavior indicative of cognitive changes in *H. sapiens* can be detected from the time our species appeared in Africa, with social, technological, ecological, economic, and symbolic manifestations. Second, that the increase in the complexity of *H. sapiens* culture was gradual. Third, that only by leaving aside African evidence is it possible to be surprised by the "sudden" appearance of the Aurignacian artistic explosion. Artistic activity took place throughout the whole existence of our species and left some early

samples, but in Africa, not in Europe. The murals of southern France and northern Spain represent the final manifestation of the new cognitive capacities. Those magnificent paintings may be due to new rituals as well as to the possibility of occupying the caves after the disappearance of animals capable of disputing them with humans, but they need not indicate any last moment mental change. The model of the early African origin of these cognitive processes accommodates better to the rules of parsimony.

The gradual explanation elaborated by McBrearty and Brooks (2000) has little to do with the gradualist model we referred to in previous chapters. McBrearty and Brooks accept the taxonomic distinction between the species *H. sapiens*—a term they use to designate only anatomically modern humans—and *H. neanderthalensis*. They also separate *H. sapiens* sensu stricto and "archaic" *H. sapiens* as two different species. Furthermore, they do not agree on the unity of the latter and—in accordance with the meaning of a grade—they argue that "archaic sapiens" probably encompasses several species. Consequently, their gradual model of the appearance of symbolism does not involve the existence of a single hominin species since the middle Pleistocene until our days. McBrearty and Brooks (2000) do not favor either

Figure 10.18 Left: piece from Blombos Cave (South Africa). Picture from http://www.svf.uib.no/sfu/blombos/Picture_Gallery.html. Right: current decorative uses.

hybridization between Neanderthals and anatomically modern humans. They touch on a tender spot when noting the most evident inconsistency of the model of sudden cognitive change: the enormous lapse of time between the appearance of anatomically modern humans and the artistic explosion identified in Western Europe. As they say, within our species, human anatomical and behavioral patterns have developed over a period of close to 200,000 years.

The only thing this hypothesis rejects, on very firm grounds and supported by a broad range of indications, from archaeological to molecular, is the notion of a "revolution" as it is suggested by the European models: that is, a revolution caused by a sudden and late change around 40,000–30,000 years ago. The model presented by McBrearty and Brooks seems much more reasonable than the one positing a late sudden change, which is very hard to explain in view of the available evidence.

10.3 The origin of language

Language is one of our species' most distinctive behavioral traits. No other animal speaks like we do, by means of symbolic creative languages. When dealing with its evolution we are faced with two certainties. First, it is an apomorphy, a derived trait, that we can be absolutely certain evolved within the human lineage. Second, it is hard to offer any evidence of how and when this evolution took place. There are plenty of hypothetical speculations and philosophical models of the evolution of human language. They usually move between the pragmatic and genetic conceptions of language. From the pragmatic point of view, human speech is the result of a learning process that starts from scratch. From the genetic point of view, the faculty of language requires a prior substrate available only to humans; thus, trying to teach human language to an animal is senseless.

The starting point for any explanation for the evolution of human speech is the so-called "Plato's paradox" or "poverty of stimulus". In a very short time and based on disperse and confused information, children manage to understand and produce correct syntactic constructions. The best starting point to work out an explanation for this paradox of human language is the genetic or nativist hypothesis. Noam Chomsky stands out among the supporters of this perspective. To justify Plato's paradox Chomsky postulated the existence of genetic baggage that makes the capacity to speak an innate asset of our species (Chomsky, 1966, 1980, 1989). Applying the Chomskyan model to the evolution of language, this faculty would be considered as a product of phylogenesis, as a human apomorphy.

Ullin Place (2000) qualified this Chomskyan idea as a *"deus ex machina"*. According to Place, Chomsky's hypothesis requires positing a sudden appearance of the innate capacities for speech. Place believes it is not possible to imagine a single mutation of that kind. Rather, a series of cumulative mutations must have occurred over millions of years, each one affording new selective advantages. The earliest mutations would have been useful for noncommunicative activities and would be related with tool making. At that time communication would have taken place through gestures. Ullin Place cited the work of such ethologists as Savage-Rumbaugh to document a similar kind of communication in chimpanzees.

As we will soon see, we are faced with yet another instance of polysemy, now referred to by the term language. If it is understood as any kind of sufficiently informative communication, there is no doubt that chimpanzees use language, and that ants, bees, and other insects, use sophisticated languages. But Chomsky (1966) was referring to a specific kind of communication that is symbolic and allows generating virtually infinite messages: what in Cartesian terms is known as the creative aspect of language.

This specific faculty is innate, in the same way that the capacity to produce and interpret gestures or recognize faces is innate. This does not imply that *any* kind of sign communication system could have appeared in virtue of the same mechanisms and at the same time as the linguistic competence alluded by Chomsky. Rather, it is reasonable to assume that this is not the case. The path to language is the combination of a very diverse series of

---

Box 10.18 Plato's paradox

Terrence Deacon (1997) agreed with Plato's paradox. Deacon argued that the explanation for that paradox lies in a characteristic shared by all languages that makes them accessible to humans. The model, which Deacon illustrated with computer examples, suggests that language is an external phenomenon: it appeared, it organized itself, and evolved separately from humans, though, naturally, in interaction with them. Such an autonomous language could, in fact, occur independently from our species. It could be achieved by means of other information-processing structures not available to animals

but, maybe, they could be available to computers. According to Deacon (1997, p. 109), languages need children more than children need languages. However, languages are constructs of the individuals who speak them, and they are not straightforwardly comparable to computational designs. Deacon's model explains, in a complicated way, an evolutionary piece of evidence: given the genetic constitution of our species, fixed by natural selection, we speak in a such a way that our languages follow rules which children can easily interiorize.

---

communicative aptitudes whose phylogenesis must have extended for 2 million years. But it is also important to know that, at a certain moment, communicative capacities took a completely new direction owing to the appearance of three novelties: first, an organ that produced sounds capable of modulating vowels and consonants; second, means of phonetic/semantic identification which associates the combinations of vowels and consonants with meanings; and third, combination of phonetic/semantic units capable of generating an unlimited number of messages subjected to syntactic rules.

Such a combination of capacities is exclusively human. It may have been generated by very specific mutations that turned the previous communicative abilities into a new and unique kind of language.

As we noted above, during ontogenesis linguistic competence is achieved rapidly and without systematic learning tasks. Hence, it is necessary to posit the existence of an innate genetic baggage. But which specific baggage? Which genes control this human capacity for speech?

FOXP2 was the first gene ever identified with a function related to language (Lai *et al.*, 2001). *FOXP2* was isolated while studying a family with an inherited severe language and speech impairment. The affected members of the family had difficulties in selecting and sequencing fine orofacial movements, and exhibited grammatical

deficits as well as slight nonverbal cognitive impairments. It was soon determined that the disorder was transmitted as an autosomal dominant monogenetic trait, associated with a point mutation in a gene situated on chromosome 7 (Lai *et al.*, 2000).

FOXP2 is not exclusively human, but is part of the genome of animals as evolutionarily distant as humans and mice. The *FOXP2* protein is highly conserved among mammals: it has undergone only one amino acid replacement during the 70 million years or so that have elapsed between the last common ancestor of primates and mice and the last common ancestor of humans and chimpanzees. Since the divergence of the human and chimpanzee lineages, 6–8 Ma, the human protein has undergone two amino acid changes, while the chimpanzee form has not changed. If the gene participates in laying down the neural circuits involved in speech and language, it seems possible that the last two mutations that occurred in the human lineage were crucial for the development of language. Wolfgang Enard and colleagues (2002) have suggested that these events happened during the last 200,000 years. We do not know whether language would have been possible with a chimpanzee or earlier hominid version of the gene, but the estimated age of the mutated gene fits well with the estimated age for the appearance of modern humans. One possibility is that language appeared more-or-less suddenly in

Box 10.19 The dual patterning of human language

A small shift in the phonetic chain created by combining consonants and vowels may produce a completely different semantic content. This is what happens with the words *fake* and *fate*. A few phonetic units generate, in virtue of this first articulation, a very large number of words.

By means of a second articulation words combine to form sentences. Sentences follow each language's syntactic rules, such that varying the syntactic order of the units—words—changes the message. "Peter killed John" is not the same message as "John killed Peter". The second articulation allows an infinite number of messages.

Box 10.20 *FOXP2* acts in the subcortical region

Brain-imaging techniques revealed that a subcortical region, the caudate nucleus, was reduced bilaterally in the affected members of the family presenting a defective allele of the gene *FOXP2* (Watkins *et al.*, 2002). This observation is striking, because the traditional viewpoint

tends to restrict language-related areas to the cortex. It seems that *FOXP2* is involved in the regulation of the development of subcortical neural circuitry critical for language and speech.

the hominid lineage with the advent of modern *H. sapiens*. However, the high degree of conservation of the protein, and of the pattern of the gene's expression in the brain, suggests that language and speech are, at least in part, supported by neural structures present in other species, which would support a gradual emergence of the capacity for language through the recruitment or fine-tuning of pre-existing neural pathways.

We have to acknowledge, following Phillip Tobias (1997b), that humans speak with their brains. But human language cannot be reduced to a matter of genetically controlled neural processes. Our species' children must be submerged in the environment created by human language for their brain to mature during exterogestation: the development stage that occurs outside the maternal uterus. This remarkable combination of innate and acquired elements operating in a feedback loop leads to a language that can express an infinite number of sentences. A slow and gradual maturation of cognitive capacities and a last leap in the adjustment of the phonation organs, which occurred in our species and maybe in Neanderthals and allowed phonetic/semantic associations with dual patterning, are the minimum

components of any language-evolution model compatible with the Chomskyan proposal.

Any proposal concerning the phylogenesis of language is necessarily speculative. However, it is worth trying to constrain speculation by means of available evidence and keeping it to the indispensable minimum necessary to offer a coherent model of the phylogenesis of our linguistic competence.

10.3.1 Animal languages?

Linguists, philosophers, psychologists, ethologists, primatologists, paleontologists, sociobiologists, and archaeologists have spent much time and effort discussing whether language is a human apomorphy. The means of communication of animals that are phylogenetically close to us (such as chimpanzees), some that are further away (such as humpback whales), and even some that are very remote (such as bees), can be considered languages. A very long-running controversy, which we will not review here, has offered all possible arguments for and against the efforts of many authors (Gardner, Premack, Terrace, Savage-Rumbaugh) to teach human language to

Box 10.21 Evolution of genes expressed in the human brain

Hill and Walsh (2005) have pointed out that human brain evolution is associated "with changes in gene expression specifically within the brain as opposed to other tissues such as liver." Current evidence indicates that some genes expressed in the brain have evolved faster in the human lineage than in the ape lineages (Caceres *et al.*, 2003; Uddin *et al.*, 2004): "although the studies differ in design and principal conclusions, they share support for an increase in expression level in a subset of brain-expressed genes in the lineage leading to humans." Genetic changes in the evolution of the human lineage have resulted in adaptive advantages, but some allelic combinations may have pathological consequences. Hill and Walsh (2005) refer to *FOXP2* as well as to two other loci, *ASPM* and *MCPH1*, which "show strong evidence for positive evolutionary selection" and yet have alleles that cause microcephally.

Box 10.22 The language-learning capacity of children

Some experiments suggest that children have a greater learning capacity than Chomskyans usually admit (Saffran *et al.*, 1996; Jusczyk and Hohne, 1997). This has led authors such as Elisabeth Bates and Jeffrey Elman (1996) to minimize the problem of the poverty of stimuli. Young children, according to this interpretation, learn language through statistically guided inductive processes. But the experiments do not suggest that inductive capability works in isolation. Gary Marcus *et al.* (1999) have shown that young children learn in at least two different ways: one that uses statistical relations between words in sentences; and another that deduces abstract rules between variables in almost an algebraic way. This second way fits better with a Chomskyan scheme. Marcus *et al.* (1999) warned that the learning of a language cannot be accounted for exclusively by these two kinds of learning. These authors believe that these are necessary prerequisites, but they are not sufficient.

chimpanzees, gorillas, and orangutans. Apes are incapable of vocalizing like we do. But, in any case, deaf and dumb humans are also incapable of expressing themselves through an oral language, and they use a sign language as an alternative. The resource of deaf and dumb people has been used by primatologists and anthropologists to test whether other species close to ours are capable of learning these signs and communicate in that fashion with us.

Sign-language teaching projects such as those of the Yerkes Primate Research Centre, the Language Research Center (LRC; at Atlanta, Georgia State, and Emory Universities), the Central Washington University's Chimpanzee and Human Communication Institute (CHCI), and the Orangutan Language Project (OLP) have made some animals famous, such as Lana, Kanzi, Washoe (Figure 10.19), Loulis, Nim, and Koko. The significance of the results achieved depends, as usual, on the way they are interpreted.

The cognitive processes involved in the learning of sign language by deaf humans are very similar, if not identical, to those that allow people with no hearing problems to communicate. However, there is more. Hiroshi Nishimura and colleagues (1999) found that in addition to visual areas, sign-language gestures activate auditory brain regions, which one would expect to be inactive, given the absence of sounds to be processed. Unless chimpanzees possess a similar system, these findings suggest, once again, that human language is an exclusive trait of our species. If this is the case, there is no way that trying to teach an ape will allow it to "talk", not even using human signs.

The argument that other primates are capable of achieving a linguistic competence similar to ours—which, by the way, none of the aforementioned

Figure 10.19 Washoe, a female chimpanzee flicking through a magazine. Photograph from CHCI, http://www.cwu.edu/~cwuchci/research.html.

> **Box 10.23 Teaching sign language to apes**
>
> Chapter 5 of Steven Mithen's *The Prehistory of the Mind* (Mithen, 1996) includes a synthesis of the achievements of sign-language teaching to apes.

authors suggests—has failed continuously. But the proposition that can readily be accepted is that chimpanzees and other primates have a highly developed intelligence. Chimpanzees are capable of applying their natural skills to learn tasks that are adaptively unrelated to their environment, such as that of learning human symbols.

The communicative competence of some primates in their natural environment is comparable to ours in some respects, according to Dorothy Cheney and Robert Seyfarth's work on vervet monkeys (*Cercopithecus aethiops*; Seyfarth and Cheney, 1984, 1992; Cheney *et al.*, 1986; Cheney and Seyfarth, 1992). These studies have shown that vervet monkeys possess a whole repertoire of vocal signals in the form of alarm calls with very precise semantic content. This is an essential evolutionary adaptation of *C. aethiops*, which are a usual prey of many different kinds of predators. It is obvious that a general alarm call would be of little use to a group of vervets. The ways to escape successfully from a cobra and an eagle are completely different.

Cheney and Seyfarth used playback techniques, recording and reproducing vocal signals to study the production of *C. aethiops* selective alarm calls in Amboseli National Park (Kenya). They found four acoustically discrete alarm calls, which corresponded to four main kinds of predator: felines, birds of prey, snakes, and primates. Cheney and Seyfarth concluded that each alarm class is discrete and different from the rest. This induces different escape responses. For instance, after an "eagle alarm", the individuals of the group of vervets run under bushes. But if the "leopard alarm" has been produced, they climb on to the thinnest branches of a tree. Mistaking an alarm call can be fatal, so *C. aethiops* have established very clear, precise, and discrete phonetic/semantic identifications. Each signal serves a specific function. This implies that vervets have a very complex cognitive representational system. The duration and volume of the warning call does not affect the escape response, which only depends on the particular class of call (Figure 10.20).

Cheney and Seyfarth argued that vervet vocalizations include semantic signals that are completely unrelated to the expression of the emotional state. But the sense in which these authors use the term semantic signal deserves clarification. We could contend that the alarm calls are mere releasing signals. Ethologists (Eibl-Eibesfeldt, 1967, for instance) have defined these as signals capable of eliciting an innate response. But, how do the vervet alarm signals work? Are they

Box 10.24 A chat with Koko the gorilla

Here is a partial transcript of an internet chat with Koko the gorilla (http://www.geocities.com/RainForest/Vines/4451/KokoLiveChat.html). HaloMyBaby is the moderator of the chat on AOL, DrPPatrsn is Koko's friend and trainer, and LiveKOKO is Koko the gorilla.

Welcome, Dr. here!	Patterson and Koko, we're so happy you're
DrPPatrsn:	You're welcome!
HaloMyBaby:	Is Koko aware that she's chatting with thousands of people now?
LiveKOKO:	Good here.
DrPPatrsn:	Koko is aware.
HaloMyBaby:	I'll start taking questions from the audience now, our first question is: MInyKitty asks, Koko are you going to have a baby in the future?
LiveKOKO:	Pink.
DrPPatrsn:	We've had earlier discussion about colors today.
LiveKOKO:	Listen, Koko loves eat.
HaloMyBaby:	Me too!
DrPPatrsn:	What about a baby? She's thinking . . .
LiveKOKO:	Unattention.
DrPPatrsn:	She covered her face with her hands . . . .which means it's not happening, basically, or it hasn't happened yet.
LiveKOKO:	I don't see it.
HaloMyBaby:	That's sad!

DrPPatrsn:	In other words, she hasn't had one yet, and she doesn't see it happening. She needs several females and one male to have a family. In our setting it really isn't possible for her to have a baby.
HaloMyBaby:	Do you see that situation changing when you get the Gorilla preserve on Maui?
DrPPatrsn:	Yes, we do.
LiveKOKO:	Listen.
DrPPatrsn:	Koko wants to hear on the phone as we're doing this.
HaloMyBaby:	Hi Koko! I can hear her! She breathed at me! This is so cool! In case you're curious, here's how Koko is able to participate in this chat: Dr. Penny Patterson is signing the questions to Koko from the online audience and a typist is entering for her.
DrPPatrsn:	I'm working to create a family. In Hawaii, we'll have the ability to do that; she's almost assured to have a family of her own.
HaloMyBaby:	So she really is looking forward to this!
DrPPatrsn:	She's making happy sounds now . . .
Question:	EFRN asks, Would Koko like to have a kitten, a dog, or another Gorilla as a friend?
LiveKOKO:	Dog.
DrPPatrsn:	She actually has two dog friends right now, one kitty and two gorillas.

Box 10.25 The adaptive nature of alarm calls in vervet monkeys

In 1967, that is, at the same time R. Gardner and B. Gardner (1969) were trying to teach sign language to chimpanzees, Struhsaker (1967) described different alarm calls in vervet monkeys, which Marler (1978) later identified as a special kind of symbolic communication, rejecting the hypothesis of a simple emotional response to attacks. The need for a high-level semantic communication is a consequence of the adaptive situation of *C. aethiops*, which is prey to the following predators: leopard, lion, hyena, cheetah, jackal, baboon, python, cobra, black mamba, green mamba, viper, martial eagle, tawny eagle, hawk-eagle, and Verreaux's owl.

efficient because they produce an adaptive behavior such as automatic escape or because these monkeys "understand" what each "word" means?

Using playback techniques—recording free-ranging vervet sounds and reproducing them in other circumstances—Seyfarth and Cheney assessed the communicative scope of these primates. After an alarm signal (one that sounds more or less like *wrr*, for instance, or the one that sounds like *chatter*) belonging to a specific individual had been

Figure 10.20 Dorothy Cheney together with a baboon (*Papio cynocephalus* or *Papio hamadryas*). Photograph from http://www.bio.upenn.edu/faculty/cheney/.

played repeatedly, the other monkeys became habituated: they stopped attending to that call. But most interesting is the psychological mechanism that saturated the signals. The habituation to the *wrr* call of a given individual *x* also produced habituation to the *chatter* call of the same individual *x*. But, habituation to the *wrr* call of individual *x* did not transfer to the *chatter* call of another individual *y*. Hence, vervets become habituated to the meaning, not to the sound. Presumably *C. aethiops* infer this meaning by means of deep psychological processes. The question whether *wrrs* and *chatters* can be considered to be "words" is mostly a philosophical matter.

10.3.2 The phylogenesis of the supralaringeal vocal tract

Different human languages use different vocal expressions, although they all consist of an ordered succession of vowels and consonants. A question that is far from trivial is determining when hominins acquired the capability of pronouncing vowels, and consonants in particular, in a way similar to modern humans. We have

mentioned that chimpanzees are incapable of doing this, and thus it is reasonable to assume that this capacity was acquired during human evolution. But at what moment? Language does not fossilize, and written language does not appear until the last split second of our species' history.

Language requires certain anatomical features, relative to the brain and the supralaringeal vocal tract, the part of the throat that goes from the larynx to the oral cavity. It is generally accepted that these features are only present in our species (Laitman, 1984). The anatomical arrangement of the human supralaringeal vocal tract allows a very particular kind of modulation of air flowing out. Through the coordination of the tongue, palate, teeth, and lips we are able to pronounce a multitude of vowels and consonants.

But whereas vocalizing requires a larynx placed in a relatively low position, certain brain mechanisms are essential for sequencing the phonemes that make up words according to precise rules. Hence, it is possible that something helpful might be said about the evolution of language by studying the evolution of the necessary anatomical elements: the supralaringeal vocal tract and the

---

Box 10.26 Prerequisites for speech

Horst Steklis (1988) has argued that field and laboratory research, as well as neurological comparisons, show that there are important similarities between the primate system of vocalizing and human language. In some species warning calls also have semantic qualities, given that they code information about events or external objects. The decoding of the calls could be governed by simple syntactic rules. In Steklis' opinion, such structural and functional complexity is also reflected at the neural control level, because there is strong evidence of volitional components that involve neocortical mechanisms. Neural control mechanisms are different for each kind of call. These data suggest, according to Steklis, that the vocal–auditory

mechanisms of early hominins was better prepared to carry out important language functions than previously assumed. The argument presented by Steklis is close to those suggesting "prerequisites" for speech. Believing such mechanisms existed is different from accepting that language had developed at a certain time in human evolution. Recent research (Fitch and Hauser, 2004) has revealed that other primates can master some simple grammars, but not certain types of grammatical construction, such as phrase-structure grammar. These sorts of result suggest that human language is a combination of novel and primitive language-related capacities.

---

brain. Let us turn to the first one: which of our ancestors had a phonating apparatus similar to ours?

Jeffrey Laitman (1984, 1986) and Philip Lieberman (1973, 1984, 1989, 1994, 1995) have devoted considerable attention to the supralaryngeal vocal tract as a possible indication of the origin of language. Lieberman has noted that human speech sounds require that the length of the kind of tube formed by the mouth be equivalent to that of the other tube descending behind the tongue. Thus, only a relatively low larynx will allow human vocalization (Figure 10.21). Such a position creates some complications for breathing and swallowing. The risk of choking increases because the pipes leading to the esophagus and the lungs cross. Human babies are capable of breast-feeding and breathing at the same time, because their larynx is still in a high position, similar to that of chimpanzees. But towards the age of 2, the descent has already begun (Laitman, 1984). Notice that 2 years is also the age at which children begin articulating words.

Laitman's (1984) reconstruction of the base of fossil hominin crania suggested that the larynx of australopiths was in a high position, similar to that of chimpanzees. The descent would have begun, according to Laitman, in *H. erectus*. Based on the study of the marks left by muscles on the basicrania, and the computational interpolation of

chimpanzee and human anatomies, Lieberman went further by arguing that speech was a very late phenomenon, distinctive of anatomically modern humans, and partially, of Neanderthals. According to Lieberman, the latter were capable of producing some human speech sounds, but not the full range. Gover Krantz (1988) was even more precise when he suggested that the descent of the larynx did not take place until 40,000 years ago, as a second phase of the evolution of our species. The first phase, which Krantz believes began 200,000 years ago, would have modernized the cranium and, partially, the larynx, leading to a cavity about half of its current length and would, thus, allow an imperfect vocal behavior. Only the second phase would lead to speech like ours.

What kind of empirical evidence can be presented for or against such hypotheses as Kranz's? The discovery of a hyoid bone—which is involved in the movements of the larynx—in the Mousterian site of Kebara, Near East (Bar-Yosef *et al.*, 1986), provided new information about the early anatomy of that part of the neck. The Kebara hyoid, attributed to a Neanderthal, has a similar shape to that of current humans, but the associated mandible is broader and more robust (Arensburg, 1989). Its discoverers suggested that the vocal apparatus of Neanderthals could have been similar to ours (Arensburg and Tillier, 1990). However, Lieberman (1999) criticized this hypothesis,

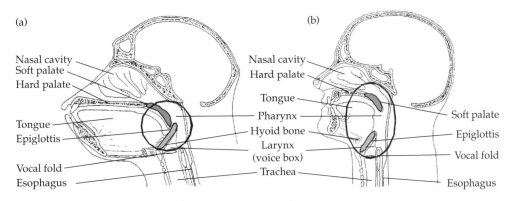

Figure 10.21 The relative positions of the supralaryngeal vocal tract in (a) chimpanzees and (b) humans.

Box 10.27 Factors in the evolution of speech and language

David Lieberman and colleagues (Lieberman and McCarthy, 1999; Lieberman et al., 2001) have challenged the assumption that phonation constraints are virtually the sole determinants of the configuration of the modern human vocal tract. They have shown how factors relating to swallowing, respiration, and the ontogenetic growth of the facial region must also be taken into account to obtain an accurate picture of the evolution of speech and language.

---

arguing that the hyoid is not involved in the ability to modulate human speech sounds.

The attribution of a complete linguistic behavior to Neanderthals is usually based on the morphology of the Kebara hyoid. However, most of the speculations on the possible language of Neanderthals—or even of *erectus*-grade hominins—are based on another kind of evidence: symbolic behavior. As Tobias (1997b) cautioned, we do not speak with our throat, but with our brain.

10.3.3 Evolutionary changes in brain areas

Primate brain organization is fairly homogeneous. An Old World monkey, such as a macaque, is in this sense similar to human beings. But the human brain exhibits a conspicuous development of the temporal and prefrontal areas related, precisely, with the processing of verbal communication and semantic processes. Determining how this evolution occurred might provide a solid base for speculations on the origin of language (Figure 10.22).

What happened to the frontal and temporal areas throughout the human lineage? Do they exhibit traces of a relative expansion that could be associated with the phylogenesis of linguistic mechanisms? For more than two decades Tobias (1975, 1995, 1997b) has argued that such early expansion existed. By examination of *H. habilis* endocrania (OH 16 and OH 24 from Olduvai, KNM-ER 1470 from Koobi Fora), and their comparison with those of *A. africanus*, Tobias (1987) found that Broca's and Wernicke's areas in the left hemisphere had already begun to develop in *H. habilis*. These areas are thought to be responsible for the better part of the brain processes related with language. Tobias also identified an enlargement of the frontal lobes that seemed to anticipate, especially in the Koobi Fora specimen, the increase of that area's volume in *erectus*-grade hominins. Tobias suggested that a *structural* change of the brain that began with *A. africanus*, which exhibited an incipient development of Broca's area, would have consolidated with *H. habilis*.

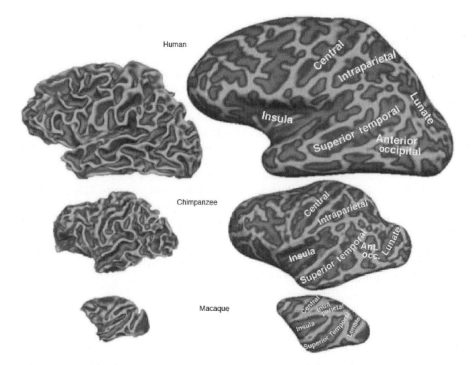

Figure 10.22 Folded (left) and unfolded (right) reconstructions of the left hemisphere. Picture from Sereno and Tootell (2005). Reprinted from *Current Opinion in Neurobiology*, Vol 15: 2, M.L. Sereno and R. Tootell 'From monkeys to humans', 135–144, 2005, with permission from Elsevier.

Box 10.28 The origin of language and developed speech

Tobias has often insisted in his lectures on something that is not always acknowledged: language at its origins—attributed by Tobias to *H. habilis*—and human languages as we know them today are very different. The origin of language and developed speech are different issues in phylogenesis of language.

Tobias' ideas were received with hostility. Tobias (1997b, p. 39–40) himself has described his discovery as "premature", in a similar way to Dart's initial remarks on *A. africanus* (1925) and Leakey and colleagues' (1964) on *H. habilis*. However, as time passed, his model of an early development of brain areas related to language has been admitted even by Lieberman (1999), a true adversary of the hypothesis of a developed language before current humans.

What was the language of *H. habilis* like? Hardly anything can be said beyond the indications on endocranial casts. Lieberman (1995) argued that *H. habilis* possessed, in the best of cases, a language

that was not quite modern. Tobias (1997b) agreed with him on this point, but suggested that we need not consider two phases in language evolution, one which was "not completely human" and another which was "completely human". Hence, Tobias believes that the evolution of language was gradual, by means of a series of stages of increasing conceptual and syntactic complexity, together with an increase of phonetic range. He argued that all these stages would qualify as "spoken human language".

Other authors who have devoted a great deal of effort to research on fossil endocasts are Ralph Holloway (1973, 1974, 1983, 1985), Harry Jerison

---

Box 10.29 Relative increases in brain size

The very popular hypothesis of a differential expansion of brain areas related to language in the genus *Homo* has been questioned by Katherina Semendeferi and Hanna Damasio. The volumetric study of the frontal lobes of some monkeys and apes (macaque, gibbon, orangutan, gorilla, and chimpanzee) shows that their absolute size was smaller than that of modern humans. But, in relative terms, the size of the frontal lobe and the distribution of its sectors are very similar in monkeys, apes, and humans (Semendeferi *et al.*, 1997). The immediate conclusion is that there was growth in brain size throughout the human lineage, but not a relative development of the frontal lobe, which plays an active role in some important processes related to language, such as creative thinking, planning future actions, artistic expression, and semantic analysis.

Semendeferi and Damasio (2000) obtained a similar result when comparing extra-allometric expansions of three broad cortical areas: the frontal, temporal, and parieto-occipital lobes. They found significant differences in absolute volume, but the relative size of the frontal lobe does not distinguish humans from other hominoids. Our frontal area is, in relative terms, the expected size for a primate with a brain the size of ours. Among hominoids, only gibbons have an allometrically smaller frontal area. Temporal and parieto-occipital areas do not show significant relative differences either. The conclusion presented by the authors was that evolutionary modifications of the frontal cortex occurred prior to the chimpanzee/human divergence.

(1975a, 1975b, 1977a, 1977b, 1988) and Dean Falk (1975, 1980, 1983, 1986, 1987, 1998; Falk *et al.*, 2000). These authors did not reach identical conclusions, but generally noted two frontiers, so to speak, in the evolution of brain complexity. The first one is the appearance of an "essentially human" neurological organization in *H. habilis* (Holloway, 1974). The second one is the continuous and rapid increase in the encephalization quotient within the genus *Homo*. The significance of this increase is, however, difficult to specify in terms of volumetric measurements and encephalization indexes. For instance, Wynn (1979, 1993) has argued that cognitive capacities underwent a leap forward with *H. erectus*, and that this species had a remarkable capacity to manage diverse data from the environment. According to Wynn, *H. erectus* might have been able to carry out complex representations of the world better than we can. Or, at least, their everyday use of those abilities was greater than ours.

According to Semendeferi and Damasio (2000), although global frontal volume did not vary greatly throughout human evolution, it is possible that some areas, such as the prefrontal cortex, could have expanded in our species. This suggestion is compatible with the observations of Falk

et al. (2000) in a comparative study of current human, gorilla, and chimpanzee endocranial measurements, which showed the same relative size of the frontal lobe in the three cases (the measurements were taken by projecting on the horizontal plane the maximum width and length dimensions). However, this study revealed certain significant differences favoring humans in certain sub-regions of the frontal lobe. This suggests that the human frontal lobe has undergone a process of reorganization in comparison to that of great African apes. Falk *et al.* (2000) also analyzed the endocrania of *A. africanus* and three paranthropine species with reference to the parameters found for *Homo*, *Pan*, and *Gorilla*. Their results suggest that the frontal lobes of paranthropines were "apelike", whereas the frontal and temporal expansion observed in *A. africanus* brings its brain anatomy closer to that of *Homo*.

10.3.4 The costs of the brain

Any hypothesis about increasing encephalization must take into account metabolic requirements. Brain tissue consumes a great amount of oxygen and glucose, and it does so continuously, independently of physical and mental states. This is

---

Box 10.30 Cortical reorganization

As Falk *et al.* (2000) noted, paleoneurologists have been interested in the frontal lobes because of their functions related to language, abstract thinking, and motor planning and execution. Semendeferi (1994), Deacon (1997), and Passingham (1998) mention the expansion of the prefrontal cortex as a human derived trait, although they all observed that this is the result of a cortical reorganization already present in our ancestors. Such a reorganization is suggested by James Rilling and Thomas Insel (1999) in their study of the primate cortex gyrification, which showed extra-allometric increases in human frontal and parietal cortex. However, further detailed studies are needed to identify precisely the brain functions associated with cognitive processes related to speech.

---

true for all mammals, but the human brain's metabolic needs are enormous. In humans, the ratio between the oxygen consumed by the brain and the whole body is 20%, double that of macaques (*Macaca mulata*; Hofman, 1983). Additionally, the metabolic index for the human cortex is 43% greater than for the rest of the brain. In comparative terms across primates, the "excessive" cost of the cortex does not depend on the size of the body, but the degree of evolutionary development of grey matter (Hofman, 1983).

The existence of high metabolic requirements is an important consideration when assessing how natural selection favors an expanding cortex. Katherine Milton (1988) suggested that the only way to meet the brain's metabolic demands for the genus *Homo* was a dietary change towards richer nutrients, particularly meat. The intestine of modern humans is relatively smaller than the intestine of other primates. It would seem strange that throughout evolution, one organ—the brain—has required a greater amount of nutrients, while the intestines responded by reducing their size. Milton (1988) believed that this is explained by the fact that the human digestive system is specialized, with a relatively long small intestine compared to the considerable length of the colon of apes.

Aiello and Wheeler's (1995) study of the relation between brain size and the length of the intestines led to similar conclusions. Herbivorous primates, with larger intestines, have relatively smaller brains than frugivorous primates. Aiello and Wheeler's explanation is known as the expensive-tissue hypothesis. Two different animal species with a similar metabolic rate have had to "chose"

between intestinal and cerebral tissues, given that both have very high energetic requirements. Herbivorous diets require very large intestines for digestion, so this expensive system would constitute a barrier for high encephalization.

Robert Martin (1996) provided a different explanation for the mechanism that allowed the phylogenetic increase of brain size. According to Martin, the correlation between carnivorous diet and high encephalization is contradicted by the small brain in insectivorous bats, whose intestine is relatively smaller than that of bats which feed on fruits. The alternative large brain/small intestine–small brain/large intestine, which grounds Aiello and Wheeler's hypothesis, does not seem to be a general law. In all fairness, though, it is true that whereas the expensive-tissue hypothesis establishes that if the intestines are voluminous the brain cannot also be large, it says nothing about what to expect with short digestive systems. But Martin (1996) offered an alternative way to establish a relation between energetic resources and brain development, known as the maternal-energy hypothesis. This hypothesis argues that the growth of a primate brain initially occurs within the fetus, during pregnancy, and later during exterogestation with breast feeding. Hence, the energetic resources available to the mother while pregnant and breast feeding are an essential variable. A mother can produce large-brained offspring by increasing the gestation period or by increasing the metabolic rate. Martin (1996, 2000) underlined the risks involved in establishing statistical correlations and deducing causal relations from them when metabolic rates, body weights, brain masses, and gestation periods are being compared. He concluded that the hypothesis of the maternal energy can explain a "passive" increase in encephalization. This can be achieved because of dietary changes—such as the increase in meat consumption—that allow higher metabolic rates, or because of an increase of the gestation period, or because of both. However, Martin also noted the possibility of an "active increase" in relative brain size due to selective pressures towards larger and more complex brains.

10.3.5 The appearance of modern language

Aiello and Dunbar (1993) noted the existence among primates of a correlation between encephalization and group size. Taking this correlation into account could allow determination of the moment at which hominin groups were large enough for language to develop as an adequate instrument of social cohesion. After studying the hominin fossil record, Aiello and Dunbar concluded that the earliest members of the genus *Homo* were the first ones to have some sort of language. They also suggested that during the second half of the middle Pleistocene a rapid increase would have occurred within the *erectus* grade.

Martin (2000) has expressed a different opinion. He believes that the notion of a brain increase in *H. habilis* is biased by calculations of the body size of gracile australopiths (*A. africanus* and *A. afarensis*). It has been generally assumed that gracile australopiths were small individuals, weighing between 25 and 35 kg. If this were the case, their encephalization would be similar to that of *H. habilis*, whose body was a similar size. But the thesis of a considerable sexual dimorphism in australopiths—an issue we dealt with in an earlier chapter—raised the estimates for average body size, which reduced their encephalization index. However, Martin (2000) believes that the *A. africanus* sample speaks of a rather monomorphic single species (with little sexual dimorphism). Consequently, its encephalization would already be considerable. The increase in encephalization would have taken place before *H. habilis*, and

would probably be related to locomotor activity and the search for a different kind of food. Both circumstances are related intimately with the requirements of a brain undergoing an expansion phase.

Falk and colleagues (2000) reexamined the available endocrania of *A. africanus* and the different paranthropine species. Their results suggested a similar brain evolution to that proposed by Robert Martin. As previously mentioned in Falk and colleagues' reinterpretation, the brain morphology of gracile australopiths turned out to be closer to that of humans than to the paranthropines.

What do we know about the origin of language beyond endocranial studies? We have already talked about the appearance of symbolic thought when we discussed Mousterian culture. If there is any available evidence of a complex language, it would certainly be symbols. The relationship between thought and language is unquestionable: advanced symbolism represents firm support for the presence of speech similar to ours, with writing as the last and most obvious example. How was this capacity for symbolism acquired?

If we knew how our mental capacities evolved, the problem of the origin of language would be much easier to resolve. The examination of manufactured objects may be helpful. It seems that the cognitive capacities necessary to produce objects with a high symbolic content, such as those that appear at a certain moment of our lineage's history, are not very different to those required for speech.

In the previous chapter we presented archaeological and paleontological evidence related to symbolic thought. Although there are early symbolic manifestations in Africa (McBrearty and Brooks, 2000), the European artistic explosion 40,000–30,000 years ago seems to have represented a gigantic qualitative and quantitative change. However, according to Mithen (1996) the first early modern humans achieved a certain degree of integration of their specialized intelligence, though they did not reach a complete cognitive fluidity. In Mithen's opinion, their minds were halfway between a Swiss army knife and a cognitively fluid mentality. Mithen's scenario does not constitute an

---

Box 10.31 Enkapune Ya Muto

The Enkapune Ya Muto site (Kenya) has afforded an interesting identification between objects and spoken communication. This is a very recent finding, and the indications that appeared there have still to be broadly considered. However, in 1998 *Science* (vol. 282, pp. 1441–1448) provided a report on transitions in prehistory, which includes the Enkapune Ya Muto and Katanda sites as significant places for finding clues about the origin of language.

---

explanation; it just interprets the meaning of Aurignacian art in a different way. Accepting it does not answer a key question: why did things happen that way? An interpretation of the reasons for that event requires a model of the evolution of the mind which is compatible with the observed facts.

Let us assume that we accept that the continuity theses for the evolution of our cognitive abilities are inadequate, so that European Aurignacian culture should be seen as a revolution. Under this supposition there are several possible interpretations. The first implies that there was some kind of change in the brains of anatomically modern humans. It suggests a sort of anagenetic evolution that, although it did not lead to a new species, increased cognitive capacity. However, there is no direct evidence of such a transformation. Indeed, we are still uncertain about the neural correlates of the so-called higher cognitive processes, such as aesthetic experience. We can say much less about how the brain of the first anatomically modern humans was with respect to those processes, not to mention *erectus*-grade hominins and Neanderthals.

Faced with the lack of direct anatomical evidence, we have to resort to behavioral interpretations. A behavioral hypothesis would argue that anatomically modern humans changed their behavior when they became producers of an art that still amazes us today. We already said that merely noting the existence of European Aurignacian art constitutes no explanation. However, the correlation of the artistic accomplishments with the behavioral habits of modern humans renders a kind of interpretation—hypothetical, certainly—of the appearance of cave paintings and decorative objects. The transition from an occasional use of caves to their continuous occupation might be an explanation for why paintings on their ceilings appeared at a particular time. The reasons for the permanent occupation of caves might be related with the disappearance of the great predators, which would have turned any permanently used cave into a trap for humans. Another interpretation, which is also behavioral, though of a different kind, refers to artistic taste. This taste is related to the production of objects lacking an immediate functional use or for the decoration of the body, which, as we saw, happened prior to the Sistine chapels of Paleolithic art. In fact, these behavioral explanations are complementary. It is very possible that modern humans went from decorating their bodies to doing the same with their homes when caves became permanent living places.

Given that there is no solid evidence in favor of an anagenesis leading to an improvement of cognitive capacities in anatomically modern humans, or of a marked behavioral change, the alternative account by McBrearty and Brooks (2000) seems the most reasonable one, as we have noted. The new mind, and maybe language, are part of our species' exclusive heritage, though not only for the most recent period, but for the whole of its existence. Davidson and Noble's (1989; Noble and Davidson, 1996) evolutionary model established a relation between language and our species' biological prerequisites following a line of thought that can be traced back to Chomsky. Based on those prerequisites, the appearance of anatomically modern humans would represent what William Calvin (1998) called "a quantum leap" in cleverness and foresight during the evolution of humans from apes. A leap that was related to the advantages of a language characterized by its dual patterning.

We can undoubtedly imagine other interpretations. But the key word here is imagine. Only the identification of the brain's processing centers related to the mental tasks of artistic production and appreciation would provide a solid starting point beyond imagined stories and speculative hypotheses. We would also require finding supporting evidence in the fossil record to know the make-up of our ancestors' brains.

10.4 The evolution of moral behavior

I fully subscribe to the judgment of those writers who maintain that of all the differences between man and the lower animals the moral sense or conscience is by far the most important. Charles Darwin, *The Descent of Man* (1871), Chapter III

Morality is a human universal characteristic. People have moral values; that is, they accept standards according to which their conduct is judged either right or wrong, good or evil. The particular norms by which moral actions are judged vary to some extent from individual to individual, and from culture to culture (although some norms, like not to kill, not to steal, and to honor one's parents are widespread and perhaps universal), but value judgments are passed in all cultures. This universality raises the question whether the moral sense is part of human nature, one more dimension of our biological make-up; and whether ethical values may be the product of biological evolution, rather than simply given by religious and cultural traditions.

Aristotle and other philosophers of classical Greece and Rome, as well as many other philosophers throughout the centuries, held that humans hold moral values by nature. A human is not only *Homo sapiens*, but also *Homo moralis*. But biological evolution brings about two important issues: timing and causation. We do not attribute ethical behavior to animals (surely, not to all animals and not to the same extent as to humans in any case). When did ethical behavior come about in human evolution? Did modern humans have an ethical sense from the beginning? Did Neandertals hold moral values? What about *H. erectus* and *H. habilis*?

And how did the moral sense evolve? Was it directly promoted by natural selection? Or did it come about as a by-product of some other attribute (such as intelligence, for example) that was the direct target of selection? Alternatively, is the moral sense an outcome of cultural evolution rather than of biological evolution?

10.4.1 Moral sense and moral codes

The question whether moral behavior can be understood as a phenomenon that came about by means of natural selection was posed by Darwin. He pointed out, as an observation that might favor this idea, the existence in other animals of behaviors that seem similar to those of human moral conduct. In Chapter III of the *Descent of Man*, Darwin (1871) writes:

Brehm encountered in Abyssinia a great troop of baboons who were crossing a valley: some had ascended the opposite mountain, and some were still on the valley: the latter were attacked by the dogs, but the old males immediately hurried down from the rocks, and with mouths widely opened, roared so fearfully, that the dogs precipitately retreated. They were again encouraged to the attack; but by this time all the baboons had reascended the heights, excepting a young one, about six months old, who, loudly calling for aid, climbed on a block of rock and was surrounded. Now one of the largest males, a true hero, came down again from the mountain, slowly went to the young one, coaxed him, and triumphantly led him away—the dogs being too much astonished to make an attack.

This is just one of the many examples given by Darwin of animals that help a distressed group member. However, in this particular case, Darwin uses a word that deserves attention: the baboon that comes down from the mountain is called "a true hero". Heroism is an ethical concept. Is Darwin using it in this sense or only metaphorically?

Darwin belongs to an intellectual tradition, originating in the Scottish Enlightenment of the eighteenth and nineteenth centuries, which uses the moral sense as a behavior that, based on sympathy, leads human ethical choice. In his account of the evolution of cooperative behavior, Darwin states that an animal, with well-defined social instincts—like parental and filial affections—"would inevitably acquire a moral sense or conscience, as soon as its intellectual powers had become as well, or nearly as well developed, as in man." (Darwin, 1871, p. 472). This is a hypothetical issue, because no other animal has ever reached the level of human mental faculties, language included. But this is an important statement, because Darwin is affirming that the moral sense, or conscience, is a necessary consequence of high intellectual powers, such as exist in modern humans. Therefore, if our intelligence is an outcome of natural selection, so it would be the moral sense. Darwin's statement further implies that the moral sense is not by itself directly conscripted by natural selection, but only indirectly as a consequence of high intelligence.

Darwin also states that even if some animal could achieve a human-equivalent degree of development of its intellectual faculties, we cannot conclude that it would also acquire exactly the same moral sense as ours. "I do not wish to maintain that any strictly social animal, if its intellectual faculties were to become as active and as highly developed as in man, would acquire the same moral sense as ours.... [T]hey might have a sense of right and wrong, though led by it to follow widely different lines of conduct" (Darwin, 1871, Chapter III). These statements imply that, according to Darwin, having a moral sense does not by itself determine what the moral code would be: which sorts of actions might by sanctioned by the code and which ones would be condemned.

This distinction is important. Much of the historical controversy, particularly between scientists and philosophers, as to whether the moral sense is or not biologically determined has arisen owing to a failure to make the distinction. Scientists often affirm that morality is a human biological attribute because they are thinking of the predisposition to pass moral *judgment*: that is, to judge some actions as good and others as evil. Some philosophers argue that morality is not biologically determined, but rather comes from cultural traditions or from religious beliefs, because they are thinking about moral *codes*, the set of norms that determine which actions are judged to be good and which are evil. They point out that moral codes vary from culture to culture and, therefore, are not biologically predetermined.

In this book, we consider this distinction fundamental. Thus, we argue that the question of whether ethical behavior is biologically determined may refer to either one of the following two issues. First, is the capacity for ethics—the proclivity to judge human actions as either right or wrong—determined by the biological nature of human beings? Second, are the systems or codes of ethical norms accepted by human beings biologically determined? A similar distinction can be made with respect to language. The question whether the capacity for symbolic creative language is determined by our biological nature is different from the question of whether the particular language we speak—English, Spanish, Chinese, etc.—is biologically determined, which in the case of language obviously it is not.

10.4.2 Moral behavior as a biological attribute

The first question asks whether or not the biological nature of humans is such that we are necessarily inclined to make moral judgments and to accept ethical values, to identify certain actions as either right or wrong. It asks whether moral behavior is an outcome of biological evolution. Affirmative answers to this first question do not necessarily determine what the answer to the second question should be. Independent of whether or not humans have a biologically determined moral sense, it remains to be determined whether particular moral prescriptions are in fact determined by the biological nature of humans, or whether they are chosen by society, or by individuals. Even if we were to conclude that people cannot avoid having moral standards of conduct, it might be that the choice of the particular standards used for judgment would be arbitrary. The need for having moral values does not necessarily tell us what the moral values should be, like the capacity for language does not determine which language we shall speak.

The first argument that we will advance is that humans are ethical beings by their biological nature; that humans evaluate their behavior as either right or wrong, moral or immoral, as a consequence of their eminent intellectual capacities that include self-awareness and abstract thinking. These intellectual capacities are products of the evolutionary process, but they are distinctively human. Thus, we will also maintain that ethical behavior is not causally related to the social behavior of animals, including kin selection and so-called reciprocal altruism.

A second argument that we will put forward is that the moral norms according to which we evaluate particular actions as either morally good or morally bad (as well as the grounds that may be used to justify the moral norms) are products of cultural evolution, not of biological evolution. The norms of morality belong, in this respect, to the same category of phenomena as political and religious institutions, or the arts, sciences, and technology. The moral codes, like these other products of human culture, are often consistent with the biological predispositions of the human species, and of other animals. But this consistency between ethical norms and biological tendencies is not necessary or universal: it does not apply to all ethical norms in a given society, much less in all human societies.

Moral codes, like any other cultural system, depend on the existence of human biological nature and must be consistent with it in the sense that they could not counteract it without promoting their own demise. Moreover, the acceptance and persistence of moral norms is facilitated whenever they are consistent with biologically conditioned human behaviors. But the moral norms are independent of such behaviors in the sense that some norms may not favor, and may hinder, the survival

and reproduction of the individual and its genes, which survival and reproduction are the targets of biological evolution. Discrepancies between accepted moral rules and biological survival are, however, necessarily limited in scope or would otherwise lead to the extinction of the groups accepting such discrepant rules.

10.4.3 Darwin and the moral sense

Two years after the end of the *Beagle*'s voyage, Darwin gathered the current literature on human moral behavior, including such works as William Paley's *The Principles of Moral and Political Philosophy* (1785), which he had encountered earlier while a student at the University of Cambridge, and the multivolume *Illustrations of Political Economy* by Harriet Martineau, published recently in 1832–1834. These two authors, like other philosophers of the time, maintained that morality was a conventional attribute of humankind, rather than a naturally determined human attribute, using an argument often exploited in our days: the diversity of moral codes. The proliferation of ethnographic voyages had brought to light the great variety of moral customs and rules. This is something Darwin had observed in South American Indians. But this apparent dispersion had not distracted him. Rather, he saw the diversity as an adaptive response to the environmental and historical conditions, unique in every different place, without necessarily implying that morality was an acquired, rather than natural, human trait.

A variable adaptive response could very well derive from some fundamental capacity, a common substrate, unique for the whole human race, but capable of becoming expressed in diverse

---

Box 10.32 Mills' 'powerful natural sentiment'

In the *Descent of Man*, Chapter III, footnote 5, Darwin refers to the famous philosopher John Stuart Mill, who "speaks in his celebrated work, 'Utilitarianism' (1864, p. 46), of the social feelings as a 'powerful natural sentiment' . . . but, on the previous page, he says, 'if, as is my own belief, the moral feelings are not innate, but acquired, they are not for that reason less natural.'" Darwin adds: "It is with hesitation that I venture to differ from so profound a thinker."

directions. Darwin did not attribute this universality to supernatural origin, but rather saw it as a product of evolution by natural selection. The presence of a universal and common foundation, endowing humans with an ethical capacity, was for Darwin compatible with different cultures manifesting different stages of a moral evolution, with different sets of moral norms.

Humans, but no other animals, have a moral sense. This attribute would be able to turn a rescue carried out by a human, but not a primate, into a heroic act, although both species are acting altruistically. Darwin does not always make clear what the differences are between an altruist baboon and an altruist human. He associates the moral sense with the idea that human beings carry out an *assessment* in order to make decisions, whereas animals do not. Human altruism appears as a similar behavior to that of other animals, but it is different in the important sense of being accompanied by this assessment, which is what makes it moral. Moral behavior is not of a kind with the automatic responses of biological altruism that happen in hymenopterans (see section 1.2). But neither could we ignore a component of predisposition due to fixed impulses, which arose during the phylogenetic process, when our ancestors exhibited altruistic behaviors similar to those of other primates. Hominids developed sophisticated social habits, including a new moral behavior, under circumstances that included biological altruistic traits emerging from their non-hominid ancestors. The final result of the process, what we now refer to as "moral altruism", exhibits some features resembling biological altruism, but also distinctive features exclusive to the human species, which include the moral sense, the assessment of a behavior as being good or evil.

Darwin shares with philosophers of his time and the previous century an effort to seek a justification of the moral codes. The *Descent of Man* (Darwin, 1871) sees human beings as endowed with a moral sense, which, with the help of biological predispositions, has constructed societies in which ethical behaviors and codes of approval for such behaviors have emerged. Initially the set of actions and codes is small, but gradually, by means of

intellectual, material, and moral progress, morality's scope grows. At first, behaviors emerge that respect and help the individual's closest relatives; then these behavioral patterns extend to the tribe; later to a whole village. In time, Darwin concludes, the whole human race might evolve a single code of morality that is universal and encompasses the whole species.

Darwin's theory of biological heredity, as is well known, accepts the heritability of acquired characters. This component of his theory, proceeding from Lamarck, provides a formula for the joint development of sympathetic instincts and ethical codes, given that the new generations benefit from each advance made in previous generations. In *Descent of Man*, the gradual development of ethics shapes humans, gradually transforming the prehuman savage into a modern citizen. Darwin's moral sense at times seems to conflate the two components of morality: the capacity to *evaluate* actions as moral or immoral (which is the moral sense, in its strict meaning) and the predisposition to behave in certain ways, to accept certain behaviors and not others as morally correct.

10.4.4 Evolutionary theories of morality after Darwin

There are many theories concerned with the rational grounds for morality, such as deductive theories that seek to discover the axioms or fundamental principles that determine what is morally correct on the basis of direct moral intuition; or theories like logical positivism or existentialism that negate rational foundations of morality, reducing moral principles to emotional decisions or to other irrational grounds. After the publication of Darwin's theory of evolution by natural selection, several philosophers as well as biologists attempted to find in the evolutionary process the justification for moral norms. The common ground to such proposals is that evolution is a natural process that achieves goals that are desirable and thereby morally good; indeed, it has produced humans. Proponents of these ideas see that only the evolutionary goals can give moral value to human action: whether a human deed is morally right depends on whether it directly or indirectly promotes the evolutionary process and its natural objectives.

Herbert Spencer was perhaps the first philosopher seeking to find the grounds of morality in biological evolution. More recent attempts include those of the distinguished evolutionists Julian S. Huxley (1953; Huxley and Huxley, 1947), C.H. Waddington (1960), and Edward O. Wilson (1975, 1978, 1998), founder of sociobiology as an independent discipline engaged in discovering the biological foundations of all social behavior.

In *The Principles of Ethics* (1893), Spencer seeks to discover values that have a natural foundation. Spencer argues that the theory of organic evolution implies certain ethical principles. Human conduct must be evaluated, like any biological activity whatsoever, according to whether it conforms to the life process; therefore, any acceptable moral code must be based on natural selection, the law of struggle for existence. According to Spencer, the most exalted form of conduct is that which leads to a greater duration, extension, and perfection of life; the morality of all human actions must be measured by that standard. Spencer proposes that, although exceptions exist, the general rule is that pleasure goes with that which is biologically useful, whereas pain marks what is biologically harmful. This is an outcome of natural selection: thus, while doing what brings them pleasure and avoiding what is painful, organisms improve their chances for survival. With respect to human behavior, we see that we derive pleasure from virtuous behavior and pain from evil actions, associations which indicate that the morality of human actions is also founded on biological nature.

Spencer proposes as the general rule of human behavior that anyone should be free to do anything that they want, so long as it does not interfere with the similar freedom to which others are entitled. The justification of this rule is found in organic evolution: the success of an individual, plant or animal, depends on its ability to obtain that which it needs. Consequently, Spencer reduces the role of the state to protecting the collective freedom of individuals so that they can do as they please. This *laissez faire* form of government may seem ruthless, because individuals would seek their own welfare without any consideration for others' (except for respecting their freedom), but Spencer believes that it is consistent with traditional Christian values. It may be added that, although Spencer sets the grounds of morality on biological nature and on nothing else, he admits that certain moral norms go beyond that which is biologically determined; these are rules formulated by society and accepted by tradition.

Social Darwinism, in Spencer's version or in some variant form, was fashionable in European and American circles during the latter part of the nineteenth century and the early years of the twentieth century, but it has few or no distinguished intellectual followers at present. Spencer's critics include the evolutionists J.S. Huxley and C.H. Waddington who, nevertheless, maintain that organic evolution provides grounds for a rational justification of ethical codes. For Huxley (1953; Huxley and Huxley, 1947), the standard of morality is the contribution that actions make to evolutionary progress, which goes from less to more "advanced" organisms. For Waddington (1960), the morality of actions must be evaluated by their contribution to human evolution.

Huxley and Waddington's views are based on value judgments about what is or is not progressive in evolution. But, contrary to Huxley's claim, there is nothing objective in the evolutionary process itself (i.e. outside human considerations; see Ayala, 1982a) that makes the success of bacteria, which have persisted as such for more than 2 billion years and in enormous numbers, less desirable than that of the vertebrates, even though the latter are more complex. Are the insects, of which more than 1 million species exist, less desirable or less successful from a purely biological perspective than humans or any other mammal species? Waddington fails to demonstrate why the promotion of human biological evolution by itself should be the standard to measure what is morally good.

10.4.5 The naturalistic fallacy

A more fundamental objection against the theories of Spencer, Huxley, and Waddington—and against any other program seeking the justification of a moral code in biological nature—is that such theories commit the "naturalistic fallacy" (Moore, 1903), which consists of identifying what "is" with what "ought to be." This error was pointed out already by the philosopher David Hume (1740; 1978, p. 469): "In every system of morality which I have hitherto met with I have always remarked that the author proceeds for some time in the ordinary way of reasoning . . . when of a sudden I am surprised to find, that instead of the usual copulations of propositions *is* and *is not*, I meet with no proposition that is not connected with an *ought* or *ought not*. This change is imperceptible; but is, however, of the last consequence. For as this *ought* or *ought not* express some new relation or affirmation, it is necessary that it should be observed and explained; and at the same time a reason should be given, for what seems altogether inconceivable, how this new relation can be a deduction from others, which are entirely different from it."

The naturalistic fallacy occurs whenever inferences using the terms ought or ought not are derived from premises that do not include such terms but are rather formulated using the connections is or is not. An argument cannot be logically valid unless the conclusions only contain terms that are also present in the premises. In order to proceed logically from that which is to what ought to be, it would be necessary to include a premise that justifies the transition between the two expressions. But his transition is what is at stake, and one would need a previous premise to justify the validity of the one making the transition, and so on in a regression *ad infinitum*. In other words, from the fact that something *is* the case, it does not follow that it *ought to be* so in the ethical sense; *is* and *ought* belong to disparate logical categories.

Because evolution has proceeded in a particular way, it does not follow that that course is morally right or desirable. The justification of ethical norms on biological evolution, or on any other natural processes, can only be achieved by introducing value judgments, human choices that prefer one rather than another object or process. Biological nature is in itself morally neutral.

It must be noted, moreover, that using natural selection or the course of evolution to determine the morality of human actions will lead to paradoxes. Evolution has produced the smallpox and AIDS viruses. But it would seem unreasonable to accuse the World Health Organization of immorality because of its campaign for total eradication of the smallpox virus; or to label unethical the efforts to control the spread of the AIDS virus. Human hereditary diseases are conditioned by mutations that are natural events in the evolutionary process. But we do not think it immoral to cure or alleviate the pain of persons with such diseases. Natural selection is a natural process that increases the frequency of certain genes and the elimination of others, that yields some kinds of organisms rather than others; but it is not a process moral or immoral in itself or in its outcome, in the same way as gravity is not a morally laden force. To consider some evolutionary events as morally right and others wrong, we must introduce human values; moral evaluations cannot be reached simply on the basis that certain events came about by natural processes.

10.4.6 Sociobiology

Edward O. Wilson (1975, p. 562) has urged that "scientists and humanists should consider together the possibility that the time has come for ethics to be removed temporarily from the hands of the philosophers and biologicized." Wilson (1978, 1994), like other sociobiologists (Barash, 1977; Alexander, 1979; see also Kitcher, 1985; Sober and Wilson, 1998; Ruse, 2000, 2006), sees that sociobiology may provide the key for finding a naturalistic basis for ethics. Sociobiology is "the systematic study of the biological basis of all forms of social behavior in all kinds of organisms" (Wilson, in the Foreword to Barash, 1977) or, in Barash's concise formulation, "the application of evolutionary biology to social behavior" (1977, p. ix). Its purpose is "to develop general laws of the evolution and biology of social behavior, which might then be extended in a disinterested manner to the study of human beings" (Wilson, in the Foreword to Barash, 1977). The program is ambitious: to discover the biological basis of human social behavior, starting from the investigation of the social behavior of animals.

The sociobiologist's argument concerning normative ethics is not that the norms of morality can be grounded in biological evolution, but rather that evolution predisposes us to accept certain moral norms, namely those that are consistent with the "objectives" of natural selection. It is because of this predisposition that human moral codes sanction patterns of behavior similar to those encountered in the social behavior of animals. The sociobiologists claim that the agreement between moral codes and the goals of natural selection in social groups was discovered when the theories of kin selection, inclusive fitness, and reciprocal altruism were formulated. According to sociobiologists, the commandment to honor one's parents, the incest taboo, the greater blame usually attributed to the wife's adultery than to the husband's, and the banning or restriction of divorce, are among the numerous ethical precepts that endorse behaviors that are also endorsed by natural selection.

The sociobiologists reiterate their conviction that science and ethics belong to separate logical realms; that one may not infer what is morally right or wrong from a determination of how things are or are not in nature. In this respect they claim to avoid committing the naturalistic fallacy. According to Wilson, "To devise a naturalistic description of human social behavior is to note a set of facts for further investigation, not to pass a value judgment or to deny that a great deal of the behavior can be deliberately changed if individual societies so wish" (in Barash, 1977, p. xiv). Barash (1977, p. 278) puts it so: "Ethical judgments have no place in the study of human sociobiology or in any other science for that matter. What is biological is not necessarily good." Alexander (1979, p. 276) asks what is it that evolution teaches us about normative ethics or about what we *ought* to do, and responds "Absolutely nothing."

There is nevertheless some question as to whether the sociobiologists are always consistent with the statements just quoted. Wilson (1975, p. 564), for example, writes that "the requirement for an evolutionary approach to ethics is self-evident. It should also be clear that no single set of moral standards can be applied to all human populations, let alone all sex-age classes within each population. To impose a uniform code is therefore to create complex, intractable moral dilemmas." Moral pluralism is, for Wilson, "innate." It seems, therefore, that according to Wilson, biology helps us at the very least to decide that certain moral codes (e.g. all those pretending to be universally applicable) are incompatible with human nature and therefore unacceptable. This is not quite an argument in favor of the biological determinism of ethical norms, but it does approach determinism from the negative side: because the range of valid moral codes is delimited by the claim that some are not compatible with biological nature.

However, Wilson goes further when he writes: "Human behavior—like the deepest capacities for emotional response which drive and guide it—is the circuitous technique by which human genetic material has been and will be kept intact. *Morality has no other demonstratable ultimate function*" (Wilson, 1978, p. 167; our italics). How is one to interpret this statement? It is possible that Wilson is simply giving the reason why ethical behavior exists at all; his proposition would be that humans are prompted to evaluate morally their actions as a means to preserve their genes, their biological nature. But this proposition is erroneous. Human beings are by nature ethical beings in the sense we will expound below: they judge morally their actions because of their innate ability for anticipating the consequences of their actions, for formulating value judgments, and for free choice. Human beings exhibit ethical behavior by nature and necessity, rather than because such behavior would help to preserve their genes or serve any other purpose.

Wilson's statement may alternatively be read as a justification of human moral codes: the function of these would be to preserve human genes. But this would entail the naturalistic fallacy and, worse yet, would seem to justify a morality that most people detest. If the preservation of human genes (be those of the individual or of the species) is the purpose that moral norms serve, Spencer's Social

Darwinism would seem right; racism or even genocide could be justified as morally correct if they were perceived as the means to preserve those genes thought to be good or desirable and to eliminate those thought to be bad or undesirable. Surely, Wilson is not intending to justify racism or genocide, but this is one possible interpretation of his words.

We shall now turn to the sociobiologists' proposition that natural selection favors behaviors similar to the behaviors sanctioned by the moral codes endorsed by most humans.

10.4.7 Altruism and inclusive fitness

Evolutionists had for years struggled to find an explanation for the apparently altruistic behavior of animals. When a predator attacks a herd of zebras, adult males attempt to protect the young in the herd, even if they are not their progeny, rather than fleeing. When a prairie dog sights a coyote, it will warn other members of the colony with an alarm call, even though by drawing attention to itself this increases its own risk. Darwin tells the story of adult baboons protecting the young. Examples of altruistic behaviors of this kind can be multiplied.

A dictionary definition of altruism is "Regard for, and devotion to, the interests of others" (*Webster's New Collegiate Dictionary*, 2nd edn). To speak of animal altruism is not to claim that explicit feelings of devotion or regard are present in them, but rather that animals act for the welfare of others at their own risk just as humans are expected to do when behaving altruistically. The problem is precisely how to justify such behaviors in terms of natural selection. Assume, for illustration, that in a certain species there are two alternative forms of a gene (two alleles), of which one but not the other promotes altruistic behavior. Individuals possessing the altruistic allele will risk their life for the benefit of others, whereas those possessing the non-altruistic allele will benefit from altruistic behavior without risking themselves. Possessors of the altruistic allele will be more likely to die and the allele for altruism will therefore be eliminated more often than the non-altruistic allele. Eventually, after some generations, the altruistic allele will be completely replaced by the non-altruistic one. But then, how is it that altruistic behaviors are common in animals without the benefit of ethical motivation?

One major contribution of sociobiology to evolutionary theory is the theory of kin selection, as well as the notion of inclusive fitness (see section 1.2). To ascertain the consequences of natural selection it is necessary to take into account a gene's effects not only on a particular individual but also on all individuals possessing that gene. When considering altruistic behavior, one must take into account not only the risks for the altruistic individual, but also the benefits for other possessors of the same allele. Zebras live in herds where individuals are blood relatives. This is also the case for baboon troops. An allele prompting adults to protect the defenseless young would be favored by natural selection if the benefit (in terms of saved carriers of that allele) is greater than the cost (due to the increased risk of the protectors). An individual that lacks the altruistic allele and carries instead a non-altruistic one, will not risk its life, but the non-altruistic allele is partially eradicated with the death of each defenseless relative.

It follows from this line of reasoning that the more closely related the members of a herd, troop, or animal group are, the more altruistic behavior should be present. This seems to be generally the case. Consider the following two examples: parental care and the castes of social insects. Parental care is most obvious in the genetic benefits it entails. Parents feed and protect their young because each child has half the genes of each parent: the genes are protecting themselves, as it were, when they prompt a parent to care for its young. Parental care is widespread among animals.

The second example is more subtle: the social organization and behavior of many species of hymenopterans: bees, wasps, and ants. Consider Meliponinae bees, with hundreds of species across the tropics. These stingless bees have typically single-queen colonies with hundreds to thousands of workers. The queen generally mates only once. The worker bees toil, building the hive and feeding and caring for the larvae even though they themselves are sterile and only the queen produces progeny. Assume that in some ancestral hive, an allele arises that prompts worker bees to behave as they now do. It would seem that such an allele would not be passed on to the following generation because such worker bees do not reproduce. But such inference is erroneous.

Queen bees produce two kinds of eggs: some remain unfertilized develop into males (which are therefore haploid; i.e. they carry only one set of genes); others are fertilized (hence, are diploid, carry two sets of genes) and develop into worker bees and occasionally into a queen. W.D. Hamilton (1964) demonstrated that with such a reproductive system the queen's daughters share in two-thirds of their genes among them, whereas the queen's daughters and their mother share in only one-half of their genes (see section 1.2 and Figure 1.19). Hence, the worker-bee genes are more effectively propagated by workers caring for their sisters than if they would produce and care for their own daughters. Natural selection can thus explain the existence in social insects of sterile castes, which exhibit a most extreme form of apparently altruistic behavior by dedicating their life to care for the progeny of another individual, the queen.

Sociobiologists point out that many of the moral norms commonly accepted in human societies sanction behaviors also promoted by natural selection (promotion of which becomes apparent only when the inclusive fitness of genes is taken into account). Examples of such behaviors are the commandment to honor one's parents, the incest taboo, the greater blame attributed to the wife's than to the husband's adultery, and many others. The sociobiologists' argument is that human ethical norms are sociocultural correlates of behaviors fostered by biological evolution. Ethical norms protect such evolution-determined behaviors as well as being specified by them.

The sociobiologists' argument, however, is misguided and does not escape the naturalistic fallacy (see Ayala 1980, 1982b). Consider altruism as an example. Altruism in the biological sense (altruism$_b$) is defined in terms of the population genetic consequences of a certain behavior.

Altruism$_b$ is explained by the fact that genes prompting such behavior are actually favored by natural selection (when inclusive fitness is taken into account), even though the fitness of the behaving individual is decreased. But altruism in the moral sense (altruism$_m$) is explained in terms of *motivations*: a person chooses to risk his/her own life (or incur some kind of cost) for the benefit of somebody else. The similarity between altruism$_b$ and altruism$_m$ is only with respect to the consequences: an individual's chances are improved by the behavior of another individual who incurs a risk or cost. The underlying causations are completely disparate: the ensuing genetic benefits in altruism$_b$; regard for others in altruism$_m$. As Darwin put it, the behavior of a baboon and a human are similar in that they both save an infant (from the dogs or from drowning), but they differ in that humans carry out an assessment, which baboons do not. We will propose below that we make moral judgments as a consequence of our eminent intellectual abilities, not as an innate way to achieve biological gain.

10.4.8 Human rationality and moral behavior

We turn now to the question of why the moral sense in its strict meaning is an attribute of our species determined by our biology and, hence, an attribute that came about by natural selection. The moral sense in this strict meaning refers to the evaluation of some actions as virtuous, or morally good, and others as evil, or morally bad. Morality in this sense is the urge or predisposition to judge human actions as either right or wrong in terms of their consequences for other human beings.

In this sense, humans are moral beings by nature because their biological constitution determines the presence in them of the three necessary conditions for ethical behavior. These conditions are: (i) the ability to anticipate the consequences of one's own actions; (ii) the ability to make value judgments; and (iii) the ability to choose between alternative courses of action. These abilities exist as a consequence of the eminent intellectual capacity of human beings.

The ability to anticipate the consequences of one's own actions is the most fundamental of the three conditions required for ethical behavior. Only if I can anticipate that pulling the trigger will shoot the bullet, which in turn will strike and kill my enemy, can the action of pulling the trigger be evaluated as nefarious. Pulling a trigger is not in itself a moral action; it becomes so by virtue of its relevant consequences. My action has an ethical dimension only if I do anticipate these consequences.

The ability to anticipate the consequences of one's actions is closely related to the ability to establish the connection between means and ends; that is, of seeing a means precisely as a means, as something that serves a particular end or purpose. This ability to establish the connection between means and their ends requires the ability to anticipate the future and to form mental images of realities not present or not yet in existence.

The ability to establish the connection between means and ends happens to be the fundamental intellectual capacity that has made possible the development of human culture and technology. As we have discussed throughout this book, the remote evolutionary roots of this capacity may be found in the evolution of bipedalism, which transformed the anterior limbs of our ancestors from organs of locomotion into organs of manipulation. The hands thereby gradually became organs adept for the construction and use of objects for hunting and other activities that improved survival and reproduction; that is, which increased the reproductive fitness of their carriers. The construction of tools depends not only on manual dexterity, but on perceiving them precisely as tools, as objects that help to perform certain actions; that is, as means that serve certain ends or purposes: a knife for cutting, an arrow for hunting, an animal skin for protecting the body from the cold. Natural selection promoted the intellectual capacity of our bipedal ancestors because increased intelligence facilitated the perception of tools as tools, and therefore their construction and use, with the ensuing amelioration of biological survival and reproduction.

The development of the intellectual abilities of our ancestors took place over several million years,

gradually increasing the ability to connect means with their ends and, hence, the possibility of making ever-more complex tools serving more remote purposes. The ability to anticipate the future, essential for ethical behavior, is therefore closely associated with the development of the ability to construct tools, an ability that has produced the advanced technologies of modern societies and that is largely responsible for the success of humans as a biological species.

The second condition for the existence of ethical behavior is the ability to make value judgments, to perceive certain objects or deeds as more desirable than others. Only if I can see the death of my enemy as preferable to his survival (or vice versa) can the action leading to his demise be thought of as moral. If the consequences of alternative actions are neutral with respect to value, an action cannot be characterized as ethical. Values are of many sorts: not only ethical, but also aesthetic, economic, gastronomic, political, and so on. But in all cases, the ability to make value judgments depends on the capacity for abstraction; that is, on the capacity to perceive actions or objects as members of general classes. This makes it possible to compare objects or actions with one another and to perceive some as more desirable than others. The capacity for abstraction requires an advanced intelligence such as it exists in humans and apparently in them alone.

The third condition necessary for ethical behavior is the ability to choose between alternative courses of actions. Pulling the trigger can be a moral action only if you have the option not to pull it. A necessary action beyond conscious control is not a moral action: the circulation of the blood or the process of food digestion are not moral actions. Whether there is free will is a question much discussed by philosophers and the arguments are long and involved. (For a brief but insightful discussion of free will in the context of evolution, see Ruse, 2006, chapter 12.) Here, we advance two considerations that are common-sense evidence of the existence of free will. One is personal experience, which indicates that the possibility to choose between alternatives is genuine rather than only apparent. The second consideration is that when we confront a given situation that requires action on our part, we are able mentally to explore alternative courses of action, thereby extending the field within which we can exercise our free will. In any case, if there were no free will, there would be no ethical behavior; morality would only be an illusion. The point is, however, that free will is dependent on the existence of a well-developed intelligence, which makes it possible to explore alternative courses of action and to choose one or another in view of the anticipated consequences.

10.4.9 Beyond rationality?

Two issues concerning the explanation of moral behavior just developed are: (1) is morality an adaptation directly favored by natural selection rather than simply a by-product of high intelligence? and (2) does morality occur in other animals, either as directly promoted by natural selection or as a consequence of animal intelligence, even if only as a rudiment?

The answer to the first question is negative. Morality consists of *judging* certain actions as either right or wrong; not of choosing and carrying out some actions rather than others, or evaluating them with respect to their practical consequences. It seems unlikely that making moral judgments would promote the reproductive fitness of those judging an action as good or evil. Nor does it seem likely that there might be some form of "incipient" ethical behavior that would then be further promoted by natural selection. The three necessary conditions for there being ethical behavior are manifestations of advanced intellectual abilities.

It rather seems that the target of natural selection was the development of these advanced intellectual capacities. This was favored by natural selection because the construction and use of tools improved the strategic position of our biped ancestors. Once bipedalism evolved and after tool-using and tool-making became practical, those individuals more effective in these functions had a greater probability of biological success. The biological advantage provided by the design and use of tools persisted long enough so that intellectual abilities continued to increase, eventually yielding

the eminent development of intelligence that is characteristic of *Homo sapiens*.

A related question is whether morality would benefit a social group within which it is practiced, and, indirectly, individuals as members of the group. This seems likely to be the case, if indeed moral judgment would influence individuals to behave in ways that increase cooperation, or benefit the welfare of the social group in some way; for example, by reducing crime or protecting private property. This brings about again the issue of whether there is group selection and the related issues of kin selection and inclusive fitness (section 1.2). We'll briefly return to these issues below when considering the codes of morality.

Group selection based on altruistic behavior is generally not an evolutionary stable strategy (ESS), because mutations that favor selfish over altruistic behavior will be favored by natural selection within a given population, so that selfish alleles will drive out altruistic alleles. Of course, it may be the case that populations with a preponderance of altruistic alleles will survive and spread better than populations consisting of selfish alleles. This would be group selection. But typically there are many more individual organisms than there are populations; and individuals are born, procreate, and die at rates much higher than populations. Thus, the rate of multiplication of selfish individuals over altruists is likely to be much higher than the rate at which altruistic populations multiply relative to predominantly selfish populations.

There is, however, an important difference between animals and humans that is relevant in this respect. Namely, the fitness advantage of selfish over altruistic behavior does not apply to humans, because humans can *understand* the benefits of altruistic behavior (to the group and indirectly to them) and thus adopt altruism and protect it, by laws or otherwise, against selfish behavior that harms the social group. As Darwin wrote in *The Descent of Man* (1871, chapter V): "It must not be forgotten that, although a high standard of morality gives but a slight or no advantage to each individual man and his children over the other men of the same tribe, yet that an advancement in the standard of morality and an increase in the number of well-endowed men will certainly give an immense advantage to one tribe over another."

The theory of sociobiology provides a ready answer to the second question raised above, whether morality occurs in other animals, even if only as a rudiment. The theory of kin selection, they argue, explains altruistic behavior, to the extent that it exists in other animals as well as in humans. However, as we have argued, moral behavior properly so does not exist, even incipiently, in nonhuman animals. The reason is that the three conditions required for ethical behavior depend on an advanced intelligence that has the capacity for free will, abstract thought, and anticipation of the future events, such as it exists in *H. sapiens* and not in any other living species. It is the case that certain animals exhibit behaviors analogous with those resulting from ethical actions in humans, such as the loyalty of dogs or the appearance of compunction when they are punished. But such behaviors are either genetically determined or elicited by training (conditioned responses). Genetic determination and not moral evaluation is also what is involved in the altruistic behavior of social insects and other animals. Biological altruism (altruism$_b$) and moral altruism (altruism$_m$) have disparate causes: kin selection in altruism$_b$, regard for others in altruism$_m$.

The capacity for ethics is an outcome of gradual evolution, but it is an attribute that only exists when the underlying attributes (i.e. the intellectual capacities) reach an advanced degree. The necessary conditions for ethical behavior only come about after the crossing of an evolutionary threshold. The approach is gradual, but the conditions only appear when a degree of intelligence is reached such that the formation of abstract concepts and the anticipation of the future are possible, even though we may not be able to determine when the threshold was crossed. Thresholds occur in other evolutionary developments—for example, in the origins of life, multicellularity, and sexual reproduction—as well as in the evolution of abstract thinking and self-awareness. Thresholds also occur in the inorganic world; for example, water heats gradually, but at 100°C boiling begins and the transition from liquid to gas starts suddenly. Surely, human intellectual capacities came about by gradual evolution, but when looking at the world of life as it exists today, it would seem that there is a radical breach between human intelligence and that of other animals. The rudimentary cultures that exist in chimpanzees (Whiten *et al.*, 1999, 2005) do not imply advanced intelligence as it is required for moral behavior.

10.4.10 Whence moral codes?

We have distinguished between moral behavior—judging some actions as good, others as evil—and moral codes—the precepts or norms according to which actions are judged. Moral behavior, we have proposed, is a biological attribute of *H. sapiens*, because it is a necessary consequence of our biological make-up, namely our high intelligence. But moral codes, we argue, are not products of biological evolution, but of cultural evolution.

It must, first, be stated that moral codes, like any other cultural systems, cannot survive for long if they run in outright contrast to our biology. The norms of morality must be consistent with biological nature, because ethics can only exist in human individuals and in human societies. One might therefore also expect, and it is the case, that accepted norms of morality will promote behaviors that increase the biological fitness of those who behave according to them, such as child care.

Box 10.33 When did morality emerge?

Did *H. habilis* or *H. erectus* have morality? What about *H. neanderthalensis*? When in hominin evolution morality emerged is a difficult question to answer. It seems likely that the emergence of morality may have been associated with the emergence of creative language, an issue that we discussed in section 10.3, and which need not be reviewed here.

But this is neither necessary nor indeed always the case: some moral precepts common in human societies have little or nothing to do with biological fitness and some moral precepts are contrary to fitness interest.

Before going any further, it seems worthwhile to consider briefly the proposition that the justification of the codes of morality derives from religious convictions and only from them. There is no necessary, or *logical*, connection between religious faith and moral principles, although there usually is a motivational or psychological connection. Religious beliefs do explain why people accept particular ethical norms, because they are motivated to do so by their religious convictions. But in following the moral dictates of one's religion, an individual is not rationally justifying the moral norms that one accepts. It may, of course, be possible to develop such rational justification; for example, when a set of religious beliefs contains propositions about human nature and the world from which a variety of ethical norms can be logically derived. Indeed, religious authors, including, for example, Christian theologians, do often propose to justify their ethics on rational foundations concerning human nature. But in this case, the logical justification of the ethical norms does not come from religious faith as such, but from a particular conception of the world; it is the result of philosophical analysis grounded on certain premises.

It may well be that the motivational connection between religious beliefs and ethical norms is the decisive one for the religious believer. But this is true in general: most people, religious or not, accept a particular moral code for social reasons, without trying to justify it rationally by means of a theory from which the moral norms can be logically derived. They accept the moral codes that prevail in their societies, because they have learned such norms from parents, school, or other authorities. The question therefore remains, how do moral codes come about?

The short answer is, as already stated, that moral codes are products of cultural evolution, a distinctive human mode of evolution that has surpassed the biological mode, because it is faster and because it can be directed. Cultural evolution is based on cultural heredity, which is Lamarckian, rather than Mendelian, so that acquired characteristics are transmitted. Most important, cultural heredity does not depend on biological inheritance, from parents to children, but is transmitted also horizontally and without biological bounds. A cultural mutation, an invention (think of the laptop computer, the cell phone, or rock music) can be extended to millions and millions of individuals in less than one generation.

Box 10.34 Morality and tribal success

Chapter V of Darwin's *The Descent of Man* is entitled, "On the Development of the Intellectual and Moral Faculties During Primeval and Civilized Times." He writes:

"There can be no doubt that a tribe including many members who, from possessing in a high degree the spirit of patriotism, fidelity, obedience, courage, and sympathy, were always ready to give aid to each other and to sacrifice themselves for the common good, would be victorious over most other tribes; and this would be natural selection. At all times throughout the world tribes have supplanted other tribes; and as morality is one element in their success, the standard of morality and the number of well-endowed men will thus everywhere tend to rise and increase."

Darwin is making two important assertions. First, that morality may contribute to the success of some tribes over others, which is natural selection in the form of group selection. Second, that the standards of morality will tend to improve over human history, because the higher the standards of a tribe, the more likely the success of the tribe. This assertion depends on which standards are thought to be "higher" than others. If the higher standards are defined by their contribution to the success of the tribe, then the assertion is circular. But Darwin asserts that there are some particular standards that, in his view, would contribute to tribal success: patriotism, fidelity, obedience, courage, and sympathy.

Box 10.35 Human behaviors and biological correlates

Parental care is a behavior generally favored by natural selection, which may also be present in virtually all codes of morality, from primitive to more advanced societies. There are other human behaviors sanctioned by moral norms that have biological correlates favored by natural selection. One example is monogamy, which occurs in some animal species but not in many others. It is also sanctioned in many human cultures, but surely not in all. Polygamy is sanctioned in some current human cultures and surely was more so in the past. Food sharing outside the mother–offspring unit rarely occurs in primates, with the exception of chimpanzees and capuchin monkeys, although even in chimpanzees food sharing is highly selective and often associated with reciprocity. A more common form of mutual aid among primates is coalition formation; alliances are formed in fighting other conspecifics, although these alliances are labile, with partners readily changing partners.

One interesting behavior, associated with a sense of justice, or equal pay for equal work, has been described by Sarah Brosnan and Frans de Waal (2003; see also de Waal, 1996) in the brown capuchin monkey, *Cebus apella*. Monkeys responded negatively to unequal rewards in exchanges with a human experimenter. Monkeys refused to participate in an exchange if they witnessed that a conspecific had obtained a more attractive reward for equal effort.

Is the capuchin behavior phylogenetically related to the human virtue of justice? This seems unlikely, since similar behavioral patterns have not been observed in other primates, including apes, phylogenetically closer to humans. Cannibalism is practiced by chimps, as well as by human cultures of the past. Do we have a phylogenetically acquired predisposition to cannibalism as a morally acceptable behavior? This seems unlikely. More generally, we have proposed in this book that moral codes arise in human societies by cultural evolution. Those moral codes tend to be widespread that lead to successful societies. But this assertion leaves open to question the issue of how extensively humans are predisposed to prevailing codes of morality.

Marc Hauser in *Moral Minds* (2006) has proposed that humans are born with a sort of moral "organ" or moral "grammar." Hauser does not believe, of course, that we are born with specific moral rules. Moral norms differ from one culture to another and evolve from one time to another, even within a given society. Rather, Hauser draws an analogy with Noam Chomsky's theory that humans possess a kind of language organ that contains a universal grammar. This grammar consists of universal syntactic rules that underlie the syntax of any specific language. The innate universal grammar explains, for example, why children quickly learn the syntax of the language to which they are exposed, without formal explanation of the rules. Hauser analogously claims that we have an innate universal moral grammar, which underlies all codes of morality but allows variation among them, just as there is syntax variation among different languages.

Since time immemorial, human societies have experimented with moral systems. Some have succeeded and spread widely through humankind, like the Ten Commandments, although other moral systems persist in different human societies. Many moral systems of the past have surely become extinct because they were replaced or because the societies that held them became extinct. The moral systems that currently exist in humankind are those that were favored by cultural evolution. They were propagated within particular societies for reasons that might be difficult to fathom, but that surely must have included the perception by individuals that a particular moral system was beneficial for them, at least to the extent that it was beneficial for their society by promoting social stability and success. Acceptance of some precepts in many societies is reinforced by civil authority (e.g. those who kill or commit adultery will be punished) and by religious beliefs (God is watching and you'll go to hell if you misbehave). Legal and political systems as well as belief systems are themselves outcomes of cultural evolution.

Glossary

Acheulean A lower-Paleolithic lithic technology. It evolved from the *Oldowan* and is characterized by handaxes and cleavers. Widespread in Africa, Europe, and parts of Asia from around 1.5 million to 150,000 years ago. Associated with *Homo erectus* and archaic *Homo sapiens*.

adaptation A structural or functional characteristic of an organism that allows it to cope better with its environment; the evolutionary process by which organisms become adapted to their environment.

adaptive radiation Diversification of a group of related *species* associated with the colonization of novel ecological niches.

adaptive value A measure of the reproductive efficiency of an organism (or *genotype*) compared with other organisms (or *genotypes*); also called *selective value*.

allele Each of the two or more different forms of a *gene*, such as the alleles for *A, B,* and *O* at the *ABO* blood-group gene *locus*.

allometry Relative growth relationships between two parts of an organism or between two *species*.

allopatric Geographically separated *populations* or *species* (see also *sympatric*).

amino acid The building blocks of *proteins*. Several hundred are known, but only 20 are normally found in proteins.

anagenesis The evolutionary change of a single lineage in the course of time (see also *cladogenesis*).

analogy (adj. **analogous**) Resemblance in function but not in structure, due to independent evolutionary origin; for example, the wings of a bird and of an insect.

anatomically modern human Human beings that appeared about 150,000 years ago and which share conspicuous morphological traits with current humans, despite cultural and symbolic differences.

anthropoid A *primate* belonging to one of the following superfamilies: Ceboidea (New World monkeys), Cercopithecoidea (*Old World monkeys*), or Hominoidea (lesser apes, great apes, and humans).

antibody A *protein*, synthesized by the immune system of a higher organism, that binds specifically to the foreign molecule (*antigen*) that induced its synthesis.

antigen see *antibody*

anvil A hard surface used as a base when extracting *flakes* from a *core*. It is usually a flat stone, although chimpanzees have been observed using roots as anvils to open nuts.

ape A member of the *primate* group that includes the gibbons, orangutans, gorillas, and chimpanzees; gibbons are often called the lesser apes and the others great apes.

apomorphy A trait that has appeared after the node where a certain *clade* originated. It is also known as a *derived trait*. A distinction is made between *synapomorphy* and *autapomorphy*.

archaic *Homo sapiens* The name given to *fossil* humans that lived in Africa, Europe, and Asia from about 400,000 to 200,000 or 100,000 years ago with features intermediate between *Homo erectus* and modern *Homo sapiens*, and which may represent separate *species*.

Ardipithecus An early *genus* of the *hominin* tribe, even though some authors have questioned this description based on the thinness of its enamel.

argon/argon A variation of the *potassium/argon* dating method.

arthropod A member of the animal *phylum* Arthropoda (insects, arachnids, millipedes, crustaceans, etc.).

articulation The joint between two bones.

assemblage A group of objects found together in an archaeological setting.

Australopithecus—An extinct *genus* of the tribe *Hominini*. Species commonly assigned to this genus include *A. anamensis, A. afarensis, A. bahrelghazali* and *A. africanus*

autapomorphy A *derived trait* found only in a single *clade* of a cladogram.

autosome A *chromosome* other than a sex chromosome.

base pair Two nitrogenous bases that pair by hydrogen-bonding in double-stranded *DNA* or *RNA*.

bed A geologic layer.

biface A flat stone tool produced by extracting *flakes* from both sides of a *core* until an edge is obtained along the whole perimeter. The most common form of bifacial tools are *handaxes*.

biogeography The geographic distribution of plants and animals.

biostratigraphy The study of the sequence of appearance and disappearance of *fossil species* throughout a series of deposits.

biota All plants and animals of a given region or time.

bipedalism A mode of locomotion involving a vertical position of the body and walking by use of only the hindlimbs.

bottleneck A period when a *population* becomes reduced to only a small number of individuals.

brachiation A mode of arboreal locomotion involving swinging alternate forelimbs to move from branch to branch.

brain case Refers to the set of bones that surround the brain; also known as the *cranium*.

breccia Cave sediments that have been calcified by filtering lime solutions.

brow ridge A ridge of bone that arches above the eye sockets, the main function of which is to protect the ocular cavities.

burin A piercing tool commonly used to engrave materials such as antler, ivory, or bone.

calcaneus The largest of the *hominin* tarsal bones.

calcareous Composed of, or containing lime (calcium carbonate).

calotte The roof of the skull.

canine Each of the two lower and two upper large teeth at the corners of the mouth, between the *incisors* and *premolars*, and whose function is to pierce food.

carbon 14 (^{14}C) see *radiocarbon dating*

carpal bone Each of the eight bones that constitute the wrist. They are placed in two rows; the proximal one articulates with the *radius* and *ulna* while the distal one articulates with the *metacarpals*.

catarrhine A member of the *primate* infraorder Catarrhini (*Old World monkeys*, apes, and humans).

category Each of the levels in the Linnaean classification system (*kingdom, phylum, class, order, family, genus, species,* and their intermediates).

ceboid A member of the *primate* superfamily Ceboidea (New World monkeys).

Cenozoic The era from 65 Ma to the present, also called the Age of Mammals. It is divided into the Tertiary and Quaternary periods.

cercopithecoid A member of the *primate* superfamily Cercopithecoidea (*Old World monkeys*).

cerebral cortex The outer layer of grey matter of the cerebral hemispheres, comprising layers of nerve cells and their interconnections.

character A trait or feature of an organism.

Chatêlperronian A tool industry that exhibits certain upper-Paleolithic features. Such instruments are found in Western Europe and they are thought to have been developed by late *Neanderthals.*

chin The anterior projection of the *mandible* at its midline that begins below the alveolar bone of the central *incisors* and extends down and out to form a raised inverted T.

chopper A stone tool which has been flaked irregularly to produce a cutting edge on one side.

chromosome A thread-shaped structure visible in the cell's *nucleus* during cell division. Chromosomes contain most of the hereditary material or genes.

chronospecies A *species* that gradually changes through time such that the original organisms and resulting ones are too different to be classified within the same species, although there is no clear cutting point along the lineage at which they can be differentiated.

chronostratigraphy The study of the temporal aspect of sediments aiming to establish a temporal reference for all elements related with rocks and *fossils* found at a site.

clade A complete group of organisms derived from a common ancestor; or the branches that separate in a cladistic event.

cladistic event see *cladogenesis*

cladistics A classification system proposed by Hennig based on phylogenetic hypotheses and common ancestry that only admits speciation by means of *cladogenesis*.

cladogenesis An evolutionary process whereby a *species* gives rise to two different ones, after which it disappears. In a *cladogram*, the divergence point is known as a node.

cladogram The graphic representation of the branching relations between *species*, genera, families, and so on, which are represented as *clades*.

class A category formed by a set of *orders*.

clavicle Each of the two collarbones that articulate with the scapulae and sternum.

cleaver A large bifacial stone tool with a transverse sharp edge on one side.

codon A group of three adjacent *nucleotides* in an *mRNA* molecule that code either for a specific *amino acid* or for *polypeptide* chain termination during *protein* synthesis.

conspecific Belonging to the same *species*.

continental drift The slow movement of the continents and their crustal plates over the Earth's surface.

convergence The parallel development of the same feature in unrelated organisms.

core What remains of a stone after *flakes* have been removed from it.

cranium The set of bones that constitute the skull except for the *mandible* (also known as the *brain case*).

Cretaceous The last geological period of the Mesozoic era, about 145–65 Ma.

Cro-Magnon The earliest anatomically modern human *populations* in Europe.

cuboid The tarsal bone placed on the lateral side of the foot. It articulates with the *calcaneous*, the lateral *cuneiform*, and the fourth and fifth *metatarsals*.

cuneiform Each of three footbones (medial, intermediate, and lateral cuneiforms) that link the *navicular* and the medial *metatarsals*.

dating methods see *radiocarbon dating, paleomagnetism, potassium/argon* and *argon/argon, fission-track dating, uranium series, thermoluminescence, electronic spin resonance*

deciduous dentition The first set of teeth to appear, before the *mandible* is capable of accommodating the *permanent dentition*.

deletion A chromosomal mutation due to the loss of a chromosomal segment.

dental arcade The shape of the lower and upper rows of teeth.

derived trait see *apomorphy*

diastema The gap between *incisors* and *canines*. It is very marked in rodents and archaic *primates*.

digit A toe or finger.

diploid A cell, tissue, or organism having two *chromosome* sets.

DNA Deoxyribonucleic acid; *nucleic acid* composed of units consisting of a deoxyribose sugar, a phosphate group, and the nitrogen bases adenine, guanine, cytosine, and thymine. The self-replicating genetic material of all living cells; it is made up of a double helix of two complementary strands of *nucleotides*.

dominant A character that is manifest in the *phenotype* of heterozygous individuals.

duplication A chromosomal mutation characterized by two copies of a *chromosome* segment.

effective population size The number of reproducing individuals in a *population*.

electronic spin resonance A dating method based in determining the amount of electrons trapped in defects of crystal lattices, which are caused by the decay of radioactive elements, through the measurement of their absorption of microwave radiation.

electrophoresis A technique for separating molecules based on their differential mobility in an electric field.

enamel A hard substance that forms a layer around the crown dentine of teeth.

encephalization The increase in brain size relative to body size.

endemic Applied to *species* restricted to a certain region or part of a region; in epidemiology, applied to diseases that are constantly present at relatively low levels in a particular *population*.

endocast or **endocranial cast** A natural or artificial mold of the inner surface of the *cranium*.

enzyme A biochemical catalyst based on specialized *protein* molecules that speeds up biochemical processes.

Eocene The second epoch of the Tertiary period, about 56.5–35.4 Ma, during which the second wave of *primates* appeared.

epithelium A tissue consisting of one or more layers of tightly bound cells that covers the external and internal surfaces of the body.

epoch A subdivision of a geological period.

era The largest division of geological time, including one or more periods.

eukaryote An organism whose cells contain a distinct *nucleus* as well as *mitochondria* and other *organelles*.

exon The *DNA* of a eukaryotic transcription unit whose transcript becomes a part of the *mRNA* produced by splicing out introns.

extant Living; the opposite of extinct.

family The category composed by a set of *genera*.

fault A fracture of the Earth's crust, across which there has been observable displacement.

fauna The animals of a region, country, special environment, or period.

femoral head Rounded upper part of the *femur* that forms the joint with the hip.

femoral neck The section of the *femur* that extends medially, separating the *femoral head* from the shaft.

femur The proximal bone of the lower limbs. It is the largest bone in the human body. It is usually divided into three sections: the upper end, which includes the *femoral head* and *femoral neck*, the femoral shaft, and the lower or condylar end.

fibula The smaller of the two distal hindlimb bones.

fission-track dating A method used for dating volcanic rocks associated with *fossil* remains. The age of the rocks is estimated as a function of the amount of uranium in them and the density of the damage trails left by the spontaneous fission of uranium (^{238}U).

fitness The reproductive contribution of an organism or *genotype* to future generations.

flake Long, small, and cutting stone fragment obtained from cobbles by percussion or pressure.

flora The plants of a region, country, particular environment, or period.

folivore An animal whose diet is composed mainly of leaves.

forager One who collects wild animal or plant food.

foramen magnum The large opening in the back of the skull through which the spinal chord passes to join the brainstem; where the vertebral column connects with the *cranium*.

formation In geology, a fundamental unit of stratigraphic classification. Often formations are given geographic names; for example, the Hadar Formation.

fossil Any preserved remains or traces of past life, more than about 10,000 years old, embedded in rock as either mineralized remains or impressions, casts, or tracks.

frugivore An animal whose diet is composed mainly of fruit.

gamete A mature reproductive cell capable of fusing with a similar cell of opposite sex to give a *zygote*; also called a *sex cell*.

gene A genetic unit found on a specific *locus* of a *chromosome*. It consists of a sequence of *DNA* that codes for an *enzyme* or a *protein*, or that regulates activity of other genes.

genetic drift Chance fluctuations in gene frequency observed especially in small *populations*.

genome The genetic content of a cell; in *eukaryotes*, it sometimes refers to only one complete (*haploid*) *chromosome* set.

genotype The genetic information of an individual.

genus The category formed by a set of closely related *species* (plural: *genera*).

glabella A prominence above the nose in the midline of the external surface of the frontal bone.

glacial period A period of cold climate, during which time a certain amount of oceanic water is deposited on the continents as glaciers.

Gorilla The *genus* to which gorillas belong.

gracile Slender or light in build, often used to characterize the *Australopithecus species A. afarensis* and *A. africanus*.

grade A group of organisms that do not necessarily form a *clade* but share a set of characteristics which set them apart from others.

great apes Chimpanzees and gorillas (in Africa) and orangutans (in Asia).

habitat The natural home or environment of a plant or animal.

half-life Statistical average time in which half of a certain amount of a radioactive isotope disappears.

hammerstone An unaltered stone used for hammering. It is commonly considered as the simplest stone tool.

handaxe The most common *biface* stone tool found in Acheulean archaeological sites. It is usually oval or teardrop-shaped.

haploid Cells, such as *gametes*, that in *eukaryotes* have half as many *chromosome* sets as the somatic cells.

heterozygote An organism with two different *alleles* at a certain *locus*.

holotype The example of a certain organism which is used for its classification. Ideally it should be typical of its *taxon*, although in *fossil* taxa the holotype is often only a partial specimen.

hominin An individual belonging to the tribe *Hominini*.

Hominini A tribe composed of current humans and their direct and lateral ancestors that are not also ancestral to chimpanzees.

hominoid An individual belonging to the superfamily *Hominoidea*.

Hominoidea The superfamily composed of lesser apes, great apes, and humans, as well as their direct and lateral ancestors that are not also ancestral to *Old World monkeys*.

Homo The *genus* to which the human *species* belongs. The species *H. habilis*, *H. ergaster*, *H. antecessor*, *H. erectus*, *H. heidelbergensis*, *H. neanderthalensis*, and *H. sapiens* are usually included within this genus.

homologous (noun **homology**) A trait that is shared by two taxa which inherited it from their closest common ancestor. Can also be used to refer to *chromosomes*.

homoplasy A similar trait in two or more taxa that is not due to inheritance from their closest common ancestor, but rather has evolved independently in the different taxa.

homozygote A cell or organism having the same *allele* at a given *locus* on *homologous chromosomes* (adj. *homozygous*).

humerus The arm's largest bone. Its proximal end articulates with the *scapula* at the shoulder joint, and at the distal end it articulates with the *radius* and the *ulna*.

hunter-gatherer One who lives by hunting and scavenging wild animals, gathering plants and, in some places, collecting shellfish and fishing, often moving in small groups (bands) from place to place.

hybridization hypothesis The proposition that the appearance of modern humans involved the admixing of modern humans with *Neanderthals* or other early *species*.

Hylobates The *genus* including gibbons and siamangs.

hyoid bone A U-shaped bone in the neck lying just above the larynx and below the tongue.

hypodigm The whole available set of *fossil* material belonging to a given *species*.

incisor Frontal teeth whose primary function is to cut food. All *hominoids* have four upper and four lower incisors.

insectivore An animal that feeds mostly on insects.

interglacial period A warm interval between two glaciations.

intron a length of *DNA* within a functional gene in *eukaryotes*, separating two segments of coding DNA (*exons*).

inversion A chromosomal mutation characterized by the reversal of a *chromosome* segment.

K/Ar see *potassium/argon*

Kenyanthropus The *hominin genus* introduced in 2001 to accommodate a *cranium* and other findings from Lomekwi (northern Kenya), which have been attributed to the *species Kenyanthropus platyops*.

kingdom The category of classification formed by a set of phyla; plants, animals, and fungi are kingdoms.

knuckle-walking   A kind of quadrupedal loco-motion typical of gorillas and chimpanzees, in which the weight of the upper body leans on the finger knuckles, bent into the palm of the hands and placed on the floor.

lesser apes   The Asian apes, gibbons, and siamangs.

Levallois technique   A tool-making technology involving significant pre-shaping of the *core*.

Levantine corridor   A migration and settlement area between Africa and Eurasia on the eastern coast of the Mediterranean Sea.

locus   A *gene's* specific place on a *chromosome*; sometimes used to refer to the gene itself (plural: *loci*).

magnetostratigraphy   A dating method based on the polarity inversions of Earth's magnetic field. Calibration using other techniques has allowed charting many such episodes, which can constrain the estimated age of *fossils*.

mandible   The lower jaw.

mandibular symphysis   The midline separating the left and right halves of the *mandible*.

maxilla   A facial bone that forms the floor of the orbital cavity, the lateral wall of the nasal cavity, and the support for the upper teeth.

member   In geology, a rock unit which is a sub-division of a broader *formation*.

mentum   see *chin* (also known as a mental protuberance)

messenger RNA (mRNA)   an *RNA* molecule whose *nucleotide* sequence is translated into an *amino acid* sequence on ribosomes during *polypep-tide* synthesis.

metacarpal   Each of a set of five bones in each hand that form the palm and the thumb's first bone. They articulate proximally with the wrist's carpal bones and distally with the fingers' *phalanges*.

metatarsal   Each of a set of five bones in each foot that form the instep.

Miocene   The fourth epoch of the Tertiary period, about 23.3–5.2 Ma.

mitochondrial DNA (mtDNA)   Genetic informa-tion contained in *mitochondria* in the form of a single circular strand which is inherited only through the maternal line.

mitochondrion   An *organelle* in a eukaryotic cell that is involved in energy metabolism; each mito-chondrion has its own small circular genome (plural: *mitochondria*).

molar   Each of the teeth with large occlusal sur-faces located at the back of the jaws for grinding and crushing food. Humans have three molars on each side of the jaw.

molecular clock   The estimated regularity of changes in *DNA* and *proteins* through time, which can be used to estimate the timing of evolutionary episodes.

monophyletic   Describing a group of organisms, including their common ancestral stock and all its descendants.

Mousterian   A middle-Paleolithic stone culture characterized by small and precise instruments, such as small *handaxes*, side-scrapers, and tri-angular points. It is commonly associated with *Neanderthals*.

multiregional   An hypothesis suggesting that the appearance of modern humans occurred by independent evolution from earlier *hominins* in different geographical regions.

mutant   An *allele* different from the wild type; or an individual carrying such an allele.

mutation   An inheritable modification of genetic material.

nasal   Of or relating to the nose.

natural selection   The differential reproduction of alternative *genotypes* due to variable *fitness*.

navicular   A footbone located on the medial side of the foot that articulates posteriorly with the *talus*, laterally with the *cuboid*, and distally with the three *cuneiform* bones.

Neanderthal   The common designation for *Homo neanderthalensis*, a *hominin species* closely related to *Homo sapiens* which inhabited southern Europe and the Levant during the Würm glaciation. They are generally associated with the *Mousterian* material culture.

Neolithic   The New Stone Age, usually asso-ciated with the beginnings of agriculture, pottery, and settlements in the Old World. In parts of western Asia, farming began as early as 10,000 years ago (although without pottery).

neoteny   The persistence of a juvenile character in adulthood.

niche   The place of an organism in its environment, including the resources it exploits and its association with other organisms.

nitrogen base   An organic compound composed of a ring containing nitrogen. Used here to refer to each of the complementary molecules that keep the two *DNA* strands together transversally or form *RNA* strands (the bases are adenine, cytosine, guanine, thymine, and uracil).

node   The point where two *clades* diverge in a *cladogram*.

nomad   One who moves continually from place to place to find food.

nucleic acid   see *DNA* and *RNA*

nucleotide   A nucleic-acid unit, composed of a sugar molecule, a nitrogen base, and a phosphate group. A set of three nucleotides constitutes a *triplet*, or *codon*, and each triplet codes for an *amino acid* or represents a stop signal during *protein* synthesis.

nucleus   A membrane-enclosed *organelle* of *eukaryotes* that contains the *chromosomes*.

occipital torus   A protuberance found on the posterior part of the *cranium*.

Oldowan   The oldest known *hominin* lithic cultural tradition, usually associated with the appearance of *Homo habilis*. The oldest Oldowan remains are about 2.5 million years old.

Old World monkey   Relating to monkeys in all geographical areas except South and Central America, in the superfamily Cercopithecoidea.

Oligocene   The third epoch of the Tertiary period, lasting about 35.4–23.3 Ma, during which there was a great reduction of the *primate order*.

omnivore   An eater of both animal and plant food.

orbit   The bony socket for the eye.

order   The taxonomic category formed by a set of *families* and a division of a *class*.

organelle   A functional membrane-enclosed body inside cells (e.g. a *nucleus* or a *mitochondrion*).

Orrorin   The *genus* including late Miocene *hominin* remains found in the Tugen Hills region of Kenya.

orthognathic   With little facial projection, opposite to *prognathic*.

orthograde   The upright position of the body.

orthologous   Referring to *genes* or *chromosomes* of different *species* which are similar because they derive from a common ancestor.

out of Africa   An hypothesis proposing that the appearance of modern humans occurred solely in Africa, from where they colonized the other continents.

ovum   A female *gamete*.

palate   The roof of the mouth. There are two different parts: a hard bony anterior one, and a soft posterior one.

Paleocene   The first epoch of the Tertiary period, about 65–56.5 Ma, during which archaic, or first-wave, *primates* appeared.

Paleolithic   The Old Stone Age, the first and longest part of the stone age that began some 2.6 Ma in Africa with the first recognizable stone tools belonging to the *Oldowan* industrial tradition and ended some 12,000–10,000 years ago.

paleomagnetism   An important dating method based on the history of changes in the Earth's magnetic polarity throughout a stratigraphic column.

Pan   The *genus* to which chimpanzees and bonobos belong.

Paranthropus   An extinct *hominin genus* including Pliocene and Pleistocene robust individuals. The *species P. boisei, P. robustus*, and *P. aethiopicus* are usually assigned to this genus.

paraphyletic   A group of organisms including some, but not all, of the descendants of the group's common ancestor.

paratype   A set of specimens in the type series other than the *holotype*.

parsimony   In evolution, the principle proposing that evolution has followed the most economical route, involving the assumption that closely related *species* (those that diverged more recently) will consistently have fewer differences than species that diverged longer ago.

patella   The sesamoid bone located at the knee joint.

pelvis   The structure composed of two pelvic bones and the sacrum. Each pelvic bone is composed of three bones, the ilium (the upper blade), ischium (the lower part), and pubis (the ventral part).

permanent dentition   The second generation of teeth that appear once the *deciduous dentition* has been lost.

phalange   One of the set of 14 finger bones in each hand and foot. There are five proximal phalanges that articulate proximally with the five *metacarpals* in the hand, and the five *metatarsals* in the foot. There are four middle phalanges in each hand and foot (absent in the thumb and the toe). There are five distal phalanges in each hand and foot.

phenetics   A system of classification of organisms based principally on the similarity of morphological traits, also known as numerical *taxonomy*.

phenotype   An organism's observable traits.

phyletic   Applied to a group of *species* with a common ancestor; a line of direct descent.

phylogenesis   The process of evolution and differentiation of organisms.

phylogenetic tree   The graphic representation of the evolutionary relations among living and extinct organisms.

phylogeny   The evolutionary history of a group of living or extinct organisms.

phylum   The category formed by a set of *classes* (plural: *phyla*).

Pithecanthropus   A *genus* created by Eugéne Dubois to include the specimens discovered by his expedition to Java. Those specimens are currently included within *Homo erectus*.

Pleistocene   The first epoch of the Quaternary period, which lasted from about 1.64 million to 10,000 years ago, and saw the radiation of the *genus Homo*.

plesiadapiform   A group of diverse *fossil* primatelike mammals that lived during the Paleocene and early Eocene. Traditionally, they have been considered as a suborder of Primates, but this view has gradually lost support in favor of their consideration as a different *order* of mammals.

plesiomorphy   A trait that is already present in the ancestral group of the *taxon* being studied.

Pliocene   The final epoch of the Tertiary period, about 5.2–1.64 Ma.

polymorphism   The existence of alternative allelic forms at a *locus* within a *population*. In

humans, there is a polymorphism for the ABO blood groups.

polypeptide   A chain of *amino acids* bound covalently by peptide linkages.

polyphyletic   The grouping of organisms derived from at least two different ancestral stocks.

Pongo   The *genus* to which orangutans belong.

population   A set of individuals belonging to the same *species* that constitute an effective reproductive community.

postcranial   Referring to any skeletal element except those forming the *cranium*.

potassium/argon   A dating method that estimates the age of volcanic terrains based on the amount of radioactive potassium (^{40}K) remaining in a sample.

premolar   Each of the teeth located between the canines and the molars. Humans have two premolars in each half of each jaw.

primate   The *order* to which the human *species* belongs, together with prosimians, tarsoids, and the rest of anthropoids.

primitive trait   A feature exhibited by an organism or set of organisms that appeared before the node representing the origin of their *clade*, opposite of *derived trait* or *apomorphy*.

prognathic   Describing the outward facial projection beyond the vertical plane that passes through the orbital cavities, opposite to *orthognatic*.

prosimian   Any *primate* in the suborder Prosimii (lemurs, lorises, and tarsiers).

protein   A molecule composed of one or more *polypeptide* subunits and possessing a characteristic three-dimensional shape imposed by the sequence of its component *amino acid* residues.

quadrupedal   An arboreal or terrestrial mode of locomotion that involves the use of the four limbs, forelimbs and hindlimbs.

Quaternary   The period of the Cenozoic era that began about 1.64 Ma.

radioactivity   The emission of ionizing radiation from an unstable chemical isotope.

radiocarbon dating   Radiometric dating method for organic materials based on the rate of decay of ^{14}C to ^{14}N.

radiometric dating   Dating methods, such as *radiocarbon dating*, based on the measurement of radioactive decay.

radius   A bone of the forearm, somewhat shorter than the *ulna*, but broader at its distal side.

ramus   An ascending backward projection from each side of the *mandible's* body (plural: *rami*).

recessive   An *allele*, or the corresponding trait, that is manifest only in *homozygotes*.

replacement hypothesis   The proposal that the modern humans that had dispersed from Africa did not admix with earlier *populations* living in the territories they colonized.

RNA   Ribonucleic acid; *nucleic acid* composed of units consisting of a ribose sugar, a phosphate group, and the nitrogen bases adenine, guanine, cytosine, and uracil.

robust   A heavily built *fossil* specimen, especially the face and jaw, normally with large masticatory apparatus.

sacrum   The continuation of the vertebral column into the pelvis; composed of several fused vertebrae.

sagittal crest   A bony protuberance along the skull's superior midline for the attachment of large muscles, observed in some robust *hominins*.

Sahelanthropus   A *genus* created to accommodate late-Miocene *fossil* specimens retrieved in Tchad, included within the *hominin* tribe, although this has been contested by some researchers.

savanna   Subtropical or tropical grassland with scattered trees and shrubs and a pronounced dry season.

scapula   The shoulder blade.

sedimentary rock   A rock formed by the accumulation and hardening of rock particles (sediments) derived from existing rocks and/or organic debris and deposited by agents such as wind, water, and ice at the Earth's surface; the source of *fossils*.

selection   see *natural selection*

selective value   see *adaptive value*

sensu lato   Latin expression meaning "in a broad sense".

sensu stricto   Latin expression meaning "in a strict sense".

sex cell   see *gamete*

sex chromosome   A *chromosome* that differs between the two sexes and is involved in sex determination (see *autosome*).

sexual dimorphism   A set of traits that distinguish male and female individuals from the same *species*.

shared derived character   A feature shared by descendants from an ancestral stock that was not present in the remote common ancestor.

simian   Any member of the *primate* suborder Anthropoidea (monkeys, apes, and humans); a higher primate.

single-species hypothesis   The notion that all *hominin* specimens after *Homo habilis* can be accommodated adequately within a single *species*.

sister group   Each one of the *clades* that separates at a node (see *clade*, *cladogenesis*, and *cladogram*).

skull   A set formed by the cranial bones and the *mandible*.

soft-hammer   A tool-making technique in which the hammer is made of a softer material than the *core*. It allows a greater precision in the production of *flakes*.

speciation   The process of evolution of a new *species*.

species   The basic unit of Linnaean classification, always expressed by two Latin names (such as *Homo sapiens*), the first of which specifies the *genus*; defined as a group of interbreeding natural *populations* that is reproductively isolated from other such groups.

stratigraphy   In geology, the study of a site's strata; usually represented by stratigraphic columns.

stratum   A layer of sedimentary terrain that corresponds to a specific sedimentary period and is differentiable from other layers above or below it.

supraorbital torus   A bar of bone extending over the superior margins of the *orbits*.

suspensory behavior   A locomotion or posture that involves hanging from branches or clinging to them.

suture   In anatomy, the junction of two parts immovably connected.

symbol   A word, behavior, or object that conveys meaning.

sympatric   Referring to *species* that share the same territory (see also *allopatric*).

synapomorphy   A *derived trait* that is shared by two or more *clades*.

systematics   The discipline that studies the classification of organisms and their evolutionary relationships.

talus   A footbone that links the leg and the rest of the foot. Proximally, it articulates with the *tibia*; anteriorly with the *navicular*; and inferiorly with the *calcaneus*.

taphonomy   A discipline that studies the processes involved in fossilization.

tarsal bone   One of the set of seven ankle bones in each foot: *calcaneus, talus, navicular, cuboid,* and three *cuneiforms*.

taxon   A defined unit in the classification of organisms. For example, *Homo* is a taxon of the *genus* category; *Homo sapiens* is a taxon of the *species* category (plural: *taxa*).

taxonomy   The rules and procedures used in the classification of organisms.

temporal bone   One of a pair of bones that form part of the side wall and base of the human *brain case* and cover the auditory ossicles in the middle ear.

Tertiary   The first period of the Cenozoic era, from 65 to 1.64 Ma.

tetrapod   A four-footed animal: any amphibian, reptile, bird, or mammal.

thalamus   A large mass of grey matter deep in the cerebral hemispheres in the middle of the forebrain.

thermoluminescence   A dating method based on the calculation of the amount of electrons trapped in defects of crystal lattices, caused by the decay of radioactive elements; it measures the intensity of the light emitted by electrons escaping as the sample is heated to a high temperature.

tibia   The largest of the two distal leg bones.

triplet   In genetics, set of three contiguous *DNA* or *RNA nucleotides* that specifies a particular *amino acid* in a *protein*, or indicates its end signal (also called *codon*).

tuff, volcanic   Rock formed by the cementing or compression of volcanic ashes.

type specimen   An individual exemplar used as the reference for naming a new *species*.

ulna   A bone of the forearm.

uranium series   A dating method based on the radioactive decay of uranium in a geological sample.

vertebrate   An animal with a backbone or vertebral column.

Villafranchian   Fauna of the early Pleistocene, after the Italian village of Villafranca.

Wernicke's area   The region of the human brain involved in the comprehension of speech, lying in the upper part of the temporal cortex and extending into the parietal cortex in the left cerebral hemisphere.

Y   The designation of the shape of the lower molars with five cusps, typical of *hominoids*.

zygomatic arch   The arch of bone that extends along the front or side of the skull beneath the *orbit*.

zygomatic bone   The lateral facial bone below the eye that in mammals forms part of the *zygomatic arch* and part of the *orbit*.

zygote   The *diploid* cell formed by the union of egg and sperm nuclei in the cell.

* The authors are grateful to Mr. Marcos Nadal for his important contribution in the preparation of the Glossary.

References

Abbate, E., Albianelli, A., Azzaroli, A., Benvenutti, M., Tesfamariam, B., Bruni, P., Cirpiani, N., Clarke, R.J., Ficcarelli, G., Macchiarelli, R. *et al.* (1998) A one-million-year-old *Homo* cranium from the Danakil (Afar) Depression of Eritrea. *Nature* **393**, 458–460.

Abbott, A. (2003) Anthropologists cast doubt on human DNA evidence. *Nature* **423**, 468.

Adcock, G.J., Dennis, E.S., Easteal, S., Huttley, G.A., Jermiin, L.S., Peacock, W.J., and Thorne, A. (2001) Mitochondrial DNA sequences in ancient Australians: implications for modern human origins. *Proceedings of the National Academy of Sciences USA* **98**, 537–542.

Agnew, N. and Demas, M. (1998) Conservación de las huellas de Laetoli. *Investigación y ciencia* **266**, 8–18.

Aguirre, E. (1970) Identificación de "Paranthropus" en Makapansgat. In *Crónica* (X.C. N.d. Arqueología, ed.), pp. 97–124. Merida, Spain: Cronica del XI Congreso Nacional de Arqueologia.

Aguirre, E. (1995) Los yacimientos de Atapuerca. *Investigación y ciencia* **229**, 42–51.

Aguirre, E. (1996) Antecedentes y contextos del bipedismo vertical. In *Significado de la postura y de la marcha humana* (A.V. Pericé, ed.), pp. 52–67. Madrid: Editorial Complutense.

Aguirre, E. (1997) Relaciones entre humanos fósiles del Pleistoceno antiguo. In *Senderos de la evolución humana* (C.J. Cela-Conde, R. Gutiérrez Lombardo, and J. Martínez Contreras, eds), pp. 53–81. México: Ludus Vitalis.

Aguirre, E. (2000) *Evolución humana. Debates actuales y vías abiertas*. Madrid: Real Academia de Ciencias Exactas, Físicas y Naturales.

Aguirre, E. and de Lumley, M.-A. (1977) Fossil men from Atapuerca, Spain: their bearing on human evolution in the middle Pleistocene. *Journal of Human Evolution* **6**, 681–688.

Aguirre, E., de Lumley, M.-A., Basabe, J.M., and Botella, M. (1980) Affinities between the mandibles from Atapuerca and L'Arago, and some East African fossil hominids. In *Proceedings of the 8th Panafrican Congress of Prehistory and Quaternary Studies* (R.E. Leakey and B.A. Ogot, eds), pp. 171–174. Nairobi: The International Louis Leakey Memorial Institute for African Prehistory.

Agustí, J. (2000) Viaje a los orígenes del bipedismo y una escala en la isla de los simios. In *Antes de Lucy. El agujero negro de la evolución humana* (J. Agustí, ed.), pp. 97–134. Barcelona: Tusquets.

Agustí, J.M., Köhler, M., Moyà-Solà, S., Cabrera, L., Garcés, M., and Parés, J.M. (1996) Can Llobateres: the pattern and timing of the Vallesian hominoid radiation reconsidered. *Journal of Human Evolution* **31**, 143–155.

Aiello, L.C. (1981) Locomotion in the Miocene Hominoidea. In *Aspects of Human Evolution* (C.B. Stringer, ed.), pp. 63–97. London: Taylor and Francis.

Aiello, L.C. (1992) Allometry and the analysis of size and shape in human evolution. *Journal of Human Evolution* **22**, 127–147.

Aiello, L.C. (1994) Variable but singular. *Nature* **368**, 399–400.

Aiello, L.C. and Dunbar, R.I.M. (1993) Neocortex size, group size and the evolution of language. *Current Anthropology* **34**, 184–193.

Aiello, L.C. and Wheeler, P. (1995) The expensive tissue hypothesis: the brain and the digestive system in human and primate evolution. *Current Anthropology* **36**, 199–221.

Aiello, L. and Collard, M. (2001) Our newest oldest ancestor? *Nature* **410**, 526–527.

Akazawa, T., Muhesen, M., Dodo, Y., Kondo, O., and Mizoguchi, Y. (1995) Neanderthal infant burial. *Nature* **377**, 585–586.

Alba, D.M., Moyà-Solà, S., and Köhler, M. (2001) Canine reduction in the Miocene hominoid *Oreopithecus bambolii*: behavioural and evolutionary implications. *Journal of Human Evolution* **40**, 1–16.

Alemseged, Z., Coppens, Y., and Geraads, D. (2002) Hominin cranium from Omo: description and taxonomy of Omo-323-1976–896. *American Journal of Physical Anthropology* **117**, 103–112.

Alemseged, Z., Spoor, F., Kimbel, W.H., Bobe, R., Geraads, D., Reed, D., and Wynn, J.G. (2006) A juvenile early hominin skeleton from Dikika, Ethiopia. *Nature* **443**, 296–301.

Alexander, R.D. (1979) *Darwinism and Human Affairs*. Seattle: University of Washington Press.

Alexeev, V.P. (1986) *The Origin of the Human Race*. Moscow: Progress Publishers.

Alpagut, B., Andrews, P., Fortelius, M., Kappelman, J., Temizsoy, I., Çelebi, H., and Lindsay, W. (1996) a new specimen of *Ankarapithecus meteai* from the Sinap Formation of Central Anatolia. *Nature* **382**, 349–351.

Ambrose, S.H. (2001) Paleolithic technology and human evolution. *Science* **291**, 1748–1753.

Anderson, S., Bankier, A.T., Barrell, B.G., de Bruijn, M.H., Coulson, A.R., Drouin, J., Eperon, I.C., Nierlich, D.P., Roe, B.A., Sanger, F. *et al.* (1981) Sequence and organisation of the human mitochondrial genome. *Nature* **290**, 457–465.

Andrews, P. (1981) Species diversity and diet in monkeys and apes during the Miocene. In *Aspects of Human Evolution* (C.B. Stringer, ed.), pp. 25–61. London: Taylor and Francis.

Andrews, P. (1983) Small mammal faunal diversity at Olduvai Gorge, Tanzania. In *Animals and Archaeology: I. Hunters and Their Prey* (J. Clutton-Brock and C. Grigson, eds), pp. 77–85. Oxford: BAR International Series **163**.

Andrews, P. (1984) An alternative interpretation of the characters used to define *Homo* erectus. In *The Early Evolution of Man with Special Emphasis on Southeast Asia and Africa* (P. Andrews and J. Frazen, eds), vol. **69**, pp. 167–175. Frankfurt: Courier Forschungsinstitut Senckenberg.

Andrews, P. (1990) Lining up ancestors. *Nature* **345**, 664–665.

Andrews, P. (1992a) Evolution and environment in the Hominoidea. *Nature* **360**, 641–646.

Andrews, P. (1992b) Reconstructing past environments. In *The Cambridge Encyclopedia of Human Evolution* (S. Jones, R. Martin, and D. Pilbeam, eds), pp. 191–195. Cambridge: Cambridge University Press.

Andrews, P. (1995) Ecological apes and ancestors. *Nature* **376**, 555–556.

Andrews, P. (1996) Palaeoecology and hominoid palaeoenvironments. *Biological Review* **71**, 257–300.

Andrews, P. and Cronin, J.E. (1982) The relationships of *Sivapithecus* and *Ramapithecus* and the evolution of the orang-utan. *Nature* **297**, 541–546.

Andrews, P.J. and Martin, L.B. (1987) Cladistic relationships of extant and fossil hominoids. *Journal of Human Evolution* **16**, 101–118.

Andrews, P. and Martin, L. (1991) Hominid dietary evoution. *Philosophical Transactions of the Royal Society of London Series B* **334**, 199–209.

Anon (2005) The Chimpanzee Genome. *Nature* **437**, 47–108.

Antón, S.C. (2002) Evolutionary significance of cranial variation in Asian *Homo* erectus. *American Jornal of Physical Anthropology* **118**, 301–323.

Antón, S.C. (2003). Natural history of *Homo* erectus. *Yearbook of Physical Anthropology* **46**, 126–170.

Antón, S.C. (2004) The face of Olduvai Hominid 12. *Journal of Human Evolution* **46**, 337–347.

Appenzeller, T. (1998) Art: evolution or revolution? *Science* **282**, 1451–1454.

ApSimon, A.M. (1980) The last Neanderthal in France? *Nature* **287**, 271–272.

Arambourg, C. (1954) L'hominien fossile de Ternifine (Algérie). *Comptes rendus de l'Académie des Sciences Paris* **239**, 893–895.

Arambourg, C. (1955) A recent discovery in human paleontology: *Atlanthropus* of Ternifine (Algeria). *American Journal of Physical Anthropology* **13**, 191–202.

Arambourg, C. and Coppens, Y. (1968) Découverte d'un Australopithécien nouveau dans les gisements de l'Omo (Ethiopie). *South African Journal of Science* **64**, 58–59.

Arensburg, B. (1989) New skeletal evidence concerning the anatomy of middle paleolithic populations in the Middle East: the Kebara skeleton. In *The Human Revolution: Behavioural and Biological Perspectives on the Origins of Modern Humans* (P. Mellars and C. Stringer, eds), pp. 165–171. Princeton, NJ: Princeton University Press.

Arensburg, B. and Tillier, A.-M. (1990) El lenguaje del hombre de Neanderthal. *Mundo científico* **107**, 1144–1146.

Arensburg, B., Bar-Yosef, O., Chech, M., Goldberg, P., Laville, H., Meignen, L., Rak, Y., Rchernov, E., Tillier, A.M., and Vandermeersch, B. (1985) Une sépulture néandertalienne dans la grotte de Kébara (Israël). *Comptes-Rendus de l'Académie des Scenices de Paris (Série D)* **300**, 227–230.

Argue, D., Donlon, S., Groves, C., and Wright, R. (2006) Homo floresiensis: microcephalic, pygmoid, Australopithecus, or Homo? *Journal of Human Evolution* **51**, 360–3744.

Arsuaga, J.L. (1994) Los hombres fósiles de la sierra de Atapuerca. *Mundo científico* **143**, 167–168.

Arsuaga, J.L. (1999) *El collar del neandertal*. Madrid: Temas de Hoy.

Arsuaga, J.L., Martínez, I., Gracia, A., Carretero, J.M., and Carbonell, E. (1993) Three new human skulls from the Sima de los Huesos Middle Pleistocene site in Sierra de Atapuerca, Spain. *Nature* **362**, 534–537.

Arsuaga, J.L., Martínez, I., Lorenzo, C., Gracia, A., Muñoz, A., Alonso, O., and Gallego, J. (1999) The human cranial remains from the Gran Dolina Early Pleistocene site (Sierra de Atapuerca, Spain). *Journal of Human Evolution* 37, 431–457.

Ascenzi, A., Biddittu, I., Cassoli, P.F., Segre, A.G., and Segre-Naldini, E. (1996) A calvarium of late *Homo* erectus from Ceprano, Italy. *Journal of Human Evolution* 31, 409–423.

Ascenzi, A., Mallegni, F., Manzi, G., Segre, A.G., and Segre Naldini, E. (2000) A re-appraisal of Ceprano calvaria affinities with *Homo* erectus, after the new reconstruction. *Journal of Human Evolution* 39, 443–450.

Asfaw, B. (1983) A new hominid parietal from Bodo, Middle Awash Valley, Ethiopia. *American Journal of Physical Anthropology* 61, 387.

Asfaw, B. (1987) The Belohdelie frontal: new evidence of early hominid cranial morphology from the Afar of Ethiopia. *Journal of Human Evolution* 16, 611–624.

Asfaw, B., Beyene, Y., Suwa, G., Walter, R.C., White, T.D., WoldeGabriel, G., and Yemane, T. (1992) The earliest Acheulan from Konso-Gardula. *Nature* 360, 732–735.

Asfaw, B., White, T., Lovejoy, O., Latimer, B., Simpson, S., and Suwa, G. (1999) *Australopithecus garhi*: a new species of early hominid from Ethiopia. *Science* 284, 629–635.

Asfaw, B., Gilbert, W.H., Beyene, Y., Hart, W.K., Renne, P.R., WoldeGabriel, G., Vrba, E.S., and White, T.D. (2002) Remains of *Homo* erectus from Bouri, Middle Awash, Ethiopia. *Nature* 416, 317–320.

Avise, J.C. (2006). *Evolutionary Pathways in Nature. A Phylogenetic Approach.* Cambridge: Cambridge University Press.

Ayala, F.J. (1980) *Origen y Evolución del Hombre.* Madrid: Alianza Editorial.

Ayala, F.J. (1982a) The evolutionary concept of progress. In *Progress and Its Discontents* (G.A. Almond, M. Chodorow, and R.H. Pearce, eds), pp. 106–124. Berkeley: University of California Press.

Ayala, F.J. (1982b) La naturaleza humana a la luz de la evolutión. *Estudios Filosóficos* 31, 397–441.

Ayala, F.J. (1995) The myth of Eve: molecular biology and human origins. *Science* 270, 1930–1936.

Bahn, P.G. (1992) Ancient art. In *The Cambridge Enciclopedia of Human Evolution* (S. Jones, R. Martin and D. Pilbeam, eds), pp. 361–364. Cambridge: Cambridge University Press.

Bahn, P. (1996) New developments in Pleistocene art. *Evolutionary Anthropology* 4, 204–215.

Bahn, P.G. (1998) Neanderthals emancipated. *Nature* 394, 719–720.

Bailey, W.J., Hayasaka, K., Skinner, C.G., Kehoe, S., Sieu, L., Slightom, J.L., and Goodman, M. (1992) Reex-amination of the African hominoid trichotomy with additional sequences from the primate beta-globin gene cluster. *Molecular Phylogenetics and Evolution* 1, 97–135.

Baird, D.M., Coleman, J., Rosser, Z.H., and Royle, N.J. (2000) High levels of sequence polymorphism and linkage disequilibrium at the telomere of 12q: implications for telomere biology and human evolution. *American Journal of Human Genetics* 66, 235–250.

Balter, M. (2004a) Skeptic to take possession of Flores hominid bones. *Science* **306, 1450.**

Balter, M. (2004b) Skeptics question whether Flores hominid is a new species. *Science* **306, 1116.**

Baltimore, D. (2001) Our genome unveiled. *Nature* **409,** 814–816.

Barash, D.P. (1977) *Sociobiology and Behavior.* Yew York: Elsevier.

Barbujani, G., Magagni, A., Minch, E., and Cavalli-Sforza, L.L (1997) An apportionment of human DNA diversity. *Proceedings of the National Academy of Sciences USA* **94,** 4516–4451.

Barreiro, L.B., Patin, E., Neyrolles, O., Cann, H.M., Gicquel, B., and Quintana-Murci, L. (2005) The heritage of pathogen pressures and ancient demography in the human innate-immunity CD209/CD209L region. *American Journal of Human Genetics* 77, 869–886.

Bar-Yosef, O. (1988) Evidence for Middle Paleolithic symbolic behaviour: a cautionary note. In *L'Homme de Néandertal* (O.B. Yosef, ed.), vol. 5, pp. 11–16. Liège: ERAUL.

Bar-Yosef, O. (2000) The middle and early Upper Paleolithic in Southwest Asia and neighboring regions. In *The Geography of Neandertals and Modern Humans in Europe and the Greater Mediterranean* (O. Bar-Yosef and D. Pilbeam, eds), pp. 107–156. Cambridge, MA: Peabody Museum of Archaeology and Ethnology, Harvard University.

Bar-Yosef, O. and Vandermeersch, B. (1981) Notes concerning the possible age of the Mousterian layers in Qafzeh Cave. In *Préhistoire du Levant* (P. Sanlaville and J. Cauvin, eds), pp. 281–285. Paris: CNRS.

Bar-Yosef, O. and Vandermeersch, B. (1991) *Le Squelette Mousterien de Kebara.* Paris: CNRS.

Bar-Yosef, O. and Vandermeersch, B. (1993) El hombre moderno de Oriente medio. *Investigación y ciencia* 201, 66–73.

Bar-Yosef, O., Vandermeersch, B., Arensburg, B., Goldberg, P., Laville, H., Meginen, L., Rak, Y., Tchernov, E., and Tillier, A.M. (1986) New data on the origins of Modern man in the Levant. *Current Anthropology* 27, 63–64.

Bates, E. and Elman, J. (1996) Learning rediscovered. *Science* **274**, 1849–1850.

Beaumont, P.B., de Villiers, H., and Vogel, J.C. (1978) Modern man in sub-Saharan Africa prior to 49,000 years B.P.: a review and reevaluation with particular reference to Border Cave. *South African Journal of Science* 74, 409–419.

Beauval, C., Maureille, B., Lacrampe-Cuyaubère, F., Serre, D., Peressinotto, D., Bordes, J.G., Cochard, D., Couchoud, I., Dubrasquet, D., Laroulandie, V. *et al.* (2005) A late Neandertal femur from Les Rochers-de-Villeneuve, France. *Proceedings of the National Academy of Sciences USA* **192**, 7085–7090.

Bednarik, R.G. (1992) Paleoart and Archeological Myths. *Cambridge Archaeological Journal* **2**, 27–57.

Bednarik, R.G. (1995) Concept-mediated marking in the lower Paleolithic. *Current Anthropology* **36**, 605–634.

Bednarik, R.G. (1997) The global evidence of early human symboling behaviour. *Human Evolution* **12**, 147–168.

Begun, D.R. (1992) Miocene fossil hominoids and the chimp-human clade. *Science* **257**, 1929–1933.

Begun, D.R. and Ward, C.V. (2005) Comment on "Pierolapithecus catalaunicus, a New Middle Miocene Great Ape from Spain". *Science* **308**, 203c.

Begun, D., Moyà-Solà, S., and Köhler, M. (1990) New Miocene hominoid specimens from Can Llobateres (Vallès-Penedès, Sapin) and their geological and paleoecological context. *Journal of Human Evolution* **19**, 255–268.

Begun, D.R., Ward, C.V., and Rose, M.D. (eds) (1997) *Miocene Hominoid Evolution and Adaptations*, vol. 19. New York, NY: Plenum Press.

Begun, D.R., Güleç, E., and Geraads, D. (2003) Dispersal patterns of Eurasian hominoids: implications from Turkey. *Deinsea* **10**, 23–39.

Behrensmeyer, A.K. and Laporte, L.F. (1981) Footprints of a Pleistocene Hominid in Northern Kenya. *Nature* **289**, 167–169.

Berckhemer, F. (1936) Der Urmenscheschädel von Steinheim. *Zeitschrift für Morphologie und Anthropologie* **35**, 463–516.

Berezikov, E, Cuppen, E., and Plasterk, R.H.A. (2006) Approaches to microRNA discovery. *Nature Genetics Supplement* **38**, S2-S7.

Berge, C. (1991) Quelle est la signification fonctionelle du pelvis très large de *Australopithecus afarensis* (AL 288–1)? In *Origine(s) de la bipédie chez les hominidés* (Y. Coppens and B. Senut, eds), pp. 113–119. Paris: Editions du CNRS.

Berger, L.R. and Parkington, J.E. (1995) A new Pleistocene hominid-bearing locality at Hoedjiespunt, South Africa. *Anerican Journal of Physical Anthropology* **98**, 601–609.

Berger, L.R. and Clarke, R.J. (1996) The load of the Taung Child. *Nature* **379**, 778–779.

Berger, L.R., Keyser, A.W., and Tobias, P.V. (1993) Brief communication: Gladysvale: first early hominid site in South Africa since 1948. *Anerican Journal of Physical Anthropology* **92**, 107–111.

Berger, L.R., Lacruz, R., and de Ruiter, D.J. (2002) Revised age estimates of Australopithecus-bearing deposits at Sterkfontein, South Africa. *American Journal of Physical Anthropology* **119**, 192–197.

Berggren, W.A., Kent, D.V., Swisher, C.C., and Aubry, M. (1995) A revised Cenozoic geochronology and chronostratigraphy. In *Geochronology, Time Scales and Global Stratigraphic Correlation*, SEPM Special Publication no. 54 (W.A. Berggren, D.V. Kent, C.C. Swisher, M. Aubry, and J. Hardenbol, eds), pp. 129–212. Tulsa, OK: Society for Sedimentary Geology.

Bermúdez de Castro, J.M. (1995) Nuevos datos sobre la biología del hombre de Atapuerca. *Fronteras de la ciencia y la tecnología* **9**, 52–55.

Bermúdez de Castro, J.M. and Nicolas, M.E. (1996) Changes in the Lower premolar-size sequence during Hominid evolution. Phylogenetic implications. *Human Evolution* **11**, 107–112.

Bermúdez de Castro, J.M., Arsuaga, J.L., Carbonell, E., Rosas, A., Martínez, I., and Mosquera, M. (1997) A hominid from the Lower Pleistocene of Atapuerca, Spain: possible ancestor to Neandertals and modern humans. *Science* **276**, 1392–1395.

Bermúdez de Castro, J.M., Carbonell, E., Cáceres, I., Díez, J. C., Fernández-Jalvo, Y., Mosquera, M., Ollé, A., Rodríguez, J., Rodríguez, X.P., Rosas, A. *et al.* (1999) The TD6 (Aurora stratum) hominid site. Final remarks and new questions. *Journal of Human Evolution* **37**, 695–700.

Bever, M.R. (2001) An overview of Alaskan Late Pleistocene archaeology: historical themes and current perspectives. *Journal of World Prehistory* **15**, 125–191.

Bigazzi, G., Balestrieri, M.L., Norelli, P., Oddone, M., and Tecle, T.,M,. (2004) Fission-track dating of a Tephra layer in the Alat Formation of the Dandiero Group (Danakil Depression, Eritrea). *Rivista Italiana di Paleontologia e Stratigrafia* **110** (supplement), 45–49.

Binford, L.R. (1981) *Bones: Ancient Men and Modern Myths*. New York, NY: Academic Press.

Binford, L. and Ho, C. (1985) Taphonomy at a distance: Zhokoudian, 'The Cave Home of Beijing Man'? *Current Anthropology* **26**, 413–442.

Binford, L. and Stone, N. (1986) Zhoukoudian: a closer look. *Current Anthropology* **27**, 453–475.

Black, D. (1927) On a lower molar hominid tooth from the Chou Kou Tien deposit. *Palaeontologia Sinica, series D7,* 1–29.

Black, D. (1930) On an adolescent skull of *Sinanthropus pekinensis* in comparison with an adult skull of the same species and with other hominid skulls, recent and fossil. *Palaeontologia Sinica, series D* VII(ii), 1–145.

Black, D. (1931) Evidences for the use of fire by *Sinanthropus.* Bulletin of the Geological Society of China **11,** 107–198.

Black, D., Teilhard de Chardin, P., Young, C.C., and Pei, W.C. (1935) Fossil man in China. The Choukoutien cave deposits with a synopsis of our present knowledge of the late Cenozoic of China. *American Anthropologist,* New Series **37(3:1),** 514–515.

Blake, C.C. (1862) On the cranium of the most ancient races of man. *Geologist* June 206.

Blake, C.C. (1867) On a human jaw from the cave of La Naulette near Dinant, Belgium. *Anthropological Review* July and October, 295–395.

Bloch, J.I. and Gingerich, P.D. (1998) *Carpolestes simpsoni,* new species (Mammalia, Proprimates) from the Late Paleocene of the Clarks Fork Basin, Wyoming. *Contributions from the Museum of Paleontology, The University of Michigan* **30,** 131–162.

Bloch, J.I. and Boyer, D.M. (2002) Grasping primate origins. *Science* **298,** 1606–1610.

Blumenschine, R. (1987) Characteristics of an early hominid scavenging niche. *Current Anthropology* **28,** 383–408.

Blumenschine, R.J., Peters, C.R., Masao, F.T., Clarke, R.J., Deino, A.L., Hay, R.L., Swisher, C.C., Stanistreet, I.G., Ashley, G.M., McHenry, L.J.*et al.* (2003) Late Pliocene *Homo* and hominid land use from Western Olduvai Gorge, Tanzania. *Science* **299,** 1217–1221.

Boaz, N.T. and Howell, F.C. (1977) A gracile hominin from Upper Member G of the Shungura Formation, Ethiopia. *American Journal of Physical Anthropology* **46,** 93–108.

Boaz, N.T., Ciochon, R.L., Xuc, Q., and Liu, J. (2004) Mapping and taphonomic analysis of the *Homo* erectus loci at Locality 1 Zhoukoudian, China. *Journal of Human Evolution* **46,** 519–549.

Bobe, R. and Behrensmeyer, A.K. (2004) The expansion of grassland ecosystems in Africa in relation to mammalian evolution and the origin of the genus *Homo.* *Palaeogeography, Palaeoclimatology, Palaeoecology* **207,** 399–420.

Boesch, C. and Tomasello, M. (1998) Chimpanzee and human cultures. *Current Anthropology* **39,** 591–595.

Bordes, F. (1953) Nodules de typologie paléolithique I. Outils musteriens à fracture volontaire, vol. 1. *Bulletin de la Societé Préhistorique Française* **50** Paris: Société préhistorique française.

Bordes, F. (1979) *Typologie du Paléolithique ancien et moyen,* vol. **1.** Paris: CNRS.

Boule, M. (1908) L'Homme fossile de La Chapelle-aux-Saints (Corrèze). *Comptes rendus de l'Académie des Sciences Paris* **147,** 1349–1352.

Boule, M. (1911–1913) L'Homme fossile de La Chapelle-aux-Saints. *Annales de Paléontologie* **6,** 7–8.

Bouyssonie, A., Bouyssonie, J., and Bardon, L. (1908) Découverte d'un squelette humain mousterian à la bouffia de la Chapelle-aux-Saints (Corrèze). *L'Anthropologie* **19,** 513–519.

Bowen, D.Q. and Sykes, G.A. (1994) How old is 'Boxgrove man'? *Nature* **371,** 751.

Bowen, G.J., Clyde, W.C., Koch, P.L., Ting, S., Alroy, J., Tsubamoto, T., Wang, Y., and Wang, Y. (2002) Mammalian dispersal at the Paleocene/Eocene boundary. *Science* **295,** 2062–2065.

Bowler, J.M. and Thorne, A.G. (1976) Human remains from Lake Mungo. In *The Origin of the Australians* (R.L. Kirk and A.G. Thorne, eds), pp. 127–138. Canberra: Australian Institute of Aboriginal Studies.

Bowler, J.M. and Magee, J. (2000) Redating Australia's oldest human remains: a sceptic's view. *Journal of Human Evolution* **38,** 719–726.

Brace, C.L. (1964) The fate of the 'Classic' Neanderthals: a consideration of hominid catastrophism. *Current Anthropology* **5,** 3–43.

Brace, C.L. (1965) *The Stages of Human Evolution.* Englewood Cliffs, NJ: Prentice Hall. (references are from the 3rd edn, 1988)

Brace, C.L., Mahler, P.E., and Rosen, R.B. (1973) Tooth measurements and the rejection of the taxon 'Homo habilis'. *Yearb. Phys. Anthropol.* 1972, **16,** 50–58.

Brace, C.L., Ryan, A.S., and Smith, B.H. (1981) Tooth wear in La Ferrassie man: comment. *Current Anthropology* **22,** 426–430.

Brain, C.K. (1958) The Transvaal ape-man bearing cave deposits. *Transvaal Museum Memoir* **11,** 1–125.

Brain, C.K. (1970) New finds at the Swartranks site. *Nature* **225,** 1112–1119.

Brain, C.K. (1982) *The Swartranks site: stratigraphy of the fossil hominids and a reconstruction of the environment of early Homo.* Paper presented at the Congrès International de Paléontologie Humaine. Premier Congrés. Prétirage. Tome 2, Nice.

Brain, C.K. and Sillen, A. (1988) Evidence from the Swartranks cave for the earliest use of fire. *Nature* **336**, 464–466.

Bramble, D.M. and Lieberman, D.E. (2004) Endurance running and the evolution of *Homo*. *Nature* **432**, 345–352.

Bräuer, G. (1989) The evolution of modern humans: a comparison of the African and non-African evidence. In *The Human Revolution: Behavioural and Biological Perspectives on the Origins of Modern Humans* (P. Mellars and C. Stringer, eds), pp. 123–154. Princeton, NJ: Princeton University Press.

Bräuer, G. (1994) How different are Asian and African *Homo* erectus? *Courier Forschungsinstitut Senckenberg* **171**, 301–318.

Bräuer, G. and Mehlman, M.J. (1988) Hominid molars from a Middle Stone Age level at Mumba Rock Shelter, Tanzania. *American Journal of Physical Anthropology* **75**, 69–76.

Bräuer, G. and Rimbach, K.W. (1990) Late archaic and modern *Homo* sapiens from Europe, Africa, and Southwest Asia: craniometric comparisons and phylogenetic implications. *Journal of Human Evolution* **19**, 789–807.

Bräuer, G. and Mbua, E. (1992) *Homo* erectus features used in cladistics and their variability in Asian and African hominids. *Journal of Human Evolution* **22**, 79–108.

Bräuer, G. and Schultz, M. (1996) The morphological affinities of the Plio-Pleistocene mandible from Dmanisi, Georgia. *Journal of Human Evolution* **39**, 445–481.

Bräuer, G., Yokoyama, Y., Falguères, C., and Mbua, E. (1997) Modern human origins backdated. *Nature* **386**, 337.

Bridgland, D.R., Gibbard, P.L., Harding, P., Kemp, R.A., and Southgate, G. (1985) New information and results from recent excavations at Barnfield Pit, Swanscombe. *Quarterly Newsletter* **46**, 25–39.

Bromage, T., Schrenk, F., and Zonneveld, F.S. (1995) Paleoanthropology of the Malawi Rift: an early hominid mandible from the Chiwondo Beds, northern Malawi. *Journal of Human Evolution* **28**, 71–103.

Brooks, A.S., Helgren, D.M., Cramer, J.S., Franklin, A., Hornyak, W., Keating, J.M., Klein, R.G., Rink, W.J., Schwarcz, H., Leith Smith, J.N. *et al.* (1995) Dating and Context of Three Middle Stone Age Sites with Bone Points in the Upper Semliki Valley, Zaire. *Science* **268**, 553–556.

Broom, R. (1936) A new fossil anthropoid skull from South Africa. *Nature* **138**, 486–488.

Broom, R. (1938) The Pleistocene anthropoid apes of South Africa. *Nature* **142**, 377–379.

Broom, R. (1939) Another new type of fossil ape-man. *Nature* **163**, 57.

Broom, R. (1950) The genera and species of the South African fossil ape man. *American Journal of Physical Anthropology* **8**, 1–13.

Broom, R. and Robinson, J.T. (1949) A new type of fossil man. *Nature* **164**, 322–323.

Broom, R. and Robinson, J.T. (1950) Man contemporaneous with the Swartranks ape-man. *American Journal of Physical Anthropology* **8**, 151–155.

Broom, R., and Robinson, J.T. (1952) Swartranks ape-man *Paranthropus crassidens*. *Transvaal Museum Memoir* **6**, 1–123.

Broom, R., Robinson, J.T., and Schepers, G.W.H. (1950) Sterkfontein Ape-man *Plesianthropus*. *Transvaal Museum Memoir* **4**, 1–117.

Brose, D.S. and Wolpoff, M.H. (1971) Early Upper Paleolithic man and Late Middle Paleolithic tools. *American Anthropologist* **73**, 1156–1194.

Brosnan, S.F. and de Waal, F.B.M. (2003) Monkeys reject unequal pay. *Nature* **425**, 297–299.

Brown, F.H., McDougall, I., Davies, T., and Maier, R. (1985a) An integrated Plio-Pleistocene chronology for the Turkana Basin. In *Ancestors: The Hard Evidence* (E. Delson, ed.), pp. 82–90. New York, NY: Alan R. Liss.

Brown, F., Harris, J., Leakey, R., and Walker, A. (1985b) Early *Homo* erectus skeleton from West Lake Turkana, Kenya. *Nature* **316**, 788–792.

Brown, P., Sutikna, T., Morwood, M.J., Soejono, R.P., Jatmiko, Saptomo, W.E., and Awe Due, R. (2004) A new small-bodied hominin from the Late Pleistocene of Flores, Indonesia. *Nature* **431**, 1055–1061.

Brown, W.M. (1980) Polymorphism in mitochondrial DNA of humans as revealed by restriction endonuclease analysis. *Proceedings of the National Academy of Sciences USA* **77**, 3605–3609.

Bruce, E.J. and Ayala, F.J. (1979) Phylogenetic relationships between man and the apes: electrophoretic evidence. *Evolution* **33**, 1040–1056.

Brumm, A., Aziz, F., van den Bergh, G.D., Morwood, M.J., Moore, M.W., Kurniawan, I., Hobbs, D.R., and Fullagar, R. (2006) Early stone technology on Flores and its implications for Homo floresiensis. *Nature* **441**, 624–628.

Brunet, M., Beauvilain, A., Coppens, Y., Heintz, E., Moutaye, A.H.E., and Pilbeam, D. (1995) The first Australopithecine 2.500 kilometres west of the Rift Valley (Chad). *Nature* **378**, 273–275.

Brunet, M., Beauvilain, A., Coppens, Y., Heintz, E., Moutaye, A.H.E., and Pilbeam, D. (1996) *Australopithecus bahrelghazali*, une nouvelle espèce d'Hominidé ancien de la région de Koro Toro (Chad). *Comptes rendus de l'Académie des Sciences Paris* **322** (series II a), 907–913.

Brunet, M., Guy, G., Pilbeam, D., Mackaye, T.H., Likius, A., Ahounta, D., Beauvilain, B., Blondel, C., Bocherensk, H., Boisserie, J.R.*et al.* (2002) A new hominid from the Upper Miocene of Chad, Central Africa. *Nature* **418**, 145–151.

Brunet, M., Guy, F., Pilbeam, D., Lieberman, D.E., Likius, A., Mackaye, H.T., Ponce de Leon, M.S., Zollikofer, C. P.E., and Vignaud, P. (2005) New material of the earliest hominid from the Upper Miocene of Chad. *Nature* **434**, 752–755.

Bunn, H.T. (1981) Archaeological evidence for meat-eating by Plio-Pleistocene hominids from Koobi Fora and Olduvai Gorge. *Nature* **291**, 574–577.

Bush, M.E. (1980) The thumb of *Australopithecus afarensis. American Journal of Physical Anthropology* **52**, 210.

Bush, M.E., Lovejoy, C.O., Johanson, D.C., and Coppens, Y. (1982) Hominid carpal, metacarpal and phalangeal bones recovered from the Hadar formation: 1974–1977 collections. *American Journal of Physical Anthropology* **57**, 651–667.

Butler, D. (2001) The battle of Tugen Hills. *Nature* **410**, 508–509.

Butzer, K.W. (1971) *Recent History of an Ethiopian Delta.* Chicago, IL: Chicago University Press.

Bye, B.A., Brown, F.H., Cerling, T.E., and McDougall, I. (1987) Increased age estimate for the Lower Paleolithic hominid site at Olorgesailie, Kenya. *Nature* **329**, 237–239.

Caceres, M., Lachuer, J., Zapala, M.A., Redmond, J.C., Kudo, L., Geschwind, D.H., Lockhart, D.J., Preuss, T. M., and Barlow, C. (2003) Elevated gene expression levels distinguish human from non-human primate brains. *Proceedings of the National Academy of Sciences USA* **100**, 13030–13035.

Calvin, W. (1998) The emergence of intelligence. *Scientific American Presents* **9**, 44–51.

Campbell, B. (1964) Quantitative taxonomy and human evolution. In *Classification and Human Evolution* (S.L. Washburn, ed.), pp. 50–74. London: Methuen and Co.

Cande, S. and Kent, D.V. (1995) Revised calibration of the geomagnetic polarity timescale for the Late Cretaceous and Cenozoic. *Journal of Geophysical Research* **100**, 6093–6095.

Cann, R.L., Stoneking, M., and Wilson, A.C. (1987) Mitochondrial DNA and human evolution. *Nature* **325**, 31–36.

Caramelli, D., Lalueza-Fox, C., Vernesi, C., Martina Lari, S., Casoli, A., Mallegni, F., Chiarelli, B., Dupanloup, I.,

Bertranpetit, J., Barbujani, G., and Bertorelle, G. (2003) Evidence for a genetic discontinuity between Neandertals and 24,000-year-old anatomically modern Europeans. *Proceedings of the National Academy of Sciences USA* **100**, 6593–6597.

Caramelli, D., Lalueza-Fox, C., Condemi, S., Longo, L., Milani, L., Manfredini, A., de Saint Pierre, M., Adoni, F., Lari, M., Giunti, P., *et al.*, (2006). A highly divergent mtDNA sequence in a Neanderthal individual from Italy. *Current Biology,* **16**, R650–652.

Carbonell, E., Bermúdez de Castro, J.M., Arsuaga, J.L., Díez, J.C., Rosas, A., Cuenca-Bescós, G., Sala, R., Mosquera, M., and Rodríguez, X.P. (1995) Lower Pleistocene hominids and artifacts from Atapuerca-TD6 (Spain). *Science* **269**, 830–832.

Carbonell, E., Bermudez de Castro, J.M., Arsuaga, J.L., Allue, E., Bastir, M., Benito, A., Caceres, I., Canals, T., Diez, J.C., van der Made, J. *et al.* (2005) An Early Pleistocene hominin mandible from Atapuerca-TD6, Spain. *Proceedings of the National Academy of Sciences USA* **102**, 5674–5678.

Carroll, S.B. (2003) Genetics and the making of *Homo sapiens. Nature* **422**, 849–857.

Cavalli-Sforza, L.L and Feldman, M.W. (2003) The application of molecular genetic approaches to the study of human evolution. *Nature Genetics* **33** (supplement), 266–275.

Cavalli-Sforza, L.L , Menozzi, P., and Piazza, A. (1994) *The History and Geography of Human Genes.* Princeton, NJ: Princeton University Press.

Cela-Conde, C.J. (1988) ¿Está bien definida la especie? La especie biológica como conjunto borroso. *Anthropos* **81–82**, 104–110.

Cela-Conde, C.J. (1989) Los homínidos del plio-pleistoceno: ¿un nuevo árbol evolutivo? *Mundo científico* **91**, 504–509.

Cela-Conde, C.J. (1998) The problem of hominoid systematics, and some suggestions for solving it. *South African Journal of Sciences* **94**, 255–262.

Cela-Conde, C.J. (2001) Hominid taxon and Hominoidea systematics. In *Abstracts of Contributions to the Dual Congress 1998* (A.A. Raath, H. Soodyall, D. Barkhan, K. L. Kuykendall, and P.V. Tobias, eds), pp. 14. Johannesburg: International Association for the Study of Human Paleontology.

Cela-Conde, C.J. and Altaba, C.R. (2002) Multiplying genera versus moving species: a new taxonomic proposal for the family Hominidae. *South African Journal of Science* **98**, 229–232.

Cela-Conde, C.J. and Ayala, F.J. (2003) Genera of the human lineage. *Proceedings of the National Academy of Sciences USA* **100**, 7684–7689.

Cerling, T.E., Brown, F.H., Cerling, B.W., Curtis, G.H., and Drake, R.E. (1979) Preliminary correlations between the Koobi Fora and Shungura Formations, East Africa. *Nature* **279**, 118–121.

Chaimanee, Y., Jolly, D., Benammi, M., Tafforeau, P., Duzer, D., Moussa, I., and Jaeger, J.J. (2003) A Middle Miocene hominoid from Thailand and orangutan origins. *Nature* **422**, 61–65.

Chaline, J., Dutrillaux, B., Couturier, J., Durand, A., and Marchand, D. (1991) Un modèle chomosomique et paléobiogéographique d'évolution des primates supérieurs. *Géobios* **24**, 105–110.

Chaline, J., Durand, A., Marchand, D., Dambricourt Malassé, A., and Deshayes, M.J. (1996) Chromosomes and the origins of Apes and Australopithecines. *Human Evolution* **11**, 43–60.

Chamberlain, A.T. (1987) *A Taxonomic Review and Phylogenetic Analysis of Homo habilis.* PhD Thesis, University of Liverpool.

Chang, K. (2000) A breed apart. DNA tests: humans not descended from Neanderthals. ABCNews–Science 29 March 2000. http://cogweb.ucla.edu/Abstracts/Goodwin_00.html.

Chase, P.G. and Dibble, H.L. (1987) Middle Paleolithic symbolism: a review of current evidence and interpretations. *Journal of Anthropological Archaeology* **6**, 263–296.

Chen, S. and Wei, Q. (2004) Lithic technological variability of the Middle Pleistocene in the Eastern Nihewan Basin, Northern China. *Asian Perspectives* **43**, 281–301.

Chen, T., Yang, Q., and Wu, E. (1993) Electro spin resonance dating of teeth enamel smaples from Jingniushan paleoanthropological site. *Acta Anthropologica Sinica* **12**, 346.

Cheney, D.L. and Seyfarth, R.M. (1992) Précis of How monkeys see the world. *Behavioral and Brain Sciences* **15**, 135–182.

Cheney, D., Seyfarth, R., and Smuts, B. (1986) Social relationships and social cognition in nonhuman primates. *Science* **234**, 1361–1366.

Chiarelli, B. (1962) Comparative morphometric analysis of primate chromosomes of the anthropoid apes and man. *Caryologia* **15**, 99–121.

Chimpanzee Sequencing and Analysis Consortium (2005) Initial sequence of the chimpanzee genome and comparison with the human genome. *Nature* **437**, 69–87.

Chomsky, N. (1966) *Cartesian Linguistics.* New York, NY: Harper and Row.

Chomsky, N. (1980) *Rules and Representations.* Oxford: Blackwell Publishing.

Chomsky, N. (1989) *El lenguaje y los problemas del conocimiento.* Madrid: Visor.

Ciochon, R.L. (1983) Hominoid cladistic and the ancestry of modern apes and humans. In *New Interpretations of Ape and Human Ancestry* (R.L. Ciochon and R.S. Corruccini, eds), pp. 783–837. New York, NY: Plenum Press.

Cipollari, M., Galderisi, U., and di Bernardo, G. (2005) Ancient DNA as a multidisciplinary experience. *Journal of Cellular Physiology* **202**, 315–322.

Clark, G.A. (1992) Continuity or replacement? Putting modern human origins in an evolutionary context. In *The Middle Paleolithic: Adaptation, Behavior, and Variability* (H. Dibble and P. Mellars, eds), pp. 183–205. Pennsylvania: University of Pennsylvania Museum.

Clark, G.A. (1997a) Through a glass darkly. In *Conceptual Issues in Modern Human Origins*, vol. 30 (G.A Clark and C.M Willermet, eds), pp. 60–76. New York, NY: Aldine de Gruyter.

Clark, G.A. (1997b) The Middle-Upper Paleolithic transition in Europe: an American perspective. *Norwegian Archaeological Review* **30**, 25–53.

Clark, G.A. and Lindly, J.M. (1989) The case for continuity: observations on the biocultural transition in Europe and Western Asia. In *The Human Revolution: Behavioural and Biological Perspectives on the Origins of Modern Humans* (P. Mellars and C. Stringer, eds), pp. 626–676. Princeton, NJ: Princeton University Press.

Clark, J.D. (1995) Introduction to research on the Chiwondo Beds, northern Malawi. *Journal of Human Evolution* **28**, 3–5.

Clark, J.D. and Kurashina, H. (1979) Hominid occupation of the east-central highlands of Ethiopia in the Plio-Pleistocene. *Nature* **382**, 33–39.

Clark, J.D., Asfaw, B., Assefa, G., Harris, J.W.K., Kurashina, H., Walter, R.C., White, T.D., and Williams, M.A.J. (1984) Paleoanthropological discoveries in the Middle Awash Valley, Ethiopia. *Nature* **307**, 423–428.

Clark, J.D., de Heinzelin, J., Schick, K.D., Hart, W.K., White, T.D., WoldeGabriel, G., Walter, R.C., Suwa, G., Asfaw, B., Vrba, E., and Selassie, Y.H. (1994) African *Homo erectus*: old radiometric ages and young Oldowan assemblages in the Middle Awash Valley, Ethiopia. *Science* **264**, 1907–1920.

Clarke, R.J. (1985) *Australopithecus* and Early *Homo* in Southern Africa. In *Ancestors: The Hard Evidence* (E. Delson, ed.), pp. 171–177. New York, NY: Alan R. Liss.

Clarke, R.J. (1988) A new *Australopithecus* cranium from Sterkfontein and its bearing on the ancestry of *Paranthropus*. In *Evolutionary History of the Robust*

Australopithecines (F.E Grine, ed.), pp. 285–292. New York, NY: Aldine de Gruyter.

Clarke, R.J. (1994) On some new interpretations of Sterkfontein stratigraphy. *South African Journal of Science* 90, 211–214.

Clarke, R.J. (1998) First ever discovery of a well-preserved skull and associated skeleton of *Australopithecus*. *South African Journal of Science* 94, 460–463.

Clarke, R.J. (1999) Discovery of complete arm and hand of the 3.3 million-year-old Australopithecus skeleton from Sterkfontein. *South African Journal of Science* 95, 477–480.

Clarke, R.J. (2000) A corrected reconstruction and interpretation of the *Homo* erectus calvaria from Ceprano, Italy. *Journal of Human Evolution* 39, 433–442.

Clarke, R.J. and Tobias, P.V. (1995) Sterkfontein Member 2 Foot Bones of the Oldest South African Hominid. *Science* **269, 521–524.**

Clarke, R.J., Howell, F.C., and Brain, C.K. (1970) More evidence of an advanced hominid at Swartranks. *Nature* 225, 1219–1222.

Close, A. (1984) Current research and recent radiocarbon dates from North Africa, II. *Journal of African History* 25, 1–24.

Cohen, M.N. and Armelagos, G.J. (1984) *Paleopathology at the Origins of Agriculture.* Orlando: Academic Press.

Collard, M. and Wood, B. (1999) Grades among the African Early Hominids. In *African Biogeography. Climate Change & Human Evolution* (T.G.Bromage and F. Schrenk, eds), pp. 316–327. New York, NY: Oxford University Press.

Collard, M. and Aiello, L.C. (2000) From forelimbs to two legs. *Nature* 404, 339–340.

Condemi, S. (1988) A review and analysis of the Riss-Würm Saccopastore skulls, can they provide evidence in regard to the origin of Near Eastern Neandertahls? In *L'Homme de Neandertal*, vol. 3. *L'anatomie* (E. Trinkaus, ed.), pp. 39–48. Liège: Etudes et Recherches Archéologiques de l'Université de Liège.

Conroy, G.C. (1997) *Reconstructing Human Origins: a Modern Synthesis.* New York, NY: W.W. Norton and Company.

Conroy, G.C., Jolly, C.J., Cramer, D., and Kalb, J.E. (1978) Newly discovered fossil hominid skull from the Afar Depression, Ethiopia. *Nature* 275, 339–406.

Conroy, G.C., Pickford, M., Senut, B., Van Couvering, J., and Mein, P. (1992) *Otavipithecus namibiensis*, first Miocene hominoid from southern Africa. *Nature* 356, 144–148.

Conroy, G.C., Weber, G.W., Seidler, H., Tobias, P.V., Kane, A., and Brundsen, B. (1998) Endocranial capacity

in an early hominid cranium form Sterkfontein, South Africa. *Science* **280**, 1730–1731.

Cook, J., Stringer, C., Currant, A., Schwarcz, H., and Wintle, A. (1982) A review of the chronology of the European middle Pleistocene hominid record. *Yearbook of Physical Anthropology* 25, 19–65.

Cooke, B. (1976) Suidae from Plio-Pleistocene Strata of the Rudolph Basin. In *Earliest Man and Environement in the Lake Rudolf Basin* (Y. Coppens, F.C Howell, G.L. Isaac, and R.E.F. Leakey, eds), pp. 251–263. Chicago, IL: Chicago University Press.

Cooke, H.B.S. (1964) Pleistocene mammal faunas of Africa, with particular reference to South Africa. In *African Ecology and Human Evolution* (F.C. Howell and F. Bourlière, eds), pp. 65–116. London: Methuen and Co.

Cooper, A., Rambaut, A., Macaulay, V., Willerslev, E., Hansen, A.L., and Stringer, C. (2001) Human origins and ancient human DNA. *Science* 292, 1655–1656.

Coppens, Y. (1972) Tentative de zonation du Pliocene et du Pléistocène d'Afrique par les grands Mammifères. *Comptes rendus des séances de l'Academie des Sciences 274*, Série D, 181–184.

Coppens, Y. (1975) Evolution des Mammifères, de leurs fréquences et de leurs associations, au cours du Plio-Pléistocène dans la basse vallée de l'Omo en Ethiopie. *Comptes rendus des séances de l'Academie des Sciences 281*, Série D, 1571–1574.

Coppens, Y. (1978a) Evolution of the hominids and of their environment during the Plio-Pleistocene in the Lower Omo Valley, Ethiopia. In *Geological Background to Fossil Man* (W.W. Bishop, ed.), pp. 499–506. Edinburgh: Scottish Academic Press.

Coppens, Y. (1978b) Les Hominidés du Pliocene et du Pléistocène d'Ethiopie; Chronologie, systématique, environnement. In *Les origines humaines et les époques de l'intelligence* (F. Singer-Polignac, ed.), pp. 79–106. Paris: Masson.

Coppens, Y. (1980) The differences between *Australopithecus* and *Homo*: preliminary conclusions from the Omo Research Expedition's studies. In *Current Argument on Early Man* (L.K. Koniggson, ed.), pp. 207–225. Stockholm: Pergamon Press.

Coppens, Y. (1983a) Les plus anciens fossiles d'hominidés. *Pontifical Academy of Science. Scripta Varia* 50, 1–9.

Coppens, Y. (1983b) Systématique, phylogénie, environnement et culture des australopithèques, hypothèses et synthèse. *Bulletin et Memoires de la Societé d'Anthropologie Paris* 10, 273–284.

Coppens, Y. (1991) L'évolution des hominidés, de leur locomotion et de leurs environnements. In *Origine(s) de* la bipédie chez les hominidés (Y. Coppens and B. Senut, eds), pp. 295–301. Paris: Editions du CNRS.

Coppens, Y. (1994) East Side Story: the origin of mankind. *Scientific American* **270**, 62–69.

Coppens, Y. and Howell, F.C. (1976) Mammalian faunas of the Omo group: distributional and biostratigraphical aspects. In *Earliest Man and Environments in the Lake Rudolf Basin* (Y. Coppens, F.C Howell, G.L. Isaac, and R.E F. Leakey, eds), pp. 177–192. Chicago, IL: Chicago University Press.

Cracraft, J. (1974) Phylogenetic models and classification. *Systematic Zoology* **23**, 71–90.

Crusafont, M. and Hürzeler, J. (1961) Les Pongides fossiles d'Espagne. *Comptes rendus de l'Académie des Sciences Paris* **252**, 582–584.

Culotta, E. (1999) Neanderthals were cannibals, bone show. *Science* **286**, 18–19.

Curnoe, D. and Thorne, A. (2003) Number of ancestral human species: a molecular perspective. *Homo* **53**, 201–224.

Curnoe, D. and Tobias, P.V. (2006) Description, new reconstruction, comparative anatomy, and classification of the Sterkfontein Stw 53 cranium, with discussions about the taxonomy of other southern African early Homo remains. *Journal of Human Evolution* **50**(1), 36–77.

Curtis, G.H., Drake, R.E., Cerling, T., and Hampel, J. (1975) Age of KBS Tuff in Koobi Fora Formation, East Rudolf, Kenya. *Nature* **258**, 395–398.

Dainton, M. and Macho, G.A. (1999) Did knuckle walking evolve twice? *Journal of Human Evolution* **36**, 171–194.

Dalton, R. (2005) Looking for the ancestors. *Nature* **434**, 432–434.

Dalton, R. (2006) Palaeoanthropology: The history man. *Nature* **443**, 268–269.

Dart, R. (1925) *Australopithecus africanus*: the man-ape of South Africa. *Nature* **115**, 195–199.

Dart, R.A. (1948) The Makapansgat proto-human *Australopithecus prometheus*. *American Journal of Physical Anthropology* **6**, 259–284.

Dart, R. (1949a) Innominate fragments of *Australopithecus prometeus*. *American Journal of Physical Anthropology* **7**, 301–334.

Dart, R. (1949b) The predatory implemental technique of *Australopithecus*. *American Journal of Physical Anthropology* **7**, 1–16.

Dart, R. (1953) The predatory transition from ape to man. *International Anthropological and Linguistic Review* **1**, 201–218.

Dart, R.A. (1957) *The Osteodontokeratic culture of Australopithecus prometheus*. Pretoria: Transvaal Museum Memoir 10.

Dart, R.A. (1962) The Makapansgat pink breccia australopithecine skull. *American Journal of Physical Anthropology* **20**, 119–126.

Dart, R.A. and Del Grande, N. (1931) The ancient iron smelting cave at Mumbwa. *Transactions of the Royal Society of South Africa* **19**, 379–427.

Darwin, C.R. (1871) *The Descent of Man, and Selection in Relation to Sex*. London: John Murray.

Davidson, I. (1990) Bilzingsleben and early marking. *Rock Art Research* **7**, 52–56.

Davidson, I. and Noble, W. (1989) The archaeology of perception. *Current Anthropology* **30**, 125–155.

Davis, P.R. (1964) Hominid fossils from Bed I, Olduvai Gorge, Tanganyika: a Tibia and a Fibula. *Nature* **201**, 967–970.

Day, M.H. (1969) Omo human skeletal remains. *Nature* **223**, 1234–1239.

Day, M.H. (1971a) Postcranial remains of *Homo* erectus from Bed IV, Olduvai Gorge, Tanzania. *Nature* **232**, 383–387.

Day, M.H. (1972) The Omo human skeletal remains. In *The Origin of Homo sapiens* (F. Bordes, ed.), pp. 31–35. Paris: UNESCO.

Day, M.H. (1973) The development of *Homo sapiens*. In *Darwin Centenary Symposium on the Origin of Man* (pp. 87–95). Rome: Accadémia Nazionale dei Lincei.

Day, M.H. (1982) The *Homo* erectus pelvis: punctuation or gradualism? *le Congres International de Paleontologie Humaine, I, Nice, Prétirage* 411–421.

Day, M.H. (1984) The postcranial remains of *Homo* erectus from Africa, Asia and possibly Europe. In *The Early Evolution of Man with Special Emphasis on Southeast Asia and Africa*, vol. 69 (P. Andrews and J.L. Franzen, eds), pp. 113–121. Frankfurt: Courier Forschungsinstitut Senckenberg.

Day, M.H. (1986) *Guide to Fossil Man*, 4th edn. Chicago, IL: University of Chicago Press.

Day, M.H. (1992) Posture and childbirth. In *The Cambridge Encyclopedia of Human Evolution* (S. Jones, R. Martin, and D. Pilbeam, eds), p. 88. Cambridge: Cambridge University Press.

Day, M.H. (1995) Remarkable delay. *Nature* **376**, 111.

Day, M.H. and Napier, J.R. (1964) Hominid fossils from Bed I, Olduvai gorge, Tanganyika: fossil foot bones. *Nature* **201**, 967–970.

Day, M.H. and Wood, B. (1968) Functional affinities of the Olduvai Hominid 8 talus. *Man* **3**, 440–445.

Day, M.H. and Wickens, E.H. (1980) Laetoli Pliocene hominid footprints and bipedalism. *Nature* **286**, 385–387.

Day, M.H. and Stringer, C.B. (1982) A reconsideration of the OmoKibish remains and the *erectus-sapiens* transition. In *L'Homo erectus et la Place de l'Homme de Tautavel parmi les Hominidés Fossiles*, vol. 2 (H. de Lumley, ed.), Vol. Prétirage, pp. 814–846. Nice: Louis-Jean Scientific and Literary Publications.

Deacon, H.J. (1989) Late Pleistocene palaeoecology and archaeology in the Southern Cape, South Africa. In *The Human Revolution: Behavioural and Biological Perspectives on the Origins of Modern Humans* (P. Mellars and C. Stringer, eds), pp. 547–564. Princeton, NJ: Princeton University Press.

Deacon, T. (1997) The symbolic species. New York, NY: W.W. Norton and Company.

Dean, D. and Delson, E. (1995) *Homo* at the gates of Europe. *Nature* **373**, 472–473.

Dean, D., Hublin, J.J., Holloway, R., and Ziegler, R. (1998) On the phylogenetic position of the pre-Neandertal specimen from Reilingen, Germany. *Journal of Human Evolution* **34**, 485–508.

Debénath, A. (1994) L'Atérien du nord de l'Afrique du Sahara. *Sahara* **6**, 21–30.

de Bonis, L., Bouvrain, G., Geraads, D., and Melentis, J. (1974) Première decouverte d'un primate hominoïde dans le Miocène supérieur de Macédoine (Grèce). *Comptes rendus de l'Académie des Sciences Paris (Serie D)* **278**, 3063–3066.

de Bonis, L., Bouvrain, G., Geraads, D., and Koufos, G. (1990) New hominid skull material from the Late Miocene of Macedonia in Northern Greece. *Nature* **345**, 712–714.

Defleur, A., White, T., Valensi, P., Slimak, L., and Crégut-Bonnoure, E. (1999) Neanderthal cannibalism at Moula-Guercy, Ardèche, France. *Science* **286**, 128–131.

de Heinzelin, J., Clark, J.D., White, T., Hart, W., Renne, P., WoldeGabriel, G., Beyene, Y., and Vrba, E. (1999) Environment and behavior of 2.5-million-year-old Bouri Hominids. *Science* **284**, 625–629.

de Heinzelin, J., Clark, J.D., Schick, K.D., and Gilbert, W.H. (eds) (2000) *The Acheulean and the Plio-Pleistocene Deposits of the Middle Awash Valley, Ethiopia*. Tervuren, Belgium: Royal Museum of Central Africa. *Annales–sciences géologiques* 104.

Deinard, A. and Kidd, K. (1999) Evolution of a HOXB6 intergenic region within the great apes and humans. *Journal of Human Evolution* **36**, 687–703.

Deino, A.L. and Potts, R. (1990) Single-crystal 40Ar/39Ar dating of the Olorgesailie Formation, southern Kenya Rift. *Journal of Geophysical Research, B, Solid Earth and Planets* **95**, 8453–8470.

Deino, A. and McBrearty, S. (2002) 40Ar/39Ar chronology for the Kapthurin Formation, Baringo, Kenya. *Journal of Human Evolution* **42**, 185–210.

Deino, A.L., Tauxe, L., Monaghan, M., and Hill, A. (2002) 40Ar/39Ar geochronology and paleomagnetic stratigraphy of the Lukeino and lower Chemeron Formations at Tabarin and Kapcheberek, Tugen Hills, Kenya. *Journal of Human Evolution* **42**, 117–140.

Deloison, Y. (1991) Les Australopithèques marchaient-ils comme nous? In *Origine(s) de la bipédie chez les hominidés* (Y. Coppens and B. Senut, eds), pp. 177–186. Paris: Editions du CNRS.

Deloison, Y. (1996) El pie de los primeros homínidos. *Mundo científico* **164**, 20–22.

Delson, E. (1986) Human phylogeny revised again. *Nature* **322**, 496–497.

Delson, E. (1988) Chronology of South African Australopith site units. In *Evolutionary History of the "Robust" Australopithecines* (F.E Grine, ed.), pp. 317–324. New York, NY: Aldine de Gruyter.

Delson, E. (1997) One skull does not a species make. *Nature* **389**, 445–446.

Delson, E., Eldredge, N., and Tattersall, I. (1977) Reconstruction of hominid phylogeny: a testable framework based on cladistic analysis. *Journal of Human Evolution* **6**, 263–278.

de Lumley, H. (1979) L'homme de Tautavel. *Les Dossiers de l'Archeologie* **36**.

de Lumley, H. and de Lumley, M.-A. (1971) Découverte de restes humains anténéandertaliens datés du début de Riss à la Caune de l'Arago à Tautavel (Pyrénées-Orientales). *Comptes rendus de l'Académie des Sciences Paris* **272**, 1729–1742.

de Lumley, H. and de Lumley, M.A. (1973) Pre-neanderthal human remains from Arago cave in Southeastern France. *Yearbook of Physical Anthropology* **17**, 162–168.

Dennell, R.W. (1983) *European Economic Prehistory: a New Approach*. London: Academic Press.

Dennell, R. and Roebroeks, W. (1996) The earliest colonization of Europe: the short chronology revisited. *Antiquity* **70**, 535–542.

De Queiroz, K. and Donoghue, M.J. (1988) Phylogenetic systematics and the species problem. *Cladistics* **4**, 317–338.

d'Errico, F. (1991) Carnivore traces or mousterian skiffle? *Rock Art Research* **8**, 61–63.

d'Errico, F. and Villa, P. (1997) Holes and grooves: the contribution of microscopy and taphonomy to the problem of art origins. *Journal of Human Evolution* **33**, 1–31.

d'Errico, F., Zilhão, J., Julien, M., Baffier, M., and Pelegrin, J. (1998) Neanderthal acculturation in western Europe? A critical review of the evidence and its interpretation. *Current Anthropology* **39** (supplement), 1–44.

de Villiers, H. (1973) Human skeletal remains from Border Cave, Ingwavumu District, KwaZulu, South Africa. *Annals of the Transvaal Museum* **28**, 229–256.

de Vos, J. and Sondaar, P. (1994) Dating hominid sites in Indonesia. *Science* **266**, 1726–1727.

de Waal, F.B.M. (1996) *Good Natured: the Origins of Right and Wrong in Humans and Other Animals*. Cambridge, MA: Harvard University Press.

de Waal, F.B.M. (1999) Cultural primatology comes of age. *Nature* **399**, 635–636.

Diamond, J. (2004) The astonishing micropygmies. *Science* **306**, 2047–2048.

Dobzhansky, Th. (1935) A critique of the species concept in biology. *Philosophy of Science* **2**, 344–355.

Dobzhansky, Th. (1937) *Genetics and the Origin of Species*. New York: Columbia University Press.

Dominguez-Rodrigo, M., Serrallonga, J., Juan-Tresserras, J., Alcala, L., and Luque, L. (2001) Woodworking activities by early humans: a plant residue analysis on Acheulian stone tools from Peninj (Tanzania). *Journal of Human Evolution* **40**, 289–299.

Dorit, R.L., Akashi, H., and Gilbert, W. (1995) Absence of polymorphism at the ZFY locus on the human Y chromosome. *Science* **268**, 1183–1185.

Doronichev, V. and Golovanova, L. (2003) Bifacial tools in the lower and middle Paleolithic of the Caucasus and their contexts. In *Multiple Approaches to the Study of Bifacial Technologies* (M. Soressi and H.L.Dibble, eds), pp. 77–107. Philadelphia, PA: Museum of Archaeolgy and Anthropology.

Duarte, C., Maurício, J., Pettitt, P.B., Souto, P., Trinkaus, E., van der Plicht, H., and Zilhão, J. (1999) The early Upper Paleolithic human skeleton from the Abrigo do Lagar Velho (Portugal) and modern human emergence in Iberia. *Proceedings of the National Academy of Sciences USA* **96**, 7604–7609.

Dubois, E. (1894) *Pithecanthropus erectus. Eine menschanähnliche Übergangsform aus Java*. Batavia: Landsdruckerei.

Duckworth, W.J.H. (1925) The fossil anthropoid ape from Taungs. *Nature* **115**, 234–235.

Eccles, J.C. (1977) Evolution of the brain in relation to the development of the self-conscious mind. In Evolution and Lateralization of the Brain, vol. 299 (S.J.Dimond and D.A.Blizard, eds), pp. 161–179. New York, NY: New York Academy of Sciences.

Eckhardt, R.B. (1977) Hominid origins: the Lothagam mandible. *Current Anthropology* **18**, 356.

Eckhardt, R.B. (2000) *Human Paleobiology*. Cambridge: Cambridge University Press.

Eglinton, G. and Logan, G.A. (1991) Molecular preservation. *Philosophical Transactions of the Royal Society of London Series B* **333**, 315–328.

Eibl-Eibesfeldt, I. (1967) *Grundriss der vergleichenden Verhaltensforschung*. Munich: R. Piper.

Elango, N., Thomas, J.W., Program, N.C.S., and Yi, S.V. (2006) Variable molecular clocks in hominoids. *Proceedings of the National Academy of Sciences USA* **103**, 1370–1375.

Elliott Smith, G. (1925) The fossil anthropoid ape from Taungs. *Nature* **115**, 234.

Enard, W., Przeworski, M., Fisher, S.E., Lai, C.SL., Wiebe, V., Kitano, T, Monaco, A.P., and Pääbo, S. (2002) Molecular evolution of *FOXP2*, a gene involved in speech and language. *Nature* **418**, 869–872.

Ennouchi, E. (1963) Les Néanderthaliens du Jebel Irhoud (Maroc). *Comptes rendus de l'Académie des Sciences Paris* **256**, 2459–2460.

Enquist, M. and Arak, A. (1994) Symmetry, beauty and evolution. *Nature* **372**, 169–172.

Evernden, J.F. and Curtiss, G.H. (1965) Potassium argon dating of Late Cenozoic rocks in East Africa and Italy. *Current Anthropology* **6**, 348–385.

Excoffier, L. (2002) Human demographic history: refining the recent African origin model. *Current Opinion in Genetics and Development* **12**, 675–682.

Falk, D. (1975) Comparative anatomy of larynx in man and the chimpanzee: implications for language in Neanderthal. *American Journal of Physical Anthropology* **43**, 123–132.

Falk, D. (1980) Language, handedness, and primate brains: did the australopithecines sign? *American Anthropologist* **82**, 72–73.

Falk, D. (1983) Cerebral cortices of East Africa early hominids. *Science* **221**, 1072–1074.

Falk, D. (1986) Endocranial casts and their significance for primate brain evolution. In *Comparative Primate Biology* (D.R.Swindler and J. Erwin, eds), pp. 477–490. New York, NY: Alan R. Liss.

Falk, D. (1987) Hominid paleoneurology. *Annual Review of Anthropology* **16**, 13–30.

Falk, D. (1998) Hominid brain evolution: looks can be deceiving. *Science* **280**, 1714.

Falk, D., Redmond, J.CJ., Guyer, J., Conroy, C., Recheis, W., Weber, G.W., and Seidler, H. (2000) Early hominid brain evolution: a new look at old endocasts. *Journal of Human Evolution* **38**, 695–717.

Falk, D., Hildebolt, C., Smith, K., Morwood, M.J., Sutikna, T., Brown, P., Jatmiko, Saptomo, E.W., Brunsden, B., and Prior, F. (2005) The brain of LB1, *Homo floresiensis*. *Science* **308**, 242–245.

Falk, D., Hildebolt, C., Smith, K., Morwood, M.J., Sutikna, T., Jatmiko, Saptomo, E.W., Imhof, H., Seidler, H., and Prior, F. (2007). Brain shape in human microcephalics and Homo floresiensis. *Proceedings of the National Academy of Sciences*, **104** 2513–2518.

Farrand, W.R. (1979) Chronology and paleoenvironment of Levantine prehistoric sites as seen from sediment studies. *Journal of Archaeological Science* **6**, 369–392.

Fay, J.C. and Wu, C.I. (1999) A human population bottleneck can account for the discordance between patterns of mitochondrial versus nuclear DNA variation. *Molecular Biology and Evolution* **16**, 1003–1005.

Feibel, C.S., Brown, F.H., and McDougall, I. (1989) Stratigraphic context of fossil hominids from the Omo group deposits: Northern Turkana Basin, Kenya and Ethiopia. *American Journal of Physical Anthropology* **78**, 595–622.

Ferembach, D. (1976) Les reste humaines de la Grotte de Dar-es-Soltane 2 (Maroc) Campagne 1975. *Bulletin et Memoires de la Societé d'Anthropologie Paris* **3**, 183–193.

Fernández-Jalvo, Y., Carlos Díez, J., Bermúdez de Castro, J.M., Carbonell, E., and Arsuaga, J.L. (1996) Evidence of early cannibalism. *Science* **271**, 269–270.

Finlayson, C., Giles Pacheco, F., Rodriquez-Vidal, J., Fa, D.A., Maria Gutierrez Lopez, J., Santiago Perez, A., Finlayson, G., Allue, E., Baena Preysler, J., Caceres, I., *et al.* (2006). Late survival of Neanderthals at the southernmost extreme of Europe. *Nature*, **443**, 850–853.

Fischman, J. (1994) Putting our oldest ancestors in their proper place. *Science* **265**, 2011–2012.

Fitch, F.J. and Miller, J.A. (1970) Radioisotopic age determinations of Lake Rudolf artifact site. *Nature* **231**, 241–245.

Fitch, F.J., Hooker, P.J., and Miller, J.A. (1976) 40Ar/39Ar dating of the KBS Tuff in Koobi Fora Formation, East Rudolf, Kenya. *Nature* **263**, 740–744.

Fitch, W.M. and Margoliash, E. (1967) The construction of phylogenetic trees—a generally applicable method utilizing estimates of the mutation distance obtained from cytochrome c sequences. *Science* **155**, 279–84.

Fitch, W.T. and Hauser, M.D. (2004) Computational constraints on syntactic processing in a nonhuman primate. *Science* **303**, 377–380.

Fleagle, J.G. (1988) *Primate Adaptation and* Evolution. San Diego, CA: Academic Press.

Fleagle, J.G. (1992) Primate locomotion and posture. In *The Cambridge Encyclopedia of Human Evolution* (S. Jones, R. Martin, and D. Pilbeam, eds), pp. 75–79. Cambridge: Cambridge University Press.

Foley, R.A. and Lahr, M.M. (1997) Mode 3 technologies and the evolution of modern humans. *Cambridge Archaeological Journal* **7**, 3–36.

Formicola, V. and Giannecchini, M. (1999) Evolutionary trends of stature in Upper Paleolithic and Mesolithic Europe. *Journal of Human Evolution* **36**, 319–333.

Forrer, R. (1908) *Urgescichte des Europäers von der Menschwerdung bis zum Anbruch der Geschichte*. Stuttgart: Spemann.

Gabunia, L. and Vekua, A. (1995) A Plio-Pleistocene hominid from Dmanisi, East Georgia, Caucasus. *Nature* **373**, 509–512.

Gabunia, L., Vekua, A., Lordkipanidze, D., Swisher, III, C.C., Ferring, R., Justus, A., Nioradze, M., Tvalchrelidze, M., Anto, S.C., Bosinski, G. *et al.* (2000) Earliest Pleistocene hominid cranial remains from Dmanisi, Republic of Georgia: taxonomy, geological setting, and age. *Science* **288**, 1019–1025.

Gabunia, L., de Lumley, M.A., Vekua, A., Lordkipanidze, D., and de Lumley, H. (2002) Découverte d'un nouvel hominidé à Dmanissi (Transcaucasie, Géorgie). *Comptes rendus Palévol* **1**, 243–253.

Galik, K., Senut, B., Pickford, M., Gommery, D., Treil, J., Kuperavage, A.J., and Eckhardt, R.B. (2004) External and Internal Morphology of the BAR 1002'00 Orrorin tugenensis Femur. *Science* **305**, 1450–1453.

Gamble, C.S. (1989) Comment on R. Gargett 'Grave shortcomings: the evidence for Neanderthal burial'. *Current Anthropology* **30**, 181–182.

Gamble, C. (1993) *Timewalkers*. Stroud: Alan Sutton.

Gamble, C. (1994) Time for Boxgrove man. *Nature* **369**, 275–276.

Ganopolski, A. and Rahmstorf, S. (2001) Rapid changes of glacial climate simulated in a coupled climate model. *Nature* **409**, 153–158.

Gantt, D.G. (1979) Comparative enamel histology of primate teeth. *Journal of Dental Research* **58**, 1002–1003.

Gardner, R.A. and Gardner, B.T. (1969) Teaching sign language to a chimpanzee. *Science* **165**, 664–672.

Gargett, R. (1989) Grave shortcoming: the evidence for Neandertal burial. *Current Anthropology* **30**, 157–190.

Gargett, R.H. (1999) Middle Palaeolithic burial is not a dead issue: the view from Qafzeh, Saint-Césaire, Kebara, Amud, and Dederiyeh. *Journal of Human Evolution* **37**, 27–90.

Garralda, M.D. and Vandermeersch, B. (2000) Les néandertaliens de la grotte de Combe-Grenal. *Paleo* **12**, 213–259.

Garrigan, D. and Hammer, M.F. (2006) Reconstructing human origins in the genomic era. *Nature Reviews Genetics* **7**, 669–680.

Garrigan, D., Mobasher, Z., Severson, T., Wilder, J.A., and Hammer, M.F. (2005) Evidence for archaic Asian ancestry on the human X chromosome. *Molecular Biology and Evolution* **22**, 189–192.

Garrod, D.A.E. (1962) The Middle Palaeolithic of the Near East and the problem of Mount Carmel man. *The Journal of the Royal Anthropological Institute of Great Britain and Ireland* **92**, 232–259.

Garrod, D.A.E. and Bate, D.M.A. (1937) *The Stone Age of Mount Carmel*, vol. I. *Excavations at the Wady el-Mughara*. Oxford: Clarendon Press.

Gauss, G.F. (1934) *The Struggle for Existence*. Baltimore, MD: Williams and Wilkins.

Gebo, D.L., MacLatchy, L., Kityo, R., Deino, A., Kingston, J., and Pilbeam, D. (1997) A hominoid genus from the Early Miocene of Uganda. *Science* **276**, 401–404.

Gee, H. (1995) Uprooting the human family tree. *Nature* **373**, 15.

Gee, H. (1996) Box of bones 'clinches' identity of Piltdown paleontology hoaxer. *Nature* **381**, 261–262.

Ghiselin, M.T. (1987) Species concepts, individuality, and objectivity. *Biology and Philosophy* **2**, 127–143.

Gibbar, P.L. (2003) Definition of the Middle–Upper Pleistocene boundary. *Global and Planetary Change* **36**, 201–208.

Gibbons, A. (1995) Old dates for modern behavior. *Science* **268**, 495–496.

Gibbons, A. (1996) *Homo erectus* in Java: a 250,000-year anachronism. *Science* **274**, 1870–1874.

Gibbons, A. (1998) Old, old skull has a new look. *Science* **280**, 1525.

Gibbons, A. (2000) Chinese stone tools reveal high-tech *Homo* erectus. *Science* **287**, 1566.

Gibbons, A. (2004) Oldest human femur wades into controversy. *Science* **305**, 1885a.

Gibbons, A. (2006) Lucy's 'child' offers rare glimpse of an ancient toddler. *Science* **313**, 1716.

Gibbons, A. and Culotta, E. (1997) Miocene primates go ape. *Science* **276**, 355–356.

Gibbs, S., Collard, M., and Wood, B. (2000) Soft-tissue characters in higher primate phylogenetics. *Proceedings of the National Academy of Sciences USA* **97**, 1130–1132.

Gibert, J., Arribas, A., Martínez, B., Albaladejo, S., Gaete, R., Gibert, L., Oms, O., Peña, C., and Torrico, R. (1994) Biostratigraphie et magnétostratigraphie des gisements à présence humaine et action anthropique du Pleistocène inférieur de la région d'Orce (Granada, Espagne). *Comptes rendus de l'Académie des Sciences Paris* **318** (series II), 1277–1282.

Gibert, J., Gibert, L., Iglesias, A., and Maestro, E. (1998) Two 'Oldowan' assemblages in the Plio-Pleistocene deposits of the Orce region, southeast Spain. *Antiquity* **72**, 17–25.

Gilbert, W.H., White, T.D., and Asfaw, B. (2003) *Homo erectus, Homo ergaster, Homo "cepranensis"*, and the Daka cranium. *Journal of Human Evolution* **45**, 255–259.

Gillespie, R. (2002) Dating the First Australians. *Radiocarbon* **44**, 455–472.

Gillespie, R. and Roberts, R.G. (2000) On the reliability of age estimates for human remains at Lake Mungo. *Journal of Human Evolution* **38**, 727–730.

Gingerich, P.D. (1976) Cranial anatomy and evolution of early Tertiary Plesiadapidae (Mammalia, Primates). *University of Michigan Papers on Paleontology* no. 15, 1–116.

Gingerich, P.D. (1981). Cranial Morphology and Adaptations in Eocene Adapidae. 1. Sexual Dimorphism in Adapis magnus and Adapis parisiensis. American Journal of Physical Anthropology, 56, 217-234.

Gingerich, P.D., & Gunnell, G.F. (2005). Brain of Plesiadapis cookei (Mammalia, propirimates): Surface morphology and encephalization compared to those of the Primates and Dermoptera. The University of Michigan. Contributions from the Museum of Paleontology, 31, 185-1195.

Giorgini, M.S. (1971) *Soleb II. Les Nécropoles*. Florence: Sansoni.

Gleadow, A.J.W. (1980) Fission track age of the KBS Tuff and associated hominin remains in northern Kenya. *Nature* **284**, 225–230.

Goldstein, D.B., Ruiz-Linares, A., Cavalli-Sforza, L.L , and Feldman, M.W. (1995) Microsatellite loci, genetic distances, and human evolution. *Proceedings of the National Academy of Sciences USA* **92**, 6723–6727.

Goodall, J.M. (1964) Tool-using and aimed throwing in a community of free-living chimpanzees. Nature **201**, 1264–1266.

Goodman, M. (1962) Evolution of the immunologic species specificity of human serum proteins. *Human Biology* **34**, 104–150.

Goodman, M. (1963) Man's place in the phylogeny of the primates as reflected in serum proteins. In *Classification and Human Evolution* (S.L Washburn, ed.), pp. 204–234. Chicago, IL: Aldine.

Goodman, M. (1975) Protein sequence and immunological specifity. In *Phylogeny of the Primates* (W.P.Luckett and F.S.Szalay, eds), pp. 219–248. New York, NY: Plenum Press.

Goodman, M. (1976) Towards a genealogical description of the primates. In *Molecular Anthropology* (M. Goodman and R.E.Tashian, eds), pp. 321–353. New York, NY: Plenum Press.

Goodman, M., Poulik, E., and Poulik, M.D. (1960) Variations in the serum specifities of higher primates detected by two-dimensional starch-gel electrophoresis. *Nature* **188**, 78–79.

Goodman, M., Bailey, W.J., Hayasaka, K., Stanhope, M.J., Slightom, J., and Czelusniak, J. (1994) Molecular evidence on primate phylogeny from DNA sequences. *American Journal of Physical Anthropology* **94**, 3–24.

Goodman, M., Porter, C.A., Czelusniak, J., Page, S.L., Schneider, H., Shoshani, J., Gunnell, G., and Groves, C. P. (1998) Toward a phylogenetic classification of primates based on DNA evidence complemented by fossil evidence. *Molecular Phylogenetics and Evolution* **9**, 585–598.

Goren-Inbar, N. (ed.) (1990) *Quneitra: A Mousterian Site on the Golan Heights.* Jerusalem: Monographs of the Institute of Archaeology, Hebrew University 31.

Goren-Inbar, N., Feibel, C., Cerosub, K.L., Melamed, Y., Kislev, M.E., Techernov, E., and Saragusti, I. (2000) Pleistocene milestones on the Out-of-Africa corridor at Gesher Benot Ya'aqov, Israel. *Science* **289**, 944–947.

Goren-Inbar, N., Alperson, N., Kislev, M.E., Simchoni, O., Melamed, Y., Ben-Nun, A., and Werker, E. (2004) Evidence of hominin control of fire at Gesher Benot Ya'aqov, Israel. *Science* **304**, 725–727.

Green, H.S., Stringer, C.B., Collcutt, S.N., Currant, A.P., Huxtable, J., Schwarcz, H.P., Debenham, N., Embleton, C., Bull, P., Molleson, T.I., and Bevins, R.E. (1981) Pontnewydd Cave in Wales-a new Middle Pleistocene hominid site. *Nature* **294**, 707–713.

Green, R.E., Krause, J., Ptak, S.E., Briggs, A.W., Ronan, M.T., Simons, J.F., Du, L., Egholm, M., Rothberg, J.M., Paunovic, M., and Paabo, S. (2006). Analysis of one million base pairs of Neanderthal DNA. *Nature*, **444,** 330–336.

Greenfield, L.O. (1983) Toward the resolution of discrepancies between phenetic and paleontological data bearing on the question of human origins. In *New Interpretations of Ape and Human Ancestry* (R.L.Ciochon and R.S.Corruccini, eds), pp. 659–703. New York, NY: Plenum Press.

Griffiths, R.C. (1980) Lines of descent in the diffusion approximation of neutral Wright-Fisher models. *Theoretical Population Biology* **17**, 37–50.

Grine, F.E. (1981) Trophic differences between gracile and robust *Australopithecus*: a scanning electron microscope analysis of occlusal events. *South Africa Journal of Science* **77**, 203–230.

Grine, F.E. (1985) Australopithecine evolution: the decidous dental evidence. In *Ancestors: the Hard Evidence* (E. Delson, ed.), pp. 153–167. New York, NY: Alan R. Liss.

Grine, F.E. (1987) The diet of South African australopithecines based on a study of dental microwear. *L'Anthropologie* **91**, 467–482.

Grine, F.E. (1988) Evolutionary history of the 'robust' australopithecines: a summary and historical perspective. In *Evolutionary History of the "Robust" Australopithecines* (F.E Grine, ed.), pp. 509–520. New York, NY: Aldine de Gruyter.

Grine, F.E. (2000) Middle Stone Age human fossils from Die Kelders Cave 1, Western Cape Province, South Africa. *Journal of Human Evolution* **38**, 129–145.

Grine, F.E. and Klein, R.G. (1985) Pleistocene and Holocene human remains from Equus cave, South Africa. *Anthropology* **8**, 55–98.

Grine, F.E. and Martin, L.B. (1988) Enamel thickness and development in *Australopithecus* and *Paranthropus*. In *Evolutionary History of the "Robust" Australopithecines* (F.E Grine, ed.), pp. 3–42. New York, NY: Aldine de Gruyter.

Grine, F.E., Klein, R.-G., and Volman, T.P. (1991) Dating, archaeology and human fossils from the Middle Stone Age levels of Die Kelders, South Africa. *Journal of Human Evolution* **21**, 363–395.

Grine, F.E., Henshilwood, C.S., and Sealy, J.C. (2000) Middle Stone Age human fossils from Die Kelders Cave 1, Western Cape Province, South Africa. *Journal of Human Evolution* **38**, 755–765.

Groves, C.P. (1986) Systematics of the great apes. In *Comparative Primate Biology* (D.R.Swindler and J. Erwin, eds), pp. 187–217. New York, NY: Alan R. Liss.

Groves, C.P. (1989) *A Theory of Human and Primate Evolution.* Oxford: Clarendon Press.

Groves, C.P. and Mazák, V. (1975) An approach to the taxonomy of the Hominidae: gracile Villafranchian hominids of Africa. *Casopis pro Nineralogii Geologii* **20**, 225–247.

Groves, C.P. and Paterson, J.F. (1991) Testing hominoid phylogeny with the PHYLIP programs. *Journal of Human Evolution* **20**, 167–183.

Grün, R. and Stringer, C.B. (1991) Electron spin resonance dating and the evolution of modern humans. *Archeometry* **33**, 153–199.

Grün, R., Stringer, C., and Schwarcz, H. (1991) ESR dating of teeth from Garrod's Tabun cave collection. *Journal of Human Evolution* **20**, 231–248.

Grün, R., Stringer, C., McDermott, F., Nathan, R., Porat, N., Robertson, S., Taylor, L., Mortimer, G., Eggins, S., and McCulloch, M. (2005) U-series and ESR analyses of bones and teeth relating to the human burials from Skhul. *Journal of Human Evolution* **49**, 316–334.

Haeckel, E. (1868) *Natürliche Schopfungsgeschichte*. Berlin: Reimer.

Haeckel, E. (1883) *The History of Creation, or the Development of the Earth and its inhabitants by Natural Causes*. New York, NY: Appleton.

Haeckel, E. (1905) *The Wonders of Life*. New York, NY: Harper.

Haile-Selassie, Y. (2001) Late Miocene hominids from the Middle Awash, Ethiopia. *Nature* **412**, 178–181.

Haile-Selassie, Y., Suwa, G., and White, T.D. (2004) Late Miocene teeth from Middle Awash, Ethiopia, and early hominid dental evolution. *Science* **303**, 1503–1505.

Hamilton, W.D. (1964) The genetical evolution of social behavior. *Journal of Theoretical Biology* **7**, 1–51.

Hammer, M.F., Garrigan, D., Wood, E., Wilder, J.A., Mobasher, Z., Bigham, A., Krenz, J.G., and Nachman, M.W. (2004) Heterogeneous patterns of variation among multiple human X-linked loci: the possible role of diversity-reducing selection in non-Africans. *Genetics* **167**, 1841–1853

Harcourt-Smith, W.E.H. and Aiello, L.C. (2004) Fossils, feet and the evolution of human bipedal locomotion. *Journal of Anatomy* **204**, 403–416.

Harding, R.M. and McVean, G. (2004) A structured ancestral population for the evolution of modern humans. *Current Opinion in Genetics and Development* **14** (**6**), 667–674.

Harding, R.M., Fullerton, S.M., Griffiths, R.C., Bond, J., Cox, M.J., Schneider, J.A., Moulin, D.S., and Clegg, J.B. (1997) Archaic African and Asian lineages in the genetic ancestry of modern humans. *American Journal of Human Genetics* **60(4)**, 772–789.

Hardy, J. Pittman, A., Myers, A., Gwinn-Hardy, K., Fung, H.C., de Silva, R., Hutton, M., and Duckworth, J. (2005) Evidence suggesting that *Homo* neanderthalensis contributed the H2 *MAPT* haplotype to *Homo* sapiens. Biochemical Society Transactions 33, 582–585.

Harland, W.B., Armstrong, R.L., Cox, A.V., Craig, L.E., Smith, A.G., and Smith, D.G. (1990) *A Geologic Time Scale*. Cambridge: Cambridge University Press.

Harpending, H. and Rogers, A. (2000) Genetic perspectives on human origins and differentiation. *Annual Review of Genomics and Human Genetics* **1**, 361–385.

Harris, J.M. (1985) Age and paleoecology of the Upper Laetoli Beds, Laetoli, Tanzania. In *Ancestors: the Hard Evidence* (E. Delson, ed.), pp. 76–81. New York, NY: Alan R. Liss.

Harris, E.E. and Hey, J. (1999) X chromosome evidence for ancient human histories. *Proceedings of the National Academy of Sciences USA* **96**, 3320–3324.

Harrison, T. (1991) Some observations on the Miocene hominoids from Spain. *Journal of Human Evolution* **20**, 515–520.

Hartwig-Scherer, S. and Martin, R.D. (1991) Was "Lucy" more human than her "child"? Observations on early hominid post-cranial skeletons. *Journal of Human Evolution* **21**, 439–449.

Hattori, M., Fujiyama, A., Taylor, T.D., Watanabe, H., Yada, T., Park, H.-S., Toyoda, A., Ishii, K., Totoki, Y., Choi, D.-K. *et al.* (2000) The DNA sequence of human chromosome 21. *Nature* **405**, 311–319.

Hauser, M. (2006) *Moral Minds: How Nature Designed Our Universal Sense of Right and Wrong*. New York: Ecco/ HarperCollins.

Häusler, M. and Schmid, P. (1995) Comparison of the pelvis of Sts 14 and AL 288–1: implications for birth and sexual dimorphism in australopithecines. *Journal of Human Evolution* **29**, 363–383.

Hawks, J., Hunley, K., Lee, S.-H., and Wolpoff, M. (2000) Population bottlenecks and Pleistocene human evolution. *Molecular Biology and Evolution* **17**, 2–22.

Hay, R.L. and Leakey, M.D. (1982) Fossil footprints of Laetoli. *Scientific American Presents* February, 50–57.

Hayakawa, T., Aki, I., Varki, A., Satta, Y., and Takahata, N. (2006) Fixation of the human-specific CMP-*N*-acetylneuraminic acid hydroxylase pseudogene and implications of haplotype diversity for human evolution. *Genetics* **172**, 1139–1146.

Hayden, B. (1993) The cultural capacities of Neandertals: a review and re-evaluation. *Journal of Human Evolution* **24**, 113–146.

Hedenström, A. (1995) Lifting the Taung's Child. *Nature* **378**, 670.

Heim, J.L. (1997) Lo que nos dice la nariz. *Mundo científico* **177**, 526–534.

Hein, J. (2004) Human evolution: pedigrees for all humanity. *Nature* **431**, 518–519.

Heizmann, E.P.J. and Begun, D.R. (2001) The oldest Eurasian hominoid. *Journal of Human Evolution* **41**, 463–481.

Hellman, M. (1928) Racial characters in human dentition. *Proceedings of the American Philosophical Society* **67**, 157–174.

Hemmer, H. (1972) Notes sur la position phylétique de l'homme de Petralona. *Anthropologie* **76**, 155–162.

Hennig, W. (1950) *Grundzüge einer Theorie der phylogenetischen Systematik*. Berlin: Aufbau.

Hennig, W. (1966) Phylogenetic systematics. *Annual Review of Entomology* **10**, 97–116.

Hey, J. (1997) Mitochondrial and nuclear genes present conflicting portraits of human origins. *Molecular Biology and Evolution* **14**, 166–172.

Higgs, E.S. (1961) Some Pleistocene faunas of the Mediterranean coastal areas. *Proceedings of the Prehistory Society* **27**, 144–154.

Higuchi, R., Bowman, B., Freiberger, M., Ryder, O.A., and Wilson, A.C. (1984) DNA sequences from the quagga, an extinct member of the horse family. *Nature* **312**, 282–284.

Hill, A. (1999) The Baringo Basin, Kenya: from Bill Bishop to BPRP. In *Late Cenozoic Environments and Hominid Evolution: a Tribute to Bill Bishop* (P. Andrews and P. Banham, eds), pp. 85–97. London: Geological Society of London.

Hill, A. (2002) Paleoanthropological research in the Tugen Hills, Kenya. *Journal of Human Evolution* **42**, 1–10.

Hill, A., Drake, R., Tauxe, L., Monaghan, M., Barry, J.C., Behrensmeyer, A.K., Curtis, G., Fine Jacobs, B., Jacobs, L., Johnson, N., and Pilbeam, D. (1985) Neogene palaeontology and geochronology of the Baringo Basin, Kenya. *Journal of Human Evolution* **14**, 759–773.

Hill, A., Ward, S., and Brown, B. (1992) Anatomy and age of the Lothagam mandible. *Journal of Human Evolution* **22**, 439–451.

Hill, R.S. and Walsh, C.A. (2005) Molecular insights into human brain evolution. *Nature* **437**, 64–67.

Hillson, S. (1996) *Dental Anthropology*. Cambridge: Cambridge University Press.

Hinds, D.A., Stuve, L.L , Nilsen, G.B., Halperin, E., Eskin, E., Ballinger, D.G., Frazer, K.A., and Cox, D.R. (2005) Whole-genome patterns of common DNA variation in three human populations. *Science* **307**, 1072–1079.

Hofman, M.A. (1983) Energy metabolism, brain size and longevity in mammals. *Quarterly Review of Biology* **58**, 495–512.

Hofreiter, M., Serre, D., Poinar, H.N., Kuch, M., and Pääbo, S. (2001) Ancient DNA. *Nature* **Genetics 2**, 353–359.

Holden, C. (1999) A new look into Neandertals' noses. *Science* **285**, 31–33.

Holden, C. (2001) Oldest human DNA reveals Aussie oddity. *Science* **291**, 230–231.

Holliday, T.W. (1995) Body size and proportions in the Late Pleistocene Western Old World and the origins of modern humans. PhD Thesis, University of New Mexico, Albuquerque.

Holloway, R. (1973) New endocranial values for the East African early hominids. *Nature* **243**, 97–99.

Holloway, R. (1974) The casts of fossil hominid brains. *Scientific American* **23**, 106–116.

Holloway, R. (1983) Human paleontological evidence relevant to language behavior. *Human Neurobiology* **2**, 105–114.

Holloway, R. (1985) The poor brain of *Homo* sapiens neanderthalensis: see what you please… In *Ancestors: the Hard Evidence* (E. Delson, ed.), pp. 319–324. New York, NY: Alan R. Liss.

Holloway, R. (1996) Evolution of the human brain. In *Handbook of Human Symbolic Evolution* (A. Lock and C. R.Peteres, eds), pp. 74–107. Oxford: Clarendon Press.

Hooker, J.J., Russell, D.E., and Phélizon, A. (1999) A new family of Plesiadapiformes (Mammalia) from the Old World Lower Paleogene. *Palaeontology* **42**, 377–407.

Hopwood, A.T. (1933) Miocene primates from British East Africa. *Annals and Magazine of Natural History, Series 10* **11**, 96–98.

Horai, S., Hayasaka, K., Kondo, R., Tsugane, K., and Takahata, N. (1995) Recent African origin of modern humans revealed by complete sequences of hominoid mitochondrial DNAs. *Proceedings of the National Academy of Sciences USA* **92**, 532–536.

Höss, M. (2000) Neanderthal population genetics. *Nature* **404**, 453–454.

Howell, F.C. (1952) Pleistocene glacial ecology and the evolution of 'Classic Neanderthal' man. *Southwestern Journal of Anthropology* **8**, 377–410.

Howell, F.C. (1957) Evolutionary significance of variation and varieties of 'Neanderthal Man'. *Quarterly Review of Biology* **37**, 340–347.

Howell, F.C. (1959) Upper Pleistocene stratigraphy and early man in the Levant. *Proceedings of the American Philosophical Society* **103**, 1–65.

Howell, F.C. (1960) European and northwest African Middle Pleistocene Hominids. *Current Anthropology* **1**, 195–232.

Howell, F.C. (1978) Hominidae. In *Evolution of African Mammals* (V.J.Maglio and H.B.S. Cooke, eds), pp. 154–248. Cambridge, MA: Harvard University Press.

Howell, F.C. (1981) Some views of *Homo* erectus with special reference to its occurrence in Europe. In *Homo erectus: Papers in honor of Davidson Black* (B.A Sigmon and J.S Cybulski, eds), pp. 153–157. Toronto: University of Toronto Press.

Howell, F.C. and Coppens, Y. (1976) An overview of Hominidae from the Omo succession, Ethiopia. In *Earliest Man and Environments in the Lake Rudolf Basin* (Y. Coppens, F.C Howell, G.L.Isaac, and R.E F. Leakey, eds), pp. 522–532. Chicago, IL: Chicago University Press.

Howell, F.C., Haesaerts, P., and de Heinzelin, J. (1987) Depositional environments, archeological occurences and hominids from Members E and F of the Shungura Formation (Omo basin, Ethiopia). *Journal of Human Evolution* 1987, 665–700.

Howells, W.W. (1973) *Cranial Variation in Man: a Study by Multivariate Analysis of Pattern of Differences Among Recent Human Populations*. Cambridge, MA: Harvard University Press.

Howells, W.W. (1974) Neanderthals: names, hypotheses, and scientific method. *American Anthropologist* 76, 24–38.

Howells, W.W. (1980) *Homo* erectus -who, when and where: a survey. *Yearbook of Physical Anthropology* 23, 1–23.

Howells, W.W. (1989) *Skull Shapes and the Map: Craniometric Analysis of Modern Homo*. Cambridge, MA: Harvard University Press.

Hublin, J.J. (1985) Human fossils from the North African Middle Pleistocene and the origin of *Homo* sapiens. In *Ancestors: the Hard Evidence* (E. Delson, ed.), pp. 283–288. New York, NY: Alan R. Liss.

Hublin, J.J. and Tillier, A.M. (1981) The Mousterian juvenile mandible from Irhoud (Morocco): a phylogenetic interpretation. In *Aspects of Human Evolution* (C.B Stringer, ed.), pp. 167–185. London: Taylor and Francis.

Hublin, J.J., Spoor, F., Braun, M., Zonneveld, F., and Condemi, S. (1996) A late Neanderthal associated with Upper Palaeolithic artefacts. *Nature* 381, 224–226.

Hudson, R.R. (1990) Gene genealogies and the coalescent process. *Oxford Surveys in Evolutionary Biology* 7, 1–44.

Huffman, O.F. (2001) Geologic context and age of the Perning/Mojokerto Homo erectus, East Java. *Journal of Human Evolution* 40, 353–362.

Huffman, O.F., Shipman, P., Hertlerc, C., de Vos, J., and Aziz, F. (2005) Historical Evidence of the 1936 Mojokerto Skull Discovery, East Java. *Journal of Human Evolution* 48, 321–363.

Huffman, O.F., Zaim, Y., Kappelman, J., Ruez, Jr, D.R., de Vos, J., Rizal, Y., Aziz, F., and Hertler, C. (2006) Relocation of the 1936 Mojokerto skull discovery site near Perning, East Java. *Journal of Human Evolution* 50, 431–451.

Hughes, A.R. and Tobias, P.V. (1977) A fossil skull probably of the genus *Homo* from Sterkfontein, Transvaal. *Nature* 265, 310–312.

Hull, D. (1977) The ontological status of species as evolutionary units. In *Foundational Problems in the Special Sciences* (R. Butts and J. Hintikka, eds), pp. 91–102. Dordrecht: Reidel Publishing Company.

Hume, D. (1740; 1978) *Treatise of Human Nature*. Oxford: Oxford University Press.

Hunt, K. and Vitzthum, V.J. (1986) Dental metric assessment of the Omo Fossils: implications for the phylogenetic position of *Australopithecus africanus*. *American Journal of Physical Anthropology* 71, 141–155.

Hurford, A.J., Gleadow, A.JW., and Naeser, C.W. (1976) Fission-track dating of pumice from the KBS Tuff, East Rudolf, Kenya. *Nature* 263, 738–740.

Hürzeler, J. (1958) *Oreopithecus bambolii* Gervais: a preliminary report. *Verhandlungen der Naturforchenden Gesellschaft. Basel* 69, 1–48.

Huxley, J.S. (1953) *Evolution* in Action. New York: Harper.

Huxley, J. (1958) Evolutionary processes and taxonomy with special reference to grades. *Uppsala University Arsskrift* 6, 21–38.

Huxley, T.H. and Huxley, J.S. (1947) *Touchstone for Ethics*. New York: Harper.

Huynen, L., Millar, C., Scofield, R.P., and Lambert, D.M. (2003) Nuclear DNA sequences detect species limits in ancient moa. *Nature* 425, 175–178.

Hyodo, M., Watanabe, N., Sunata, W., and Susanto, E.E. (1993) Magnetostratigraphy of hominid fossil bearing formations in Sangiran and Mojokerto, Java. *Anthropological Science* 101, 157–186.

Ingman, M., Kaessmann, H., Pääbo, S., and Gyllensten, U. (2000) Mitochondrial genome variation and the origin of modern humans. *Nature* 408, 708–713.

International Human Genome Sequencing Consortium (2001) Initial sequencing and analysis of the human genome. *Nature* 409, 860–921.

Isaac, G.L. (1969) Studies of early cultures in East Africa. *World Archaeology* 1, 1–28.

Isaac, G.L. (1978a) The food-sharing behavior of proto-human hominids. *Scientific American* 238, 90–106.

Isaac, G.L. (1978b) The archaeological evidence for the activities of early African hominids. In *Early Hominids of Africa* (C.J Jolly, ed.), pp. 219–254. London: Duckworth.

Isaac, G. (1980) Palaeanthropology in the New China. In *Current Argument on Early Man* (L.K Königsson, ed.), pp. 226–251. Oxford: Pergamon Press.

Ishida, H. and Pickford, M. (1997) A new Late Miocene hominoid from Kenya: *Samburupithecus kiptalami* gen.

et sp. nov. *Comptes rendus de l'Académie des sciences. Série 2. Sciences de la terre et des planètes* **325**, 823–829.

Ishida, H., Pickford, M., Nakaya, H., and Nakano, Y. (1984) Fossil Anthropoids from Nachola and Samburu Hills, Samburu District, Kenya. *African Study Monograph* (supplement) **2**, 73–85.

Ishida, H., Kunimatsu, Y., Nakatsukasa, M., and Nakano, Y. (1999) New hominoid genus from the Middle Miocene of Nachola, Kenya. *Anthropological Science* **107**, 189–191.

Ishida, H., Kunimatsu, Y., Takano, T., Nakano, Y., and Nakatsukasa, M. (2004) Nacholapithecus skeleton from the Middle Miocene of Kenya. *Journal of Human Evolution* **46**, 69–103.

Jacob, T. (1976) Solo Man and Peking Man. In *Papers in honor of Davidson Black* (B.A Sigmon and J.S Cybulski, eds), pp. 87–104. Toronto: University of Toronto Press.

Jacob, T., Indriati, E., Soejono, R.P., Hsu, K., Frayer, D.W., Eckhardt, R.B., Kuperavage, A.J., Thorne, A., and Henneberg, M. (2006) Pygmoid Australomelanesian *Homo* sapiens skeletal remains from Liang Bua, Flores: population affinities and pathological abnormalities. *Proceedings of the National Academy of Sciences USA* **103**, 13421–13426.

Jaeger, J.J. (1975) Decouverte d'un Crâne d'Hominidé dans le Pleistocene Moyen du Maroc. In *Problémes actuels de Paléontologie*, Colloque internationaux CNRS no. 218, pp. 897–902. Paris: CNRS.

James, S.R. (1989) Hominid use of fire in the Lower and Middle Pleistocene: a review of the evidence. *Current Anthropology* **30**, 1–26.

Jerison, H.J. (1975a) Fossil evidence of the evolution of human brain. *Annual Review of Anthropology* **4**, 27–58.

Jerison, H.J. (1975b) Paleoneurology and the Evolution of Mind. *Scientific American* **234**, 90–101.

Jerison, H.J. (1977a) Evolution of the brain. In *The Human Brain* (M.C. Wittrock, J. Beatty, J.E. Bogen, M.S. Gazzaniga, H.J. Jerison, S.D. Krashen, R.D. Nebes, and T.J. Teyler, eds), pp. 39–62. Englewood Cliffs, NJ: Prentice Hall.

Jerison, H.J. (1977b) The Theory of Encephalization. In *Evolution* and Lateralization of the Brain, vol. 299 (S.J. Dimond and D.A. Blizard, eds), pp. 146–160. New York, NY: New York Academy of Sciences.

Jerison, H.J. (1988) Evolutionary neurology and the origin of language as a cognitive adaptation. In *The Genesis of Language* (M.E Landsberg, ed.), pp. 3–9. Berlin: Mouton de Gruyter.

Johanson, D.C. (1976) Ethiopia yields first 'family' of early man. *National Geographic* **150**, 790–811.

Johanson, D.C. and Coppens, Y. (1976) A preliminary anatomical diagnosis of the first Plio-Pleistocene hominid discoveries in the Central Afar, Ethiopia. *American Journal of Physical Anthropology* **45**, 217–234.

Johanson, D.C. and Taieb, M. (1976) Plio-Pleistocene hominid discoveries in Hadar, Ethiopia. *Nature* **260**, 293–297.

Johanson, D.C. and White, T.D. (1979) A systematic assessment of Early African hominids. *Science* **202**, 322–330.

Johanson, D.C. and Edey, M.A. (1981) *Lucy: The Beginnings of Humankind*. New York, NY: Simon and Schuster.

Johanson, D. and Edgar, B. (1996) *From Lucy to Language*. New York, NY: Simon and Schuster.

Johanson, D., White, T., and Coppens, Y. (1978) A New Species of the Genus *Australopithecus* (Primates: Hominidae) from the Pliocene of Eastern Africa. *Kirtlandia* **28**, 1–14.

Johanson, D.C., Masao, F.T., Eck, G.G., White, T.D., Walter, R.C., Kimbel, W.H., Asfaw, B., Manega, P., Ndessokia, P., and Suwa, G. (1987) New partial skeleton of *Homo* habilis from Olduvai Gorge, Tanzania. *Nature* **327**, 205–209.

Johnstone, R.A. (1994) Female preference for symmetrical males as a by-product of selection mate recognition. *Nature* **372**, 172–175.

Jolly, C.J. (1970) The seed-eaters: a new model of hominid differentiation. *Man* (New Series) **5**, 5–26.

Jorde, L.B., Rogers, A.R., Bamshad, M., Watkins, W.S., Krakowiak, P., Sung, S., Kere, J., and Harpending, H.C. (1997) Microsatellite diversity and the demographic history of modern humans. *Proceedings of the National Academy of Sciences USA* **94**, 3100–3103.

Jouffroy, F. (1991) La 'main sans talon' du primate bipède. In *Origine(s) de la bipédie chez les hominidés* (Y. Coppens and B. Senut, eds), pp. 21–35. Paris: Editions du CNRS.

Jungers, W.L. (1994) Ape and hominid limb lenght. *Nature* **369**, 194.

Jusczyk, P.W. and Hohne, E.A. (1997) Infants' memory for spoken words. *Science* **277**, 1984–1986.

Kaessmann, H., Heissig, F., von Haeseler, A., and Pääbo, S. (1999) DNA sequence variation in a non-coding region of low recombination on the human X chromosome. *Nature Genetics* **22**, 78–81.

Kahn, P. and Gibbons, A. (1997) DNA from an extinct human. *Science* **277**, 176–178.

Kalb, J.E., Jolly, C.J., Mebrate, A., Tebedge, S., Smart, C., Oswald, E.B., Cramer, D., Witehead, P., Wood, C.B., Conroy, G.C. *et al.* (1982) Fossil mammals and artefacts from the Middle Awash Valley, Ethiopia. *Nature* **298**, 25–29.

Kappelman, J.J. (1991) The paleoenvironment of *Kenyapithecus* at Fort Ternan. *Journal of Human Evolution* **20**, 95–129.

Kappelman, J. (1997) They might be giants. *Nature* **387**, 126–127.

Karavanic, I. and Smith, F.H. (1998) The Middle/Upper Paleolithic interface and the relationship of Neanderthals and early modern humans in Hrvatsko Zagorje, Croatia. *Journal of Human Evolution* **34**, 223–248.

Kay, R.F. (1981) The nut-crackers—a new theory of the adaptations of the Ramapithecine. *American Journal of Physical Anthropology* **55**, 141–151.

Kay, R.F. and Grine, F.E. (1985) Tooth morphology, wear and diet in *Australopithecus* and *Paranthropus* from Southern Africa. In *Evolutionary History of the 'Robust' Australopithecines* (F.E Grine, ed.), pp. 427–447. New York, NY: Aldine de Gruyter.

Kay, R.F., Ross, C., and Williams, B.A. (1997) Anthropoid origins. *Science* **275**, 797–804.

Keith, A. (1903) The extent to which the posterior segments of the body have been transmuted and suppressed in the evolution of man and allied primates. *Journal of Anatomy and Physiology* **37**, 271–273.

Keith, A. (1911) The early history of the Gibraltar cranium. *Nature* **87**, 314.

Keith, A. (1925a) The fossil anthropoid ape from Taungs. *Nature* **115**, 234.

Keith, A. (1925b) The Taungs skull. *Nature* **116**, 11.

Keith, A. (1947) Australopithecines or Dartians. *Nature* **159**, 377.

Kelley, J. (1992) Evolution of apes. In *The Cambridge Enciclopedia of Human Evolution* (S. Jones, R. Martin and D. Pilbeam, eds), pp. 223–230. Cambridge: Cambridge University Press.

Kennard, A.S. (1942) Faunas of the high terrace at Swanscombe. *Proceedings of the Geological Association of London* **53**, 105.

Kennedy, G.E. (1980) *Paleoanthropology*. New York, NY: McGraw-Hill.

Kennedy, G.E. (1984) The emergence of *Homo* sapiens: the post-cranial evidence. *Man* **19**, 94–110.

Kennedy, G.E. (1999) Is "*Homo* rudolfensis" a valid species? *Journal of Human Evolution* **36**, 119–121.

Keyser, A.W. (2000) The Drimolen skull: the most complete australopithecine cranium and mandible to date. *South African Journal of Science* **96**, 189–193.

Khaitovich, P., Hellmann, I., Enard, W., Nowick, K., Leinweber, M., Franz, H., Weiss, G., Lachmann, M., and Pääbo, S. (2005) Parallel patterns of evolution in the genomes and transcriptomes of humans and chimpanzees. *Science* **309**, 1850–1854.

Kimbel, W.H., White, T.D., and Johanson, D.C. (1988) Implications of KNM-WT 17000 for the evolution of "robust" *Australopithecus*. In *Evolutionary History of the "Robust" Australopithecines* (F.E Grine, ed.), pp. 259–268. New York, NY: Aldine de Gruyter.

Kimbel, W.H., Johanson, D.C., and Rak, Y. (1994) The first skull and other new discoveries of *Australopithecus afarensis* at Hadar, Ethiopia. *Nature* **368**, 449–451.

Kimbel, W.H., Walter, R.C., Johanson, D.C., Reed, K.E., Aronson, J.L., Assefa, Z., Marean, C.W., Eck, G.G., Hovers, E., Rak, Y. *et al.* (1996) Late Pliocene *Homo* and Oldowan tools from the Hadar Formation (Kada Hadar Member), Ethiopia. *Journal of Human Evolution* **31**, 549–561.

Kimbel, W.H., Johanson, D.C., and Rak, Y. (1997) Systematic assessment of a maxilla of *Homo* from Hadar, Ethiopia. *American Journal of Physical Anthropology* **103**, 235–262.

Kimura, I., Okada, M., and Ishida, H. (1985) Kinesiological characteristics of primate walking: its significance in human walking. In *Environment, Behavior and Morphology: Dynamic Interactions in Primates* (M.R.Morbeck, H. Preuschoft, and N. Gomberg, eds), pp. 297–311. New York, NY: Gustav Fischer.

King, W. (1864) The reputed fossil man of the Neanderthal. *Quarterly Journal of Science* **1**, 88–97.

Kingman, J.F.C. (1982a) The coalescent. *Stochastic Processes and Their Applications* **13**, 235–248.

Kingman, J.F.C. (1982b) On the genealogy of large populations. *Journal of Applied Probability* **19A**, 27–43.

Kingston, J.D., Marino, B.D., and Hill, A. (1994) Isotopic evidence for neogene hominid paleoenvironments in the Kenya Rift Valley. *Science* **264**, 955–959.

Kitcher, P. (1985) *Vaulting Ambition. Sociobiology and the Quest for Human Nature*. Cambridge, MA: MIT Press.

Knight, C., Powers, C., and Watts, I. (1995) The Human Symbolic Revolution: a Darwinian account. *Cambridge Archaeological Journal* **5**, 75–114.

Köhler, M. and Moyà-Solà, S. (1999) A finding of Oligocene primates on the European continent. *Proceedings of the National Academy of Sciences USA* **96**, 14664–14667.

Kokkoros, P. and Kanellis, A. (1960) Découverte d'un crane d'homme paléolithique dnas la peninsule Chalcidique. *Anthropologie* **64**, 132–147.

Kordos, L. and Begun, D.R. (2001) Primates from Rudabánya: allocation of specimens to individuals, sex and age categories. *Journal of Human Evolution* **40**, 1–16.

Kramer, A. (1986) Hominid-pongid distinctiveness in the Miocene-Pliocene fossil record: the Lothagam mandible. *American Journal of Physical Anthropology.*, **70**, 457–473.

Kramer, P.A. and Eck, G.G. (2000) Locomotor energetics and leg length in hominid bipedality. *Journal of Human Evolution* **38**, 651–666.

Krantz, G.S. (1988) Laryngeal descent in 40,000 year old fossils. In *The Genesis of Language* (M.E Landsberg, ed.), pp. 173–180. Berlin: Mouton de Gruyter.

Kraus, B., Jordan, R., and Abrams, L. (1969) *Dental Anatomy and Occlusion*. Baltimore, MD: Williams and Wilkins.

Krings, M., Stone, A., Schmitz, R.W., Krainitzki, H., Stoneking, M., and Pääbo, S. (1997) Neandertal DNA sequences and the origin of modern humans. *Cell* **90**, 19–30.

Krings, M., Capelli, C., Tschentscher, F., Geisert, H., Meyer, S., von Haeseler, A., Grossschmidt, K., Possnert, G., Paunovic, M., and Pääbo, S. (2000) A view of Neandertal genetic diversity. *Nature Genetics* **26**, 144–146.

Kullmer, O., Sandrock, O., Abel, R., Schrenk, F., Bromage, T.G., and Juwayeyi, Y.M. (1999) The first *Paranthropus* from the Malawi Rift. *Journal of Human Evolution* **37**, 121–127.

Kuman, K. and Clarke, R. (2000) Stratigraphy, artefact industries and hominid associations for Sterkfontein, Member 5. *Journal of Human Evolution* **38**, 827–847.

Kummer, B. (1991) Biomechanical foundations of the development of human bipedalism. In *Origine(s) de la bipédie chez les hominidés* (Y. Coppens and B. Senut, eds), pp. 1–8. Paris: Editions du CNRS.

Kurtén, B. (1959) New evidence on the age of Pekin Man. *Vertebrata Palasiatica* **3**, 173–175.

Lahr, M.M. and Foley, R. (1994) Multiple dispersals and modern human origins. *Evolutionary Anthropology* **3**, 48–58.

Lahr, M.M. and Foley, R. (2004) Human evolution writ small. *Nature* **431**, 1043–1044.

Lai, C.SL., Fisher, S.E., Hurst, J.A., Levy, E.R., Hodgson, S., Fox, M., Jeremiah, S., Povey, S., Jamison, D.C., Green, E.D. *et al.* (2000) The SPCH1 region on human 7q31: genomic characterization of the critical interval and localization of translocations associated with speech and language disorder. *American Journal of Human Genetics* **67**, 357–368.

Lai, C.SL., Fisher, S.E., Hurst, J.A., Vargha-Khadem, F., and Monaco, A.P. (2001) A forkhead-domain gene is mutated in a severe speech and language disorder. *Nature* **413**, 519–523.

Laitman, J. (1984) The anatomy of human speech. *Natural History* **92**, 20–27.

Laitman, J.T. (1986) El origen del lenguaje. *Mundo científico* **64**, 1182–1191.

Lalueza-Fox, C., Sampietro, M.L., Caramelli, D., Puder, Y., Lari, M., Calafell, C., Martínez-Maza, C., Bastir, M., Fortea, J., de la Rasilla, M. *et al.* (2005) Neandertal evolutionary genetics: mitochondrial DNA data from the Iberian Peninsula. *Molecular Biology and Evolution* **22**, 1077–1081.

Lamarck, J.B. (1809) *Philosophie zoologique*. Paris: Dentu.

Larick, R., Ciochon, R.L., Zaim, Y., Sudijono, Suminto, Rizal, Y., Aziz, F., Reagan, M., and Heizler M. (2001) Early Pleistocene 40Ar/39Ar ages for Bapang Formation hominins, Central Jawa, Indonesia. *Proceedings of the National Academy of Sciences USA* **98**, 4866–4871.

Lartet, E. (1856) Note sur un gran singe fossile qui se rattache au groupe des singes supérieurs. *Comptes rendus de l'Académie des Sciences Paris* **43**, 219–223.

Lartet, E. and Christy, H. (1865–1875) *Reliquiae Aquitanicae*. London: Williams and Norgate.

Latimer, B. and Lovejoy, C.O. (1989) The calcaneus of *Australopithecus afarensis* and its implications in the evolution of bipedality. *American Journal of Physical Anthropology* **78**, 369–386.

Latimer, B., Ohman, J.C., and Lovejoy, C.O. (1987) Talocrural joint in African hominoids: implications for *Australopithecus afarensis*. *American Journal of Physical Anthropology* **74**, 155–175.

Leakey, L.S.B. (1951) *Olduvai Gorge. A report on the evolution of the handaxe culture in Beds I-IV*. Cambridge: Cambridge University Press.

Leakey, L.S.B. (1958) Recent discoveries at Olduvai Gorge, Tanganyika. *Nature* **181**, 1099–1103.

Leakey, L. (1959) A new fossil skull from Olduvai. *Nature* **184**, 491–493.

Leakey, L.S.B. (1961a) The juvenile mandible from Olduvai Gorge, Tanganyika. *Nature* **191**, 417–418.

Leakey, L.S.B. (1961b) New finds at Olduvai Gorge, Tanganyika. *Nature* **189**, 649–650.

Leakey, L.S.B. (1967a) An early Miocene member of Hominidae. *Nature* **213**, 155–163.

Leakey, L.S.B. (1967b) *Olduvai Gorge 1951–61. Volume I*. Cambridge: Cambridge University Press.

Leakey, L.SB., Evernden, J.F., and Curtiss, G.H. (1961) Age of Bed I, Olduvay Gorge, Tanganyika. *Nature* **191**, 478–479.

Leakey, L.SB., Tobias, P.V., and Napier, J.R. (1964) A new species of the genus *Homo* from Olduvai. *Nature* **202**, 7–9.

Leakey, M.D. (1969) Recent discoveries of hominid remains at Olduvai Gorge, Tanzania. *Nature* **223**, 754–756.

Leakey, M.D. (1971) *Olduvai Gorge 3. Excavations in Beds I and II 1960–1963*. Cambridge: Cambridge University Press.

Leakey, M.D. (1975) Cultural patterns in the Olduvai sequence. In *After the Australopithecines* (K.W. Butzer and G.L. Isaac, eds), pp. 477–493. The Hague: Mouton.

Leakey, M.D. (1981) Tracks and tools. *Philosophical Transactions of the Royal Society of London Series B* **292**, 95–102.

Leakey, M.D. and Hay, R.L. (1979) Pliocene footprints in the Laetoli Beds at Laetoli, northern Tanzania. *Nature* **278**, 317–323.

Leakey, M.D. and Hay, R.L. (1982) The cronological positions of the fossil hominids of Tanzania. In *L'Homo erectus andla place de l'homme de Tautavel parmi les hominidés fossiles* (H. de Lumley, ed.), pp. 753–765. Paris: CNRS.

Leakey, M.D., Clarke, R.J., and Leakey, L.S.B. (1971) New hominid skull from Bed I, Olduvai Gorge, Tanzania. *Nature* **232**, 308–312.

Leakey, M.D., Curtis, G.H., Drake, R.E., Jackes, M.K., and White, T.D. (1976) Fossil hominids from the Laetolil Beds. *Nature* **262**, 460–466.

Leakey, M.G., Feibel, C.S., McDougall, I., and Walker, A. (1995) New four-million-year-old Hominid species from Kanapoi and Allia Bay, Kenya. *Nature* **376**, 565–572.

Leakey, M.G., Feibel, C.S., McDougall, I., Ward, C., and Walker, A. (1998) New specimens and confirmation of an early age for *Australopithecus anamensis*. *Nature* **393**, 62–66.

Leakey, M.G., Spoor, F., Brown, F.H., Gathogo, P.N., Kiarie, C., Leakey, L.N., and McDougall, I. (2001) New hominin genus from eastern Africa shows diverse middle Pliocene lineages. *Nature* **410**, 433–440.

Leakey, R.E.F. (1970) New hominid remains and early artefacts from Northern Kenya. *Nature* **226**, 223–224.

Leakey, R.E.F. (1973a) Skull 1470. *National Geographic* **143**, 819–829.

Leakey, R.E.F. (1973b) Evidence for an advanced Plio-Pleistocene hominid from East Rudolf, Kenya. *Nature* **242**, 447–450.

Leakey, R.E.F. (1974) Further evidence of Lower Pleistocene hominins from East Rudolf, North Kenya, 1973. *Nature* **248**, 653–656.

Leakey, R.E.F. (1976) New hominid fossil from the Koobi Fora Formation, North Kenya. *Nature* **261**, 574–576.

Leakey, R.E.F. (1981) *The Making of Mankind*. London: The Rainbird Publishing Group.

Leakey, R.E.F. and Wood, B.A. (1973) New evidence of the genus *Homo* from East Rudolf, Kenya. II. *American Journal of Physical Anthropology* **39**, 355–368.

Leakey, R.E.F. and Walker, A.C. (1976) *Australopithecus, Homo* erectus and the single species hypothesis. *Nature* **261**, 572–574.

Leakey, R.E.F. and Walker, A.C. (1980) On the status of *Australopithecus afarensis*. *Science* **207**, 1103.

Lee, S.-H. (2005) Is variation in the cranial capacity of the Dmanisi sample too high to be from a single species? *American Journal of Physical Anthropology* **127**, 263–266.

Lee-Thorp, J.A. and van der Merwe, N.J. (1993) Stable carbon isotope studies of Swartkrans fossil assemblages. In *Swartkrans: a Cave's Chronicle of Early Man* (C. K Brain, ed.), pp. 243–250. Pretoria: Transvaal Museum Monograph.

Lee-Thorp, J.A., van der Merwe, N.J., and Brain, C.K. (1994) Diet of *Australopithecus robustus* at Swartkrans from stable carbon isotopic analysis. *Journal of Human Evolution* **27**, 361–372.

Le Gros Clark, W. (1947) Observation on the anatomy of the fossil Australopithecinae. *Journal of Anatomy* **83**, 300–333.

Le Gros Clark, W.E. (1950) New paleontological evidence bearing on the evolution of the Hominoidea. *Quarterly Journal of the Geological Society of London* **105**, 225–264.

Le Gros Clark, W.E. (1955) *The Fossil Evidence for Human Evolution*. Chicago, IL: University of Chicago Press.

Le Gros Clark, W.E. (1964a) The evolution of man. *Discovery* **25**, 49.

Le Gros Clark, W.E. (1964b) *The Fossil Evidence for Human Evolution*. Chicago, IL: University of Chicago Press.

Le Gros Clark, W.E. and Leakey, L.S.B. (1951) The Miocene Hominoidea. In *Fossil Mammals of Africa*, vol. I (B. Museum, ed.), pp. 1–117. London: British Museum.

Le Gros Clark, W., Oakley, K.P., Morant, G.M., King, W.B.R., Hawkes, C.F.C. *et al.* (1938) Report of the Swanscombe Committee. *Journal of the Royal Anthropological Institute* **68**, 17–98.

Leroi-Gourhan, A. (1958) Étude des restes humains fossiles provenant des grottes d'Arcy-sur-Cure (Yonne). *Annales de Paléontologie* **44**, 87–148.

Leroi-Gourhan, A. (1961) Les fouilles d'Arcy-sur-Cure (Yonne). *Gallia Préhistoire* **4**, 3–16.

Leroi-Gourhan, A. (1964) *Le Geste et la Parole*. Paris: Albin Michel.

Leroi-Gourhan, A. (1975) The flowers found with Shanidar IV, a Neanderthal burial in Iraq. *Science* **190**, 562–564.

Lévêque, F. and Vandermeersch, B. (1980) Les découvertes de restes humains dans un niveau castelperronien à Saint-Césaire (Charente-Maritime). *Comptes rendus de l'Académie des Sciences de Paris* **291D**, 187–189.

Lewin, R. (1984) *Human Evolution*. Oxford: Blackwell Publishing.

Lewin, R. (1986) New fossil upsets human family. *Science* **233**, 720–721.

Lewin, R. (1987) *Bones of Contention. Controversies in the Search for Human Origins.* New York, NY: Simon and Schuster.

Lewis, G. (1934) Preliminary notice of the new manlike apes from India. *American Journal of Science* **27**, 161–181.

Lewis, O.J. (1972) The evolution of the hallucial tarsometatarsal joint in the Anthropoidea. *American Journal of Physical Anthropology.*, **37**, 13–34.

Lewis, O.J. (1980) The joints of the evolving foot. Part III. The fossil evidence. *Journal of Anatomy (London)* **131**, 275–298.

Lieberman, D. (1998) Sphenoid shortening and the evolution of modern human cranial shape. *Nature* **393**, 158–162.

Lieberman, D.E. (2001) Another face in our family tree. *Nature* **410**, 419–420.

Lieberman, D.E. (2005) Further fossil finds from Flores. *Nature* **437**, 957–958.

Lieberman, D.E. and McCarthy, R.C. (1999) The ontogeny of cranial base angulation in humans and chimpanzees and its implications for reconstructing pharyngeal dimensions. *Journal of Human Evolution* **36**, 487–517.

Lieberman, D.E., Wood, B.A., and Pilbeam, D.R. (1996) Homoplasy and early *Homo*: an analysis of the evolutionary relationships of *H. habilis sensu stricto* and *H. rudolfensis. Journal of Human Evolution* **30**, 97–120.

Lieberman, D.E., McCarthy, R.C., Hiiemae, K.M., and Palmer, J.B. (2001) Ontogeny of postnatal hyoid and larynx descent in humans. *Archives of Oral Biology* **46**, 117–128.

Lieberman, P. (1973) On the evolution of language: a unified view. *Cognition* **2**, 59–94.

Lieberman, P. (1984) *The Biology and Evolution of Language.* Cambridge, MA: Harvard University Press.

Lieberman, P. (1989) The origins of some aspects of human language and cognition. In *The Human Revolution: Behavioural and Biological Perspectives on the Origins of Modern Humans* (P. Mellars and C. Stringer, eds), pp. 391–414. Princeton, NJ: Princeton University Press.

Lieberman, P. (1994) Human language and human uniqueness. *Language and* Communication **14**, 87–95.

Lieberman, P. (1995) Manual versus speech motor control and the evolution of language. *Behavioral and Brain Sciences* **18**, 197–198.

Lieberman, P. (1999) Silver-tongued Neandertals? *Science* **283**, 181–182.

Lindly, J.M. and Clark, G.A. (1990) Symbolism and modern human origins. *Current Anthropology* **31**, 233–262.

Linnaeus, C. (1735) *Systema Naturae per Naturae Regna Tria, Secundum Classes, Ordines, Genera, Species cum Characteribus, Synonymis, Locis,* 10th edn (1758). Stockholm: Laurentii Sylvii.

Liu, Z. (1985) Sequence of sediments at Locality I in Zhoukoudian and correlation with loess stratigraphy in northern China and with the chronology of deep-sea cores. *Quaternary Research* **23**, 139–153.

Lockwood, C.A. and Kimbel, W.H. (1999) Endocranial capacity of early hominids. *Science* **283**, 9b.

Lockwood, C.A. and Tobias, P.V. (1999) A large male hominin cranium from Sterkfontein, South Africa, and the status of *Australopithecus africanus. Journal of Human Evolution* **36**, 637–685.

Lockwood, C.A. and Tobias, P.V. (2002) Morphology and affinities of new hominin cranial remains from Member 4 of the Sterkfontein Formation, Gauteng Province, South Africa. *Journal of Human Evolution* **42**, 389–450.

Lordkipanidze, D., Vekua, A., Ferring, R., Rightmire, G. P., Agusti, J., Kiladze, G., Mouskhelishvili, A., Nioradze, M., de Leon, M.SP., Tappen, M., and Zollikofer, C.P.E. (2005) The earliest toothless hominin skull. *Nature* **434**, 717–718.

Lowenstein, J.M. (1986) Where there's smoke. *Pacific Discovery* **39**, 30–32.

Lovejoy, C.O. (1975) Biomechanical perspectives on the lower limb of early hominids. In *Primate Functional Morphology and Evolution* (R.H Tuttle, ed.), pp. 291–326. The Hague: Mouton.

Lovejoy, C.O. (1981) The origin of man. *Science* **211**, 341–350.

Lumsden, C.J. and Wilson, E.O. (1983) *Promethean Fire. Reflections on the Origin of Mind.* Cambridge, MA: Harvard University Press.

Macchiarelli, R., Bondioli, L., Chech, M., Coppa, A., Fiore, I., Russom, R., Vecchi, F., Libsekaal, Y., and Rook, L. (2004) The late early Pleistocene human remains from Buia, Danakil depression, Eritrea. *Rivista Italiana di Paleontologia e Stratigrafia* **110**, 133–144.

Macintosh, N.W.G. and Larnach, S., L. (1972) The persistence of *Homo erectus* traits in Australian Aboriginal crania. *Archaeology and Physical Anthropology in Oceania* **7**, 1–7.

MacLatchy, L. (1998) The controversy continues. *Evolutionary Anthropology* **6**, 147–150.

MacRae, A. (1998–2004) *Radiometric Dating and the Geological Time Scale. Circular Reasoning or Reliable Tools?* The Talk Origins Archive. http:/ / www.talkorigins. org/ faqs/ dating.html.

Maglio, V.J. (1972) Vertebrate faunas and chronology of hominin-bearing sediments east of Lake Rudolf, Kenya. *Nature* **239**, 379–385.

Mallegni, F., Carnieri, E., Bisconti, M., Tartarelli, G., Ricci, S., Biddittu, I., and Segre, A. (2003) *Homo* cepranensis sp. nov. and the evolution of African-European Middle Pleistocene hominids. *Comptes rendus Palévol 2*, 153–159.

Mallory, A.C. and Vaucheret, H. (2006) Functions of microRNAs and related small RNAs in plants. *Nature Genetics Supplement* **38**, S31-S36.

Marcus, G.F., Vijayan, S., Bandi Rao, S., and Vishton, P. M. (1999) Rule learning by seven-month-old infants. *Science* **283**, 77–80.

Marett, R. (1911) Pleistocene man in Jersey. *Archaeologia* **62**, 449–480.

Maringer, J. and Verhoeven, T. (1970) Die steinartefakte aus der Stegodon-fossilschicht von Mengeruda auf Flores, Indonesien. *Anthropos* **65**, 229–247.

Marlar, R.A., Leonard, B.L., Billman, B.R., Lambert, P.M., and Marlar, J.E. (2000) Biochemical evidence of cannibalism at a prehistoric Puebloan site in southwestern Colorado. *Nature* **407**, 74–78.

Marler, P. (1978) Primate vocalizations: affective or symbolic? In *Progress in Ape Research* (G. Bourne, ed.), pp. 281–324 New York, NY: Academic Press.

Marshack, A. (1988) The Neanderthals and the human capacity for symbolic thought: Cognitive and problem-solving aspects of Mousterian symbol. In *L'homme Nèandertal: Actes du Coloque International 1986, Liège*, vol. 5, *La pensée* (O. Bar-Yosef, ed.), pp. 57–91. Liège: Université de Liège.

Marshack, A. (1990) Early hominid symbolism and the evolution of human capacity. In *The Emergence of Modern Humans* (P. Mellars, ed.), pp. 457–498. Edinburgh: Edinburgh University Press.

Marshack, A. (1995) A Middle Paleolithic symbolic composition from the Golan Heights: the earliest known depictive image. *Current Anthropology* **37**, 357–365.

Marth, G.T., Czabarka, E., Murvai, J., and Sherry, S.T. (2004) The allele frequency spectrum in genome-wide human variation data reveals signals of differential demographic history in three large world populations. *Genetics* **166**, 351–372.

Martin, L.B. (1985) Significance of enamel thickness in hominoid evolution. *Nature* **314**, 260–263.

Martin, L.B. and Andrews, P.J. (1984) The phyletic position of *Graecopithecus ferybergi* Koenigswald. In *The Early Evolution of Man with Special Emphasis on Southeast Asia and Africa*, vol. 69 (P. Andrews and J.L.Franzen, eds), pp. 25–40. Frankfurt: Courier Forschungsinstitut Senckenberg.

Martin, R.D. (1990) *Primate Origins and Evolution. A Phylogenetic Reconstruction*. London: Chapman and Hall.

Martin, R.D. (1996) Scaling of the mammalian brain: the maternal energy hypotheses. *News in Physiological Sciences* **11**, 149–156.

Martin, R. (2000) Recursos energéticos y la evolución del tamaño cerebral en los hominoideos. In *Antes de Lucy. El agujero negro de la evolución humana* (J. Agustí, ed.), pp. 217–263. Barcelona: Tusquets.

Martínez-Navarro, B., Palmqvist, P., Arribas, A., and Turq, A. (1997) La adaptación a una dieta carnívora: clave de la primera dispersión humana fuera de Africa en el pleistoceno inferior. In *Senderos de la evolución humana* (C.J.Cela-Conde, R. Gutiérrez Lombardo, and J. Martínez Contreras, eds), pp. 179–204. México: Ludus Vitalis.

Martínez-Navarro, B., Rook, L., Segid, A., Yosieph, D., Ferretti, M.P., Shoshani, J., Tecle, T.M., and Libsekal, Y. (2004) The large fossil mammals from Buia (Eritrea). *Rivista Italiana di Paleontologia e Stratigrafia* **110** (supplement), 61–88.

Martini, J.EJ., Wipplinger, P.E., Moen, H.FG., and Keyser, A. (2003) Contribution to the speleology of Sterkfontein Cave, Gauteng Province, South Africa. *International Journal of Speleology* **32**, 43–69.

Masters, P. (1982) An amino acid racemization chonology for Tabun. In *The Transition from the Lower to Middle Paleolithic and the Origin of Modern Man*, **151** (A. Ronen, ed.), pp. 43–54. Oxford: British Archaeological Reports International Series.

Mayr, E. (1942) *Systematics and the Origin of Species from the Viewpoint of a Zoologist*. New York, NY: Columbia University Press.

Mayr, E. (1944) On the concepts and terminology of vertical subspecies and species. *Committee on Common Problems of Genetics, Paleontology and Systematics Bulletin* **2**, 11–16.

Mayr, E. (1950) Taxonomic categories in fossil hominins. *Cold Spring Harbor Symposium in Quantitative Biology* **15**, 109–117.

Mayr, E. (1957) Difficulties and importance of the biological species concept. In *The Species Problem* (E. Mayr, ed.), pp. 371–388. Washington DC: American Association for the Advancement of Science.

Mayr, E. (1963) *Animal Species and Evolution*. Cambridge: Belknap Press.

Mayr, E. (1970) *Population, Species, and Evolution*. Cambridge: Harvard University Press.

Mayr, E. (1976) *Evolution* and the Diversity of Life. Cambridge, MA: Harvard University Press.

McBrearty, S. and Brooks, A.S. (2000) The revolution that wasn't: a new interpretation of the origin of modern human behavior. *Journal of Human Evolution* 39, 453–563.

McBrearty, S. and Jablonski, N.G. (2005) First fossil chimpanzee. *Nature* 437, 105–108.

McCown, T. and Keith, A. (1939) *The Stone Age of Mount Carmel*, vol. 2, *The Fossil Human Remains from the Levallois Mousterian.* Oxford: Clarendon Press.

McCrossin, M.L. (1992) Human molars from late Pleistocene deposits of Witkrnas Cave, Gaap Escarpment, Kalahari margin. *Human Evolution* 7, 1–10.

McCrossin, M. and Benefit, B. (1993) Recently recovered *Kenyapithecus* mandible and its implications for great ape and human origins. *Proceedings of the National Academy of Sciences USA* 90, 1962–1966.

McDermott, F., Grün, R., Stringer, C.B., and Hawkesworth, C.J. (1993) Mass-spectrometric U-series dates for Israeli Neanderthal/early modern hominid sites. *Nature* 363, 252–255.

McDougall, I. (1981) 40Ar/39Ar age spectra from the KBS Tuff, Koobi Fora Formation. *Nature* 294, 120–124.

McDougall, I. (1985) Potassium-argon and argon^{40}/argon^{39} dating of the hominid Bearing Sequence at Koobi Fora, Lake Turkana, Northern Kenya. *Geological Society of America Bulletin* 96, 159–175.

McDougall, I., Maier, R., Sutherland-Hawkes, P., and Gleadow, A.J.W. (1980) K-Ar age estimate for the KBS Tuff, East Turkana, Kenya. *Nature* 284, 230–234.

McHenry, H. and Corruccini, R. (1980) Late Tertiary hominoids and human origins. *Nature* 285, 397–398.

McKee, J.C. (1996) Faunal evidence and Sterkfontein Member 2 foot of early man. *Science* 271, 1301–1302.

Mead, M. (1966) *Purity and Danger. An Analysis of Concepts of Pollution and Taboo.* London: Routledge and Kegan Paul.

Mead, S., Stumpf, M.PH., Whitfield, J., Beck, J.A., Poulter, M., Campbell, T., Uphill, J.B., Goldstein, D., Alpers, M., Fisher, E.M.*et al.* (2003) Balancing selection at the prion protein gene consistent with prehistoric Kurulike epidemics. *Science* 300, 640–643.

Meier, R. and Willmann, R. (2000) The Hennigian species concept. In *Species Concept and Phylogenetic Theory* (Q.D.Wheeler and R. Meier, eds), pp. 30–43. New York, NY: Columbia University Press.

Mellars, P. (1989) Major issues in the emergence of modern humans. *Current Anthropology* 30, 349–385.

Mellars, P. (1996) *The Neanderthal Legacy.* Princeton, NJ: Princeton University Press.

Mellars, P. (2006a) Why did modern human populations disperse from Africa *ca.* 60,000 years ago? A new model. *Proceedings of the National Academy of Sciences USA* 103, 9381–9386.

Mellars, P. (2006b) Going East: new genetic and archaeological perspectives on the modern human colonization of Eurasia. *Science* 313, 796–800.

Menter, C.G., Kuykendall, K.L., Keyser, A.W., and Conroy, G.C. (1999) First record of hominid teeth from the Plio-Pleistocene site of Gondolin, South Africa. *Journal of Human Evolution* 37, 299–307.

Mercader, J., Panger, M., and Boesch, C. (2002) Excavation of a chimpanzee stone tool site in the African rainforest. *Science* 296, 1452–1455.

Mercier, N., Valladas, H., Joron, J.L., Reyss, J.L., Lévêque, F., and Vandermeersch, B. (1991) Thermoluminescence dating of the late Neanderthal remains from Saint-Césaire. *Nature* 351, 737–739.

Mercier, N., Valladas, H., Bar-Yosef, O., Vandermeersch, B., Stringer, C., and Joron, J.L. (1993) Thermoluminescence date for the Mousterian burial site of Es-Skhul, Mt. Carmel. *Journal of Archaeological Science* 20, 169–174.

Miller, G.H., Beaumont, P.B., Brooks, A.S., Deacon, H.J., Hare, P.E., and Jull, A.J.T. (1999) Earliest modern humans in South Africa dated by isoleucine epimerization in ostrich eggshell. *Quaternary Science Review* 18, 1537–1548.

Milton, K. (1988) Foraging behaviour and the evolution of primate intelligence. In *Machiavellian Intelligence* (R. Byrne and A. Whiten, eds), pp. 285–305. Oxford: Clarendon Press.

Mishra, S., Venkatesan, T.R., Rajaguru, S.N., and Somayajulu, B.L.K. (1995) Earliest Acheulian industry from Peninsular India. *Current Anthropology* 36, 847–851.

Mithen, S. (1996) *The Prehistory of the Mind.* London: Thames and Hudson.

Moore, G. E . (1903) *On Aggression.* New York: Harcourt, Brace and World.

Moore, M.W., and Brumm, A. (2007). Stone artifacts and hominins in island Southeast Asia: New insights from Flores, eastern Indonesia. *Journal of Human Evolution* 52, 85–102.

Morell, V. (1995) African origins: West Side Story. *Science* 270, 1117.

Mortillet, G. (1897) *Formation de la Nation Française.* Paris: Plon-Nourrit.

Morwood, M.J., O'Sullivan, P.B., Aziz, A., and Raza, A. (1998) Fission-track ages of stone tools and fossils on the east Indonesian island of Flores. *Nature* 392, 173–176.

Morwood, M.J., Soejono, R.P., Roberts, R.G., Sutikna, T., Turney, C.SM., Westaway, K.E., Rink, W.J., Zhao, J.-X., van den Bergh, G.D., Due, R.A.*et al*. (2004) Archaeology and age of a new hominin from Flores in eastern Indonesia. *Nature* **431**, 1087–1091.

Morwood, M.J., Brown, P., Jatmiko, Sutikna, T., Saptomo, W.E., Westaway, K.E., Awe Due, R., Roberts, R.G., Maeda, T., Wasisto, S., and Djubiantono, T. (2005) Further evidence for small-bodied hominins from the Late Pleistocene of Flores, Indonesia. *Nature* **437**, 1012–1017.

Movius, H. (1948) The Lower Paleolithic cultures of southern and eastern Asia. *Transactions of the American Philosophical Society* **38**, 330–420.

Movius, H.L. (1953) The Mousterian cave of Teshik-Tash, Southeastern Uzbekistan, Central Asia. *Bulletin of the American School of Prehistoric Research* **17**, 11–71.

Moyà Solà, S. (2000) Viaje a los orígenes del bipedismo y una escala en la isla de los simios. In *Antes de Lucy. El agujero negro de la evolución humana* (J. Agustí, ed.), pp. 171–209. Barcelona: Tusquets.

Moyà-Solà, S. and Köhler, M. (1993a) *Dryopithecus* y el origen de los grandes monos actuales. *Investigación y ciencia* **207**, 30–31.

Moyà-Solà, S. and Köhler, M. (1993b) Recent discoveries of *Dryopithecus* shed new light on evolution of great apes. *Nature* **365**, 543–545.

Moyà-Solà, S. and Köhler, M. (1996) A *Dryopithecus* skeleton and the origins of great-ape locomotion. *Nature* **379**, 123–124.

Moyà-Solà, S., Pons Moyà, J., and Köhler, M. (1990) Primates catarrinos (Mammalia) del Neógeno de la península Ibérica. *Paleontologia i evolució* **23**, 41–45.

Moya-Sola, S., Kohler, M., Alba, D.M., Casanovas-Vilar, I., and Galindo, J. (2004) *Pierolapithecus catalaunicus*, a new Middle Miocene great ape from Spain. *Science* **306**, 1339–1344.

Moya-Sola, S., Kohler, M., Alba, D.M., Casanovas-Vilar, I., and Galindo, J. (2005) Response to Comment on "*Pierolapithecus catalaunicus*, a new Middle Miocene great ape from Spain". *Science* **308**, 203d.

Mturi, A.A. (1987) The archaeological sites of Lake Natron. *Lac Natron, Sciences Geologique* **40**, 209–215.

Murrill, R.L. (1975) A comparison of the Rodhesian and Petralona upper jaws in relation to other Pleistocene hominids. *Zeitschrift für Morphologie und Anthropologie* **66**, 176–187.

Nagel, T. (1974) What is it like to be a bat? *Philosophical Review* **83**, 435–450.

Nakatsukasa, M. (2004) Acquisition of bipedalism: the Miocene hominoid record and modern analogues for bipedal protohominids. *Journal of Anatomy* **204**, 385–402.

Nakatsukasa, M., Yamanaka, A., Kunimatsu, Y., Shimizu, D., and Ishida, H. (1998) A newly discovered *Kenyapithecus* skeleton and its implications for the evolution of positional behavior in Miocene East African hominoids. *Journal of Human Evolution* **34**, 657–664.

Napier, J. (1962) Fossil hand bones from Olduvai Gorge. *Nature* **196**, 409–411.

Napier, J.R. (1963) Brachiation and brachiators. *Symposia of the Zoological Society of London 10, 183–195*.

Napier, J. (1967) The antiquity of human walking. *Scientific American* **216**, 56–66.

Napier, J.R. (1980) *Hands*. New York, NY: Pantheon Books.

Napier, J.R. and Walker, A.C. (1967) Vertical clinging and leaping—a newly recognised category of locomotor behaviour of primates. *Folia Primatologia 6*, 204–219.

Nature Publishing Group (2006) *Nature Collections: Human Genome*.

Nei, M. (1987) *Molecular Evolutionary Genetics*. New York: Columbia University Press.

Nelson, G.J. (1973) Classification as an expression of phylogenetic relationships. *Systematic Zoology* **22**, 344–359.

Nishimura, H., Hashikawa, K., Doi, K., Iwai, T., Watanaba, Y., Kusuoka, H., Nishimura, T., and Kubo, T. (1999) Sign language 'heard' in the auditory cortex. *Nature* **397**, 116.

Noble, W. and Davidson, I. (1996) *Human Evolution, Language and Mind*. Cambridge: Cambridge University Press.

Noll, M.P. and Petraglia, M.D. (2003) Acheulean bifaces and early human behavioral patterns in East Africa and South India. In *Multiple Approaches to the Study of Bifacial Technologies* (M. Soressi and H.L.Dibble, eds), pp. 31–53. Philadelphia, PA: University of Pennsylvania Museum of Archaeology and Anthropology.

Noonan, J.P., Hofreiter, M., Smith, D., Priest, J.R., Rohland, N., Rabeder, G., Krause, J., Chris Detter, J., Pääbo, S., and Rubin, E.M. (2005) Genomic sequencing of Pleistocene cave bears. *Science* **309**, 597–600.

Noonan, J.P., Coop, G., Kudaravalli, S., Smith, D., Krause, J., Alessi, J., Alessi, J., Chen, F., Platt, D., Paabo, S., Pritchard, J.K., and Rubin, E.M. (2006). Sequencing and Analysis of Neanderthal Genomic DNA. *Science*, **314**, 1113–1118.

Nordborg, M. (1998) On the probability of Neandertal ancestry. *American Journal of Human Genetics* **63**, 1237–1240.

Oakley, K.P. (1956) The earliest fire-makers. *Antiquity* **30**, 102–107.

Oakley, K. (1981) Emergence of higher thought 3.0–0.2 Ma B.P. *Philosophical Transactions of the Royal Society of London Series B* **292**, 205–211.

Olson, T.R. (1985) Cranial morphology and systematics of the Hadar Formation hominids and *"Australopithecus" africanus*. In *Ancestors: The Hard Evidence* (E. Delson, ed.), pp. 102–119. New York, NY: Alan R. Liss.

Oppennoorth, W.F.F. (1932) *Homo* (Javanthropus) soloensis, een plistocene Mensch von Java. *Wetenschappelijke medeligen Dienst van den Mijnbrouw in Nederlandsch-Indië* **20**, 49–75.

Orlando, L., Darlu, P., Toussaint, M., Bonjean, D., Otte, M., and Hanni, C. (2006). Revisiting Neandertal diversity with a 100,000 year old mtDNA sequence. *Current Biology*, **16**, R400–402.

O'Rourke, D.H., Geoffrey Hayes, M., and Carlyle, S.W. (2000) Ancient DNA studies in physical anthropology. *Annual Review of Anthropology* **29**, 17–42.

O'Sullivan, P.B., Morwood, M., Hobbs, D., Suminto, F.A., Situmorang, M., Raza, A., and Maas, R. (2001) Archaeological implications of the geology and chronology of the Soa Basin, Flores, Indonesi. *Geology* **29**, 607–610.

Otte, M. (2003) The pitfalls of using bifaces as cultural markers. In *Multiple Approaches to the Study of Bifacial Technologies* (M. Soressi and H.L.Dibble, eds), pp. 183–192. Philadelphia, PA: University of Pennsylvania Museum of Archaeology and Anthropology.

Ovchinnikov, I.V., Götherström, A., Romanova, G.P., Kharitonov, V.M., Liden, K., and Goodwin, W. (2000) Molecular analysis of Neanderthal DNA from the northern Caucasus. *Nature* **404**, 490–493.

Owen, R. (1843) *Lectures on the Comparative Anatomy and Physiology of the Invertebrate Animals, Delivered at the Royal College of Surgeons in 1843*. London: Longman, Brown, Green and Longmans.

Oxnard, C. and Lisowski, P. (1980) Functional articulation of some hominoid foot bones: implications for the Olduvai (Hominid 8) foot. *American Journal of Physical Anthropology* **52**, 107–117.

Pääbo, S. (1984) Über den Nachweiss von DNA in altägyptischen Mumien. *Das Latertum* **30**, 213–218.

Pääbo, S. (1985) Molecular cloning of ancient Egyptian mummy DNA. *Nature* **314**, 644–645.

Pääbo, S., Poinar, H., Serre, D., Jaenicke-Després, V., Hebler, J., Rohland, N., Kuch, M., Krause, J., Vigilant, L., and Hofreiter, M. (2004) Genetic analyses from ancient DNA. *Annual Review of Genetics* **38**, 645–679.

Palmqvist, P., Arribas, A., and Martínez-Navarro, B. (1999) Ecomorphological study of large canids from the Lower Pleistocene of southeastern Spain. *Lethaia* **32**, 75–88.

Parés, J.M. and Pérez-González, A. (1995) Paleomagnetic age for hominid fossils at Atapuerca archaeological site, Spain. *Science* **269**, 830–832.

Parfitt, S.A., Barendregt, R.W., Breda, M., Candy, I., Collins, M.J., Coope, G.R., Durbidge, P., Field, M.H., Lee, J.R., Lister, A.M.*et al.* (2005) The earliest record of human activity in northern Europe. *Nature* **438**, 1008–1012.

Partridge, T.C. (1975) *Stratigraphic, Geomorphological and Paleo-environmental Studies of the Makapansgat Limeworks and Sterkfontein hominid sites: a progress report on research carried out between 1965 and 1975*. Paper presented at the Third Scientific Congress, Cape Town, May–June 1975.

Partridge, T.C. (1978) Re-appraisal of lithostratigraphy of Sterkfontein hominid site. *Nature* **275**, 282–287.

Partridge, T.C. (1982) The chronological positions of the fossil hominids of southern Africa. In *L'Homo erectus et la place de l'homme de Tautavel parmi les hominids fossils*, vol. 2 (M.A De Lumley, ed.), pp. 617–675. Nice: Premier Congres International de Paleontologie Humaine.

Partridge, T.C. (2000) Hominid-bearing cave and tufa deposits. In *The Cenozoic of Southern Africa* (T.C.Partridge and R.R.Maud, eds), pp. 100–125. New York, NY: Oxford University Press.

Partridge, T.C. and Watt, I.B. (1991) The stratigraphy of the Sterkfontein hominid deposit and its relationship to the underground cave system. *Palaeont. afr.*, **28**, 35–40.

Partridge, T.C., Shaw, J., Heslop, D., and Clarke, R. (1999) The new hominid skeleton from Sterkfontein, South Africa: age and preliminary assessment. *Journal of Quaternary Science* **14**, 239–298.

Partridge, T.C., Granger, D.E., Caffee, M.W., and Clarke, R.J. (2003) Lower Pliocene hominid remains from Sterkfontein. *Science* **300**, 607–612.

Passingham, R.E. (1998) The specializations of the human neocortex. In *Comparative Neuropsychology* (A.D Milner, ed.), pp. 271–298 Oxford: Oxford University Press.

Patterson, B., Behrensmeyer, A.K., and Sill, W.D. (1970) Geology and fauna of a new Pliocene locality in North-Western Kenya. *Nature* **226**, 918–921.

Paunovic, M., Krings, M., Capelli, C., Tshentscher, F., Geisert, H., Meyer, S., von Haesler, A., Grossschmidt, K., Possnert, G., and Pääbo, S. (2001) *The Vindija hominids: a view of Neandertal genetic diversity*. Paper presented at the Paleoanthropology Society. Abstracts for the 2001 Meeting.

Pearson, O.M. (2000) Postcranial remains and the origins of modern humans. *Evolutionary Anthropology* **9**, 229–247.

Pei, W. and Zhang, S. (1985) A study on the lithic artifacts of *Sinanthropus. Palaeontologica Sinica, New Series D12*, 1–277.

Penck, A. and Brückner, E. (1909) *Die Alpen im Eiszeitalter*. Leipzig: Tauchnitz.

Pickering, T.R., White, T.D., and Toth, N. (2000) Cutmarks on a Plio-Pleistocene Hominid From Sterkfontein, South Africa. *American Journal of Physical Anthropology* **111**, 579–584.

Pickford, M. (1975) Late Miocene sediments and fossils from the Northern Kenya Rift Valley. *Nature* **256**, 279–284.

Pilbeam, D.R. (1966) Notes of *Ramapithecus*, the earliest known hominid. *American Journal of Physical Anthropology* **25**, 1–6.

Pilbeam, D.R. (1968) The earliest hominids. *Nature* **219**, 1335–1338.

Pilbeam, D.R. (1978) Rethinking human origins. *Discovery* **13**, 2–9.

Pilbeam, D., Meyer, G., Badgley, C., Rose, M., Pickford, M., Behrensmeyer, A., and Ibrahim Shah, S.M. (1977) New hominoid primates from the Siwaliks of Pakistan and their bearing on Hominoid evolution. *Nature* **270**, 689–694.

Pilbeam, D.R., Rose, M.D., Barry, J.C., and Shah, S.M.I. (1990) New *Sivapithecus* humeri from Pakistan and the relationship of *Sivapithecus* and *Pongo. Nature* **348**, 237–239.

Place, U.T. (2000) The role of the hand in the evolution of language. *Psycoloquy* **11(7)**. http://psycprints.ecs.soton.ac.uk/archive/00000007/.

Platnick, N.I. (1979) Philosophy and the transformation of cladistics. *Systematic Zoology* **28**, 537–546.

Plummer, T. (2004) Flaked stones and old bones: biological and cultural evolution at the dawn of technology. *Yearbook of Physical Anthropology* **47**, 118–164.

Poirier, F.E. (1987) *Understanding Human Evolution*, 2nd edn (1990). Englewood Cliffs, NJ: Prentice Hall.

Pollard, K.S., Salama, S.R., Lambert, N., Lambot, M.-A., Coppens, S., Pedersen, J.S., Katzman, S., King, B., Onodera, C., Siepel, A. *et al.* (2006) An RNA gene expressed during cortical development evolved rapidly in humans. *Nature* **443**, 167–172.

Pope, G.G. (1988) Recent advances in Far Eastern Paleoanthropology. *Annual Review of Anthropology* **17**, 43–77.

Popper, K.R. and Eccles, J.C. (1977) *The Self and Its Brain*. Berlin: Springer-Verlag.

Potts, R. (1992) The hominid way of life. In *The Cambridge Encyclopedia of Human Evolution* (S. Jones, R. Martin and D. Pilbeam, eds), pp. 325–334. Cambridge: Cambridge University Press.

Potts, R., Behrensmeyer, A.K., Deino, A., Ditchfield, P., and Clark, J. (2004) Small Mid-Pleistocene hominin associated with East African Acheulean technology. *Science* **305**, 75–78.

Poulianos, A.N. (1971) Petralona: a Middle Pleistocene cave in Greece. *Archaeology* **24**, 6–11.

Poulianos, A.N. (1978) Stratigraphy and age of the Petralona Archanthropus. *Anthropos* **5**, 37–46.

Prat, S., Brugal, J.P., Tiercelin, J.J., Barrat, J.A., Bohn, M., Delagnes, A., Harmand, S., Kimeu, K., Kibunjia, M., Texier, P.J., and Roche, H. (2005) First occurrence of early *Homo* in the Nachukui Formation (West Turkana, Kenya) at 2.3–2.4Myr. *Journal of Human Evolution* **49**, 230–240.

Prentice, M.L. and Denton, G.H. (1988) The deep-sea oxygen isotope record, the global ice sheet system and hominid evolution. In *Evolutionary History of the "Robust" Australopithecines* (F.E Grine, ed.), pp. 383–403. New York, NY: Aldine de Gruyter.

Qian, F. (1980) Magnetostratigraphic study on the cave deposits containing fossil Peking Man at Zhoukoudian. *Kexue Tongbao* **25**, 359.

Rak, Y. (1983) *The Australopithecine Face*. New York, NY: Academic Press.

Rak, Y. (1985) Systematic and functional implications of the facial morphology of *Australopithecus* and Early *Homo*. In *Ancestors: The Hard Evidence* (E. Delson, ed.), pp. 168–170. New York, NY: Alan R. Liss.

Rak, Y. (1988) On variation in the masticatory system of *Australopithecus boisei*. In *Evolutionary History of the "Robust" Australopithecines* (F.E Grine, ed.), pp. 193–198. New York, NY: Aldine de Gruyter.

Rak, Y., Ginzburg, A., and Geffen, E. (2002) Does *Homo neanderthalensis* play a role in modern human ancestry? The mandibular evidence. *American Journal of Physical Anthropology* **119**, 199–204.

Ramirez Rozzi, F. (1998) Can enamel microstructure be used to establish the presence of different species of Plio-Pleistocene hominids from Omo, Ethiopia? *Journal of Human Evolution* **35**, 543–576.

Ramirez Rozzi, F. and Bermúdez de Castro, J.M. (2004) Surprisingly rapid growth in Neanderthals. *Nature* **428**, 936–939.

Rayner, R.J., Moon, B.P., and Masters, J.C. (1993) The Makapansgat australopithecine environment. *Journal of Human Evolution* **24**, 219–231.

Reader, J. (1981) *Missing Links. The Hunt for Earliest Man*. London: Collins.

Reich, D.E., Cargill, M., Bolk, S., Ireland, J., Sabeti, P.C., Richter, D.J., Lavery, T., Kouyoumjian, R., Farhadian, S.

F., Ward, R. *et al.* (2001) Linkage disequilibrium in the human genome. *Nature* **411**, 199–204.

Relethford, J.H. (1997) *The Human Species. An Introduction to Biological Anthropology*, 3rd edn. Mountain View, CA: Mayfield Publishing Company.

Reynolds, T.R. (1985) Mechanics of increased support of weight by the hindlimbs in primates. *American Journal of Physical Anthropology* **67**, 335–349.

Ribot, F. and Gibert, J. (1996) A reinterpretation of the taxonomy of *Dryopithecus* from Vallès-Penedès, Catalonia (Spain). *Journal of Human Evolution* **31**, 129–141.

Richmond, B.G. and Strait, D.S. (2000) Evidence that humans evolved from a knuckle-walking ancestor. *Nature* **404**, 382–385.

Riel-Salvatore, J. and Clark, G.A. (2001) Grave markers. Middle and Early Upper Paleolithic burials and the use of chronotypology in contemporary paleolithic research. *Current Anthropology* **42**, 449–479.

Rightmire, G.P. (1976) Relationships of Middle and Upper Pleistocene from sub-Saharan Africa. *Nature* **260**, 238–240.

Rightmire, G.P. (1979) Cranial remains of *Homo* erectus from Beds II and IV, Olduvai Gorge, Tanzania. *American Journal of Physical Anthropology* **51**, 99–115.

Rightmire, G.P. (1980) Middle Pleistocene Hominids From Olduvai Gorge, Northern Tanzania. *American Journal of Physical Anthropology* **53**, 225–241.

Rightmire, G.P. (1986) Species recognition and *Homo* erectus. *Journal of Human Evolution* **15**, 823–826.

Rightmire, G.P. (1990) *The Evolution of Homo erectus. Comparative Anatomical Studies of an Extinct Human Species*. Cambridge: Cambridge University Press.

Rightmire, G.P. (1993) Variation among early *Homo* crania From Olduvai Gorge and the Koobi Fora region. *American Journal of Physical Anthropology* **90**, 1–33.

Rightmire, G.P. (1997) Deep roots for the Neanderthals. *Nature* **389**, 917–918.

Rightmire, G.P., Lordkipanidze, D., and Vekua, A. (2006) Anatomical descriptions, comparative studies and evolutionary significance of the hominin skulls from Dmanisi, Republic of Georgia. *Journal of Human Evolution* **50**, 115–141.

Rilling, J.K. and Insel, T.R. (1999) The primate neocortex in comparative perspective using magnetic resonance imaging. *Journal of Human Evolution* **37**, 191–223.

Robbins, L.M. (1987) Hominid footprints from site G. In *Laetoli, a Pliocene Site in Northern Tanzania* (M.D.Leakey and J.M.Harris, eds), pp. 407–502. Oxford: Clarendon Press.

Roberts, M.B., Stringer, C.B., and Parfitt, S.A. (1994) A hominid tibia from Middle Pleistocene sediments at Boxgrove, UK. *Nature* **369**, 311–313.

Roberts, N. (1992) Climate change in the past. In *The Cambridge Encyclopedia of Human Evolution* (S. Jones, R. Martin, and D. Pilbeam, eds), pp. 174–178. Cambridge: Cambridge University Press.

Robinson, J.T. (1954) Prehominid dentition and hominid evolution. *Evolution* **8**, 324–334.

Robinson, J.T. (1962) The origin and adaptive radiation of the Australopithecines. In *Evolution und Hominization* (G. Kurth, ed.), pp. 150–175. Stuttgart: Stuttgarter Verlagskontor.

Robinson, J.T. (1963). Adaptive radiation in the Australopithecines and the origin of man. In *African Ecology and Human Evolution* (F.C Howell and F. Bourliere, eds), pp. 385–416. Chicago, IL: Aldine.

Robinson, J.T. (1968). The origin and adaptive radiation of the australopithecines. In *Evolution* und Hominisation (G. Kurth, ed.), pp. 150–175. Stttugart: Fischer.

Robinson, J.T. and Manson, R.J. (1957). Occurence of stone artefacts with *Australopithecus* at Sterkfontein. *Nature* **180**, 521–524.

Roche, H., Delagnes, A., Brugal, J.-P., Feibel, C., Kibunjia, M., Mourre, V., and Texier, J.-P. (1999) Early hominid stone tool production and technical skill 2.34 Myr ago in West Turkana, Kenya. *Nature* **399**, 57–60.

Roebroeks, W. (1994) Updating the earliest occupation of Europe. *Current Anthropology* **35**, 301–305.

Roebroeks, W. and Van Kolfschoten, T. (1994) The earliest occupation of Europe: a short chronology. *Antiquity* **68**, 489–503.

Rogers, A.R. and Jorde, L.B. (1995) Genetic evidence on modern human origins. *Human Biology* **67**, 1–36.

Rohde, D.LT., Olson, S., and Chang, J.T. (2004) Modelling the recent common ancestry of all living humans. *Nature* **431**, 562–566.

Rolland, N. (2004) Was the emergence of home bases and domestic fire a punctuated event? A review of the Middle Pleistocene record in Eurasia. *Asian Perspectives* **43**, 248–280.

Ron, H. and Levi, S. (2001) When did hominids first leave africa? New high-resolution magnetostratigraphy from the Erk-el-Ahmar Formation, Israel. *Geology* **29**, 887–890.

Rook, L. (2000) *The Taxonomy of Dmanisi Crania: Crucial Consequences for Hominid Taxonomy*. Science Published dEbate responses. http: / /www.sciencemag.org/cgi/eletters/289/5476/55b#EL2.

Rook, L., Bondioli, L., Köhler, M., and Moyà-Solà, S. (1999) *Oreopithecus* was a bipedal after all. Evidence from the iliac cancellous architecure. *Proceedings of the National Academy of Sciences USA* **96**, 8795–8799.

Rosas, A. and Bermúdez de Castro, J.M. (1998) On the taxonomic affinities of the Dmanisi mandible (Georgia). *American Journal of Physical Anthropology* **107**, 145–162.

Rose, M.D. (1984) A hominine hip bone, KNM-ER 3228, from East Lake Turkana, Kenya. *American Journal of Physical Anthropology* **63**, 371–378.

Rossie, J.B., Simons, E.L., Gauld, S.C., and Rasmussen, D. T. (2002) Paranasal sinus anatomy of *Aegyptopithecus*: implications for hominoid origins. *Proceedings of the National Academy of Sciences USA* **99**, 8454–8456.

Ruff, C.B., Trinkaus, E., and Holliday, T.W. (1997) Body mass and encephalization in Pleistocene *Homo*. *Nature* **387**, 173–176.

Rukang, W. (1964) A newly discovered mandible of the Sinanthropus type-*Sinanthropus lantianensis*. *Scientia Sin.*, **13**, 891–911.

Rukang, W. (1966) The hominid skull of Lantian, Shensi. *Vertebrata Palasiatica* **10**, 1–16.

Rukang, W. (1980) Palaeanthropology in the New China. In *Current Argument on Early Man* (L.K Königsson, ed.), pp. 182–206. Oxford: Pergamon Press.

Rukang, W. (1985) New Chinese *Homo* erectus and recent work at Zhoukoudian. In *Ancestors: The Hard Evidence* (E. Delson, ed.), pp. 245–248. New York, NY: Alan R. Liss.

Rukang, W. (1987) A revision of the classification of the Lufeng great apes. *Acta Anthropologica Sinica* **6**, 265–271.

Rukang, W. and Dong, X. (1982) Preliminary study of *Homo* erectus remains from Hexian, Anhui. *Acta Anthropologica Sinica* **1**, 2–13.

Ruse, M. (2000) *The Evolution Wars: a Guide to the Debates*. Santa Barbara, CA: ABC-Clio.

Ruse, M. (2006) *Darwinism and Its Discontents*. Cambridge: Cambridge University Press.

Ruvolo, M. (1993) Mitochondrial COII sequences and modern human origins. *Molecular Biology and Evolution* **10**, 1115–1135.

Ruvolo, M. (1997) Molecular phylogeny of the hominoids: inferences from multiple independent DNA sequence data sets. *Molecular Biology and Evolution* **14**, 248–265.

Ruvolo, M., Zehr, S., von Dornum, M., Pan, D., Chang, B., and Lin, J. (1993) Mitochondrial COII sequences and modern human origins. *Molecular Biology and Evolution* **10**, 1115–1135.

Sabater Pi, J. (1984) *El chimpancé y los orígenes de la cultura*. Barcelona: Anthropos.

Sabater Pi, J., Véa, J.J., and Serrallonga, J. (1997) Did the first hominids build nests? *Current Anthropology* **38**, 914–916.

Saffran, J.R., Aslin, R.N., and Newport, E.L. (1996) Statistical learning by 8-month-old infants. *Science* **274**, 1926–1928.

Santa Luca, A.P. (1980) The Ngandong fossil hominids. *Yale University Publications in Anthropology* **78**, 1–175.

Sarich, V. and Wilson, A.C. (1967a) Immunological time scale for hominid evolution. *Science* **158**, 1200–1203.

Sarich, V. and Wilson, A.C. (1967b) Rates of albumin evolution in primates. *Proceedings of the National Academy of Sciences USA* **58**, 142–148.

Sarmiento, E.E. (1998) Generalized quadrupeds, committed bipeds, and the shift to open habitats: an evolutionary model of hominid divergence. *American Museum Novitates* **3250**, 1–78.

Sarmiento, E.E. and Marcus, L.F. (2000) The os navicular of humans, great apes, OH 8, Hadar, and *Oreopithecus*: function, phylogeny, and multivariate analyses. *American Museum Novitates* **3288**, 1–38.

Sartono, S. (1971) Observations on a new skull of *Pithecanthropus erectus* (Pithecanthropus VIII) from Sangiran, Central Java. *Proceedings of the Academy of Science, Amsterdam B* **74**, 185–194.

Schaaffhausen, H. (1880) *Funde in der Sipkahöhle in Mahren. Sonderbericht der niederrheinischen Gesellschaft für Natur-und Heilkunde*, 260–264.

Schaeffer, B., Hecht, M.K., and Eldredge, N. (1972) Phylogeny and paleontology. *Evolutionary Biology* **6**, 31–46.

Schick, K.D. and Toth, N. (1993) *Making Silent Stones Speak*. New York, NY: Simon and Schuster.

Schlanger, N. (1994) *Châine opératoire* for an archaeology of the mind. In *The Ancient Mind. Elements of Cognitive Archaeology* (C. Renfrew and E.B.W. Zubrow, eds), pp. 143–151. Cambridge: Cambridge University Press.

Schmitz, R.W. (2003) Interdisziplinäre Untersuchungen an den Neufunden aus dem Neandertal Johann Carl Fuhlrott (1803–1877) gewidmet. *Mitteilungen der Gesellschaft für Urgeschichte* **12**, 25–45.

Schmitz, R.W., Serre, D., Bonani, G., Feine, S., Hillgruber, F., Krainitzki, H., Pääbo, S., and Smith, F.H. (2002) The Neandertal type site revisited: interdisciplinary investigations of skeletal remains from the Neander Valley, Germany. *Proceedings of the National Academy of Sciences USA* **99**, 13342–13347.

Schoetensack, O. (1908) *Der Unterkiefer des Homo heidelbergensis aus den Sanden von Mauer bei Heidelberg*. Leipzig: Wilhelm Englemann.

Schrein, C.M. (2006) Metric variation and sexual dimorphism in the dentition of *Ouranopithecus macedoniensis*. *Journal of Human Evolution* **50**, 460–468.

Schrenk, F., Bromage, T.G., Betzler, C.G., Ring, U., and Juwayeyi, Y.M. (1993) Oldest *Homo* and Pliocene biogeography of the Malawi Rift. *Nature* **365**, 833–836.

Schultz, A.H. (1930) The skeleton of the trunk and limbs of higher primates. *Human Biology* **2**, 303–348.

Schwarcz, H.P., Simpson, J.J., and Stringer, C.B. (1998) Neanderthal skeleton from Tabun: U-series data by gamma-ray spectrometry. *Journal of Human Evolution* **35**, 635–645.

Schwartz, G.T., Francis Thackeray, J., Reid, C., and van Reenan, J.F. (1998) Enamel thickness and the topography of the enamel–dentine junction in South African Plio-Pleistocene hominids with special reference to the Carabelli trait. *Journal of Human Evolution* **35**, 523–542.

Schwartz, H., Gabunia, L., Vekua, A., and Lordkipanidze, D. (2000) Taxonomy of the Dmanisi crania. *Science* **289**, 55–56.

Schwartz, J.H. (1984) Hominoid evolution: a review and a reassessment. *Current Anthropology* **25**, 655–672.

Schwartz, J.H. (1990) *Lufengpithecus* and its potential relationship to an orang-outan clade. *Journal of Human Evolution* **19**, 591–605.

Schwartz, J.H. (1999) Homeobox genes, fossils, and the origin of species. *The Anatomical Record (New Anatomy)* **257**, 15–31.

Schwartz, J.H. (2004) Getting to know *Homo* erectus. *Science* **305**, 53–54.

Schwartz, J.H. and Tattersall, I. (1996a) Significance of some previously unrecognized apomorphies in the nasal region of *Homo* neanderthalensis. *Proceedings of the National Academy of Sciences USA* **93**, 10852–10854.

Schwartz, J.H. and Tattersall, I. (1996b) Whose teeth? *Nature* **381**, 201–202.

Schwartz, J.H., Tattersall, I., and Eldredge, N. (1978) Phylogeny and classification of the primates revisited. *Yearbook of Physical Anthropology* **21**, 95–133.

Semaw, S., Renne, P., Harris, J.WK., Feibel, C.S., Bernor, R.L., Fesseha, N., and Mowbray, K. (1997) 2.5-million-year-old Stone Tools from Gona, Ethiopia. *Nature* **385**, 333–336.

Semaw, S., Simpson, S.W., Quade, J., Renne, P.R., Butler, R.F., McIntosh, W.C., Levin, N., Dominguez-Rodrigo, M., and Rogers, M.J. (2005) Early Pliocene hominids from Gona, Ethiopia. *Nature* **433**, 301–305.

Semendeferi, K. (1994) Evolution of the hominoid prefrontal cortex: a quantitative and image analysis of area 13 and 10. PhD Thesis, University of Iowa.

Semendeferi, K. and Damasio, H. (2000) The brain and its main anatomical subdivisions in living hominoids using magnetic resonance imaging. *Journal of Human Evolution* **38**, 317–332.

Semendeferi, K., Damasio, H., and Frank, R. (1997) The evolution of the frontal lobes: a volumetric anlysis based on three-dimensional reconstructions of magnetic resonance scans of human and ape brains. *Journal of Human Evolution* **32**, 375–388.

Senut, B. (1991) Origine(s) de la bipédie humaine: approache paléontologique. In *Origine(s) de la bipédie chez les hominidés* (Y. Coppens and B. Senut, eds), pp. 245–257. Paris: Editions du CNRS.

Senut, B., Pickford, M., Gommery, D., Mein, P., Cheboi, K., and Coppens, Y. (2001) First hominid from the Miocene (Lukeino Formation, Kenya). *Comptes rendus de l'Académie des Sciences Paris* **332**, 137–144.

Sereno, M.I. and Tootell, R. (2005) From monkeys to humans: what do we now know about brain homologies? *Current Opinion in Neurobiology* **15**, 135–144.

Serre, D., Langaney, D., Chech, M., Teschler-Nicola, M., Paunovicz, M., Mennecier, P., Hofreiter, M., Possnert, G., and Pääbo, S. (2004) No evidence of Neandertal mtDNA contribution to early modern humans. *PLoS Biology* **2**, 313–317.

Sesé, C. and Gil, E. (1987) Los micromamíferos del Pleistoceno Medio del complejo cárstico de Atapuerca (Burgos). In *El hombre fósil de Ibeas y el Pleistoceno de la Sierra de Atapuerca* (E. Aguirre, E. Carbonell, and J.M. Bermúdez de Castro, eds), pp. 74–87. Valladolid: Junta de Castilla y León. Consejería de Cultura y Bienestar Social.

Sevink, J., Rammelzwaal, A., and Spaargaren, O.C. (1984) The soils of Southern Lazio and adjacent Campania. *Fysisch Geografisch en. Bodenkundig Laboratorium Universiteit van Amsterdam* **38**, 1–44.

Seyfarth, R.M. and Cheney, D.L. (1984) Grooming, alliances and reciprocal altruism in Vervet monkeys. *Nature* **308**, 541–543.

Seyfarth, R.M. and Cheney, D.L. (1992) Meaning and mind in monkeys. *Scientific American* **267**, 122–128.

Shapiro, H.L. (1976) *Peking Man: The Discovery, Disappearance and Mystery of a Priceless Scientific Treasure*. London: Allen and Unwin.

Shen, C. and Qi, W. (2004) Lithic technological variability of the Middle Pleistocene in the Eastern Nihewan Basin, Northern China. *Asian Perspectives* **43**, 281–301.

Shreeve, J. (1994) 'Lucy', crucial early human ancestor, finally gets a head. *Science* **264**, 34–35.

Shreeve, J. (1995) Sexing fossils: a boy named Lucy? *Science* **270**, 1297–1298.

Sibley, C.G. and Ahlquist, J.E. (1984) The phylogeny of the hominoid primates, as indicated by DNA-DNA hybridization. *Journal of Molecular Evolution* **20**, 2–15.

Sibley, C., Comstock, J.A., and Ahlquist, J.E. (1990) DNA hybridization evidence of hominoid phylogeny: a reanalysis of the data. *Journal of Molecular Evolution* **30**, 202–236.

Siddall, M. (1998) The follies of ancestor worship. *Nature Debates*, http://helix.nature.com/debates/fossil/fossil_3.html.

Sillen, A. (1992) Strontium-calcium ratios (Sr/Ca) of *Australopithecus robustus* and associated fauna from Swartkrans. *Journal of Human Evolution* **23**, 495–516.

Sillen, A., Hall, G., and Armstrong, R. (1995) Strontium calcium ratios (Sr/Ca) and strontium isotopic ratios (87Sr/86Sr) of *Australopithecus robustus* and *Homo* sp. in Swartkrans. *Journal of Human Evolution* **28**, 277–285.

Sillen, A., Hall, G., Richardson, S., and Armstrong, R. (1998) 87Sr/86Sr ratios in modern and fossil food-webs of the Sterkfontein Valley: implications for early hominid habitat preference. *Geochimica et Cosmochimica Acta* **62**, 2463–2473.

Simons, E.L. (1961) The phyletic position of *Ramapithecus*. *Postilla* **57**, 1–9.

Simons, E.L. (1964) On the mandible of *Ramapithecus*. *Proceedings of the National Academy of Sciences USA* **51**, 528–535.

Simons, E.L. (1965) New fossil apes from Egypt and the initial differentiation of Hominoidea. *Nature* **205**, 135–139.

Simons, E.L. and Pilbeam, D. (1965) Preliminary revision of the Dryopithecinae (Pongidae, Anthropoidea). *Folia Primatologica* **3**, 81–152.

Simpson, G.G. (1931) A new classification of mammals. *Bulletin of the American Museum of Natural History* **59**, 259–293.

Simpson, G.G. (1945) The principles of classification and a classification of mammals. *Bulletin of the American Museum of Natural History* **85**, 1–350.

Simpson, G.G. (1953) *The Major Features of Evolution*. New York, NY: Columbia University Press.

Singer, R. and Wymer, J. (1982) *The Middle Stone Age of Klasies River Mouth in South Africa*. Chicago, IL: Chicago University Press.

Skelton, R.R., McHenry, H.M., and Drawhorn, G.M. (1986) Phylogenetic analysis of early hominids. *Current Anthropology* **27**, 21–35.

Skinner, M.M., Gordon, A.D., and Collard, N.J. (2006) Mandibular size and shape variation in the hominins at Dmanisi, Republic of Georgia. *Journal of Human Evolution* **51**, 36–49.

Slatkin, M. and Hudson, R.R. (1991) Pairwise comparisons of mitochondrial DNA sequences in stable and exponentially growing populations. *Genetics* **129**, 555–562.

Smith, K. (2006) Homing in on the genes for humanity. What makes us different from chimps? *Nature* **442**, 725.

Smouse, P.E. and Li, W.H. (1987) Likelihood analysis of mitochondrial restriction-clavage patterns for the human-chimpanzee-gorilla trichotomy. *Evolution* **41**, 1162–1176.

Sober, E. and Wilson, D.S. (1998) *Unto Others. The Evolution and Psychology of Unselfish Behavior*. Cambridge, MA: Harvard University Press.

Soldevila, M., Andrés, A.M., Ramírez-Soriano, A., Marquès-Bonet, T., Calafell, C., Arcadi Navarro, A., and Bertranpetit, J. (2006) The prion protein gene in humans revisited: Lessons from a worldwide resequencing study. *Genome Research* **16**, 231–239.

Solecki, R.S. (1953) The Shanidar Cave Sounding 1953 season. With notes concerning the discovery of the first paleolithic skeleton in Iraq. *Sumer* **9**, 229–232.

Solecki, R. (1975) Shanidar IV, a Neanderthal flower burial in Northern Iraq. *Science* **190**, 880–881.

Soligo, C. and Martin, R.D. (2006) Adaptive origins of primates revisited. *Journal of Human Evolution* **50**, 414–430.

Sondaar, P.Y., Van den Bergh, G., Mubroto, B., Aziz, F., de Vos, J., and Batu, U.L. (1994) Middle Pleistocene faunal turnover and colonization of Flores (Lesser Sunda Islands, Indonesia) by *Homo* erectus. *Comptes rendus de l'Académie des Sciences Paris* **319**, 1255–1262.

Spencer, H. (1893) *The Principles of Ethics*. London: Williams and Norgate.

Spencer, F. (1990) *Piltdown: A Scientific Forgery*. Oxford: Oxford University Press.

Sponheimer, M. and Lee-Thorp, J.A. (1999) Isotopic evidence for the diet of an early hominid, *Australopithecus africanus*. *Science* **283**, 368–370.

Sponheimer, M., Passey, B.H., de Ruiter, D.J., Guatelli-Steinberg, D., Cerling, T.E., and Lee-Throp, J.A. (2006). Isotopic Evidence for Dietary Variability in the Early Hominin Paranthropus robustus. *Science*, **314,** 980–982.

Spoor, F., Wood, B., and Zonneveld, F. (1994) Implications of early hominid labyrinthine morphology for evolution of human bipedal locomotion. *Nature* **369**, 645–648.

Spoor, F., O'Higgins, P., Dean, C., and Lieberman, D.E. (1999) Anterior sphenoid in modern humans. *Nature* **397**, 572.

Steele, J. (1999) Stone legacy of skilled hands. *Nature* **399**, 24–25.

Stefansson, H., Helgason, A., Thorleifsson, G., Steinthorsdottir, V., Masson, G., Barnard, J., Baker, A., Jonasdottir, A., Ingason, A., Gudnadottir, V.G.*et al.*

(2005) A common inversion under selection in Europeans. *Nature Genetics* **37**, 129–137.

Stehlin, H.G. (1909) Remarques sur les faunules de mammiferes des couches eocenes et oligocenes du Basin de Paris. *Bulletin de la Société Géologique de France* serie 8 **9**, 488–520.

Steklis, H.D. (1988) Primate communication, comparative neurology, and the origin of language re-examined. In *The Genesis of Language* (M.E Landsberg, ed.), pp. 37–63. Berlin: Mouton de Gruyter.

Stern, J.T. and Susman, R.L. (1983) The locomotor anatomy of *Australopithecus afarensis*. *American Journal of Physical Anthropology* **60**, 279–317.

Stern, Jr, J.T. and Susman, R.L. (1991) 'Total morphological pattern' versus the 'magic trait': conflicting approaches to the study of early hominid bipedalism. In *Origine(s) de la bipédie chez les hominidés* (Y. Coppens and B. Senut, eds), pp. 99–111. Paris: Editions du CNRS.

Stewart, T.D. (1958) First views of the restored Shanidar I skull. *Sumer* **14**, 90–96.

Stoneking, M., Jorde, L.B., Bhatia, K., and Wilson, A.C. (1990) Geographic variation of human mitochondrial DNA from Papua New Guinea. *Genetics* **124**, 717–733.

Strait, D.S. and Grine, F.E. (1999) Cladistics and early hominid phylogeny. *Science* **285**, 1210–1211.

Strait, D.S., Grine, F.E., and Moniz, M.A. (1997) A reappraisal of early hominid phylogeny. *Journal of Human Evolution* **32**, 17–82.

Stern, J.T. and Susman, R.L. (1983) The locomotor anatomy of *Australopithecus afarensis*. *American Journal of Physical Anthropology* **60**, 279–317.

Stringer, C.B. (1980) The phylogenetic position of the Petralona cranium. *Anthropos* **7**, 81–95.

Stringer, C.B. (1984) The definition of *Homo* erectus and the existence of the species in Africa and Europe. In *The Early Evolution of Man with Special Emphasis on Southeast Asia and Africa*, vol. 69 (P. Andrews and J.L.Franzen, eds), pp. 131–143. Frankfurt: Courier Forschungsinstitut Senckenberg.

Stringer, C.B. (1985) Middle Pleistocene hominid variability and the origin of Late Pleistocene humans. In *Ancestors: the Hard Evidence* (E. Delson, ed.), pp. 289–295. New York, NY: Alan R. Liss.

Stringer, C.B. (1986) The credibility of *Homo* habilis. In *Major Topics in Primate and Human Evolution* (B. Wood, L. Martin, and P. Andrews, eds), pp. 266–294. Cambridge: Cambridge University Press.

Stringer, C.B. (1987) A numerical cladistic analysis for the genus *Homo*. *Journal of Human Evolution* **16**, 135–146.

Stringer, C.B. (1992) Evolution of early humans. In *The Cambridge Encyclopedia of Human Evolution* (S. Jones, R. Martin, and D. Pilbeam, eds), pp. 241–251. Cambridge: Cambridge University Press.

Stringer, C. (1993) Secrets of the pit of the bones. *Nature* **362**, 501–502.

Stringer, C.B. (1996) Current issues in modern human origins. In *Contemporary Issues in Human Evolution* (W.E Meikle, F.C Howell, and N.G Jablonski, eds), pp. 115–134. San Francisco, CA: California Academy of Sciences.

Stringer, C.B. and Trinkaus, E. (1981) The Shanidar Neanderthal crania. In *Aspects of Human Evolution* (C.B Stringer, ed.), pp. 129–165. London: Taylor and Francis.

Stringer, C.B. and Andrews, P. (1988) Genetic and fossil evidence for the origin of modern humans. *Science* **239**, 1263–1268.

Stringer, C. and Gamble, C. (1993) *In Search of the Neanderthals*. London: Thames and Hudson.

Stringer, C.B., Howell, F.C., and Melentis, J.K. (1979) The significance of the fossil hominid skull from Petralona Cave, Greece. *Nature* **292**, 81–95.

Stringer, C.B., Hublin, J.J., and Vandermeersch, B. (1984) The origin of anatomically modern humans in Western Europe. In *The Origins of Modern Humans: a World Survey of the Fossil Evidence* (F.H.Smith and F. Spencer, eds), pp. 51–135. New York, NY: Alan R. Liss.

Stringer, C., Grün, R., Schwarcz, H., and Goldberg, P. (1989) ESR dates for the hominid burial site of Es Skuhl in Israel. *Nature* **338**, 756–758.

Stringer, C.B., Trinkaus, E., Roberts, M.B., Parfitt, S.A., and Macphail, R.I. (1998) The Middle Pleistocene human tibia from Boxgrove. *Journal of Human Evolution* **34**, 509–557.

Struhsaker, T.T. (1967) Auditory communication among vervet monkeys (*Cercopithecus aethiops*). In *Social Communication Among Primates* (S.A Altmann, ed.), pp. 281–324. Chicago, IL: Chicago University Press.

Susman, R.L. and Stern, J. (1979) Telemetered electron myography of flexor digitum superficialis in *Pan troglodytes* and implications for interpretation of the OH 7 hand. *American Journal of Physical Anthropology* **50**, 565–574.

Susman, R.L. and Stern, J.T. (1991) Locomotor behavior of early hominids: epistemology and fossil evidence. In *Origine(s) de la bipédie chez les hominidés* (Y. Coppens and B. Senut, eds), pp. 121–131. Paris: Editions du CNRS.

Susman, R.L., Stern, Jr, J.T., and Jungers, W.L. (1984) Arboreality and bipedality on the Hadar hominids. *Folia Primatologica* **43**, 113–156.

Susman, R.L., de Ruiter, D., and Brain, C.K. (2001) Recently identified postcranial remains of *Paranthropus* and Early *Homo* from Swartkrans Cave, South Africa. *Journal of Human Evolution* **41**, 607–629.

Suwa, G. (1988) Evolution of the 'Robust' Australopithecines in the Omo succession: evidence from mandibular premolar morphology. In *Evolutionary History of the "Robust" Australopithecines* (F.E Grine, ed.), pp. 199–222. New York, NY: Aldine de Gruyter.

Suwa, G., Asfaw, B., Beyene, Y., White, T.D., Katoh, S., Nagaoka, S., Nakaya, H., Uzawa, K., Renne, P., and WoldeGabriel, G. (1997) The first skull of *Australopithecus boisei*. *Nature* **389**, 489–492.

Suzuki, H., and Takai, F. (eds) (1970) *The Amud Man and his Cave Site*. Tokyo: Keigaku Publishing Co.

Swisher, III, C.C., Curtis, G.H., Jacob, T., Getty, A.G., and Widiasmoro, A.S. (1994) Age of the earliest known hominids in Java, Indonesia. *Science* **263**, 1118–1121.

Swisher, III, C.C., Rink, W.J., Antón, S.C., Schwarcz, H.P., Curtis, G.H., Suprijo, A., and Widiasmoro, A.S. (1996) Latest *Homo* erectus of Java: potential contemporanity with *Homo* sapiens in Southeast Asia. Science 274, 1870–1874.

Swisher, C.C., Rink, W.J., Schwarcz, H.P., and Antón, S.C. (1997) Dating the Ngandong humans. *Science* **276**, 1575–1576.

Szalay, F.S. and Delson, E. (1979) *Evolutionary History of Primates*. New York, NY: Academic Press.

Tague, R.G. and Lovejoy, C.O. (1986) The obstetrics pelvis of AL 288–1 (Lucy). *Journal of Human Evolution* **15**, 237–255.

Taieb, M. (1974) *Évolution Quaternaire du Bassin de l'Awash*. PhD Thesis, University of Paris VI.

Taieb, M., Johanson, D.C., Coppens, Y., and Aronson, J.L. (1976) Geological and paleontological background of Hadar hominid site, Afar, Ethiopia. *Nature* **260**, 289–293.

Tajima, F. (1983) Evolutionary relationship of DNA sequences in finite populations. *Genetics* **105**, 437–460.

Takahata, N. and Nei, M. (1985) Gene genealogy and variance of interpopulational nucleotide differences. *Genetics* **110**, 325–344.

Takahata, N., Lee, S.-H., and Satta, Y. (2001) Testing multi-regionality of modern human origins. *Molecular Biology and Evolution* **18**, 172–183.

Tatsumi, Y. and Kimura, N. (1991) Secular variation of basalt chemistry in the Kenya rift: evidence of the pulsing of asthenosphere upwelling. *Earth and Planetary Science Letters* **104**, 99–113.

Tattersall, I. (1995a) *The Fossil Trail*. Oxford: Oxford University Press.

Tattersall, I. (1995b) *The Last Neanderthal*. New York, NY: Macmillan.

Tattersall, I. (1998) Neanderthal genes: what do they mean? *Evolutionary Anthropology* **6**, 157–158.

Tattersall, I. and Schwartz, J.H. (1999) Hominids and hybrids: the place of Neanderthals in human evolution. *Proceedings of the National Academy of Sciences USA* **96**, 7117–7119.

Tauxe, L., Deino, A.L., Behrensmeyer, A.K., and Potts, R. (1992) Pinning down the Brunhes/Matuyama and upper Jaramillo Boundaries: A reconciliation of orbital and isotopic time scales. *Earth and Planetary Science Letters* **109**, 561–562.

Tavare, S. (1984) Line of descent and genealogical process and their application in population genetics model. *Theoretical Population Biology* **26**, 119–164.

Tchernov, E. (1989) The age of the Ubeidiya Formation. *Israeli Journal of Earth Sciences* **36**, 3–30.

Teaford, M.F. and Walker, A.C. (1984) Quantitative differences in dental microwear between primate species with different diets and a comment on the presumed diet of *Sivapithecus*. *American Journal of Physical Anthropology* **64**, 191–200.

Templeton, A.R. (2002) Out of Africa again and again. *Nature* **416**, 45–51.

Templeton, A.R. (2005) Haplotype trees and modern human origins. *Yearbook of Physical Anthropology* **48**, 33–59.

The Chimpanzee Sequencing and Analysis Consortium (2005) Initial sequence of the chimpanzee genome and comparison with the human genome. *Nature* **437**, 69–87.

Thorne, A.G. and Wolpoff, M.H. (1992) The multiregional evolution of humans. *Scientific American Presents* **266**, 76–83.

Thorne, A., Grün, R., Mortimer, G., Spooner, N.A., Simpson, J.J., McCulloch, M., Taylor, L., and Curnoe, D. (1999) Australia's oldest human remains: age of the Lake Mungo 3 skeleton. *Journal of Human Evolution* **36**, 591–612.

Thrush, P.W. (1968) *A Dictionary of Mining, Mineral, and Related Terms*. Washington DC: US Bureau of Mines.

Tianyuan, L. and Etler, D.A. (1992) New Middle Pleistocene hominid crania from Yunxian in China. *Nature* **357**, 404–407.

Tiemei, C. and Sixun, Y. (1988) Uranium-series dating of bones and teeth from Chinese paleolithic sites. *Archaeometry* **30**, 59–76.

Tiemei, C. and Yinyun, Z. (1991) Paleolithic chronology and possible coexistence of *Homo* erectus and *Homo* sapiens in China. World Archaeology 23, 147–154.

Tillet, T. (1978) Présence de pendeloques en milieu atérien au Niger oriental. *Bulletin de la Société Préhistorique Française, Paris* **75**, 273–275.

Tobias, P.V. (1964) The Olduvai Bed I hominine with special reference to its cranial capacity. *Nature* **202**, 3–4.

Tobias, P.V. (1965) *Australopithecus, Homo* habilis, tool-using and tool-making. *South African Archaeological Bulletin* **20**, 167–192.

Tobias, P.V. (1967) *The Cranium and Maxillary Dentition of Australopithecus (Zinjanthropus) boisei. Olduvai Gorge*, vol. 2. Cambridge: Cambridge University Press.

Tobias, P.V. (1968) Middle and early Upper Pleistocene members of the genus *Homo* in Africa. In *Evolution und Hominisation* (G. Karth, ed.), pp. 176–194. Stuttgart: G. Fischer.

Tobias, P.V. (1971) *The Brain in Hominid Evolution*. New York, NY: Columbia University Press.

Tobias, P.V. (1975) Brain evolution in the Hominoidea. In *Primate Functional Morphology and Evolution* (R.H Tuttle, ed.), pp. 353–392. The Hague: Mouton Publishers.

Tobias, P.V. (1978) The earliest Transvaal members of the genus *Homo* with another look at some problems of hominid taxonomy and systematics. *Zeitschrift für Morphologie und Anthropologie* **69**, 225–265.

Tobias, P.V. (1980) '*Australopithecus afarensis*' and *A. africanus*: Critique and alternative hypothesis. *Palaeontologia africana* **23**, 1–17.

Tobias, P.V. (1982a) Man the tottering biped: the evolution of his erect posture. In *Proprioception, Posture and Emotion* (D. Garlick, ed.), pp. 1–13. Sydney: Committee in Postgraduate Medical Education. The University of New South Wales.

Tobias, P.V. (1982b) The antiquity of man: human evolution. In *Human Genetics, Part A: The Unfolding Genome* (B. Bonné-Tamir, ed.), pp. 195–214. New York, NY: Alan R. Liss.

Tobias, P.V. (1985a) Single characters and the total morphological pattern redefined: the sorting effected by a selection of morphological features of the early hominids. In *Ancestors: The Hard Evidence* (E. Delson, ed.), pp. 94–101. New York, NY: Alan R. Liss.

Tobias, P.V. (1985b) The conquest of the savannah and the attaining of erect bipedalism. In *Homo*: Journey to the Origin of Man's History (C. Peretto, ed.), pp. 36–45. Cataloghi Marsilio.

Tobias, P.V. (1985c) Punctuational and phyletic evolution in the hominids. In *Species and Speciation* (E.S Vrba, ed.), pp. 131–141. Pretoria: Transvaal Museum Monograph no. 4.

Tobias, P.V. (1986) Delineation and dating of some major phases in hominidization and hominization since the Middle Miocene. *South African Journal of Science* 82, 92–94.

Tobias, P.V. (1987) The brain of *Homo* habilis: a new level of organization in cerebral evolution. *Journal of Human Evolution* 6, 741–761.

Tobias, P.V. (1988) Numerous apparently synapomorphic features in *Australopithecus robustus, Australopithecus boisei* and *Homo* habilis: support for the Skelton-McHenry-Drawhorn Hypothesis. In *Evolutionary History of the "Robust" Australopithecines* (F.E Grine, ed.), pp. 293–308. New York, NY: Aldine de Gruyter.

Tobias, P.V. (1990) When and by whom was the Taung skull discovered? In *Para conocer al hombre: Homenaje a Santiago Genovés a los 33 años como investigador en la UNAM* (L.L Tapia, ed.), pp. 207–213. México: UNAM.

Tobias, P.V. (1991a) Relationship between apes and humans. In *Perspectives in Human Evolution* (A. Sahni and R. Gaur, eds), pp. 1–19. Delhi: Renaissance Publishing House.

Tobias, P.V. (1991b) The environmental background of hominin emergence and the appearance of the genus *Homo. Human Evolution* 6, 129–142.

Tobias, P.V. (1991c) *Olduvai Gorge, Volume IV. The Skulls, Endocasts, and Teeth of Homo habilis*. Cambridge: Cambridge University Press.

Tobias, P.V. (1991d) The age at death of the Olduvai *Homo* habilis population and the dependence of demographic patterns on prevailing environmental conditions. In *Studia Archeologica: Liber Amicorum Jacques A.E. Nenquin* (H. Thoes, J. Bourgeois, F. Vermeulen, P. Crombe, and K. Verlaeckt, eds), pp. 57–65. Gent: Universiteit Gent.

Tobias, P.V. (1992) The species *Homo* habilis: example of a premature discovery. *Annales Zoologici Fennici* **28**, 371–380.

Tobias, P.V. (1994) The evolution of early hominids. In *Companion Encyclopedia of Anthropology. Humanity, Culture and Social Life* (T. Ingold, ed.), pp. 33–78. London: Routledge.

Tobias, P.V. (1995) *The Communication of the Dead. Earliest Vestiges of the Origin of Articulate Language.* Amsterdam: Stichting Nederlands Museum voor Anthropologie en Praehistorie.

Tobias, P. (1997a) El descubrimiento de Little Foot y la luz que proporciona sobre cómo los homínidos se volvieron bípedos. *Ludus Vitalis* **8**, 3–20.

Tobias, P.V. (1997b) Orígenes evolutivos de la lengua hablada. In *Senderos de la evolución humana* (C.J.Cela-Conde, R. Gutiérrez Lombardo and J. Martínez Contreras, eds), pp. 35–52. México: Ludus Vitalis, número especial 1.

Tobias, P.V. (2003) Encore Olduvai. *Science* **299**, 1193–1194.

Tobias, P.V. and von Koegniswald, G.H.R. (1964) A comparison between the Olduvai hominines and those of Java and some implications for hominid phylogeny. *Nature* **204**, 515–518.

Tobias, P.V. and Clarke, R.J. (1996) Faunal evidence and Sterkfontein Member 2 foot of early man: response to J. C. McKee. *Science* **271**, 1301–1302.

Toth, N. (1985a) Archaeological evidence for preferential right-handedness in the lower and middle Pleistocene, and its possible implication. *Journal of Human Evolution* **14**, 607–614.

Toth, N. (1985b) The Oldowan reassessed: a close look at early stone artifact. *Journal of Archaeological Science* **12**, 101–120.

Toth, N. and Schick, K.D. (1993). Early stone industries and inferences regarding language and cognition. In *Tools, Language and Cognition in Human Evolution* (K.R Gibson and T. Ingold, eds), pp. 346–362. Cambridge: Cambridge University Press.

Toussaint, M., Macho, G.A., Tobias, P., Partridge, T., and Hughes, A.R. (2003) The third partial skeleton of a late Pliocene hominin (Stw 431) from Sterkfontein, South Africa. *South African Journal of Science* **99**, 215–223.

Trinkaus, E. (1981) Neanderthal limb proportions and cold adaptation. In *Aspects of Human Evolution* (C.B. Stringer, ed.), pp. 187–224. London: Taylor and Francis.

Trinkaus, E. (1983) *The Shanidar Neandertals*. New York, NY: Academic Press.

Trinkaus, E. (1984) Neanderthal pubic morphology and gestation length. *Current Anthropology* **25**, 509–514.

Trinkaus, E. (1988) The evolutionary origins of the Neandertals, or, why were there Neandertals? In *L'Homme de Neandertal*, vol. 3, *L'anatomie* (E. Trinkaus, ed.), pp. 11–29. Liège: Etudes et Recherches Archéologiques de l'Université de Liège.

Trinkaus, E. (2001) The Neandertal paradox. In *Neanderthals and Modern Humans in Late Pleistocene Eurasia* (Finlayson C, ed.), pp. 73–74. Gibraltar: The Gibraltar Museum.

Trinkaus, E. and Howells, W.W. (1980) Neandertales. In *Paleontología Humana* (E. Aguirre, ed.), pp. 15–26. Barcelona: Prensa Científica.

Trinkaus, E. and Smith, F.H. (1985) The Fate of Neandertals. In *Ancestors: the Hard Evidence* (E. Delson, ed.), pp. 325–333. New York, NY: Alan R. Liss.

Trinkaus, E. and Shipman, P. (1993) *The Neandertals. Changing the Image of Mankind*. New York, NY: Alfred A. Knopf.

Trinkaus, E. and Zilhao, J. (2002) Phylogenetic implications. In *Portrait of the Artist as a Child. The Gravetian Human Skeleton from the Abrigo DoLagar Velho and its Archeological Context* (pp. 497–558). Lisbon: Portuguese Institute of Archaeology.

Trinkaus, E., Zilhão, J., and Duarte, C. (1999) The Lapedo Child: Lagar Velho 1 and our Perceptions of the Neandertals. *Mediterranean Prehistory Online*, http://www.med.abaco-mac.it/articles/doc/013.htm.

Turbón, D., Pérez-Pérez, A., and Stringer, C.B. (1997) A multivariate analysis of Pleistocene hominids: testings hypotheses of European origins. *Journal of Human Evolution* **32**, 449–468.

Turner, A. (1992) Large carnivores and earliest European hominids: changing determinants of resource availability during the Lower and Middle Pleistocene. *Journal of Human Evolution* **22**, 109–126.

Turq, A., Martínez-Navarro, B., Palmqvist, P., Arribas, A., Agustí, A., and Rodríguez Vidal, J. (1996) Le Plio-Pleistocene de la région d'Orce, province de Granade, Espagne: bilan et perspectives de recherche. *Paleo* **8**, 161–204.

Tuttle, R.H. (1981) Evolution of hominid bipedalism, and prensile capacities. *Philosophical Transactions of the Royal Society of London Series B* **292**, 89–94.

Tuttle, R.H. and Basmajian, J.V. (1974) Electromyography of brachial muscles in *Pan*, *Gorilla* and hominid evolution. *American Journal of Physical Anthropology* **41**, 71–90.

Tuttle, R.H., Webb, D., Tuttle, N.I., and Baksh, M. (1990) Further progress on the Laetoli Trails. *Journal of Archaeological Science* **17**, 347–362.

Tuttle, R.H., Webb, D.M., and Tuttle, N.I. (1991) Laetoli footprint trails and the evolution of hominid bipedalism. In *Origine(s) de la bipédie chez les hominidés* (Y. Coppens and B. Senut, eds), pp. 187–198. Paris: Editions du CNRS.

Uddin, M., Wildman, D.E., Liu, G., Xu, W., Johnson, R. M., Hof, P.R., Kapatos, G., Grossman, L.I., and Goodman, M. (2004) Sister grouping of chimpanzees and humans as revealed by genome-wide phylogenetic analysis of brain gene expression profiles. *Proceedings of the National Academy of Sciences USA* **101**, 2957–2962.

Ungar, P.S. and Grine, F. (1991) Incisor size and wear in *Australopithecus africanus* and *Paranthropus robustus*. *Journal of Human Evolution* **20**, 313–340.

Ungar, P.S., Grine, F.E., Teaford, M.F., and El Zaatari, S. (2006) Dental microwear and diets of African early *Homo*. *Journal of Human Evolution* **50**, 78–95.

Valladas, H., Reyss, J.L., Arensburg, B., Belfer-Cohen, A., Goldberg, P., Laville, H., Meignen, L., Rak, Y.T., E., Tillier, A., and Vandermeersch, B. (1987) Thermoluminiscence dates for the Neanderthal burial site at Kebara in Israel. *Nature* **330**, 159–160.

Valladas, H., Reyss, J.L., Joron, J.L., Valladas, G., Bar-Yosef, O., and Vandermeersch, B. (1988) Thermoluminiscence dating of Mousterian 'Proto-Cro-Magnon' remains from Israel and the origin of modern man. *Nature* **331**, 614–616.

Vallois, H. (1935) Le *Javanthropus. Anthropologie* **45**, 71–84.

Vallois, H.V. (1960) L'Homme de Rabat. *Bulletin d'Archéologie marocaine* **3**, 87–91.

van der Merwe, N.J., Francis Thackeray, J., Lee-Thorpa, J. A., and Luyta, J. (2003) The carbon isotope ecology and diet of *Australopithecus africanus* at Sterkfontein, South Africa. *Journal of Human Evolution* **44**, 581–597.

Vandermeersch, B. (1981) *Les Hommes Fossiles de Qafzeh (Israël)*. Paris: CNRS.

Vandermeersch, B. (1989) The evolution of modern humans: recent evidence from Southwest Asia. In *The Human Revolution: Behavioural and Biological Perspectives on the Origins of Modern Humans* (P. Mellars and C. Stringer, eds), pp. 155–164. Princeton, NJ: Princeton University Press.

Vandermeersch, B. (1996) New perspectives on the 'Proto-cromagnons' from Israel. *Human Evolution* **11**, 107–112.

Vandermeersch, B. and Garralda, M.D. (1994) El origen del hombre moderno en Europa. In *Biología de poblaciones humanas: problemas metodológicos e interpretación ecológica* (C. Bernis, C. Varea, G. Robles and A. González, eds), pp. 26–33. Madrid: Ediciones de la Universidad Autónoma.

Vekua, A., Lordkipanidze, D., Rightmire, G.P., Agusti, J., Ferring, R., Maisuradze, G., Mouskhelishvili, A., Nioradze, M., de Leon, M.P., Tappen, M., Tvalchrelidze, M., and Zollikofer, C. (2002) A new skull of early Homo from Dmanisi, Georgia. *Science* **297**, 85–89.

Venter, J.C., Adams, M.D., Myers, E.W., Li, P.W., Mural, R.J., Sutton, G.G., Smith, H.O., Yandell, M., Evans, C. A., Holt, R.A. *et al.* (2001) The sequence of the human genome. *Science* **291**, 1304–1351.

Verhaegen, M. (1996) Morphological distance between australopithecine, human and ape Skulls. *Human Evolution* **11**, 35–41.

Vermeersch, P.M., Paulissen, E., Stokes, S., Charlier, C., Van Peer, P., Stringer, C., and Lindsay, W. (1998) Middle Paleolithic burial of a modern human at Taramsa Hill, Egypt. *Antiquity* **72**, 475–484.

Vigilant, L., Stoneking, M., Harpending, H., Hawkes, K., and Wilson, AC (1991) African populations and the evolution of human mitochondrial DNA. *Science* **253**, 1503–1507.

Vignaud, P., Duringer, P., Mackaye, H.T., Likius, A., Blondel, C., Boisserie, J.-R., de Bonis, L., Eisenmann, V., Etienne, M.-E., Geraads, D. *et al.* (2002) Geology and palaeontology of the Upper Miocene Toros-Menalla hominid locality, Chad. *Nature* **418**, 152–155.

Vogel, G. (1999) Chimps in the wild show stirrings of culture. *Science* **284**, 2070–2073.

Vogel, J.C. and Waterbolk, H.T. (1967) Groningen radio carbon dates VII. *Radio Carbon* **9**, 145.

Voight, B.F., Adams, A.M., Frisse, L.A., Qian, Y., Hudson, R.R., and Di Rienzo, A. (2005) Interrogating multiple aspects of variation in a full resequencing data set to infer human population size changes. *Proceedings of the National Academy of Sciences USA* **102**, 18508–18513.

von Koenigswald, G.H.R. (1935) Eine fossile Saugetierne-fauna mit simia aus Sudchina. *Proceedings, Koninklijke Akademie van Wetenschappen, Amsterdam* **38**, 872–879.

von Koenigswald, G.H.R. (1938) Ein neuer Pithecan-thropus- Schädel. *Proceeding, Royal Academy of Amsterdam* **41**, 185–192.

von Koenigswald, G.H.R. (1949) Zur stratigraphie des javanischen Pleistocän. *De Ingenieur in Nederlandsch-Indie I*, 185–201.

von Koenigswald, G.H.R. (1981) Davidson Black, Peking Man, and the Chinese Dragon. In *Homo* erectus: Papers in Honor of Davidson Black (B.A Sigmon and J.S Cybulski, eds), pp. 27–39. Toronto: University of Toronto Press.

von Koenigswald, G.H.R. and Weidenreich, F. (1939) The relationship between *Pithecanthropus* and *Sinanthropus. Nature* **144**, 926–929.

Vrba, E.S. (1974) Chronological and ecological implications of the fossil Bovidae at the Sterkfontein Australopithecine site. *Nature* **250**, 19–23.

Vrba, E.S. (1980) Evolution, species and fossils: how does life evolve? *South African Journal of Science* **76**, 61–84.

Vrba, E.S. (1982) *Biostratigraphy and chronology, based particularly on Bovidae of Southern African Hominid-associated assemblages: Makapansgat, Sterkfontein, Taung, Kromdraai, Swartranks; also Elandsfontein (Saldanha), Broken Hill (now Kabwe) and Cave of Hearths*. Paper presented at the 1er Congress International de Paleontologie Humaine, Nice.

Vrba, E.S. (1984) Evolutionary patterns and process in the sister-group Alcelaphini-Aepycerotini (Mammalia: Bovidae). In *Living Fossils* (N. Eldredge and S.M Stanley, eds), pp. 62–79. New York, NY: Springer Verlag.

Vrba, E.S. (1985) Ecological and adaptive changes associated with early hominin evolution. In *Ancestors: the Hard Evidence* (E. Delson, ed.), pp. 63–71. New York, NY: Alan R. Liss.

Waddington, C.H. (1960) *The Ethical Animal*. London: Allen and Unwin.

Waddle, D.M. (1994) Matrix correlation tests support a single origin for modern humans. *Nature* **368**, 452–454.

Walker, A. (1981) The Koobi Fora hominins and their bearing on the origins of the genus *Homo*. In *Homo erectus: Papers in Honor of Davidson Black* (B.A Sigmon and J.S Cybulski, eds), pp. 193–215. Toronto: University of Toronto Press.

Walker, A. (1993) Perspectives on the Nariokotome discovery. In *The Nariokotome Homo erectus Skeleton* (A. Walker and R. Leakey, eds), pp. 411–430. Berlin: Springer-Verlag.

Walker, A. and Leakey, R.E.F. (1978) The hominins of East Turkana. *Scientific American* **239**, 44–56.

Walker, A.C. and Leakey, R.E. (1988) The evolution of *Australopithecus boisei*. In *Evolutionary History of the "Robust" Australopithecines* (F.E Grine, ed.), pp. 247–258. New York, NY: Aldine de Gruyter.

Walker, A. and Leakey, R. (eds) (1993) *The Nariokotome Homo erectus skeleton*. Berlin: Springer Verlag.

Walker, A.C., Falk, D., Smith, R., and Pickford, M. (1983) The skull of *Proconsul africanus*: reconstruction and cranial capacity. *Nature* **305**, 525–527.

Walker, A., Leakey, R.E., Harris, J.M., and Brown, F.H. (1986) 2.5 Myr *Australopithecus boisei* from West of Lake Turkana, Kenya. *Nature* **322**, 517–522.

Wallace, J.A. (1975) Did La Ferrassie I use his teeth as a tool? *Current Anthropology* **16**, 393–401.

Walsh, J.E. (1996) *Unravelling Piltdown: the Science Fraud of the Century and Its Solution*. New York, NY: Random House.

Walter, R.C. and Aronson, J.L. (1982) Revisions of K/Ar ages for the Hadar hominid site, Ethiopia. *Nature* **296**, 122–127.

Walter, R.C. and Aronson, J.L. (1993) Age and source of the Sidi Hakoma Tuff, Hadar Formation, Ethiopia. *Journal of Human Evolution* **25**, 229–240.

Wang, E.T., Kodama, G., Baldi, P., and Moyzis, R.K. (2006) Global landscape of recent inferred Darwinian selection for *Homo* sapiens. Proceedings of the National Academy of Sciences USA **103**, 135–140.

Wang, Y., Xue, X., Yue, L., Zhao, J., and Liu, S. (1979) Discovery of Dali fossil man and its preliminary study. *Scientia Sinica* **24**, 303–306.

Wanpo, H. (1960) Restudy of the CKT *Sinanthropus* deposits. *Vertebrata Palasiatica* **4**, 45–46.

Wanpo, H., Ciochon, R., Yumin, G., Larick, R., Qiren, F., Schwarcz, H., Yonge, C., de Vos, J., and Rink, W. (1995) Early *Homo* and Associated Artefacts from Asia. *Nature* **378**, 275–278.

Ward, C.V. (2002) Interpreting the posture and locomotion of *Australopithecus afarensis*: where do we stand? *Yearbook of Physical Anthropology* **85**, 185–215.

Ward, C.V., Leakey, M.G., Brown, B., Brown, F., Harris, J., and Walker, A. (1999) South Turkwel: a new Pliocene hominid site in Kenya. *Journal of Human Evolution* **36**, 69–95.

Ward, R. and Stringer, C. (1997) A molecular handle on the Neanderthals. *Nature* **388**, 225–226.

Ward, S., Brown, B., Hill, A., Kelley, J., and Downs, W. (1999) *Equatorius*: a new hominoid genus from the Middle Miocene of Kenya. *Science* **285**, 1382–1386.

Washburn, S.L. (1957) Australopithecines; the hunters or the hunted? *American Anthropologist* **59**, 612–614.

Washburn, S.L. (1960) Tools and human evolution. In *Human Ancestors* (R.E F. Leakey, ed.), pp. 110–123. San Francisco, CA, W.H. Freeman and Co.

Washburn, S.L. (1967) Behaviour and the origin of man. *Proceedings of the Royal Anthropological Institute of Great Britain and Ireland 1967*, 21–27.

Washburn, S.L. and Lancaster, C.S. (1968) The evolution of hunting. In *Man the Hunter* (R.B.Lee and I. DeVore, eds), pp. 293–303. Chicago, IL: Aldine Publishing Co.

Watanabe, H., Fujiyama, A., Hattori, M., Taylor, T.D., Toyoda, A., Kuroki, Y., Noguchi, H., BenKahla, A., Lehrach, H., Sudbrak, R. *et al.* (2004) DNA sequence and comparative analysis of chimpanzee chromosome 22. *Nature* **429**, 382–388.

Watkins, K.E., Vargha-Khadem, F., Ashburner, J., Passingham, R.E., Connelly, A., Friston, K.J., Frackowiak, R.S.J., Mishkin, M., and Gadian, D.G. (2002) MRI analysis of an inherited speech and language disorder: structural brain abnormalities. *Brain* **125**, 465–478.

Watson, E., Esteal, S., and Penny, D. (2001) *Homo* genus: a taxonomic revision. In *Abstracts of Contributions to the Dual Congress 1998* (A.A Raath, H. Soodyall, D. Barkhan, K.L Kuykendall, and P.V Tobias, eds), p. 16. Johannesburg: International Association for the Study of Human Paleontology.

Weber, J., Czarnetzki, A., and Pusch, C.M. (2005) Comment on "The Brain of LB1, *Homo floresiensis*". *Science* **310**, 236b.

Wei, L. (1984) 200.000-year-old skeleton unearthed. *Beijing Review* **27**, 33–34.

Weidenreich, F. (1927) Der Schadel von Weimar-Ehringsdorf. *Verhandlungen der Gesellaschaft für physische Anthropologie* **2**, 34–41.

Weidenreich, F. (1933) Über pithekoide Merkmale bei *Sinanthropus pekinensis* und seine stammesgeschichtliche Beurteilung. *Zeitschrift für Anatomie und Entwicklungsgeschichte* **99**, 212–253.

Weidenreich, F. (1936) The mandibles of *Sinanthropus pekinensis*: a comparative study. *Palaeontologia Sinica*, series D7 III, 1–163.

Weidenreich, F. (1940) Some problems dealing with ancient man. *American Anthropologist* **42**, 375–383.

Weidenreich, F. (1941) The extremity bones of *Sinanthropus pekinensis*. *Palaeontologia Sinica, new series D* **5**, 1–150.

Weidenreich, F. (1943) The skull of *Sinanthropus pekinensis*: a comparative study on a primitive hominid skull. *Palaeontologia Sinica, series D* **10**, 1–298.

Weiner, J. and Stringer, C. (2003) *The Piltdown Forgery (50th anniversary edition)*. Oxford: Oxford University Press.

Weiner, S., Xu, Q., Goldberg, P., Liu, J., and Bar-Yosef, O. (1998) Evidence for the use of fire at Zhoukoudian, China. *Science* **281**, 251–253.

Weinert, H. (1950) Über die Neuen Vor- und Frühmenschenfunde aus Afrika, Java, China und Frankreich. *Zeitschrift für Morphologie und Anthropologie* **42**, 113–148.

White, R. (2001) Personal ornaments from the Grotte du Renne at Arcy-sur-Cure. *Athena Review* **2**, 41–46.

White, T.D. (1980a) Evolutionary implications of Pliocene hominid footprints. *Science* **208**, 175–176.

White, T.D. (1980b) Additional fossil hominids From Laetoli, Tanzania: 1976–1979 Specimens. *American Journal of Physical Anthropology* **53**, 487–504.

White, T.D. (1986) *Australopithecus afarensis* and the Lothagam mandible. *Anthropos* **23**, 73–90.

White, T.D. (1994) Ape and hominid limb lenght. *Nature* **369**, 194.

White, T.D. and Suwa, G. (1987) Hominid footprints at Laetoli: facts and interpretations. *American Journal of Physical Anthropology* **72**, 485–514.

White, T.D., Johanson, D.C., and Kimbel, W.H. (1981) *Australopithecus afarensis*: its phyletic position reconsidered. *South Africa Journal of Sciences* **77**, 445–470.

White, T.D., Suwa, G., Hart, W.K., Walter, R.C., Wolde-Gabriel, G., de Heinzelin, J., Clark, J.D., Asfaw, B., and Vrba, E. (1993) New discoveries of *Australopithecus* at Maka in Ethiopia. *Nature* **366**, 261–265.

White, T.D., Suwa, G., and Asfaw, B. (1994) *Australopithecus ramidus*, a new species of early hominid from Aramis, Ethiopia. *Nature* **371**, 306–312.

White, T.D., Suwa, G., and Asfaw, B. (1995) *Australopithecus ramidus*, a new species of early hominid from Aramis, Ethiopia. *Nature* **375**, 88.

White, T.D., Suwa, G., Simpson, S., and Asfaw, B. (2000) Jaws and teeth of *Australopithecus afarensis* from Maka, Middle Awash, Ethiopia. *American Journal of Physical Anthropology* **111**, 45–68.

White, T.D., Asfaw, B., DeGusta, D., Gilbert, H., Richards, G.D., Suwa, G., and Clark Howell, F. (2003) Pleistocene *Homo* sapiens from Middle Awash, Ethiopia. *Nature* **423**, 742–747.

White, T.D., WoldeGabriel, G., Asfaw, B., Ambrose, S., Beyene, Y., Bernor, R.L., Boisserie, J.-R., Currie, B., Gilbert, H., Haile-Selassie, Y. *et al.* (2006) Asa Issie, Aramis and the origin of Australopithecus. *Nature* **440**, 883–889.

Whiten, A., Goodall, J., McGrew, W.C., Nishida, T., Reynolds, V., Sugiyama, Y., Tutin, E.G., Wrangham, R. W., and Boesch, C. (1999) Cultures in chimpanzees. *Nature* **399**, 682–685.

Whiten, A., Horner, V., and de Waal, F.B.M. (2005) Conformity to cultural norms of tool use in chimpanzees. *Nature* **437**, 737–740.

Whitworth, T. (1966) A fossil hominid from Rudolf. *South African Archaeological Bulletin* **21**, 138–150.

Williams, M.AJ., Dunkerley, D.L., De Deckker, P., Kershaw, A.P., and Stokes, T. (1993) *Quaternary Environments*. London: Edward Arnold.

Wills, C. (1995) When did Eve live? An evolutionary detective story. *Evolution* **49**, 593–607.

Wilson, E.O. (1975) *Sociobiology, the New Synthesis*. Cambridge: Belknap Press.

Wilson, E.O. (1978) *On Human Nature*. Cambridge, MA: Harvard University Press.

Wilson, E.O. (1994) *Naturalist*. Washington DC: Island Press.

Wilson, E.O. (1998) *Consilience. The Unity of Knowledge*. New York: Knopf.

WoldeGabriel, G., White, T.D., Suwa, G., Renne, P., de Heinzelin, J., Hart, W.K., and Heiken, G. (1994) Ecological and temporal placement of early Pliocene hominids at Aramis, Ethiopia. *Nature* **371**, 330–333.

Wolpoff, M.H. (1971a) Competitive exclusion among Lower Pleistocene hominids: the single species hypothesis. *Man* **6**, 601–614.

Wolpoff, M.H. (1971b) The evidence for two australopithecine lineages in South Africa. *American Journal of Physical Anthropology* **39**, 375–394.

Wolpoff, M.H. (1980) *Paleoanthropology*. New York, NY: Knopf.

Wolpoff, M.H. (1982) *Ramapithecus* and Hominid Origins. *Current Anthropology* **23**, 501–510.

Wolpoff, M.H. (1991) Levantines and Londoners. *Science* **255**, 142.

Wolpoff, M.H. (1999) The systematics of *Homo*. *Science* **284**, 1774–1775.

Wolpoff, M.H. and Caspari, R. (2000) The many species of humanity. *Anthropological Review* **63**, 3–17.

Wolpoff, M.H., Spuhler, J.N., Smith, F.H., Radovcic, J., Pope, G., Frayer, D.W., Eckhardt, R., and Clark, G. (1988) Modern human origins. *Science* **241**, 772–774.

Wolpoff, M.H., Thorne, A.G., Selinek, J., and Yinyun, Z. (1994) The case for sinking *Homoerectus*: 100 years of *Pithecanthropus* is enough! *Courier Forschungsinstitut Senckenberg* **171**, 341–361.

Wolpoff, M., Spuhler, J., Smith, F., Radovcic, J., Pope, G., Frayer, D., Eckhardt, R., and Clark, G. (1988) Modern human origins. *Science* **241**, 772–774.

Wolpoff, M.H., Hawks, J., Frayer, D.W., and Hunley, K. (2001) Modern human ancestry at the peripheries: a test of the replacement theory. *Science* **291**, 293–297.

Wolpoff, M.H., Senut, B., Pickford, M., and Hawks, J. (2002) *Sahelanthropus* or '*Sahelpithecus*'? *Nature* **419**, 581–582.

Woo, J.K. (1980) Palaeanthropology in the New China. In *Current Argument on Early Man* (L.K Königsson, ed.), pp. 182–206. Oxford: Pergamon Press.

Woo, J.K. and Chao, T.K. (1959) New discovery of *Sinanthropus* mandible from Choukoutien. *Vertebrata Palasiatica 3*, 160–172.

Wood, B. (1984) The Origin of *Homo erectus*. In *The Early Evolution of Man with Special Emphasis on Southeast Asia and Africa*, vol. 69 (P. Andrews and J.L.Franzen, eds), pp. 99–111. Frankfurt: Courier Forschungsinstitut Senckenberg.

Wood, B. (1985) Early *Homo* in Kenya, and its systematic relationships. In *Ancestors: The Hard Evidence* (E. Delson, ed.), pp. 206–214. New York, NY: Alan R. Liss.

Wood, B.A. (1988) Are "robust" australopithecines a monophyletic group? In *Evolutionary History of the "Robust" Australopithecines* (F.E Grine, ed.), pp. 269–284. New York, NY: Aldine de Gruyter.

Wood, B.A. (1991) *Koobi Fora Research Project, vol. 4: Hominid Cranial Remains*. Oxford: Clarendon Press.

Wood, B. (1992a) Origin and evolution of the genus *Homo*. *Nature* **355**, 783–790.

Wood, B. (1992b) Early hominid species and speciation. *Journal of Human Evolution* **22**, 351–365.

Wood, B.A. (1992c) Evolution of the australopithecines. In *The Cambridge Encyclopedia of Human Evolution* (S. Jones, R. Martin, and D. Pilbeam, eds), pp. 325–334. Cambridge: Cambridge University Press.

Wood, B. (1997) Mary Leakey 1913–1996. *Nature* **385**, 28.

Wood, B. (1997) The oldest whodunnit in the world. *Nature* **385**, 292–293.

Wood, B. (2002) Hominid revelations from Chad. *Nature* **418**, 313–315.

Wood, B. (2006) Paleoanthropology: a precious little bundle. *Nature* **443**, 278–281.

Wood, B.A. (1992) Evolution of the Australopithecines. In *The Cambridge Encyclopedia of Human Evolution* (S. Jones, R. Martin and D. Pilbeam, eds), pp. 325–334. Cambridge: Cambridge University Press.

Wood, B. and Turner, A. (1995) Out of Africa and into Asia. *Nature* **378**, 239–240.

Wood, B. and Collard, M. (1999a) The changing face of the *Homo* genus. *Evolutionary Anthropology 8*, 195–207.

Wood, B. and Collard, M. (1999b) The human genus. *Science* **284**, 65–71.

Wood, B. and Richmond, B.G. (2000) Human evolution: taxonomy and paleobiology. *Journal of Anatomy* **196**, 19–60.

Wood, B. and Lieberman, D.E. (2001) Craniodental variation in *Paranthropus boisei*: a developmental and functional perspective. *American Journal of Physical Anthropology* **116**, 13–25.

Wood, B. and Strait, D. (2004) Patterns of resource use in early *Homo* and *Paranthropus*. *Journal of Human Evolution* **46**, 119–162.

Wood, B., Wood, C., and Konigsberg, L. (1994) *Paranthropus boisei*: an example of evolutionary stasis? *American Journal of Physical Anthropology* **95**, 117–136.

Woodward, A.S. (1921) A new cave man from Rodhesia, South Africa. *Nature* **108**, 371–372.

Woodward, A. (1925) The fossil anthropoid ape from Taungs. *Nature* **115**, 234–235.

Wu, X. (1981) A well-preserved cranium of an archaic type of early *Homo* sapiens from Dali, China. *Scientia Sinica* **241**, 530–539.

Wu, X. (1999) Investigating the possible use of fire at Zhoukoudian, China. *Science* **283**, 299.

Wynn, J.G., Alemseged, Z., Bobe, R., Geraads, D., Reed, D., and Roman, D.C. (2006) Geological and palaeontological context of a Pliocene juvenile hominin at Dikika, Ethiopia. *Nature* **443** (7109), 332–336.

Wynn, T. (1979) The intelligence of Oldowan hominids. *Journal of Human Evolution* **10**, 529–541.

Wynn, T. (1989) *The Evolution of Spatial Competence*. Urbana, IL: University of Illinois Press.

Wynn, T. (1993) Two developments in the mind of early *Homo*. *Journal of Anthropological Archaeology* **12**, 299–322.

Wynn, T. (2002) Archaeology and cognitive evolution. *Behavioral and Brain Sciences* **25**, 389–402.

Xu, Q. and You, Y. (1984) Hexian fauna: correlation with deep sea sediments. *Acta Anthropologica Sinica* **1**, 180–190.

Yamei, H., Potts, R., Baoyin, Y., Zhengtang, G., Deino, A., Wei, W., Clark, J., Guangmao, X., and Weiwen, H. (2000) Mid-Pleistocene Acheulan-like Stone Technology of the Bose Basin, South China. *Science* **287**, 1622–1626.

Yellen, J.E., Brooks, A.S., Cornelissen, E., Mehlman, M.J., and Stewart, K. (1995) A Middle Stone Age worked bone industry from Katanda, Upper Semliki Valley, Zaire. *Science* **268**, 553–556.

Yinyun, Z. (1998) Fossil human crania from Yunxian, China: morphological comparison with *Homo* erectus crania from Zhokoudian. *Human Evolution* **13**, 45–48.

Young, N.M. and MacLatchy, L. (2004) The phylogenetic position of *Morotopithecus*. *Journal of Human Evolution* **46**, 163–184.

Yu, N. and Li, W.-H. (2000) No fixed nucleotide difference between Africans and NonAfricans at the pyruvate dehydrogenase e1 alpha-subunit locus. *Genetics* **155**, 1481–1483.

Yu, N., Chen, F.-C., Ota, S., Jorde, L.B., Pamilo, P., Patthy, L., Ramsay, M., Jenkins, T., Shyue, S.-K., and Li, W.-H. (2002) Larger genetic differences within Africans than between Africans and Eurasians. *Genetics* **161**, 269–274.

Zagorski, N. (2006) Profile of Svante Pääbo. *Proceedings of the National Academy of Sciences USA* **103**, 13575–13577.

Zhang, S. (1985) The Early Palaeolithic of China. In *Palaeoanthropology and Palaeolithic Archaeology in the People's Republic of China* (W.U. Rukang and J.W. Olsen, eds), pp. 147–186. London: Academic Press.

Zhao, Z., Jin, L., Fu, Y.-X., Ramsay, M., Jenkins, T., Leskinen, E., Pamilo, P., Trexler, M., Patthy, L., Jorde, L.B. *et al.*, (2000) Worldwide DNA sequence variation in a 10-kilobase noncoding region on human chromosome 22. *Proceedings of the National Academy of Sciences USA* **97**, 11354–11358.

Zhao, Z., Yu, N., Fu, Y.-X., and Li, W.-H. (2006) Nucleotide variation and haplotype diversity in a 10-kb noncoding region in three continental human populations. *Genetics* **174**, 399–409.

Zhu, R.X., Hoffman, K.A., Potts, R., Deng, C.L., Pan, Y.X., Guo, B., Shi, C.D., Guo, Z.T., Yuan, B.Y., and Hou, Y.M. (2001) Earliest presence of humans in northeast Asia. *Nature* **413**, 413–417.

Zhu, R.X., Potts, R., Xie, F., Hoffman, K.A., Deng, C.L., Shi, C.D., Pan, Y.X., Wang, H.Q., Shi, R.P., Wang, Y.C. *et al.* (2004) New evidence on the earliest human presence at high northern latitudes in northeast Asia. *Nature* **431**, 559–562.

Ziaei, M., Schwarcz, H.P., Hall, C.M., and Grün, R. (1990) Radiometric dating of the Mousterian site at Quneitra. In *Quneitra: a Mousterian Site on the Golan Heights* (N. Goren-Inbar, ed.), pp. 232–235. Jerusalem: Monographs of the Institute of Archaeology, Hebrew University 31.

Ziętkiewicz, E., Yotova, V., Gehl, D., Wambach, T., Arrieta, I., Batzer, M., Cole, D.EC., Hechtman, P., Kaplan, F., Modiano, D. *et al.* (2003) Haplotypes in the dystrophin DNA segment point to a mosaic origin of modern human diversity. *American Journal of Human Genetics* **73(5)**, 994–1015.

Zilhao, J., d'Errico, F., Bordes, J.G., Lenoble, A., Texier, J. P., and Rigaud, J.P. (2006) Analysis of Aurignacian interstratification at the Châtelperronian-type site and implications for the behavioral modernity of Neandertals. *Proceedings of the National Academy of Sciences USA* **103**, 12643–12648.

Zimmer, C. (1999a) Kenyan skeleton shakes ape family tree. *Science* **285**, 1335–1337.

Zimmer, C. (1999b) New date for the dawn of dream time. *Science* **284**, 1243–1246.

Index

Note: page numbers in *italics* refer to Figures, Tables and Boxes, whilst those in **bold** refer to Glossary entries.